CONTROL SYSTEMS ENGINEERING

CONTROL SYSTEMS ENGINEERING

(SECOND EDITION)

S SIVANAGARAJU
Associate Professor
Department of Electrical Engineering
JNTU College of Engineering
Kakinada, Andhra Pradesh

L DEVI
Associate Professor
Department of Electrical Engineering
Intel College of Engineering
Anantapur, Andhra Pradesh

NEW AGE INTERNATIONAL (P) LIMITED, PUBLISHERS
LONDON • NEW DELHI • NAIROBI
Bangalore • Chennai • Cochin • Guwahati • Hyderabad • Kolkata • Lucknow • Mumbai
Visit us at www.newagepublishers.com

Published by New Age International (P) Ltd., Publishers
First Edition: 2008
Second Edition: 2010
Reprint: 2017

GLOBAL OFFICES

- **New Delhi** **NEW AGE INTERNATIONAL (P) LIMITED, PUBLISHERS**
7/30 A, Daryaganj, New Delhi-110002, (INDIA)
Tel.: (011) 23253771, 23253472, **Telefax:** 23267437, 43551305
E-mail: contactus@newagepublishers.com • Visit us at www.newagepublishers.com
- **London** **NEW AGE INTERNATIONAL (UK) LTD.**
27 Old Gloucester Street, London, WC1N 3AX, UK
E-mail: info@newacademicscience.co.uk • Visit us at www.newacademicscience.co.uk
- **Nairobi** **NEW AGE GOLDEN (EAST AFRICA) LTD.**
Ground Floor, Westlands Arcade, Chiromo Road (Next to Naivas Supermarket)
Westlands, Nairobi, KENYA, **Tel.:** 00-254-713848772, 00-254-725700286
E-mail: kenya@newagepublishers.com

BRANCHES

- **Bangalore** 37/10, 8th Cross (Near Hanuman Temple), Azad Nagar, Chamarajpet, Bangalore- 560 018
Tel.: (080) 26756823, **Telefax:** 26756820, **E-mail: bangalore@newagepublishers.com**
- **Chennai** 26, Damodaran Street, T. Nagar, Chennai-600 017, **Tel.:** (044) 24353401, **Telefax:** 24351463
E-mail: chennai@newagepublishers.com
- **Cochin** CC-39/1016, Carrier Station Road, Ernakulam South, Cochin-682 016, **Tel.:** (0484) 2377303, **Telefax:** 4051304
E-mail: cochin@newagepublishers.com
- **Guwahati** Hemsen Complex, Mohd. Shah Road, Paltan Bazar, Near Starline Hotel, Guwahati-781 008,
Tel.: (0361) 2513881, **Telefax:** 2543669, **E-mail: guwahati@newagepublishers.com**
- **Hyderabad** 105, 1st Floor, Madhiray Kaveri Tower, 3-2-19, Azam Jahi Road, Near Kumar Theater, Nimboliadda Kachiguda, Hyderabad-500 027, **Tel.:** (040) 24652456, **Telefax:** 24652457
E-mail: hyderabad@newagepublishers.com
- **Kolkata** RDB Chambers (Formerly Lotus Cinema) 106A, 1st Floor, S N Banerjee Road, Kolkata-700 014
Tel.: (033) 22273773, **Telefax:** 22275247, **E-mail: kolkata@newagepublishers.com**
- **Lucknow** 16-A, Jopling Road, Lucknow-226 001, **Tel.:** (0522) 2209578, 4045297, **Telefax:** 2204098
E-mail: lucknow@newagepublishers.com
- **Mumbai** 142C, Victor House, Ground Floor, N.M. Joshi Marg, Lower Parel, Mumbai-400 013
Tel.: (022) 24927869, **Telefax:** 24915415, **E-mail: mumbai@newagepublishers.com**
- **New Delhi** 22, Golden House, Daryaganj, New Delhi-110 002, **Tel.:** (011) 23262368, 23262370, **Telefax:** 43551305
E-mail: sales@newagepublishers.com

ISBN: 978-81-224-3099-8
C-16-12-9926

Printed in India at Kanishik Printing Press, Delhi.
Typeset at In-house, Delhi.

NEW AGE INTERNATIONAL (P) LIMITED, PUBLISHERS
7/30 A, Daryaganj, New Delhi-110002
Visit us at **www.newagepublishers.com**
(CIN: U74899DL1966PTC004618)

Preface

This book entitled 'Control Systems Engineering' has been written for undergraduate students of Indian universities. This book is helpful for those interested in learning about basics of control systems. The organisation of this book reflects the desire on our part to provide the reader with a thorough understanding of the basic principles and techniques of control systems. This book has been developed with the goal of making it easy for students to learn independently. This book has been presented in the simplest way and we are sure that it will be useful for learning control systems for students of any level. A number of solved and exercise problems, descriptive questions, short questions and answers as well as objective type questions are included.

The book is divided into nine chapters. The first chapter summarizes the important results in the Laplace Transform theory which are of direct relevance to control system engineering. The second chapter gives a brief introduction and historical background of the subject as well as the modeling of systems based on classical concepts such as transfer function, block diagrams and signal flow graphs.

Chapter 3 explains the sensitivity to parameter variation of closed loop control system and different control system components. Chapter 4 deals with transient response analysis and steady state error analysis. Chapter 5 presents the stability concept in time domain using Routh-Hurwitz criterion and Root locus technique.

Chapter 6 deals with frequency domain specifications and Bode plots. Chapter 7 presents investigation of stability using the Polar plots and Nyquist stability criterion. Chapter 8 deals with design of compensating techniques using frequency domain approach.

The last chapter deals with state variable formulation of the control system, both in electrical and mechanical systems. Evaluation of state transition matrix and its response, controllability as well as observability of the system are also included in this chapter.

Although great care has been exercised to check the manuscript, some errors in the text might have crept in. We would like to get suggestions for improvement of this book and we would very much appreciate hearing from users of this book. Any comments and criticism would be most helpful. This book can be used for quick reference by students of all branches of B.E., B.Tech and A.M.I.E. The diagrams, examples and small question and answers given in this book will give you a clear idea about the concepts.

We express wholehearted gratefulness to Dr.Ch. Sai Babu, Professor in Electrical and Electronics Engineering, JNTU College of Engg., Anantapur for his encouragement for completing the book and our colleagues of the Department of Electrical and Electronics Engineering in particular who have directly or indirectly helped us to bring this book in the present form. We also sincerely express thankfulness to the publishers, M/s New Age International Publishers particularly to Sri. E. Ravi Kumar, Sales Manager, Hyderabad and Sri. V.R. Damodaran, Chief Editor, who have taken initiative for publishing this book.

—Authors

Contents

1 Laplace Transformation

1.1 INTRODUCTION

The first step in control system is to get a system, after that we have to give mathematical model by writing mathematical equations. Mathematical equations consist of differential and integral terms, and to get a solution to such equation is very difficult. In order to get a solution, Laplace transformation plays a very important role. Laplace transformation will change the time domain equations into s domain equations.

Definition: Let $f(t)$ be the function of time, Laplace transformation of function $f(t)$ is

$$F(s) = L[f(t)] = \int_0^{\infty} f(t)\, e^{-st}\, dt \tag{1.1}$$

L = symbol for Laplace transform

From the equation (1.1), Laplace transform $F(s)$ of $f(t)$ is equal to integral term which is the product of $f(t)$ and e^{-st}, whose limits are zero (lower) and infinity (upper) and integral product w.r.t. time.

Mathematical expression for Laplace transform is,

$$L[f(t)] = F(s), \text{ for } t \geq 0$$

or

$$F(s) = \int_0^{\infty} f(t)\, e^{-st}\, dt \tag{1.2}$$

The time function $f(t)$ is obtained back from the Laplace transform by a process called inverse Laplace transformation and is denoted by L^{-1}.

Thus

$$L^{-1}[L\{f(t)\}] = L^{-1}[F(s)] = f(t) \tag{1.3}$$

The time function $f(t)$ and its Laplace transform $F(s)$ are called transform pair.

1.2 LAPLACE TRANSFORM OF SOME USEFUL FUNCTIONS

Laplace transform of some useful functions are discussed in this section:

(*i*) Laplace transform of exponential term, $f(t) = e^{-at}$

From the definition of Laplace transform

$$L[e^{-at}] = \int_0^{\infty} e^{-at}\, e^{-st}\, dt = \int_0^{\infty} e^{-(a+s)t}\, dt = \frac{1}{s+a}$$

$$\therefore \qquad L[e^{-at}] = \frac{1}{s+a} \tag{1.4}$$

(*ii*) Laplace transform of constant term, $f(t) = k$

By definition

$$L\,[1] = \int_0^\infty e^{-st}\,dt = \frac{1}{s}$$

$$\therefore \qquad L\,[1] = \frac{1}{s} \tag{1.5}$$

(*iii*) Laplace transform of cos ωt

By definition

$$L\,[\cos \omega t] = \int_0^\infty \left[\frac{e^{j\omega t} + e^{-j\omega t}}{2}\right] e^{-st}\,dt \qquad \left[\because\ \cos\omega t = \frac{e^{j\omega t} + e^{-j\omega t}}{2}\right]$$

$$= \frac{1}{2}\left[\int_0^\infty e^{j\omega t}\,e^{-st}\,dt + \int_0^\infty e^{-\omega t}\,e^{-st}\,dt\right]$$

$$= \frac{1}{2}\left[\int_0^\infty e^{(-j\omega+s)t}\,dt + \int_0^\infty -e^{(j\omega+s)t}\,dt\right]$$

$$= \frac{1}{2}\left[\frac{1}{s-j\omega} + \frac{1}{s+j\omega}\right]$$

$$= \frac{1}{2}\left[\frac{s+j\omega+s-j\omega}{(s+j\omega)(s-j\omega)}\right] = \frac{1}{2}\left[\frac{2s}{s^2+\omega^2}\right]$$

$$= \frac{s}{s^2+\omega^2}$$

$$\therefore \qquad L\,[\cos\omega t] = \frac{s}{s^2+\omega^2} \tag{1.6}$$

(*iv*) Laplace transform of sin ωt

By definition

$$L\,[\sin\omega t] = \int_0^\infty \sin\omega t\; e^{-st}\,dt$$

In exponential form, $\sin\omega t = \dfrac{e^{j\omega t} - e^{-j\omega t}}{2j}$

$$\therefore \qquad L\,[\sin\omega t] = \int_0^\infty \frac{e^{j\omega t} - e^{-j\omega t}}{2j}\,e^{-st}\,dt$$

$$= \frac{1}{2j}\left[\int_0^\infty e^{(j\omega - s)t}\,dt - \int_0^\infty e^{-j\omega t}\,e^{-st}\,dt\right]$$

$$= \frac{1}{2j}\left[\int_0^\infty e^{-(s-j\omega)t}\,dt - \int_0^\infty e^{-(s+j\omega)t}\,dt\right]$$

$$= \frac{1}{2j}\left[\frac{1}{s-j\omega} - \frac{1}{s+j\omega}\right]$$

$$= \frac{1}{2j}\left[\frac{s+j\omega - s + j\omega}{(s-j\omega)(s+j\omega)}\right] = \frac{1}{2j}\left[\frac{2j\omega}{s^2+\omega^2}\right]$$

$$= \frac{\omega}{s^2+\omega^2}$$

$$\therefore \qquad L\,[\sin \omega t] = \frac{\omega}{s^2+\omega^2} \qquad (1.7)$$

(v) Laplace transform of $e^{-at}\cos \omega t$

By definition

$$L\,[e^{-at}\cos \omega t] = \int_0^\infty e^{-at}\cos \omega t\, e^{-st}\, dt$$

In exponential form, $\cos \omega t = \dfrac{e^{j\omega} + e^{-j\omega}}{2}$

$$\therefore \qquad L\,[e^{-at}\cos \omega t] = \int_0^\infty e^{-at}\left[\frac{e^{j\omega t}+e^{-j\omega t}}{2}\right]e^{-st}\, dt$$

$$= \frac{1}{2}\left[\int_0^\infty (e^{-at}\, e^{j\omega t}\, e^{-st} + e^{-at}\, e^{-j\omega t}\, e^{-st})\, dt\right.$$

$$= \frac{1}{2}\left[\int_0^\infty e^{-(s+a-j\omega)t}\, dt + \int_0^\infty e^{-(s+a+j\omega)t}\, dt\right]$$

$$= \frac{1}{2}\left[\frac{1}{s+a-j\omega} + \frac{1}{s+a+j\omega}\right]$$

$$= \frac{1}{2}\left[\frac{s+a+j\omega+s+a-j\omega}{(s+a-j\omega)\,(s+a+j\omega)}\right] = \frac{1}{2}\left[\frac{2(s+a)}{(s+a)^2+\omega^2}\right]$$

$$\therefore \qquad L\,[e^{-at}\cos \omega t] = \frac{s+a}{(s+a)^2+\omega^2} \qquad (1.8)$$

(vi) Laplace transform of $e^{-at}\sin \omega t$

By definition

$$L\,[e^{-at}\sin \omega t] = \int_0^\infty e^{-at}\sin \omega t\, e^{-st}\, dt$$

In exponential form, $\sin \omega t = \dfrac{e^{j\omega t} - e^{-j\omega t}}{2j}$

$$\therefore \qquad L\,[e^{-at}\sin \omega t] = \int_0^\infty e^{-at}\left[\frac{e^{j\omega t}-e^{-j\omega t}}{2j}\right]e^{-st}\, dt$$

$$= \frac{1}{2j}\left[\int_0^\infty (e^{-at}\, e^{j\omega t}\, e^{-st} - e^{-at}\, e^{-j\omega t}\, e^{-st})\, dt\right]$$

$$= \frac{1}{2j}\left[\int_0^\infty e^{-(s+a-j\omega)t}\, dt - \int_0^\infty e^{-(s+a+j\omega)t}\, dt\right]$$

$$= \frac{1}{2j}\left[\frac{1}{s+a-j\omega} - \frac{1}{s+a+j\omega}\right]$$

$$= \frac{1}{2j}\left[\frac{s+a+j\omega-s-a+j\omega}{(s+a-j\omega)(s+a+j\omega)}\right] = \frac{1}{2j}\left[\frac{2j\omega}{(s+a)^2+\omega^2}\right]$$

$$\therefore \quad L\,[e^{-at} \sin \omega t] = \frac{\omega}{(s+a)^2+\omega^2} \tag{1.9}$$

(*vii*) Laplace transform of e^t

$$L\,[e^t\,] = \int_0^{\infty} e^t\, e^{-st}\, dt$$

$$= \int_0^{\infty} e^{-(s-1)\,t}\, dt = \frac{1}{s-1}$$

$$\because \quad e^t = 1 + t + \frac{t^2}{2!} + \frac{t^3}{3!} + \ldots\ldots \tag{1.10}$$

$$L\,[e^t\,] = \frac{1}{s-1} = L\left[1 + t + \frac{t^2}{2!} + \ldots\right]$$

$$= \frac{1}{s} + \frac{1}{s^2} + \frac{1}{s^3} + \ldots$$

$\therefore$ Comparing the terms with eqn. (1.10)

$$L[1] = \frac{1}{s} \text{ [unit step]} \tag{1.11}$$

$$L\,[t] = \frac{1}{s^2} \text{ [unit ramp]} \tag{1.12}$$

$$L\left[\frac{t^2}{2!}\right] = \frac{1}{s^2} \text{ [unit parabolic]}$$

$$L\left[\frac{t^n}{n!}\right] = \frac{1}{s^{n+1}} \quad \text{or} \quad L[t^n] = \frac{n!}{s^{n+1}}$$

$$\therefore \quad \text{General, } L\,[t^n\,] = \frac{n!}{s^{n+1}} \tag{1.13}$$

1.3 SOME PROPERTIES OF LAPLACE TRANSFORM

(*i*) Laplace transform of constant times a function is equal to the constant times the Laplace transform of function

Let $f(t)$ is the function of time

If $\quad L[f(t)] = F(s)$

Then, $\quad L[K f(t)\,] = K\,F(s) \quad$ (1.14)

where K is constant.

(*ii*) The Laplace transform of the derivative function $f(t)$ whose derivative is $\dfrac{df(t)}{dt}$.

By definition

$$L\left[\frac{df(t)}{dt}\right] = \int_0^\infty \frac{df(t)}{dt} e^{-st} dt$$

$$= \int_0^\infty df(t)\, e^{-st} = s\,F(s) - f(0) \qquad (1.15)$$

where $f(0)$ is the initial value of $f(t)$.

In general for n^{th} derivative of $f(t)$

$$L\left[\frac{d^n f(t)}{dt^n}\right] = s^n\,F(s) - s^{n-1} f(0) - s^{n-2} f(0) \ldots - f^{n-1}(0) \qquad (1.16)$$

(*iii*) Laplace transform of integral terms.

Let $f(t)$ is the function of time 't' whose integral is $\int f(t)\, dt$.

By definition

$$L\left[\int f(t)\, dt\right] = \int_0^\infty \left[\int f(t)\, dt\right] e^{-st} dt$$

$$= \frac{1}{s} \int_0^\infty f(t)\, e^{-st} dt$$

$$= \frac{1}{s} [F(s) + f^{-1}(0)] \qquad (1.17)$$

In general Laplace transform of an integral term of nth order is

$$L\left[\iint f \ldots\ldots f(t)\, dt^n\right] = \frac{F(s)}{s^n} + \frac{f^{-1}(0)}{s^n} + \ldots + \frac{f^n(0)}{s} \qquad (1.18)$$

(*iv*) Laplace transform of the sum of two functions is equal to sum of the Laplace transform of the individual functions.

Let $f_1(t)$ and $f_2(t)$ are the two functions of time t

$$L\,[f_1(t) + f_2(t)] = L\,[f_1(t)] + L\,[f_2(t)]$$

$$= F_1(s) + F_2(s) \qquad (1.19)$$

(*v*) Laplace transform of the difference of two functions is equal to differences of the Laplace transform of the individual functions.

Let $f_1(t)$ and $f_2(t)$ are the two functions of time t

$$L\,[f_1(t) - f_2(t)] = L[f_1(t)] - L[f_2(t)]$$

$$= F_1(s) - F_2(s) \qquad (1.20)$$

(*vi*) Initial value theorem and Final value theorems.

Initial value of function $f(t)$, whose Laplace transform $F(s)$ is obtained by

$$f(0) = \underset{t\to 0}{\text{Lt}}\, f(t) = \underset{s\to\infty}{\text{Lt}}\, sF(s) \qquad (1.21)$$

Final value of function $f(t)$, is obtained by

$$f(\infty) = \underset{t\to\infty}{\text{Lt}}\, f(t) = \underset{s\to 0}{\text{Lt}}\, sF(s) \qquad (1.22)$$

(*vii*) Laplace transform of shifted function

Let $f(t)$ is the function of time, shifted by 'a' internally.

$f(t-a)$ is the shifted function

$$L[f(t-a)] = \int_0^\infty f(t-a)\,e^{-st}\,dt$$

$$= e^{-as}\int_0^\infty e^{as} f(t-a)\,e^{-st}\,dt$$

$$= e^{-as}\int_0^\infty f(t-a)\,e^{-(t-a)s}\,dt$$

$$= e^{-as}\int_0^\infty f(x)\,e^{-xs}\,dx \qquad \text{(Let } x = t-a,\ dx = dt\text{)}$$

$$= e^{-as}\,F(s) \tag{1.23}$$

1.4 INVERSE LAPLACE TRANSFORM

In the inverse Laplace transformation any 's' domain function is changed into time (t) domain function.

$$L^{-1}[F(s)] = f(t) \tag{1.24}$$

where $F(s)$ is the Laplace transform of $f(t)$ and $L^{-1}[F(s)]$ is the inverse Laplace transform of $F(s)$.

(*i*) Inverse Laplace transform of $\frac{1}{s}$ is

$$L^{-1}\left(\frac{1}{s}\right) = 1 \tag{1.25}$$

Similarly $$L^{-1}\left(\frac{1}{s^{n+1}}\right) = t^n \tag{1.26}$$

(*ii*) Inverse Laplace transform of $\frac{1}{s+a}$

$$L^{-1}\left[\frac{1}{s+a}\right] = e^{-at} \tag{1.27}$$

(*iii*) Inverse Laplace transform of $\frac{s}{s^2+\omega^2}$

$$L^{-1}\left[\frac{s}{s^2+\omega^2}\right] = \cos\omega t \tag{1.28}$$

Similarly, $$L^{-1}\left[\frac{\omega}{s^2+\omega^2}\right] = \sin\omega t \tag{1.29}$$

Table 1.1 Table of Laplace Transform

Sl. No.	Function f(t)	Laplace transform F(s)	Inverse Laplace transform
1	$u(t) = 1$ (unit step)	$\frac{1}{s}$	$L^{-1}\left[\frac{1}{s}\right]$
2	$r(t) = t$ (unit ramp)	$\frac{1}{s^2}$	$L^{-1}\left[\frac{1}{s^2}\right]$
3	$r(t) = \frac{t^2}{2!}$ (unit parabolic)	$\frac{1}{s^3}$	$L^{-1}\left[\frac{1}{s^3}\right]$
4	$\delta(t) = 1$ (unit impulse)	1	$L^{-1}[\delta(t)]$
5	e^{-at}	$\frac{1}{s+a}$	$L^{-1}\left[\frac{1}{s+a}\right]$
6	$\cos \omega t$	$\frac{s}{s^2+\omega^2}$	$L^{-1}\left[\frac{s}{s^2+\omega^2}\right]$
7	$\sin \omega t$	$\frac{\omega}{s^2+\omega^2}$	$L^{-1}\left[\frac{\omega}{s^2+\omega^2}\right]$
8	$e^{-at} \cos \omega t$	$\frac{s+a}{(s+a)^2+\omega^2}$	$L^{-1}\left[\frac{s+a}{(s+a)^2+\omega^2}\right]$
9	$e^{-at} \sin \omega t$	$\frac{\omega}{(s+a)^2+\omega^2}$	$L^{-1}\left[\frac{\omega}{(s+a)^2+\omega^2}\right]$
10	$\frac{df(t)}{dt}$	$sF(s) - f(o)$	—
11	$\int f(t)\,dt$	$\frac{F(s)}{s} + \frac{f(0)}{s}$	—
12	$\frac{t^{n-1}}{(n-1)!}$	$\frac{1}{s^n}$ $(n = 1, 2, \ldots)$	$L^{-1}\left[\frac{1}{s^n}\right]$
13	$t\,e^{-at}$	$\frac{1}{(s+a)^2}$	$L^{-1}\left[\frac{1}{(s+a)^2}\right]$
14	$\frac{t^{n-1}}{(n-1)!} e^{-at}$	$\frac{1}{(s+a)^n}$	$L^{-1}\left[\frac{1}{(s+a)^n}\right]$
15	$\cosh \omega t$	$\frac{s}{s^2-\omega^2}$	$L^{-1}\left[\frac{s}{s^2-\omega^2}\right]$
16	$\sinh \omega t$	$\frac{\omega}{s^2-\omega^2}$	$L^{-1}\left[\frac{\omega}{s^2-\omega^2}\right]$

(*iv*) Inverse Laplace transform of $\dfrac{s+a}{(s+a)^2+\omega^2}$

$$L^{-1}\left[\frac{s+a}{(s+a)^2+\omega^2}\right] = e^{-at}\cos\omega t \tag{1.30}$$

Similarly, $$L^{-1}\left[\frac{\omega}{(s+a)^2+\omega^2}\right] = e^{-at}\sin\omega t \tag{1.31}$$

Problem 1.1. *Find the Laplace transform of following functions :*

(*i*) $f(t) = 20$ (*ii*) $f(t) = 2 + 4e^{-2t}$ (*iii*) $f(t) = t - 2e^{-t}$.

Solution:

(*i*) $$f(t) = 20$$

$$F(s) = L[f(t)] = \int_0^\infty f(t)\, e^{-st}\, dt$$

Substituting $$f(t) = 20$$

$\therefore$ $$F(s) = \int_0^\infty 20e^{-st}\, dt = 20\left[-\frac{1}{s}e^{-st}\right]_0^\infty = \frac{20}{s}$$

(*ii*) $$f(t) = 5 + 4e^{-2t}$$

$$F(s) = L[f(t)] = \int_0^\infty f(t)\, e^{-st}\, dt$$

Substituting $$f(t) = 2 + 4e^{-2t}$$

$$F(s) = \int_0^\infty (2 + 4e^{-2t})\, e^{-st}\, dt$$

$$= \int_0^\infty 2e^{-st}\, dt + \int_0^\infty 4e^{-(s+2)t}\, dt$$

$$= \frac{2}{s}\left[-e^{-st}\right]_0^\infty + \frac{4}{s+2}\left[-e^{(s+2)t}\right]_0^\infty$$

$$= \frac{2}{s} + \frac{4}{s+2}$$

$$= \frac{6s+10}{s(s+2)}$$

(*iii*) $$f(t) = t - 2e^{-t}$$

$$F(s) = L[f(t)] = \int_0^\infty f(t)\, e^{-st}\, dt$$

Substituting $$f(t) = (t - 2e^{-t})$$

$\therefore$ $$F(s) = -\int_0^\infty 2e^{-(s+1)t}\, dt + \int_0^\infty te^{-st}\, dt$$

$$= -\frac{2}{s+1}\left[-e^{(s+1)t}\right]_0^\infty + \left[t\int_0^\infty \frac{1(-e)^{-st}}{s}\,dt\right]$$

$$= \frac{2}{s+1} + \left[\frac{-te^{-st}}{s} - \frac{e^{-st}}{s^2}\right]_0^\infty$$

$$= -\frac{2}{s+1} + \frac{1}{s^2}$$

$$= \frac{1+s-2s^2}{s^2(s+1)}$$

Problem 1.2. *Find the Laplace transform [t – cos 2t].*

Solution:

Let $\qquad f(t) = t - \cos 2t$

By definition

$$L[f(t)] = L(t - \cos 2t)$$
$$= L[t] - L[\cos 2t]$$

From table 1.1

$$F(s) = \frac{1}{s^2} - \frac{s}{s^2+4}$$

$$= \frac{s^2+4-s^3}{s^2(s^2+4)}.$$

Problem 1.3. *Determine the Laplace transform of 'y' in the following differential equations.*

(*i*) $\frac{d^2y}{dt^2} + 5\frac{dy}{dt} + 7y = 20$ $\qquad$ (*ii*) $\frac{d^3y}{dt^3} + 10\frac{d^2y}{dt^2} + 20\frac{dy}{dt} + 6y = 9u(t)$

Assuming initial conditions are to be zero.

Solution:

(*i*)
$$\frac{d^2y}{dt^2} + 5\frac{dy}{dt} + 7y = 20$$

Taking Laplace transform on both sides

$$[s^2 Y(s) - sy(0) - y'(0)] + 5[sY(s) - y(0)] + 7Y(s) = \frac{20}{s}$$

Initial conditions are $y(0) = 0$ and $y'(0) = 0$

$$\therefore \qquad s^2Y(s) + 5s\,Y(s) + 7Y(s) = \frac{20}{s}$$

$$\therefore \qquad Y(s) = \frac{20}{s(s^2+5s+7)}$$

(*ii*)
$$\frac{d^3y}{dt^3} + 10\frac{d^2y}{dt^2} + 20\frac{dy}{dt} + 6y = 9x$$

Taking Laplace transform on both sides, we get

$$[s^3 Y(s) - s^2 y(0) - sy'(0) - y''(0)] + 10\ [s^2Y(s)(0) - sy(0) - y'(0)]$$
$$+ 20\ [sY(s) - y(0)] + 6Y(s) = 9\,X(s)$$

As initial conditions are zero,

$$s^3 Y(s) + 10s^2 Y(s) + 20sY(s) + 6Y(s) = 9X(s)$$

$$\therefore \qquad Y(s) = \frac{9X(s)}{(s^3 + 10s^2 + 20s + 6)}.$$

Problem 1.4. *Find the inverse Laplace transform for the following :*

(i) $\dfrac{1}{s+1}$ (ii) $\dfrac{1}{s} - \dfrac{1}{s+2}$ (iii) $\dfrac{(s+1)}{(s+2)(s+3)}$.

Solution: (i) $F(s) = \dfrac{1}{s+1}$

From table 1.1

$$F(t) = L^{-1}\left[\frac{1}{s+1}\right] = e^{-t}$$

(ii) $F(s) = \dfrac{1}{s} - \dfrac{1}{(s+2)}$

From table 1.1

$$F(t) = L^{-1}\left[\frac{1}{s}\right] - L^{-1}\left[\frac{1}{s+2}\right] = 1 - e^{-2t}$$

(iii) $F(s) = \dfrac{(s+1)}{(s+2)(s+3)}$

Expanding into partial fractions as

$$\frac{(s+1)}{(s+2)(s+3)} = \frac{A}{(s+2)} + \frac{B}{(s+3)}$$

$$(s + 1) = A\,(s + 3) + B\,(s + 2)$$

at $s = -3, \quad -B = -2 \quad \Rightarrow \quad B = 2$

$s = -2, \quad A = -1$

$$\therefore \qquad F(s) = \frac{-1}{(s+2)} + \frac{2}{(s+3)}$$

Taking inverse Laplace transformation on both sides from table 1.1

$$f(t) = -e^{-2t} + 2e^{-3t}$$

Problem 1.5. *Determine the inverse Laplace transform of following functions:*

(i) $F(s) = \dfrac{20}{s(s+6)(s+8)}$ (ii) $F(s) = \dfrac{10(s+6)}{s(s+6)(s^2+3s+2)(s^2+7s+12)}$

Solution: (i)

$$F(s) = \frac{20}{s(s+6)(s+8)}$$

$$= \frac{A_1}{s} + \frac{A_2}{(s+6)} + \frac{A_3}{(s+8)} \tag{1.32}$$

The partial fraction coefficients are

$$A_1 = sF(s)\Big|_{s=0} = \frac{20}{(s+6)(s+8)}\Big|_{s=0} = \frac{5}{12}$$

$$A_2 = (s+6)\,F(s)\Big|_{s=-6} = \frac{20}{s(s+8)}\Big|_{s=-6} = \frac{20}{-6\times 2} = -\frac{5}{3}$$

$$A_3 = (s+8)\,F(s)\Big|_{s=-8} = \frac{20}{s(s+6)}\Big|_{s=-8} = \frac{20}{-8\times(-2)} = \frac{5}{4}$$

Substituting these values of coefficients in eqn. (1.32)

$$F(s) = \frac{\frac{5}{12}}{s} - \frac{\frac{5}{3}}{(s+6)} + \frac{\frac{5}{4}}{(s+8)} \tag{1.33}$$

From Laplace transform Table 1.1, the inverse Laplace transform of eqn.(1.33) is

$$f(t) = \frac{5}{12} - \frac{5}{3}\,e^{-6t} + \frac{5}{4}\,e^{-8t}$$

$$= \frac{5}{12}\,[1 - 4e^{-6t} + 3\,e^{-8t}]$$

(*ii*) $$F(s) = \frac{10(s+6)}{s(s+5)(s^2+3s+1)(s^2+7s+12)}$$

$$= \frac{10(s+6)}{s(s+1)(s+2)(s+3)(s+4)(s+5)}$$

$$= \frac{A_1}{s} + \frac{A_2}{(s+1)} + \frac{A_3}{(s+2)} + \frac{A_4}{(s+3)} + \frac{A_5}{(s+4)} + \frac{A_6}{(s+5)} \tag{1.34}$$

The partial fraction coefficients are

$$A_1 = sF(s)\Big|_{s=0}\ \frac{10\,(s+6)}{(s+1)\,(s+2)\,(s+3)\,(s+4)\,(s+5)}\Big|_{s=0} = \frac{1}{2}$$

$$A_2 = (s+1)\,F(s)\Big|_{s=-1}$$

$$= \frac{10(s+6)}{s(s+2)(s+3)(s+4)(s+5)}\Big|_{s=-1} = -\frac{25}{12}$$

$$A_3 = (s+2)\,F(s)\Big|_{s=-2}$$

$$= \frac{10(s+6)}{s(s+1)(s+3)(s+4)(s+5)}\Big|_{s=-2} = \frac{10}{3}$$

$$A_4 = (s+3)\,F(s)\Big|_{s=-3}$$

$$= \frac{10(s+6)}{s(s+1)(s+2)(s+4)(s+5)}\Bigg|_{s=-3} = -\frac{5}{2}$$

$$A_5 = (s+4)\,F(s)\Big|_{s=-4}$$

$$= \frac{10(s+6)}{s(s+1)(s+2)(s+3)(s+5)}\Bigg|_{s=-4} = \frac{5}{6}$$

$$A_6 = (s+5)\,F(s)\Big|_{s=-5}$$

$$= \frac{10(s+6)}{s(s+1)(s+2)(s+3)(s+4)}\Bigg|_{s=-5} = -\frac{1}{12}$$

Substituting these coefficients in eqn. (1.34)

$$F(s) = \frac{\frac{1}{2}}{s} + \frac{\frac{-25}{12}}{(s+1)} + \frac{\frac{10}{3}}{(s+2)} + \frac{\frac{-5}{2}}{(s+3)} + \frac{\frac{5}{6}}{(s+4)} + \frac{\frac{-1}{12}}{(s+5)} \tag{1.35}$$

From the Laplace transform Table 1.1, the inverse Laplace transform of eqn. (1.35) is

$$f(t) = \left[\frac{1}{2} - \frac{25}{12}e^{-t} + \frac{10}{3}e^{-2t} - \frac{5}{2}e^{-3t} + \frac{5}{6}e^{-4t} - \frac{1}{12}e^{-5t}\right]$$

$$= \frac{1}{12}\,[6 - 25\,e^{-t} + 40\,e^{-2t} - 30\,e^{-3t} + 10\,e^{-4t} - e^{-5t}]$$

Problem 1.6. *Find the inverse Laplace transform of the following function*

$$F(s) = \frac{10}{(s+3)^2\,(s+4)^3}$$

Solution:
$$F(s) = \frac{10}{(s+3)^2(s+4)^3}$$

$$F(s) = \frac{A_2}{(s+3)^2} + \frac{A_1}{(s+3)} + \frac{B_3}{(s+4)^3} + \frac{B_2}{(s+4)^2} + \frac{B_1}{(s+4)} \tag{1.36}$$

$$A_2 = (s+3)^2\,F(s)\Big|_{s=-3} = \frac{10}{(s+4)^3}\Bigg|_{s=-3} = 10$$

$$A_1 = \frac{1}{1!}\,\frac{d}{ds}\,(s+3)^2\,F(s)\Big|_{s=-3} = \frac{-30}{(s+4)^4}\Bigg|_{s=-3} = -30$$

$$B_3 = (s+4)^3\,F(s)\Big|_{s=-4} = \frac{10}{(s+3)^2}\Bigg|_{s=-4} = 10$$

$$B_2 = \frac{1}{1!}\frac{d}{ds}(s+4)^3 F(s)\Big|_{s=-4} = \frac{-20}{(s+3)^3}\Big|_{s=-4} = 20$$

$$B_1 = \frac{1}{2!}\frac{d^2}{ds^2}(s+4)^3 F(s)\Big|_{s=-4} = \frac{30}{2(s+3)^4}\Big|_{s=-4} = 15$$

Substituting these values in eqn. (1.36)

$$F(s) = \frac{10}{(s+3)^2} - \frac{30}{(s+3)} + \frac{10}{(s+4)^3} + \frac{20}{(s+4)^2} + \frac{15}{(s+4)} \qquad (1.37)$$

From Table 1.1, inverse Laplace transform of eqn. (1.37) is

$$f(t) = 10\,te^{-3t} - 30e^{-3t} + \frac{10}{2!}t^2 e^{-4t} + 20t\,e^{-4t} + 15e^{-4t}$$

$$= 10\,[1.5\,e^{-4t} + t\,(e^{-3t} + 2e^{-4t}) + 0.5t^2\,e^{-4t} - 30e^{-3t}]$$

$$= 10\,[1.5\,(e^{-4t} - 2e^{-3t}) + t\,(e^{-3t} + 2e^{-4t}) + 0.5t^2\,e^{-4t}]$$

Problem 1.7. *Determine the inverse Laplace transform of the following function.*

$$F(s) = \frac{5}{(s^2+6s+10)}$$

Solution:

$$F(s) = \frac{5}{(s^2+6s+10)}$$

or

$$F(s) = \frac{5}{[(s+3)^2+1]} = \frac{5}{(s+3+j)\,(s+3-j)} \qquad (1.38)$$

Function (1.38) contains complex conjugate pole.

$$\therefore \quad F(s) = \frac{A}{(s+3+j)} + \frac{A^*}{(s+3-j)} \qquad (1.39)$$

The partial fraction coefficients of eqn. (1.39) are

$$A = (s+3+j)\,F(s)\Big|_{s=-3-j} = \frac{5}{(s+3-j)}\Big|_{s=-3-j}$$

$$= -\frac{5}{2j} = \frac{5}{2}e^{-j\pi/2}$$

$$A^* = (s+3-j)\,F(s)\Big|_{s=-3+j} = \frac{5}{(s+3+j)}\Big|_{s=-3+j}$$

$$= \frac{5}{2j} = \frac{5}{2}e^{-j\pi/2}$$

Substituting these coefficients in eqn. (1.39)

$$F(s) = \frac{5/2e^{j\pi/2}}{(s+3+j)} + \frac{5/2e^{-j\pi/2}}{(s+3-j)}$$

Taking inverse Laplace transform

$$F(t) = \frac{5}{2}\,[e^{-(3+j)t}\,e^{-j\,\pi/2} + e^{-(3-j)t}\,e^{-j\,\pi/2}]$$

$$= \frac{5}{2}\,e^{-3t}\,[e^{\,j(\pi/2-t)} + e^{-j(\pi/2-t)}]$$

$$= 5e^{-3t}\cos(\pi/2 - t)$$

$$= 5e^{-3t}\sin t$$

Problem 1.8 *Find the inverse Laplace transform of the following function.*

$$F(s) = \frac{3}{s(s^2+4)}$$

Solution:

$$F(s) = \frac{3}{s(s^2+4)} = \frac{3}{s(s+2j)(s-2j)}$$

$$= \frac{A}{s} + \frac{B}{(s+2j)} + \frac{B^*}{(s-2j)} \tag{1.40}$$

$$A = sF(s)\Big|_{s=0} = \frac{3}{(s+2j)\,(s-2j)} = \frac{3}{4}$$

$$B = (s+2j)\;F(s)\Big|_{s=-2j} = \frac{3}{s(s-2j)}\Big|_{s=-2j} = +\frac{3}{8j^2} = -\frac{3}{8}$$

$$B^* = (s-2j)\;F(s)\Big|_{s=2j} = \frac{3}{s(s+2j)}\Big|_{2j} = +\frac{3}{8j^2} = -\frac{3}{8}$$

Substituting these coefficients in eqn. (1.40)

$$F(s) = \frac{3}{4s} - \frac{3}{8(s+2j)} - \frac{3}{8(s-2j)}$$

$$= \frac{3}{4}\left[\frac{1}{s} - \frac{1}{2}\left\{\frac{1}{(s+2j)} + \frac{1}{(s-2j)}\right\}\right]$$

From Table 1.1, the inverse Laplace transform is

$$f(t) = \frac{3}{4}\left[1 - \frac{1}{2}(e^{-2jt} + e^{2jt})\right]$$

$$= \frac{3}{4}\,(1 - \cos 2t)$$

Problem 1.9. *Calculate the current i(t) in the integro differential equation*

$$\frac{di(t)}{dt} + 6\int_0^t i(x)\,.\,dx + 5\,i(t) = 10e^{-4t}.$$ *Assume zero initial conditions.*

Solution: Taking Laplace transform of both sides

$$[sI(s) - sI(0)] + 5I(s) + \frac{6}{s}\,[I(s) - I(0^+)] = \frac{10}{s+4}$$

Substituting initial conditions to be zero, *i.e.*, $I(0) = 0$

$$I(s)\left[s+5+\frac{6}{s}\right] = \frac{10}{(s+4)}$$

or
$$I(s) = \frac{10s}{(s+4)(s^2+5s+6)}$$

or
$$I(s) = \frac{10s}{(s+2)(s+3)(s+4)}$$

The partial fractions are

$$\frac{A_1}{(s+2)} + \frac{A_2}{(s+3)} + \frac{A_3}{(s+4)} \qquad (1.41)$$

$$A_1 = (s+2)\, I(s)\Big|_{s=-2} = \frac{10s}{(s+3)(s+4)}\Big|_{s=-2} = -10$$

$$A_2 = (s+3)\, I(s)\Big|_{s=-3} = \frac{10s}{(s+2)(s+4)}\Big|_{s=-3} = 30$$

$$A_3 = (s+4)\, I(s)\Big|_{s=-4} = \frac{10s}{(s+2)(s+3)}\Big|_{s=-4} = -20$$

Substituting these partial factor coefficients in eqn. (1.41) and taking inverse Laplace transform from Table 1.1.

$$\therefore \qquad i(t) = L^{-1}[I(s)] = -10\,[e^{-2t} - 3e^{-3t} - 2e^{-4t}].$$

Problem 1.10. *Determine the initial and final values of f(t) for the following functions:*

(*i*) $F(s) = \dfrac{8s(s+3)}{(s+1)(s+4)(s+5)}$ (*ii*) $F(s) = \dfrac{5(s+1)}{s(s+2)(s^2+s+1)}$

Solution: (*i*) $F(s) = \dfrac{8s(s+3)}{(s+1)(s+4)(s+5)} = \dfrac{8s^2+24s}{s^3+10s^2+29s+20}$

Applying initial value theorem,

$$\lim_{t\to 0} f(t) = \lim_{s\to\infty} sF(s) = \lim_{s\to\infty} \frac{8s(s^2+3s)}{(s^3+10s^2+29s+20)}$$

$$= \lim_{s\to\infty} \frac{8\left(1+\frac{3}{s}\right)}{\left(1+\frac{10}{s}+\frac{29}{s^2}+\frac{20}{s^3}\right)} = 8$$

Applying final value theorem,

$$\lim_{t\to\infty} f(t) = \lim_{s\to 0} sF(s) = \lim_{s\to 0} \frac{8s(s^2+3s)}{(s^3+10s^2+29s+20)} = 0$$

(*ii*) $$F(s) = \frac{5(s+1)}{s(s+2)(s^2+s+1)} = \frac{5(s+1)}{(s^4+3s^3+3s^2+2s)}$$

Applying initial value theorem,

$$\lim_{t\to 0} f(t) = \lim_{s\to\infty} sF(s) = \lim_{s\to\infty} \frac{5s(s+1)}{(s^4+3s^3+3s^2+2s)}$$

$$= \lim_{s\to\infty} \frac{5(s^2+s)}{(s^4+3s^3+3s^2+2s)}$$

$$= \lim_{s\to\infty} \frac{5\left(\frac{1}{s^2}+\frac{1}{s^3}\right)}{\left(1+\frac{3}{s}+\frac{3}{s^2}+\frac{2}{s^3}\right)} = 0$$

Applying final value theorem,

$$\lim_{t\to\infty} f(t) = \lim_{s\to 0} sF(s) = \frac{5s(s+1)}{(s^4+3s^3+3s^2+2s)}$$

$$= \lim_{s\to 0} \frac{5(s+1)}{(s^3+3s^2+3s+2)}$$

$$= \frac{5}{2} = 2.5$$

Problem 1.11. *Write the Laplace transform equations for a given RLC network shown in Fig. 1.1.*

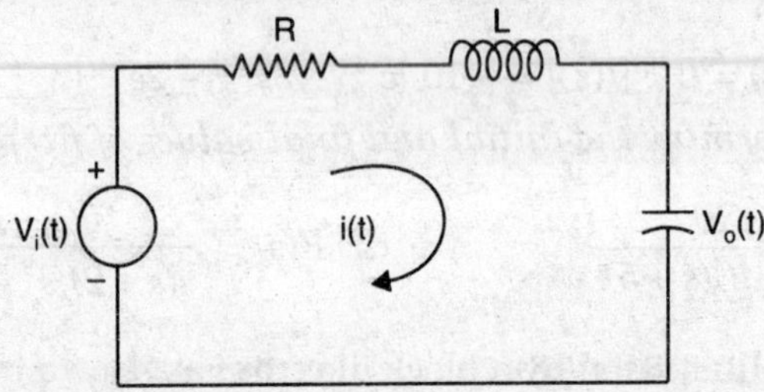

Fig. 1.1

Solution: Voltage loop equation,

$$V_i(t) = R\,i(t) + L\frac{di(t)}{dt} + \frac{1}{C}\int i(t)\,dt \tag{1.42}$$

Output voltage, $$V_0(t) = \frac{1}{C}\int i(t)\,dt \tag{1.43}$$

Taking Laplace transformation to eqns. (1.42) and (1.43)

$$V_i(s) = R\,I(s) + Ls\,I(s) + \frac{1}{Cs}I(s)$$

$$V_0(s) = \frac{1}{Cs}I(s)$$

Since Laplace transform of any derivative term is 's' times the variable. Similarly Laplace transform of any integral term is $\frac{1}{s}$ times the variable.

2 Mathematical Modelling

2.1 INTRODUCTION

Automatic control systems played an important role in the development of modern civilization. Air conditioners and furnaces with thermostatic control are examples of application of control system on domestic front. They help us to live comfortably whether we are in the tropics or cold regions. In industries, control systems are widely used to improve both the quality and quantity of products manufactured making them available for large number of people at cheap rates. In defense, most of the modern weapons are automatically controlled for accuracy by the application of control system. Thus we see that an engineer, to whatsoever branch he may belong, has to come face to face with control systems, sooner or later, either by way of its maintenance or its design and manufacture. The purpose of this book is to provide an understanding and critical appreciation of the basic principles which govern the theory and performance of the simplest type of control systems namely linear control systems, with continuous signals and constant parameters.

In order to solve the problems selecting to this, you should have a through knowledge of Laplace transform and their application as well as fundamentals of simplex variable theory.

2.2 BASIC CONTROL SYSTEM

A basic control system is illustrated by a block diagram as shown in Fig. 2.1

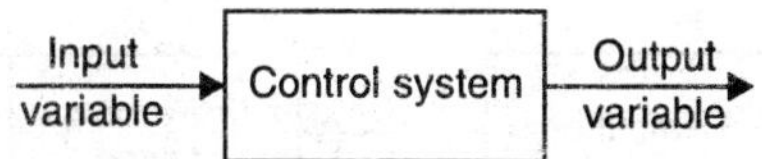

Fig. 2.1 Block diagram of basic control system

The basic control system mainly consists of reference input, control system and control output.

System: A system is collection of objectives connected in a design manner to perform the specified task.

Example: *Computer is a collection of CPU, monitor and keyboard.*

Control system: Control system is an arrangement of different objectives connected in such a manner so as to regulate, direct or command itself or some other system.

Input variable: It is a variable applied to control system from an external energy source in order to produce a specified output.

Output variable: It is a variable or response obtained from control system, when input is applied to the control system.

Control system: Control system consists of plant or process controller.

Plant or process: The portion of a system which is to be controlled.

Controller: The system which controls the plant or process.

Example for control system

Consider the steering system of an automobile. In this case, we can identify the items referred to the Fig. 2.1 as follows:

Input: Angular position of steering wheel.

Output: Position or direction of front wheels.

Control systems: Elements or components of steering mechanisms.

2.3 CLASSIFICATION OF CONTROL SYSTEMS

Control systems are classified under various heads as follows.

(*i*) Open loop and closed loop control systems

(*ii*) Linear and non-linear control systems

(*iii*) Static and dynamic systems

(*iv*) Continuous and discrete data systems

(*v*) Single input–single output and multi input–multi output systems.

(*i*) Open loop and closed loop control systems

Open loop system

Definition: Control action is independent of input, or change in input does not change the output.

In a practical control system, the output is controlled by a powering device which requires for stronger signals to operate it than those which correspond to input. As such, input signal is amplified by an amplifier before it is applied to powering device.

This situation is illustrated in Fig. 2.2.

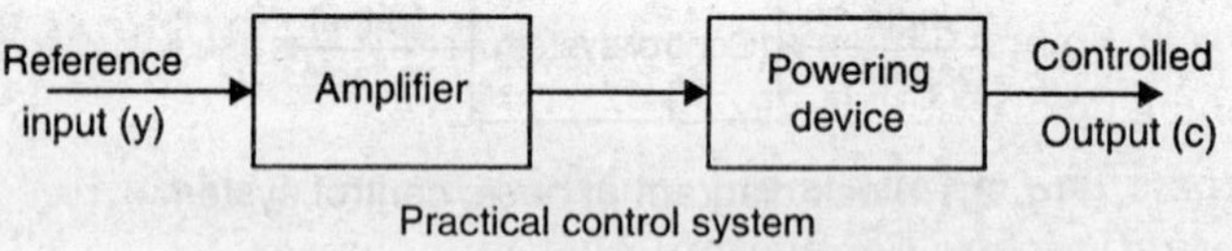

Fig. 2.2 Block diagram of open loop system

Examples of open loop control system

(*a*) *Furnace without thermostatic control but with timing device.* Human judgement is required to determine the setting of timing device in order to attain a given temperature. After preset time, the furnace is turned off and at that time, the temperature may be greater or less than the required temperature.

(*b*) *Washing machine with preset washing time.* The example given above, is not automatic because their time of operation is not based on the state of output.

Their effectiveness depends upon the experience of the person who sets the time of operation and do not adapt to the changes in environment. They also require frequent calibration.

Advantages of Open Loop System

- Simplicity
- Low cost
- Easy to construct
- Generally open loop systems are stable.

Disadvantages

- Lack of reliability and accuracy
- Change in output due to change in disturbances is not corrected automatically.

Closed Loop System

Definition. Control action somewhat depends upon input, or change in output is due to change in input.

In this type of control system, powering device is actuated by a signal proportional to the difference between the input and output. The principle of operation is illustrated by the block diagram as shown Fig. 2.3.

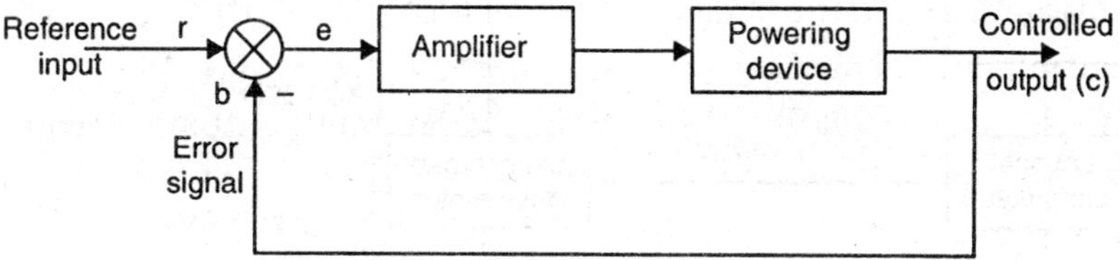

Fig. 2.3 Block diagram of closed loop control system

From Fig. 2.3, we note that powering device will be actuated only so long as there is error signal '*e*', which is the difference between input '*r*' and output '*c*'. Closed loop system is costlier than open loop system because additional components like error sensing device and feedback elements are required.

Types of Closed Loop Systems

(*a*) *Servomechanisms:* These are control systems whose output is a mechanical quantity such as position, velocity, etc.

(*b*) *Regulators:* The output is any quantity which is maintained at the desired level. The fluctuations are kept within allowable limits.

Examples: *Voltage regulators, speed regulators, frequency regulators, etc.*

(*c*) *Continuous time control systems:* Normally the plant or process which is being controlled has input and output as continuous functions of time. In case the control signal is applied continuously through a continuous feedback, it is a continuous time control system.

(*d*) *Discrete time control systems:* If the feedback is discrete in time as the output is sampled and / or the error is sampled, the controller gets the sampled error signal.

This is a discrete time or digital controller. The control system is then a digital or discrete time control system.

(*e*) *Relay Control Systems:* (On-off control systems). The controller action is effected when the relay is closed. The output is maintained at the desired level within allowable tolerances. When the output changes beyond the set limit of tolerance, the deviation is sensed and the relay is closed to enable the control action to take place.

Examples of Feed Back or Closed Loop Control Systems

(*a*) *Missile Launching and Guidance System.* In the defense applications of using missiles to hit the target enemy airplanes, feedback systems are extensively used. The probability of hitting the targets must be the highest. Fig. 2.4 shows a typical ground launching system for a surface to air missile.

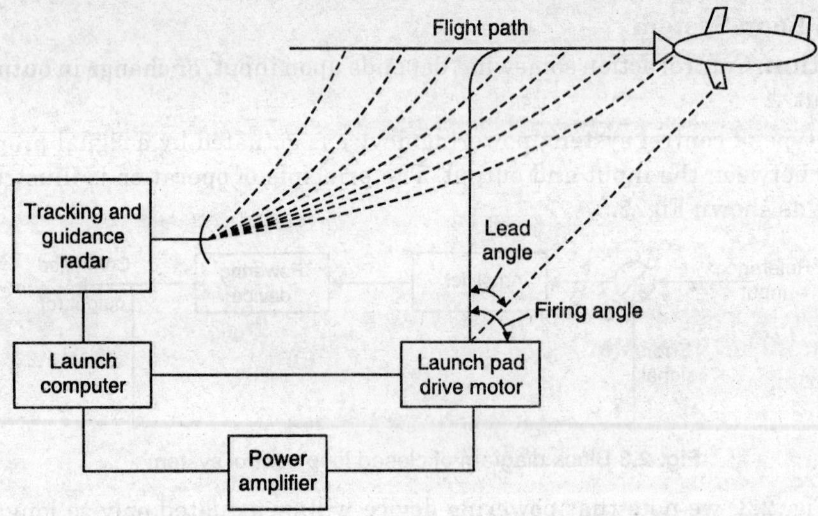

Fig 2.4

The radar senses the arrival of the target plane when one enters its zone of sensing. The location is measured in terms of the *x, y, z* co-ordinates and the respective time derivatives, accelerations, forces in the co-ordinate directions etc. The information is fed to a ground computer, which will compute the angle of lead to be given to the missile at the time of firing so that the missile takes a predetermined path and travels the distance exactly to meet the target plane in a future location. The correct angle to lead is obtained through a feedback, which senses the angle of lead and the computer takes the feedback signal, estimates the drive required and suitably gives the command signal to the drive motor through a power amplifier.

(*b*) *Temperature Control System for a Plating Bath:* In Fig. 2.5, a control system which controls the temperature of jacket water is described. The temperature of the water in the supply tank is controlled by controlling the relative amounts of steam and cold water in a mixing chamber (supply tank) and it is measured by a thermocouple. The feedback signal is then provided by the thermocouple to the controller which in turn corrects the valve settings of the steam and cold water supply lines.

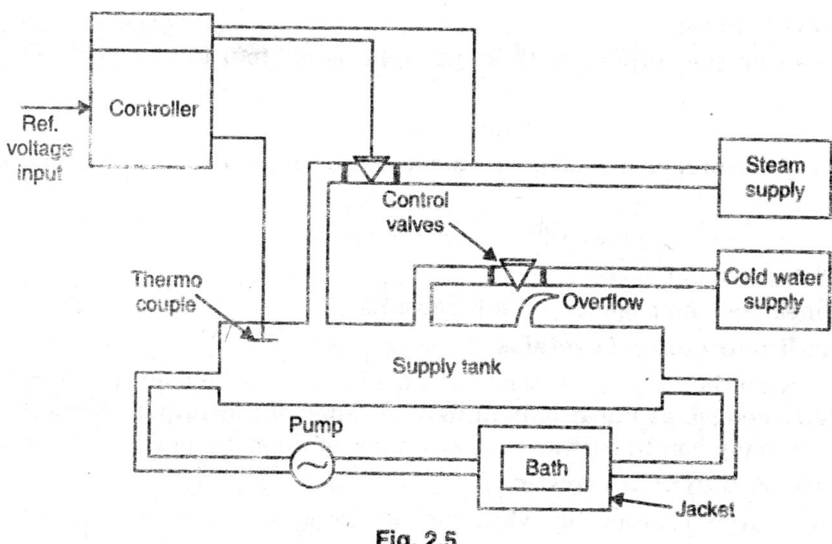

Fig. 2.5

Feedback System in Nature

(*a*) *Biological system:* A very good example of feedback system is the temperature control in the human body. The normal temperature of the body is 98.4°F or 36.8°C which is a very comfortable temperature for the human being. During the summer, when the outside temperature goes up by about 5°C, the body absorbs the excess heat of the radiation and the temperature goes up. The thermostat action is started by the skin and causes the sweating or perspiration. The moisture is produced on the skin so that the air contacting the skin and the surface temperature gets cooled. This happens as the skin expands and the pores on the skin get widely opened. On the contrary during the winter, the outside temperature goes below the normal temperature of the body. Now the skin gets contracted and all the pores are closed preventing the loss of body heat to the surrounding medium. In normal days, the skin pores are optimally conditioned so as to keep the body temperature optimum and constant within acceptable variations.

Non-Engineering Applications

The fiscal policy of a government is based on the dynamics of the economic system in the society or the state. All such systems are non-linear. However, they are modeled by making minimum assumptions. The feedback in such systems is provided through media or citizen forum etc.

(*ii*) **Linear control systems**

In most of the control systems, the output is uniformly proportional to input, the output and input are functions of time. We denote this fact by writing $c(t)$ and $r(t)$.

Linear control systems satisfy the principle of superposition and homogeneous property.

Superposition: Consider a system described by an algebraic equation.

$$y = f(x)$$

It implies that the output y with input $(x_1 + x_2)$ is equal to the sum of output y_1 with input x_1 and y_2 with input x_2

i.e.,
$$f(x_1 + x_2) = f(x_1) + f(x_2) \tag{2.1}$$

Homogeneous property

It implies that the output y with input (αx) is equal to α times y with input x for any scalar constant α,

i.e.,
$$f(\alpha x) = \alpha f(x) \tag{2.2}$$

Both the above properties can be combined into a single one, the linear property implies that

$$f(\alpha x_1 + \beta x_2) = \alpha f(x_1) + \beta f(x_2) \tag{2.3}$$

where α and β are constants.

Thus, linearity = homogeneity + superposition

(*iii*) **Non-linear control systems**

Practically all the physical systems have a non-linear behaviour. In other words, many physical systems behave as linear only in limited range of the output around the operating point. There are many non-linearities and only a few of them are modeled and defined.

(*iv*) **Static and dynamic systems**

The input-output relationship which do not involve derivation w.r.t the time '*t*' the systems are said to be static. Otherwise the system is said to be dynamic system. All resistive networks are static systems and RLC networks are dynamic systems.

(*v*) **Continuous and discrete data systems**

In Fig. 2.3, the signals r, c and e are considered as continuous functions of time. Thus Fig. 2.3 represents as continuous data control system. Position control of load is an example of continuous data control system.

On the other hand if any or all of the signals r, c and e are considered only at discrete time intervals instead of continuous, then we have discrete data or sampled data system. A radar tracking system is an example of sampled data system.

(*vi*) **Single input–single output and multi input–multi output systems**

In the examples so far considered, there is a single output controlled by single input. They are called SISO systems. On the other hand if a set of output variables $\bar{c}$ are to be controlled by another set of input variables $\bar{r}$, then we have multi input–multi output system. We call $\bar{c}$ and $\bar{r}$, as vectors, each consisting of several elements.

2.4 COMPARISON BETWEEN THE OPEN LOOP AND CLOSED LOOP CONTROL SYSTEMS

Open loop system	*Closed loop system*
1. So long as the calibration is good, an open loop system performance will be accurate.	1. Due to feedback, the performance of closed loop system is accurate.
2. Organization is simple and easy to construct	2. Complicated and difficult
3. Generally stable in operation	3. Stability depends on system components.
4. If non-linearity present, system operation degenerates.	4. Comparatively, the performance is better than open loop system, if non-linearity present.

2.5 THERMAL SYSTEM

A thermal system is made up of heat exchangers, and heat storage devices. The heat storage device is a thermal capacitance and heat exchange device is a thermal resistance. The following input-output relationship holds for these elements.

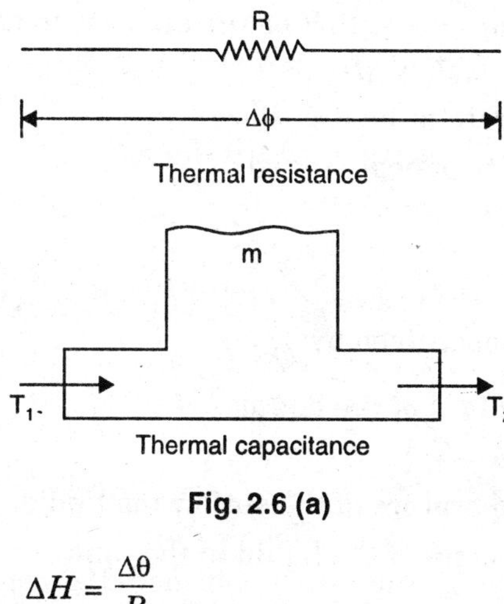

Fig. 2.6 (a)

$$\Delta H = \frac{\Delta\theta}{R}$$

where ΔH = incremental increase in heat energy

R = thermal resistance

$\Delta\theta$ = incremental rise in the temperature of the medium

Illustrate the modeling of a thermal system. The system is assumed to be a linear system. Thermal systems in general, are not linear. Hence the example is a limited case of modeling.

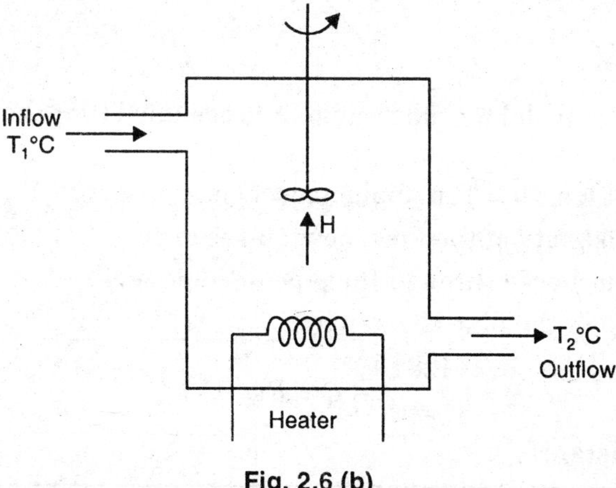

Fig. 2.6 (b)

Fig. 2.6 (*b*) shows a storage heater for a liquid. The inlet temperature of the liquid is T_1 °C and outlet T_2°C. Heat is given to the liquid in the tank through the heater. A uniform temperature is maintained for the liquid in the tank, by consistently stirring the contents of the tank. Let H be the rate at which heat is input by the heater.

Assumptions

- There is no heat gain due to external sources or heat loss for the liquid in the tank. Hence the entire liquid in the tank is at only one temperature.
- The liquid inflow and outflow of the tank is a constant.

For thermal equilibrium, the rate of heat input must be balanced by the rate of heat outflow and the rate of heat storage. Let ΔH be the increase in the rate of heat input.

Thus, $$\Delta H = \Delta H_1 + \Delta H_2$$

where ΔH_1 = increase in rate of heat outflow

ΔH_2 = increase in rate of heat storage

$$\Delta H = \frac{\Delta T}{R} + C\frac{d(\Delta T)}{dt} \tag{2.4}$$

where R is the thermal resistance, given by $\frac{1}{QS}$

Q is the steady rate of inflow of the liquid,

S is the specific heat and

ΔT is the rise in the temperature of the outflowing liquid.

$C = MS$, where M is the mass of the liquid in the tank,

$\frac{d\Delta t}{dt}$ is the rate of temperature rise of the liquid in the tank.

Rearranging of eqn. (2.4), we have

$$RC\frac{d(\Delta T)}{dt} + \Delta T = R.\,\Delta H \tag{2.5}$$

Equation (2.5) is the model of the system.

2.6 FLUID SYSTEM

In order to obtain a linear model for the system, it is assumed that the fluid is incompressible and the flow is laminar.

Consider a pipe of length = 1 m, diameter = D m.

μ = Viscosity of the flow, newton-sec/m^2,

Q = discharge through the pipe in cubic sec.

The pressure drop in the pipe is given by

$$P = \left(\frac{1281\mu}{\pi D^4}\right)Q = RQ \tag{2.6}$$

where R is the fluid resistance.

The rate of fluid storage in the tank:

$$dV = A\frac{dH}{dt} \tag{2.7}$$

where A is the area of cross-section of the tank in square-meter and

H is height of tank in meters.

Since the pressure is

$$P = \rho g H \tag{2.8}$$

At the bottom of the tank

$$H = \frac{P}{\rho g} \tag{2.9}$$

Substituting H from eqn. (2.9) in eqn. (2.7)

$$dV = \frac{A}{\rho g}\left(\frac{dP}{dt}\right) = C\,\frac{dP}{dt} \tag{2.10}$$

where C is the capacity of the tank.

Example: A liquid level system is shown in Fig. 2.7.

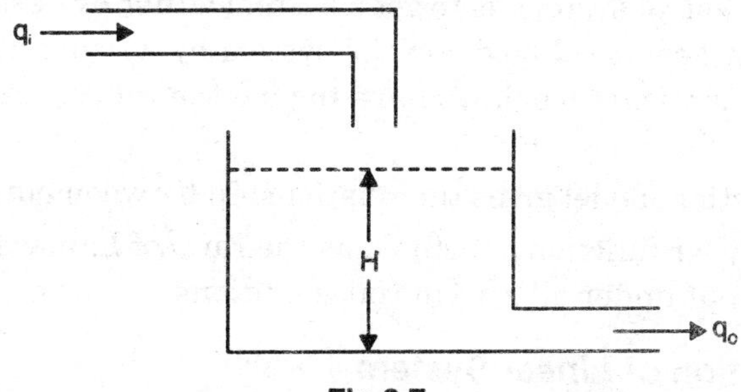

Fig 2.7

Liquid is flowing into the tank with a steady rate inlet discharge of q_1 cusecs and flowing out of the tank with a steady rate discharge of q_0 cusecs. Let H be the steady state level in the tank. Hence $q_i = q_0$ in order to maintained H steady.

Let Δq_i be the increase in q_1

Let ΔH be the increase in the height at H

Then increase in the outflow $\Delta q_0 = \dfrac{\Delta H}{R}$

where $R = \dfrac{1281\mu}{\pi D^2 \rho g}$ considering for laminar flow.

Rate of liquid storage in the tank is given by

$$= \text{Rate of inlet discharge} - \text{Rate of outlet discharge}$$

$$= \Delta q_i - \Delta q_0$$

$$= \Delta q_i - \frac{\Delta H}{R}$$

$$= C\,\frac{dH}{dt} \tag{2.11}$$

where C is the capacity of the tank.

Rearranging the terms in eqn. (2.11), we have

$$RC\,\frac{d\Delta H}{dt} + \Delta H = R\,(\Delta q_i) \tag{2.12}$$

Equation (2.12) is the model of the system.

2.7 TRANSFER FUNCTION

The nature of control system engineering involves consideration of two problems. These are analysis and design of a control system configuration.

***Analysis*:** It is the investigation of the properties of the existing system.

***Design*:** The problem in design is the choice and arrangement of control system components to perform specified task.

Design by analysis and design by synthesis are the characteristics of existing or standard system configuration.

***Design by synthesis*:** By defining the form of the system from its specifications.

Transfer Function model

First step in the control system, is to get a system either by design or by analysis. After that we have to give mathematical model of the system by writing mathematical equations. After getting the mathematical model, we have to give the relationship between output and input of the system.

The transfer function model gives the relationship between output and input.

***Definition*:** Transfer function is defined as the ratio of Laplace transform of output to Laplace transform of input under all zero initial conditions.

2.7.1 Transfer Function of Linear System

Consider a single input–single output system, where the input $r(t)$ and output $c(t)$ are the functions of time. The system has got linear time invariant parameters. The block diagram of the system in Fig. 2.1 is considered. Since $r(t)$ and $c(t)$ are continuous in time and are analytical functions, their Laplace transform can be obtained from

$$R(s) = L[r(t)] = \int_0^{\infty} r(t)\, e^{-st}\, dt$$

$$C(s) = L[c(t)] = \int_0^{\infty} c(t)\, e^{-st}\, dt \qquad (2.13)$$

The system with transformed inputs and outputs is reproduced in Fig. 2.1.

And for the system, if

$$C(s) = G(s)\, R(s)$$

the $G(s)$ is called the transfer function of the system, which is obtained when the system parameters are transformed into s-domain. We had seen already that a dynamic system is described by ordinary differential equation, connecting the input and output.

Hence for given model and input, output can be obtained by solving the differential equation. It is well known that classical methods of solution of differential equation of higher order than 2, are cumbersome and difficult. So Laplace transform is used to advantage that the differential equation is converted into an algebraic equation. To solve an algebraic equation is considered less cumbersome to differential equations. In this process, if the initial conditions are assumed to be zero, when the input is applied, the resulting ratio

$$\frac{C(s)}{R(s)} = G(s),$$

The above expression represents the transfer function of the system.

Block diagram representation of open loop system is shown in the Fig. 2.8.

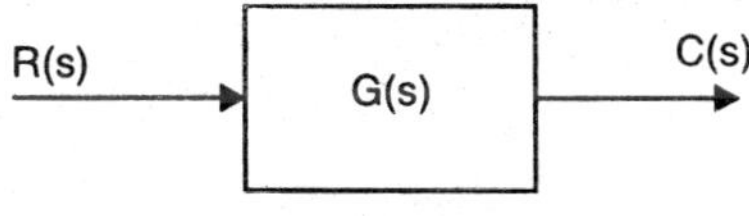

Fig. 2.8

In the block diagram, $C(s)$ is output, $R(s)$ is input.

$\therefore$ The transfer function of the system, $\dfrac{C(s)}{R(s)} = G(s)$.

PROBLEMS

Problem 2.1. *Define transfer function and determine the transfer function of RLC series circuit if voltage across the capacitor is output variable and input is voltage source V(s).*

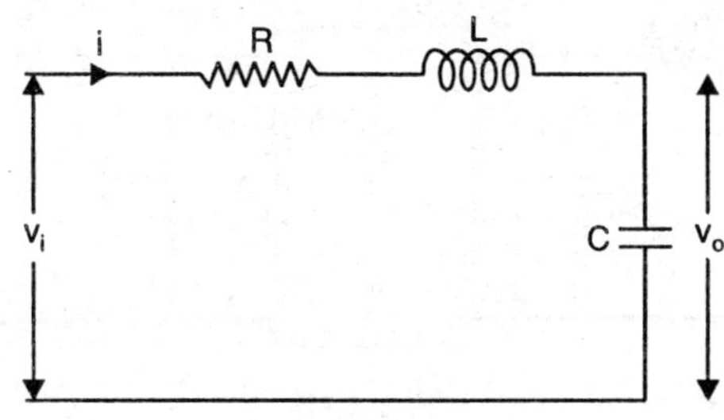

Fig 2.9

Solution:

Transfer function: It is defined as the ratio of Laplace transform of output to Laplace transform of input under initial zero conditions.

RLC series circuit

By applying KVL for the electrical network of Fig. 2.9

$$V_i(t) = R\, i(t) + L\frac{d\,i(t)}{dt} + \frac{1}{C}\int i(t)\, dt \tag{2.14}$$

And output voltage, $V_0(t) = \dfrac{1}{C}\displaystyle\int i(t)\, dt$ (2.15)

Taking Laplace transformation of eqns. (2.14) and (2.15), we get

$$V_i(s) = R\, I(s) + Ls\, I(s) + \frac{1}{Cs} I(s) \text{ (since initial conditions are zero)}$$

$$= \left[R + Ls + \frac{1}{Cs}\right] I(s) \tag{2.16}$$

and

$$V_0(s) = \frac{1}{Cs} I(s) \tag{2.17}$$

Dividing the eqn. (2.17) with eqn. (2.16)

Transfer function of the system, $\dfrac{V_0(s)}{V_i(s)}$

$$= \frac{\frac{1}{Cs} I(s)}{\left[R + Ls + \frac{1}{Cs}\right] I(s)}$$

$$= \frac{1}{LCs^2 + RCs + 1} = \frac{\frac{1}{LC}}{s^2 + \left(\frac{R}{L}\right)s + \frac{1}{LC}}$$

Problem 2.2. *Find the transfer function of a circuit as shown in Fig. 2.10.*

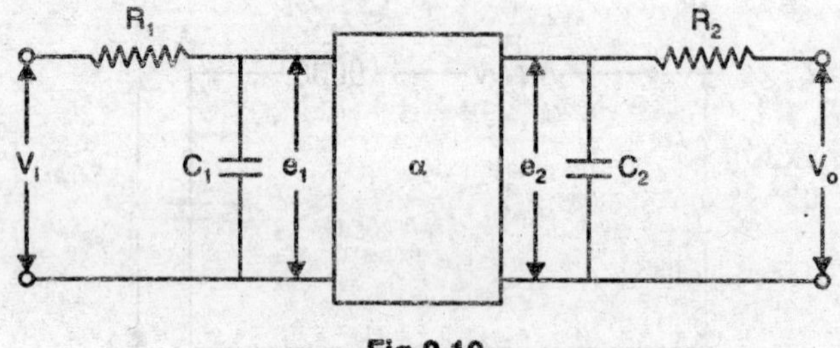

Fig 2.10

Solution: Applying KVL

$$V_i(t) = i(t)\, R_1 + \frac{1}{C_1} \int i_1 (t)\, dt \tag{2.18}$$

$$e_1 (t) = \frac{1}{C_1} \int i_1 (t)\, dt \tag{2.19}$$

$$e_2 (t) = \frac{1}{C_2} \int i_2 (t)\, dt \tag{2.20}$$

$$V_0 (t) = \frac{1}{C_2} \int i_2 (t)\, dt + i_2 (t)\, R_2 \tag{2.21}$$

Taking Laplace transform of eqns. (2.18) to (2.21)

$$V_i (s) = I_1 (s) \left[R_1 + \frac{1}{C_1 s}\right] \tag{2.22}$$

$$E_1(s) = \frac{I_1(s)}{C_1 s}$$

$$\Rightarrow \qquad I_1(s) = C_1 s\, E_i(s) \tag{2.23}$$

$$V_0(s) = \frac{1}{C_2 s} I_2(s) + I_2(s) R_2 \tag{2.24}$$

$$\Rightarrow \qquad I_2(s) = \frac{V_0(s)}{R_2 + \frac{1}{C_2 s}} \tag{2.25}$$

But $\qquad E_1(s) = E_2(s)$

Substituting eqn. (2.23) in eqn. (2.22)

$$V_i(s) = C_1 s\, E_1(s) \left[R_1 + \frac{1}{C_1 s} \right]$$

$$= E_1 (R_1 C_1 s + 1)$$

$$= E_2 (R_1 C_1 s + 1) \qquad [\because \; E_1 = E_2] \tag{2.26}$$

From eqn. (2.20),

$$E_2(s) = \frac{I_2(s)}{C_2 s}$$

$$= \frac{1}{C_2 s} \times \frac{V_0(s)}{R_2 + \frac{1}{C_2 s}} \tag{2.27}$$

Substituting eqn. (2.27) in eqn. (2.26)

$$V_i(s) = \frac{V_0(s)}{R_2 + \frac{1}{C_2 s}} \times \frac{1}{C_2 s} \times (R_1 C_1 s + 1)$$

$$= \frac{V_0(s)}{(R_2 C_2 s + 1)} (R_1 C_1 s + 1)$$

$$\frac{V_0(s)}{V_i(s)} = \frac{(R_2 C_2 s + 1)}{(R_1 C_1 s + 1)} = \frac{(1 + \tau_2 s)}{(1 + \tau_1 s)}$$

where $\quad \tau_1 = R_1 C_1$

$\tau_2 = R_2 C_2$

Problem 2.3. *Derive the transfer function of the electrical network shown as in Fig. 2.11.*

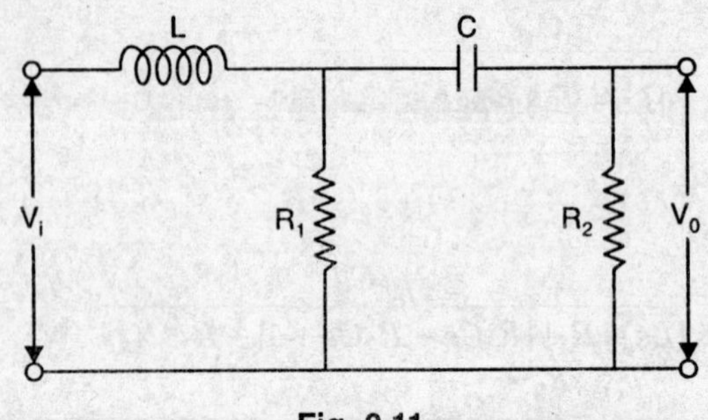

Fig. 2.11

Solution:

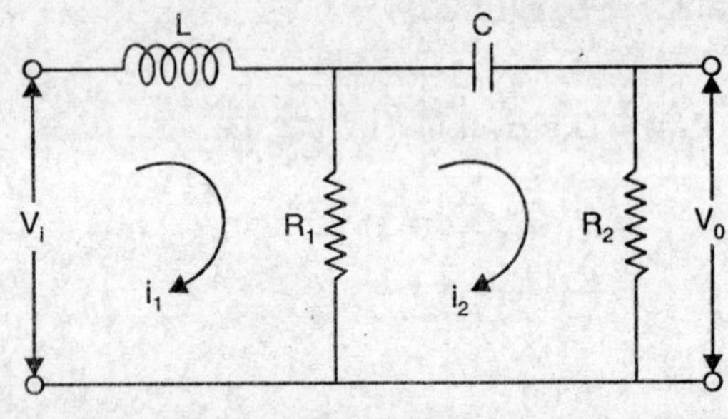

Fig. 2.12

Consider loop 1, $V_i(t) = L\dfrac{di_1(t)}{dt} + R_1[i_1(t) - i_2(t)]$ (2.28)

Loop 2, $0 = R_1(i_2(t) - i_1(t) + \dfrac{1}{C}\int i_2\,dt + R_2 i_2(t)$ (2.29)

And output voltage, $V_o(t) = R_2 i_2(t)$ (2.30)

Taking Laplace transformation on both sides of eqns. (2.28), (2.29) and (2.30)

$$V_i(s) = [Ls + R_1]\,I_1(s) - R_1 I_2(s) \quad (2.31)$$

$$0 = -R_1 I_1(s) + \left[R_1 + \frac{1}{Cs} + R_2\right] I_2(s) \quad (2.32)$$

and $$V_0(s) = R_2 I_2(s) \quad (2.33)$$

Writing $I_1(s)$ from eqn. (2.32) in terms of $I_2(s)$

$$I_1(s) = \frac{\left[R_1 + \frac{1}{Cs} + R_2\right]}{R_1} I_2(s) \quad (2.34)$$

Substituting $I_2(s)$ from eqn. (2.34) in eqn. (2.31)

$$V_i(s) = \left[\frac{[Ls + R_1]\left[R_1 + \frac{1}{Cs} + R_2\right]}{R_1} - R_1\right] I_2(s) \quad (2.35)$$

Dividing the eqn. (2.33) with eqn. (2.35)

$$\frac{V_0(s)}{V_i(s)} = \frac{R_2 I_2(s)}{\left[\dfrac{[Ls+R_1]\left[R_1+\dfrac{1}{Cs}+R_2\right]-R_i}{R_1}\right] I_2(s)}$$

$$= \frac{R_2 R_1 Cs}{[Ls+R_1]\,[R_1Cs+R_2Cs+1]-R_1^2\,Cs]}$$

$\therefore$ Transfer function, $\dfrac{V_0(s)}{V_i(s)} = \dfrac{R_1R_2Cs}{[Ls+R_1]\,[Cs\,(R_1+R_2)+1]-R_1^2\,Cs}$

Problem 2.4. *Obtain the transfer function for the electrical network as shown in Fig. 2.13.*

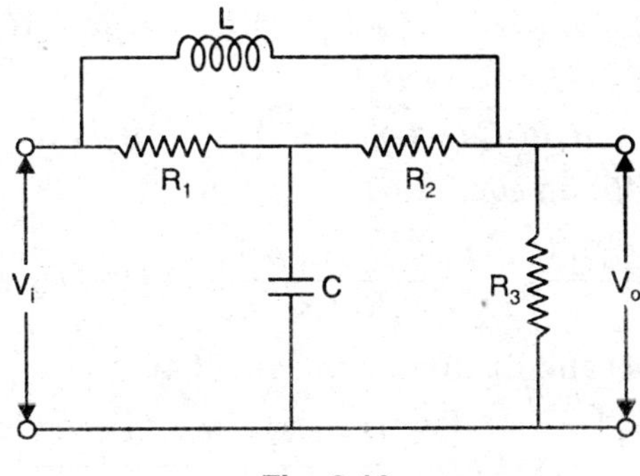

Fig. 2.13

Solution:

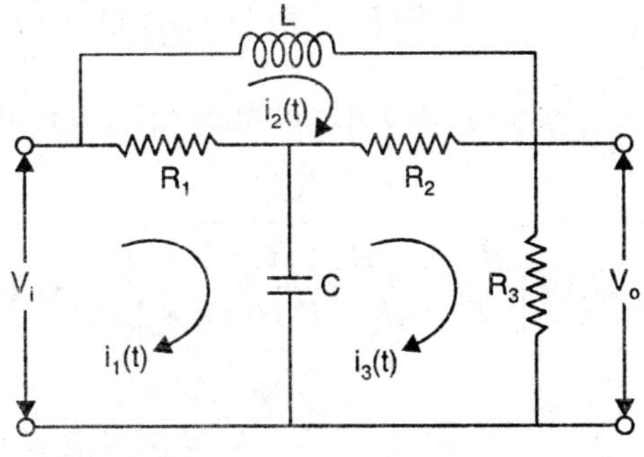

Fig. 2.14

Applying KVL for loop 1, $V_i(t) = R_1(i_1-i_2) + \dfrac{1}{C}\int (i_1-i_3)\,dt$ (2.36)

Loop 2, $0 = R_1(i_2-i_1) + L\dfrac{di_2}{dt} + R_2(i_2-i_3)$ (2.37)

Loop 3, $0 = \dfrac{1}{C}\int (i_3-i_1)\,dt + R_2(i_3-i_2) + R_3\,i_3$ (2.38)

Output equation, $V_0(t) = R_3\,i_3$ (2.39)

Taking Laplace transfer function on both sides of eqns. (2.36), (2.37), (2.38) and (2.39)

$$V_i(s) = I_1(s)\left[R_1 + \frac{1}{Cs}\right] - R_1 I_2(s) - \frac{1}{Cs} I_3(s) \qquad (2.40)$$

$$0 = -R_1 I_1(s) + [Ls + R_1 + R_2]\, I_2(s) - R_2 I_3(s) \qquad (2.41)$$

$$0 = -\frac{1}{Cs} I_1(s) - R_2 I_2(s) + \left[\frac{1}{Cs} + R_3 + R_2\right] I_3(s) \qquad (2.42)$$

and

$$V_0(s) = R_3 I_3(s) \qquad (2.43)$$

Writing $I_1(s)$ from eqn. (2.42), in terms of $I_2(s)$ and $I_3(s)$

$$I_1(s) = -R_2\, Cs\, I_2(s) + [1 + (R_2 + R_3)\, Cs]\, I_3(s) \qquad (2.44)$$

Substituting $I_1(s)$ from eqn. (2.44) in eqn. (2.41)

$$0 = -R_1 \{[-R_2\, Cs\, I_2(s) + [1 + (R_2 + R_3)\, Cs]\, I_3(s)\} + [Ls + R_1 + R_2]\, I_2(s) - R_2 I_3(s)$$

$$= R_1 R_2\, Cs\, I_2(s) - R_1 [1 + (R_2 + R_3)\, Cs]\, I_3(s) + [Ls + (R_1 + R_2)]\, I_2(s) - R_2 I_3(s)$$

$$= I_2(s)\, [R_1 R_2\, Cs + Ls + R_1 + R_2) - [R_1 (1 + (R_2 + R_3)\, Cs] + R_2]\, I_3(s)$$

$$\therefore \quad I_2(s) = \frac{[R_1 (1 + (R_2 + R_3)\, Cs) + R_2]}{(R_1 + R_2 + R_1 R_2\, Cs + Ls)} I_3(s) \qquad (2.45)$$

Substituting eqn. (2.45) in eqn. (2.44)

$$I_1(s) = \left\{-R_2\, Cs\left[\frac{R_1 (1 + (R_2 + R_3)\, Cs + R_2}{R_1 + R_2 + R_1 R_2\, Cs + Ls}\right] + [1 + (R_2 + R_3)\, Cs]\right\} I_3(s)$$

Substitute eqns. (2.44) and (2.45) and in eqn. (2.40)

$$V_i(s) = \left\{\left[R_1 + \frac{1}{Cs}\right]\left\{-R_2\, Cs\left[\frac{R_1 [1 + (R_2 + R_3)\, Cs + R_2}{R_1 + R_2 + R_1 R_2\, Cs + Ls}\right] + [1 + (R_2 + R_3)\, Cs]\right\}\right.$$

$$\left. R_1\left[\frac{R_1 [1 + (R_2 + R_3)\, Cs] + R_2}{R_1 + R_2 + R_1 R_2\, Cs + Ls}\right] - \frac{1}{Cs}\right\} I_3(s) \qquad (2.46)$$

$\therefore$ Transfer function, $\dfrac{V_o(s)}{V_i(s)}$ is the ratio of eqns. (2.43) and (2.46)

$$\frac{V_o(s)}{V_i(s)} = \frac{R_3}{\left(R_1 + \dfrac{1}{Cs}\right)\left\{-R_2 Cs\left[\dfrac{R_1 [1 + (R_2 + R_3)\, Cs] + R_2}{R_1 + R_2 + R_1 R_2\, Cs + Ls}\right] + [1 + (R_2 + R_3)\, Cs] - R_1\left[\dfrac{R_1 [1 + (R_2 + R_3)\, Cs] + R_2}{R_1 + R_2 + R_1 R_2\, Cs + Ls}\right] - \dfrac{1}{Cs}\right\}}$$

Problem 2.5. *Derive the transfer function of the network as shown in Fig. 2.15.*

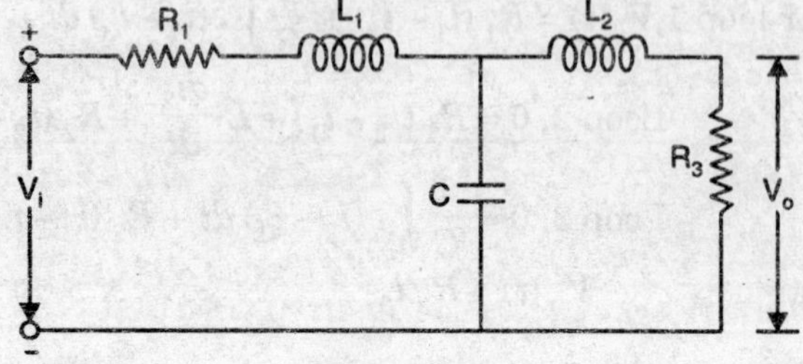

Fig. 2.15

Solution:

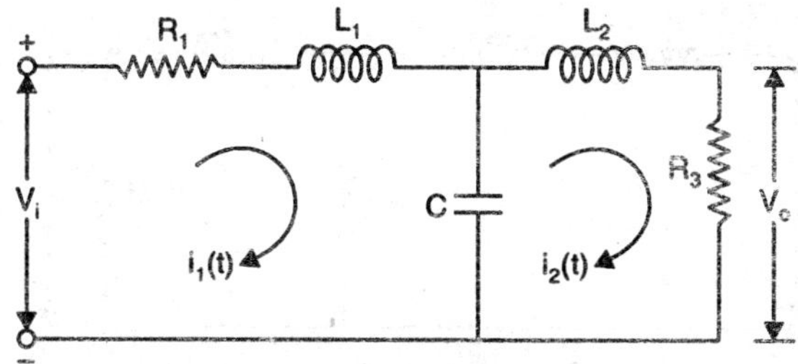

Fig. 2.15 (a)

Applying KVL for loop 1:

$$V_i(t) = R_1 i_1 + L_1 \frac{di_1(t)}{dt} + \frac{1}{C}\int (i_1(t) - i_2(t))\, dt \qquad (2.47)$$

For loop 2:
$$0 = \frac{1}{C}\int (i_2(t) - i_1(t)\, dt + L_2 \frac{di_2}{dr} + R_3 i_2(t) \qquad (2.48)$$

Output voltage, $V_o(t) = R_3 i_2(t)$ (2.49)

Taking Laplace transfer function on both sides of eqns. (2.47), (2.48) and (2.49)

$$V_i(s) = I_1(s)\left[R_1 + L_1 s + \frac{1}{Cs}\right] - \frac{1}{Cs} I_2(s) \qquad (2.50)$$

$$0 = -\frac{1}{Cs} I_1(s) + \left[L_2 s + \frac{1}{Cs} + R_3\right] I_2(s) \qquad (2.51)$$

$$V_o(s) = R_3 I_2(s) \qquad (2.52)$$

Writing $I_1(s)$ from eqn. (2.51), in terms of $I_2(s)$

$$I_1(s) = Cs\left[L_2 s + \frac{1}{Cs} + R_3\right] I_2(s)$$

$$= [L_2 Cs^2 + R_3 Cs + 1] I_2(s) \qquad (2.53)$$

Substituting eqn. (2.53) in eqn. (2.50)

$$V_i(s) = \left[R_1 + L_1 s + \frac{1}{Cs}\right][L_2Cs^2 + R_3Cs + 1]\ I_2(s) - \frac{1}{Cs} I_2(s)\Big]$$

$$= \frac{[L_1Cs^2 + R_1Cs + 1]\ [L_2Cs^2 + R_3Cs + 1] - 1]}{Cs} I_2(s) \qquad (2.54)$$

Dividing eqn. (2.52) by eqn. (2.54)

$$\frac{V_o(s)}{V_i(s)} = \frac{R_3\, I_2(s)}{[(L_1Cs^2 + R_1Cs + 1)(L_2Cs^2 + R_3Cs + 1) - 1]} \times \frac{Cs}{I_2(s)}$$

$\therefore$ Transfer function, $$\frac{V_o(s)}{V_i(s)} = \frac{R_3\, Cs}{[L_1Cs^2 + R_1Cs + 1]\ [L_2Cs^2 + R_3Cs + 1]}\,.$$

Problem 2.6. *An electrical network is shown in Fig. 2.16, derive its transfer function.*

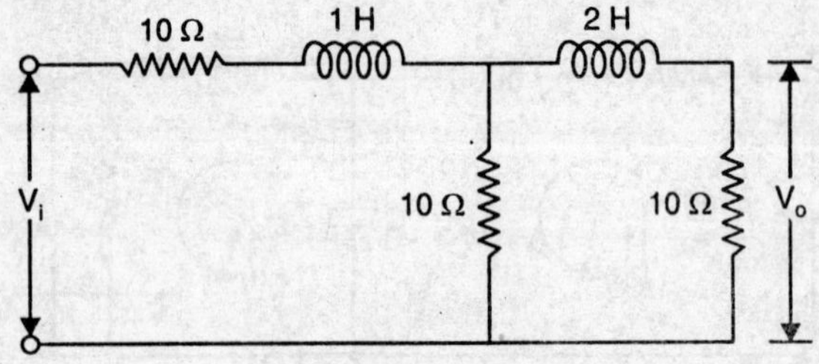

Fig. 2.16

Solution:

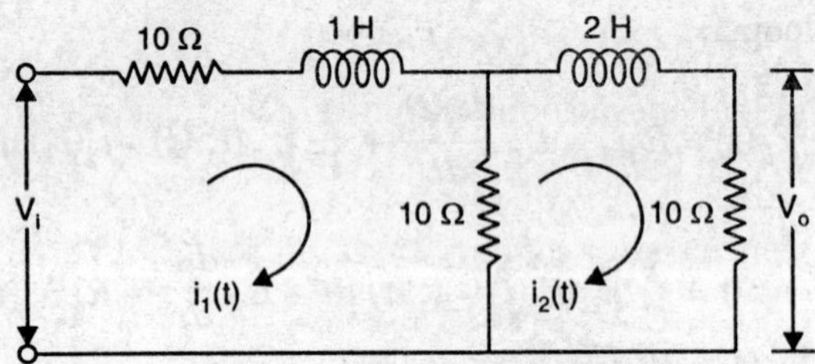

Fig. 2.16 (a)

Applying KVL for loop 1: $V_i(t) = 10\, i_1(t) + 1\, \dfrac{di_1(t)}{dt} + 10\, [i_1(t) - i_2(t)]$ (2.55)

For loop 2 : $0 = 10\, [i_2(t) - i_1(t)] + 2\, \dfrac{di_2(t)}{dt} + 10\, i_2(t)$ (2.56)

Output voltage, $V_o(t) = 10\, i_2(t)$ (2.57)

Taking Laplace transfer function on both sides of eqns. (2.55), (2.56) and (2.57)

$$V_i(s) = I_1(s)\,[10 + s + 10] - 10\, I_2(s) \quad (2.58)$$

$$0 = -\,10\, I_1(s) + [10 + 2s + 10]\, I_2(s) \quad (2.59)$$

$$V_o(s) = 10\, I_2(s) \quad (2.60)$$

Writing $I_1(s)$ from eqn. (2.59), in terms of $I_2(s)$

$$I_1(s) = 0.1\,[20 + 2s]\, I_2(s) \quad (2.61)$$

Substituting $I_1(s)$ from eqn. (2.61) in eqn. (2.58)

$$V_i(s) = [20 + s] \times 0.1\,[20 + 2s]\, I_2(s) - 10\, I_2(s)]$$

$$= [(20 + s) \times 0.2\,(s + 10) - 10\,]\, I_2(s)$$

$$= [0.2\,[20s + 10s + s^2 + 200] - 10]\, I_2(s)$$

$$= [6s + 0.2s^2 + 40 - 10]\, I_2(s) = [0.2s^2 + 6s + 30]\, I_2(s) \quad (2.62)$$

Dividing eqn. (2.60) by eqn. (2.62)

$$\frac{V_o(s)}{V_i(s)} = \frac{10\, I_2(s)}{[0.2s^2 + 6s + 30]\, I_2(s)}$$

$$= \frac{10}{\dfrac{2}{10}\,[s^2 + 30s + 150]} = \frac{50}{s^2 + 30s + 150}$$

$\therefore$ Transfer function, $\dfrac{V_o(s)}{V_i(s)} = \dfrac{50}{s^2 + 30s + 150}$.

Problem 2.7. *Find the transfer function of an operational amplifier as shown in Fig. 2.17.*

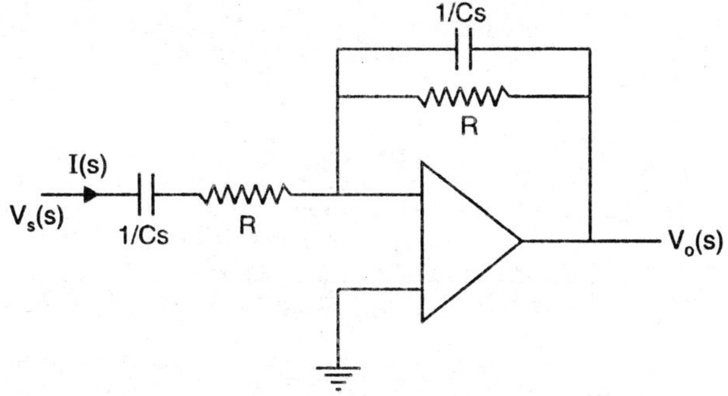

Fig. 2.17

Solution: In an ideal operational amplifier, the inverting terminal is a virtual ground.

Hence, $I_s(s) = -I_f(s)$

Let $Z_F(s)$ represent parallel combination of R and $\dfrac{1}{Cs}$

$$Z_F(s) = \frac{R}{RCs+1}$$

$$I_{f(s)} = \frac{V_o(s)}{Z_F(s)} = \frac{RCs+1}{R}\, V_o(s)$$

$$V_s(s) = I_s(s)\, Z_{in}(s)$$

$$= -I_f(s).\, Z_{in}(s)$$

$$= -\frac{(RCs+1)}{R}\, V_o(s)\left[R + \frac{1}{Cs}\right] \qquad \left(\because\ Z_{in}(s) = R + \frac{1}{Cs}\right)$$

$$= -\frac{(RCs+1)^2}{RCs}\, V_o(s)$$

$$\therefore \qquad \frac{V_o(s)}{V_s(s)} = -\frac{RCs}{(RCs+1)^2}$$

Problem 2.8. *For network shown in Fig. 2.18, determine the transfer function* $\dfrac{V_2(s)}{V_1(s)}$.

Assume zero initial conditions.

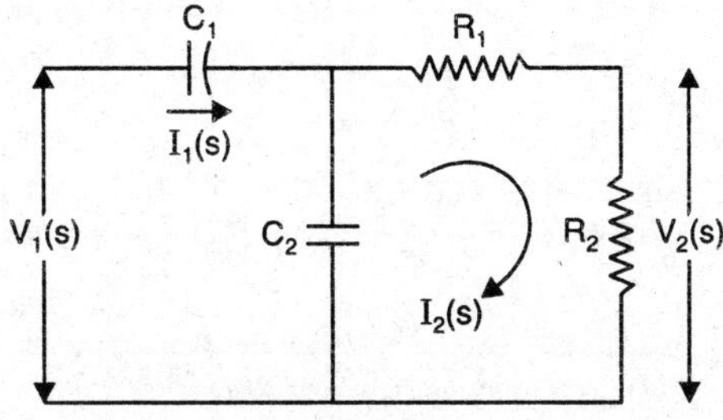

Fig. 2.18

Solution: Applying KVL

$$V_1(s) = I_1(s)\left[\frac{1}{C_1 s} + (R_1 + R_2) \,||\, \frac{1}{C_2 s}\right]$$

$$= I_1(s)\left[\frac{1}{C_1 s} + \frac{(R_1 + R_2)\dfrac{1}{C_2 s}}{R_1 + R_2 + \dfrac{1}{C_2 s}}\right]$$

$$= I_1(s)\left[\frac{R_1 + R_2 + \dfrac{1}{C_2 s} + (R_1 + R_2)\dfrac{C_1}{C_2}}{C_1 s\left(R_1 + R_2 + \dfrac{1}{C_2 s}\right)}\right] \tag{2.63}$$

$$I_2(s) = I_1(s)\,\frac{\dfrac{1}{C_2 s}}{R_1 + R_2 + \dfrac{1}{C_2 s}} \tag{2.64}$$

From eqn. (2.63),

$$I_1(s) = \frac{V_1(s)\, C_1 s\left(R_1 + R_2 + \dfrac{1}{C_2 s}\right)}{R_1 + R_2 + \dfrac{1}{C_2 s} + (R_1 + R_2)\dfrac{C_1}{C_2}} \tag{2.65}$$

Substituting $I_1(s)$ from eqn. (2.65) in eqn. (2.64)

$$I_2(s) = \frac{V_1(s)\, C_1 s \times \left(R_1 + R_2 + \dfrac{1}{C_2 s}\right)\dfrac{1}{C_2 s}}{\left[R_1 + R_2 + \dfrac{1}{C_2 s} + (R_1 + R_2)\dfrac{C_1}{C_2}\right]\left[R_1 + R_2 + \dfrac{1}{C_2 s}\right]}$$

$$= \frac{V_1(s)\dfrac{C_1}{C_2}}{\left[R_1 + R_2 + \dfrac{1}{C_2 s} + (R_1 + R_2)\dfrac{C_1}{C_2}\right]}$$

$$\therefore \quad \frac{V_2(s)}{V_1(s)} = \frac{R_2\dfrac{C_1}{C_2}}{R_1 + R_2 + \dfrac{1}{C_2 s} + (R_1 + R_2)\dfrac{C_1}{C_2}} \qquad [\because\; I_2(s) : R_2\, V_2(s)]$$

$$= \frac{R_2 C_1 s}{(R_1 + R_2)\, C_2 s + (R_1 + R_2)\, C_1 s + 1}.$$

Problem 2.9. *The pole zero plot of a linear system with gain 20 is shown in Fig 2.19. Determine the transfer function of the system.*

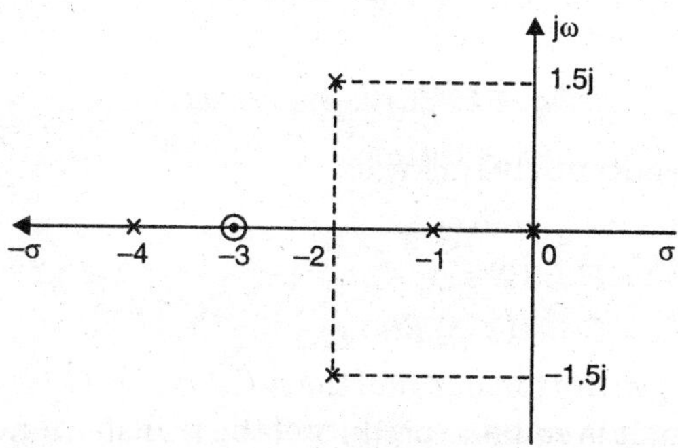

Fig. 2.19

Solution: From the pole zero plot of Fig 2.19, the system transfer function has simple pole at origin and – 4, a second order pole at – 1 and complex poles at (– 2 + 1.5 *j*) and (– 2 – 1.5 *j*). There is only one zero of the transfer function at – 3. Hence the transfer function of the system having a gain of 20 is

$$G(s) = \frac{20(s+3)}{s(s+4)\,(s+1)^2\,(s+2-1.5\,j)\,(s+2+1.5\,j)}$$

$$= \frac{20(s+3)}{s(s+1)^2\,(s+4)\,(s^2+4s+6.25)}.$$

2.8 BLOCK DIAGRAM REDUCTION TECHNIQUE

In previous section, we have discussed the transfer function model for linear time invariant control systems. When the system is small, system transfer function can be calculated easily. If the system is large, it consists large number of components. So the calculation of transfer function will be difficult. In such a case, we have introduced ficticious method to find the transfer function for large system. In this block diagram reduction technique, take unit transfer function of a component as a block, combine the blocks of each component in a designed manner, we will get the complete block diagram of entire system. Then reduce the complete block diagram until you will get a system with single block by using block diagram reduction technique.

When we are using the block diagram reduction technique, we should follow some rules.

2.8.1 Rules in Block Diagram Reduction Technique

(*i*) Cascade connections of blocks

- When the system consists of two blocks and are connected in series, it is known as cascade connections of blocks.
- Fig. 2.20 (*a*) shows cascade connections of two blocks in which the output of the block one is connected to the input of the second block.

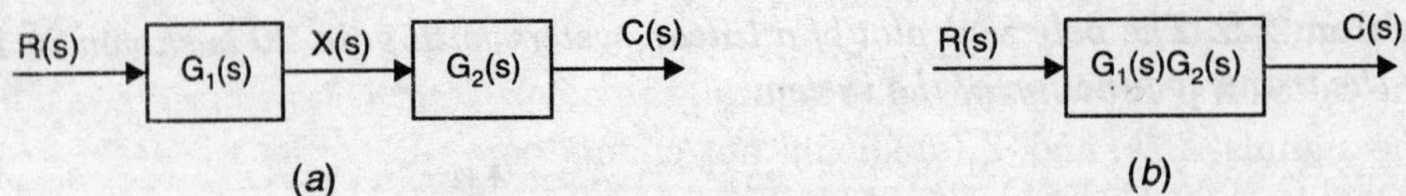

Fig. 2.20 Blocks are in cascade

- We know that the output of G_1 is $x(s)$

$$X(s) = G_1(s)\,R(s)$$
$$C(s) = G_2(s)\,X(s)$$
$$= G_1(s)\,G_2(s)\,R(s) \qquad (2.66)$$

When cascade blocks with transfer functions $G_1(s)$ and $G_2(s)$ can be replaced by single block with transfer function, it is equal to product of the transfer functions of individual blocks as shown in Fig. 2.20 (*b*).

***(ii)* Parallel connections of blocks**

- The two blocks are said to be parallel if they have common input and their outputs are connected by a summing point.

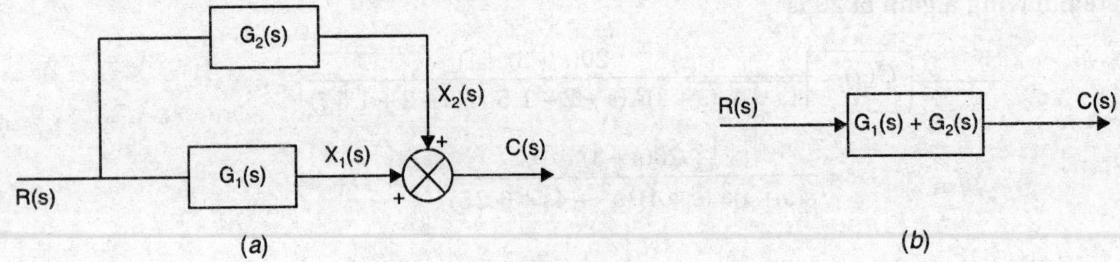

Fig. 2.21

From Fig. 2.21 (*a*),

- The output $C(s) = X_1(s) + X_2(s)$

$$C(s) = R(s)\,G_1(s) + R(s)\,G_2(s)$$
$$C(s) = [G_1(s) + G_2(s)]\,R(s) \qquad (2.67)$$

Parallel blocks with transfer functions $G_1(s)$ and $G_2(s)$ can be replaced by a single block with transfer function which is equal to the sum of the transfer functions of individual blocks as shown in Fig. 2.21 (*b*).

***(iii)* Moving a take off point to the right or in the direction of signal flow of the block**

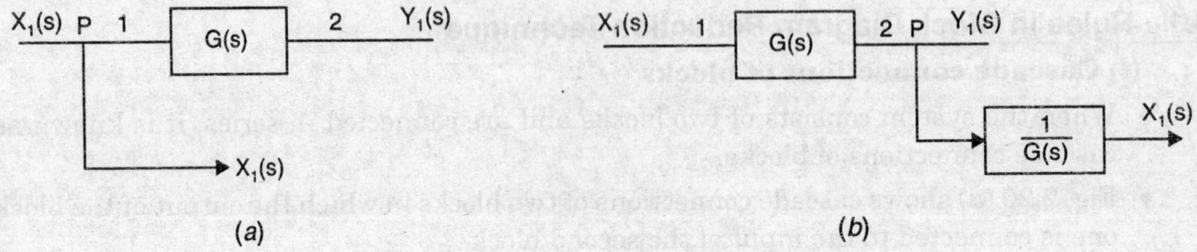

Fig. 2.22

From Fig. 2.22 (*a*),

- The take-off point *P* is moved to right of the block *G*(*s*) as shown in Fig. 2.22 (*b*).
- The signals $X_1(s)$ and $Y_1(s)$ should not change in value when the take-off point is moved to right of the block.
- Moving a take-off point from 1 to 2 is equal to multiplying the signals $x_1(s)$ by $G(s)$.
- So to compensate for this change, divide the take-off signal by *G*(*s*) after moving the take-off point.
- To shift the take-off point to the right of the block *G*(*s*) divide the take-off point branch by *G*(*s*) as shown in Fig. 2.22 (*c*).

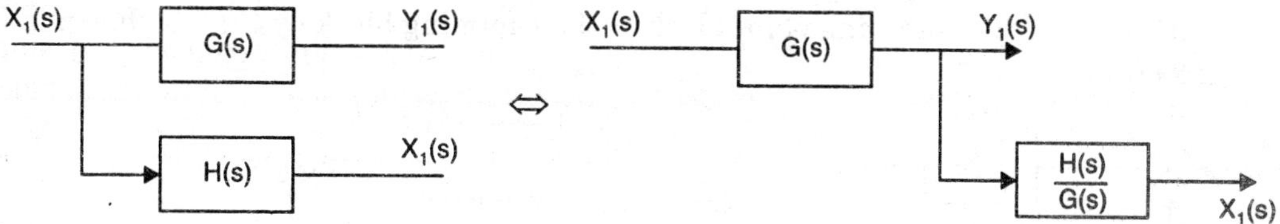

Fig. 2.22 (*c*)

(*iv*) Moving a take-off point to the left or in the direction of opposite signal flow:

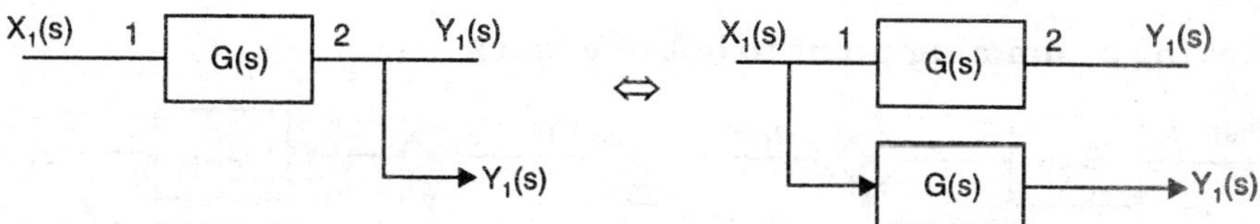

Fig. 2.23 (*a*)

From Fig. 2.23 (*a*),

- Moving a take-off point from 2 to 1 is equal to avoiding the signal from passing through *G*(*s*) to compensate this change, multiply the take-off point signal by *G*(*s*).
- To shift the take-off point to the left of the block *G*(*s*), multiply the take-off branch by *G*(*s*) as shown in Fig. 2.23 (*b*).

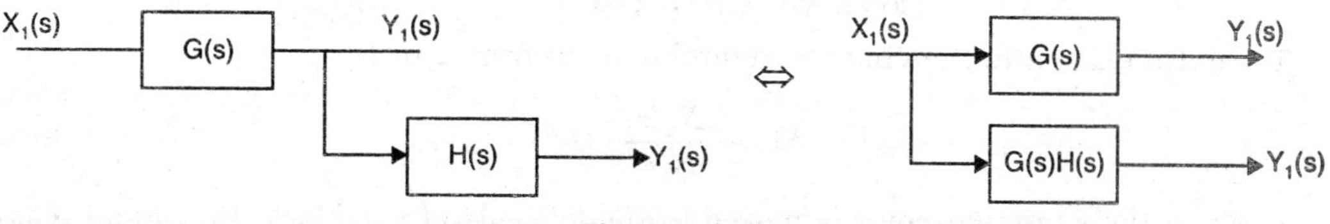

Fig. 2.23 (*b*)

(*v*) Moving a summing point to right of the block

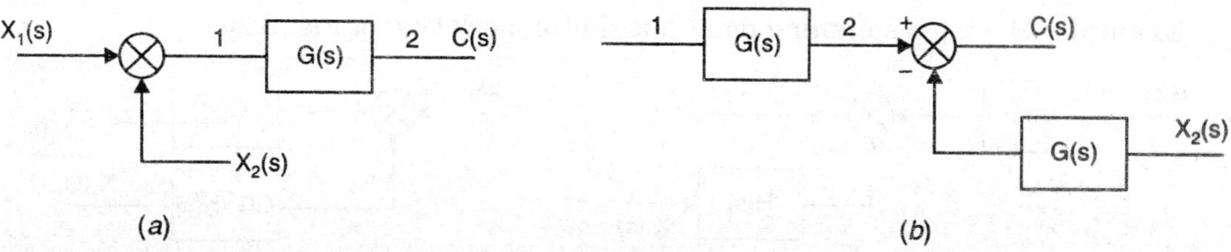

Fig. 2.24

From Fig. 2.24 (*a*),

- The summing point before $G(s)$ is moved from 1 to 2

$$C(s) = [X_1(s) - X_2(s)]\, G(s)$$

$$C(s) = G(s)\, X_1(s) - X_2(s)\, G(s) \qquad (2.68)$$

The output after shifting the summing point from 1 to 2

- When the summing point is moved to the right of the block, then the signal $X_2(s)$ must be multiplied by $G(s)$ before shifting the summing point from 1 to 2 as shown in Fig. 2.24 (*b*).
- To shift the summing point to the right side of the block $G(s)$ multiply with the signal entering at the summing point block or the summing block by $G(s)$ as shown in Fig. 2.24 (*c*).

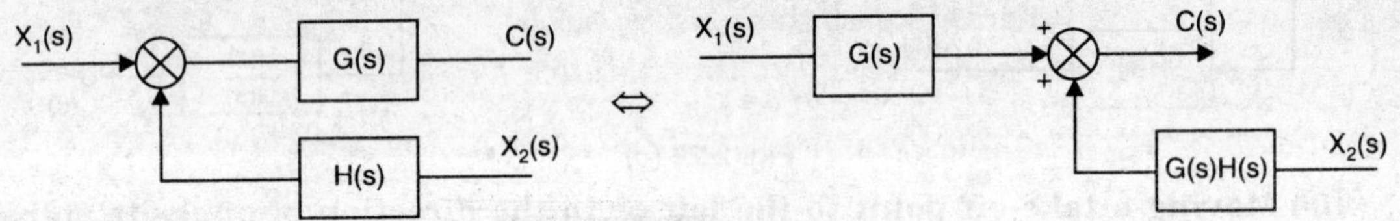

Fig. 2.24 (c)

(*vi*) Moving a summing point to left of a block

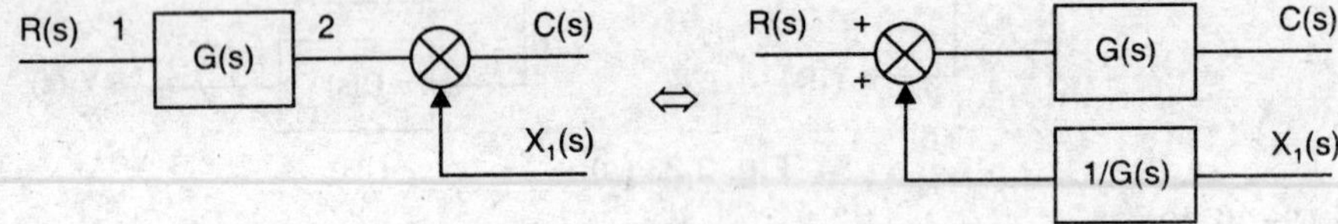

Fig. 2.25

From Fig. 2.25,

The output $C(s)$ before moving the summing point from 2 to 1 is

$$C(s) = R(s)\, G(s) - X_1(s)$$

The output $C(s)$ after moving the summing point from 2 to 1

$$C(s) = \left[R(s) - \frac{X_1(s)}{G(s)} \right] G(s)$$

- When the summing point is moved to the left side of the block, the signal $X_1(s)$ is multiplied with $\dfrac{1}{G(s)}$.

Example: Moving a summing point left side of block G(s) of Fig. 2.26.

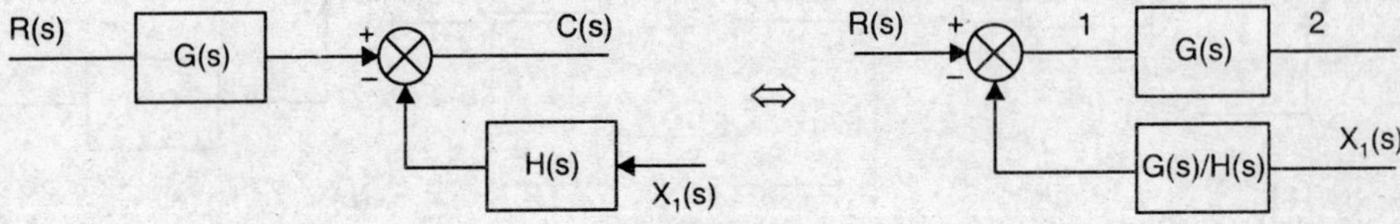

Fig. 2.26

From Fig. 2.26,

- To shift the summing point to the left of the block $G(s)$, the summing point block is multiplied with $\frac{1}{G(s)}$.

(*vii*) Feedback rule:

The closed loop control system is represented by a block diagram as shown in the Fig. 2.3.

where $B(s)$ = Feedback signal

$E(s)$ = Error signal

$$C(s) = G(s)\,E(s) \tag{2.69}$$

$G(s)$ = Forward path transfer function

$H(s)$ = Feedback path transfer function

$$H(s) = \frac{B(s)}{C(s)} \tag{2.70}$$

The overall transfer function $\frac{C(s)}{R(s)}$ is called closed loop transfer function. From Fig 2.3,

$$E(s) = R(s) - B(s)$$

$$E(s) = R(s) - H(s)\,C(s) \qquad [\because\ B(s) = H(s)\,C(s)]$$

$$E(s) = R(s) - H(s)\,G(s)\,E(s) \qquad [\because\ C(s) = G(s)\,E(s)]$$

$$E(s)\,[1 + G(s)\,H(s)] = R(s)$$

$$\therefore \qquad E(s) = \frac{R(s)}{[1 + G(s)\,H(s)]} \tag{2.71}$$

Substituting eqn. (2.71) in eqn. (2.69)

$$C(s) = G(s)\,E(s)$$

$$= G(s) \times \frac{R(s)}{1 + G(s)\,H(s)} \tag{2.72}$$

- A feedback control system whose open loop transfer function $G(s)$ has feedback transfer function $H(s)$ can be replaced by or feedback loop is eliminated by single block with transfer function, $\frac{G(s)}{1 + G(s)\,H(s)}$ (2.73)

For positive feedback loop

Block diagram for positive feedback loop is shown in the Fig. 2.27.

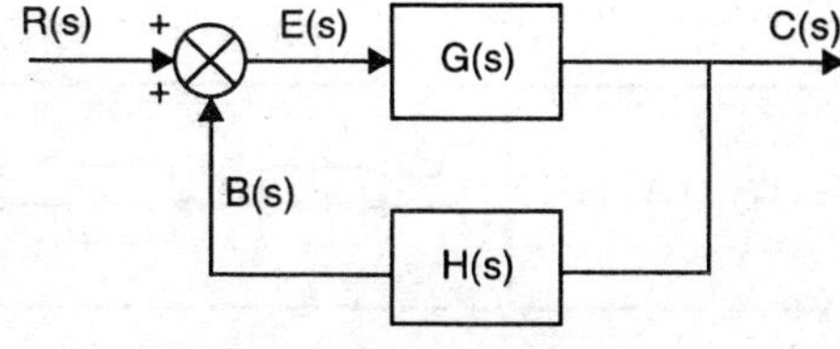

Fig. 2.27

$$C(s) = G(s)\,E(s) \tag{2.74}$$
$$E(s) = R(s) + B(s) \tag{2.75}$$
$$B(s) = H(s)\,C(s)$$

Then $E(s) = R(s) + H(s)\,C(s)$

From eqn. (2.74), $E(s) = \dfrac{C(s)}{G(s)}$

$$\frac{C(s)}{G(s)} = R(s) + H(s)\,C(s)$$
$$C(s) = G(s)\,R(s) + G(s)\,H(s)\,C(s)$$
$$C(s)\,(1 - G(s)\,H(s)) = G(s)\,R(s)$$

$\therefore$ Transfer function, $$\frac{C(s)}{R(s)} = \frac{G(s)}{1 - G(s)\,H(s)} \tag{2.76}$$

From eqns. (2.73) and (2.76), it is observed that for eliminating the negative feedback loop, we will get plus (+) in denominator and for positive feedback loop, minus (–) in denominator.

Note: (*i*) Any two summing points may be interchanged.

(*ii*) Any summing point having more than two inputs may be split into 2 summing points.

(*iii*) Any two summing points can be combined.

2.8.2 Points to be Remembered while Doing the Block Diagram Reduction

- If a block diagram consists of many feedback loops, then first find the innermost feedback loop.
- If it does not contain any take-off or summing points inside the loop, then simplify the innermost feedback loop to a single block using rule number 7.
- If it consists of take-off point or summing points inside to the innermost loop, then simplify by using rules 2 to 6.
- After eliminating such points, simplify the innermost loop using rule 7
- If any path consists of any cascade blocks, then combine by using rule 1
- If there is any parallel path then use rule 2, to simplify it.
- Repeat the above steps for obtaining the required transfer function.

Table 2.1. Block Diagram Transformations

	Transformation	*Equations*	*Original block diagram*	*Equivalent block diagram*
1	Combining blocks in cascade	$X_3(s) = G_1(s)\,G_2(s)\,X_1(s)$	$X_1(s)$ → $G_1(s)$ → $X_2(s)$ → $G_2(s)$ → $X_3(s)$	$X_1(s)$ → $G_1(s)G_2(s)$ → $X_3(s)$

2	Combining in parallel or eliminating forward loop	$X_2(s) = G_1(s)\,X_1(s) \pm G_2(s)\,X_1(s)$	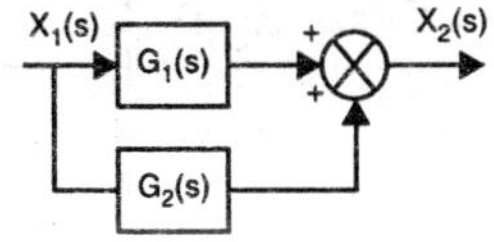	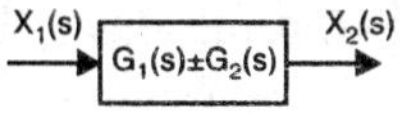
3	Removing a block from a forward path	$X_3(s) = G_1(s)\,X_1(s) \pm G_2(s)\,X_2(s)$	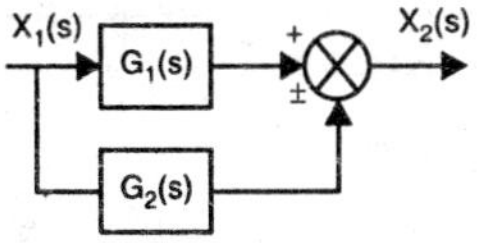	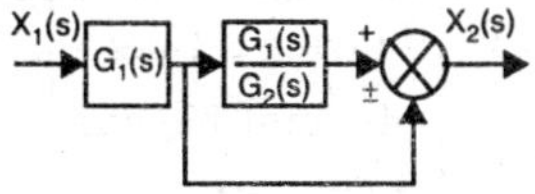
4	Eliminating a feedback loop	$X_2(s) = \dfrac{G_1(s)\,X_1(s)}{1 \mp G_1(s)\,H(s)}$	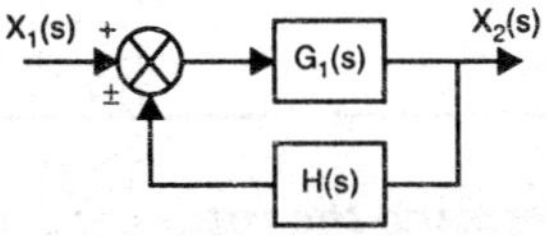	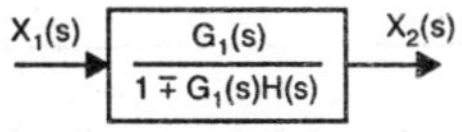
5	Removing a block from a feedback loop	$X_2(s) = \dfrac{G_1(s)\,X_1(s)}{1 \pm G_1(s)\,H(s)}$	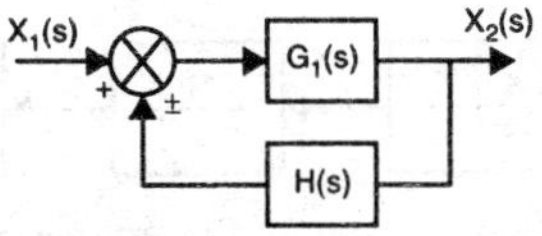	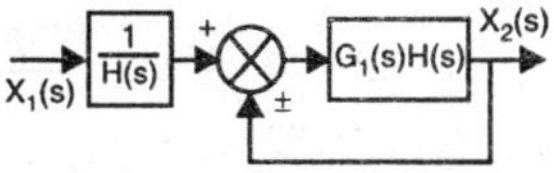
6 (*a*)	Rearranging of summing points	$X_2(s) = X_1(s) \pm X_3(s) \pm X_4(s)$	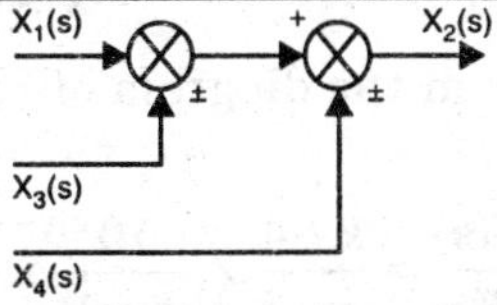	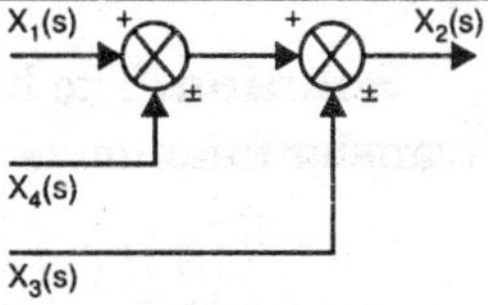
(*b*)	Rearranging of summing points	$X_2(s) = X_1(s) \pm X_3(s) \pm X_4(s)$	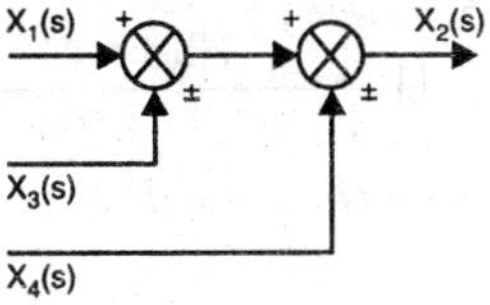	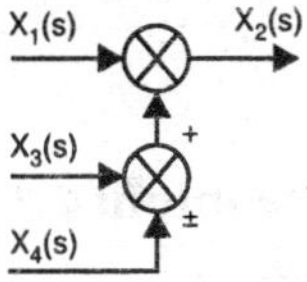
7	Moving a summing point ahead of a block	$X_2(s) = G_1(s)\,X_1(s) + X_3(s)$	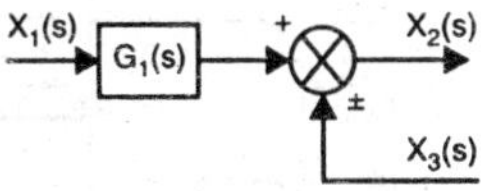	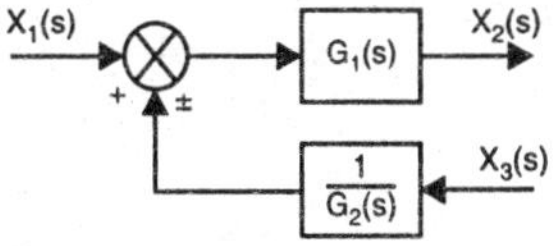
8	Moving a summing point beyond a block	$X_2(s) = G_1(s)\,[X_1(s) \pm X_3(s)]$	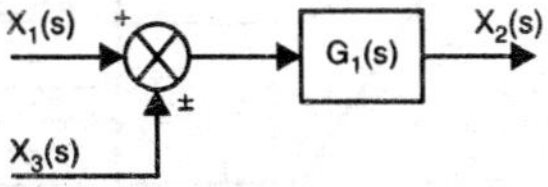	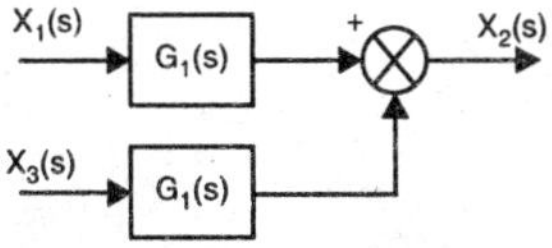
9	Moving a pick off point ahead of a block.	$X_2(s) = X_1(s)\,X_3(s)$	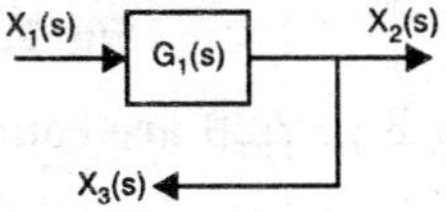	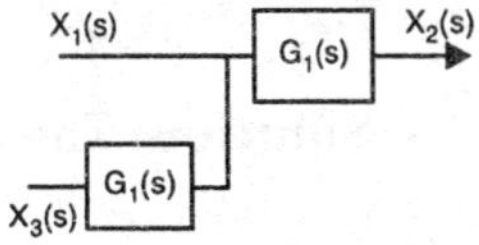
10	Moving a take off point beyond a block	$X_2(s) = X_1(s)\,X_3(s)$	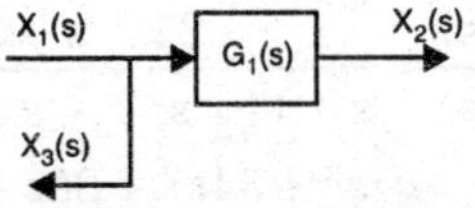	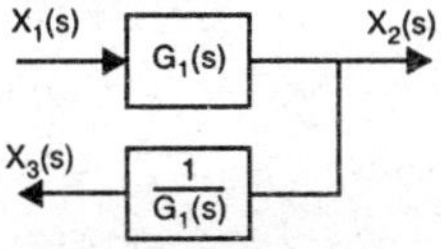

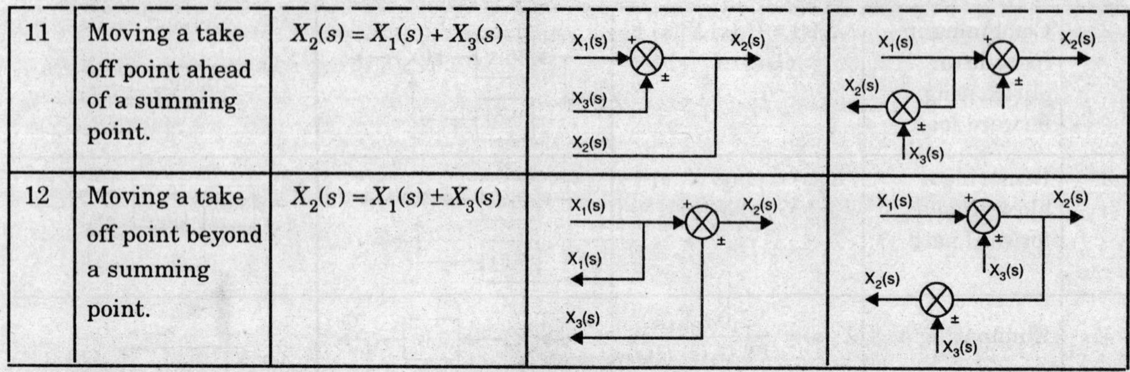

11	Moving a take off point ahead of a summing point.	$X_2(s) = X_1(s) + X_3(s)$		
12	Moving a take off point beyond a summing point.	$X_2(s) = X_1(s) \pm X_3(s)$		

Problem 2.10.: *Determine the ratio* $\frac{C(s)}{R(s)}$ *in the block diagram as shown in Fig. 2.28.*

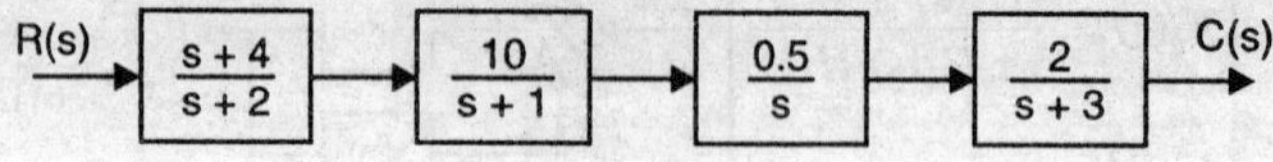

Fig. 2.28

Solution: The blocks in the diagram of Fig. 2.28 are connected in cascade. Hence over all transfer function is

$$\frac{C(s)}{R(s)} = \frac{s+4}{s+2} \times \frac{10}{(s+1)} \times \frac{0.5}{s} \times \frac{2}{s+3}$$

$$= \frac{10(s+4)}{s(s+1)(s+2)(s+3)}$$

Problem 2.11. *Determine the overall transfer function of the block diagram as shown in Fig. 2.29.*

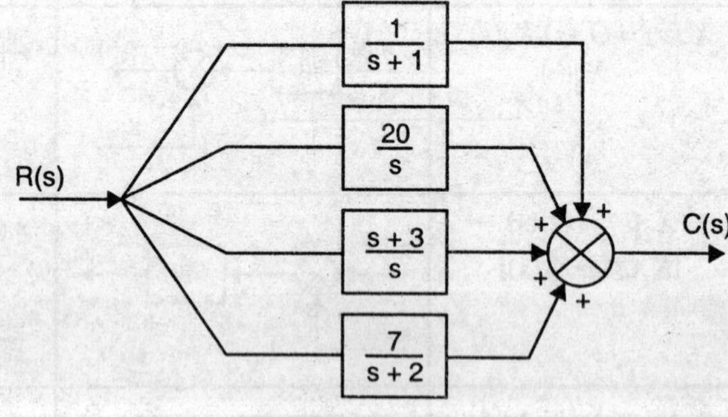

Fig. 2.29

Solution: The block in Fig. 2.29 are connected in parallel, hence the overall transfer function, $\frac{C(s)}{R(s)}$ is

$$\frac{C(s)}{R(s)} = \frac{1}{s+1} + \frac{20}{s} + \frac{s+3}{s} + \frac{7}{s+2}$$

$$= \frac{s^3 + 34s^2 + 80s + 46}{s(s+1)(s+2)}$$

Problem 2.12. *Determine the ratio* $\frac{C(s)}{R(s)}$ *in the block diagram as shown in Fig. 2.30.*

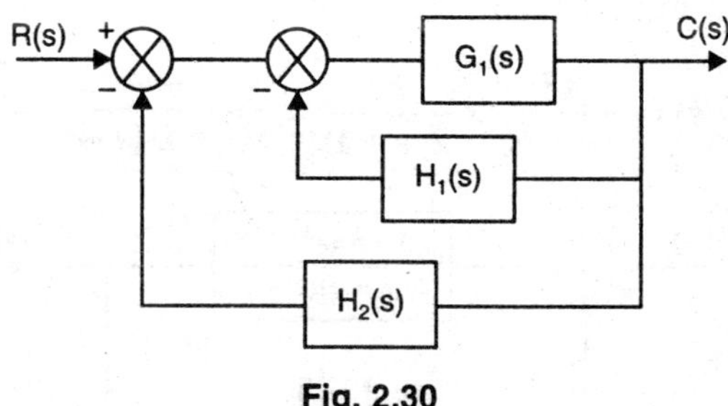

Fig. 2.30

Solution: The block diagram of Fig. 2.30 contains two loops with negative feedback.

Step 1: Eliminating the inner feedback loop, the obtained block diagram is

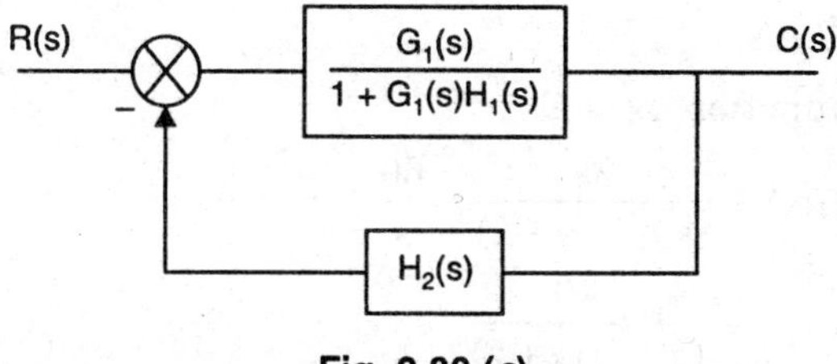

Fig. 2.30 (*a*)

Step 2: Eliminating the feedback $H_2(s)$, the obtained block diagram is

R(s)

$$\frac{\dfrac{G_1(s)}{1 + G_1(s)H_1(s)}}{1 + \dfrac{G_1(s)H_2(s)}{1 + G_1(s)H_1(s)}}$$

C(s)

Fig. 2.30 (*b*)

$\therefore$ Transfer function, $$\frac{C(s)}{R(s)} = \frac{G_1(s)}{1 + G_1H_1(s) + G_2H_2(s)}$$

Problem 2.13. *The positive feedback system is shown in Fig. 2.31. Determine the (i) loop transfer function, (ii) control ratio, (iii) feedback ratio and (iv) system characteristic equation.*

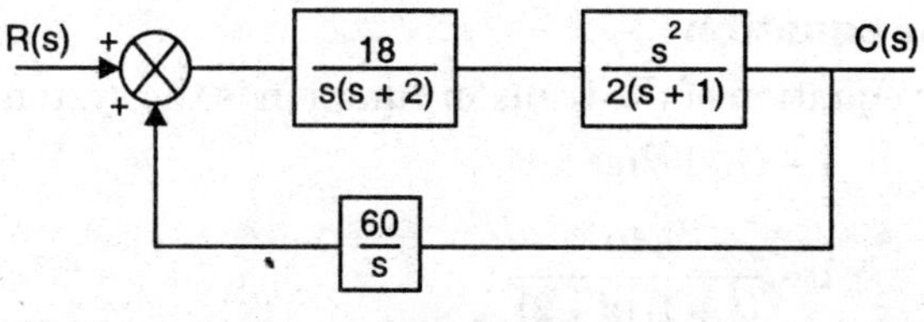

Fig. 2.31

Solution: The two blocks, $\dfrac{18}{s(s+2)}$ and $\dfrac{s^2}{2(s+1)}$ are in series. Cascading the two blocks, resultant block is shown in Fig.2.31 (*a*)

$$G(s) = \frac{18}{s\,(s+2)} \times \frac{s^2}{2\,(s+1)} = \frac{9s^2}{s(s+1)\,(s+2)} = \frac{9s}{(s+1)\,(s+2)}$$

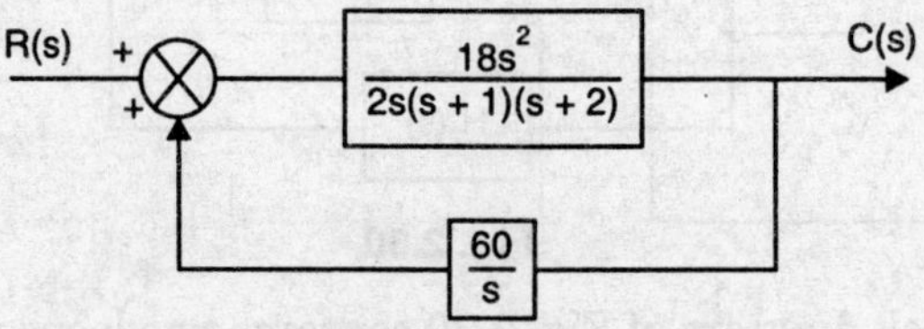

Fig. 2.31 (a)

$$H(s) = \frac{60}{s}$$

(*i*) Loop transfer function is

$$G(s)\,H(s) = \frac{9s}{(s+1)\,(s+2)} \times \frac{60}{s}$$

$$= \frac{540}{(s+1)\,(s+2)}$$

(*ii*) Control ratio of the system is

$$\frac{C(s)}{R(s)} = \frac{G(s)}{1 - G(s)\,H(s)} = \frac{\dfrac{9s}{(s+1)\,(s+2)}}{1 - \dfrac{540}{(s+1)\,(s+2)}}$$

$$\therefore \quad \frac{C(s)}{R(s)} = \frac{9(s)}{s^2 + 3s - 538}$$

(*iii*) Feedback ratio, $\quad H(s) = \dfrac{60}{s}$

(*iv*) The error ratio, $\quad \dfrac{E(s)}{R(s)} = \dfrac{1}{1 - G(s)\,H(s)}$

$$= \frac{1}{1 - \dfrac{540}{(s+1)\,(s+2)}} = \frac{s^2 + 3s + 2}{s^2 + 3s - 538}$$

System characteristic equation

The denominator equation of the transfer function is the system characteristic equation

$$1 - G(s)\,H(s) = 0$$

$$1 - \frac{540}{(s+1)\,(s+2)}$$

$$s^2 + 3s - 538 = 0.$$

Problem 2.14. *Find the transfer function for the block diagram as shown in Fig.2.32 by using block diagram reduction technique.*

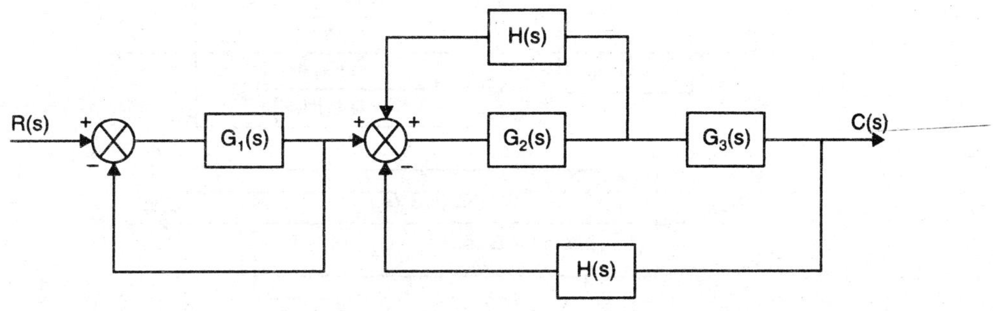

Fig. 2.32

Solution:

Step 1: Eliminating the unit negative feedback loop and H loop for G_2

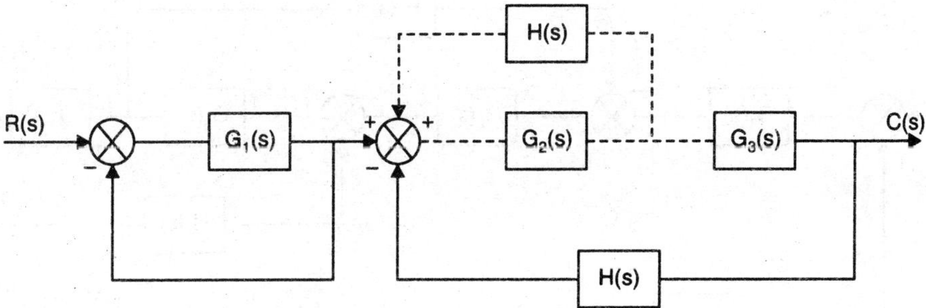

Step 2: Cascading the two blocks 1 and 2

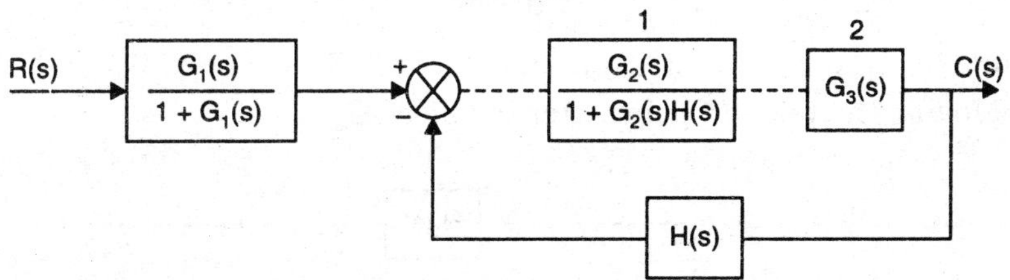

Step 3: Eliminating the negative feedback loop, H

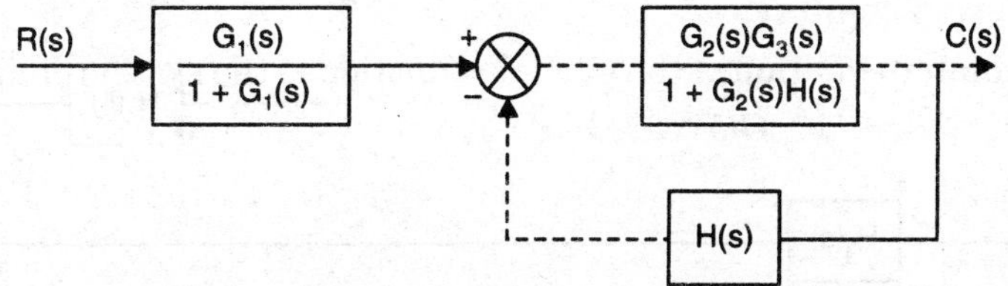

Step 4: Cascading the two blocks, the combined block is

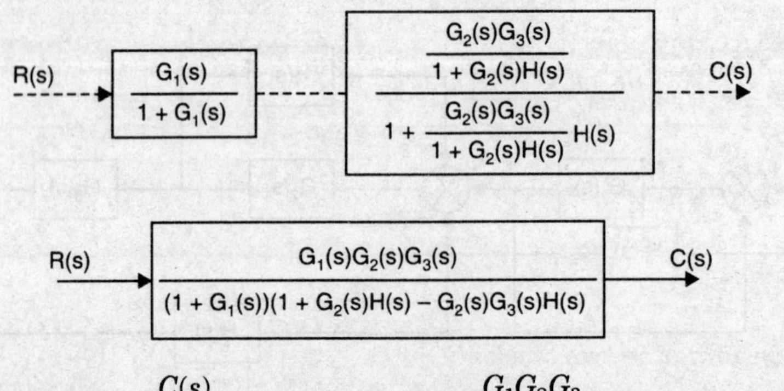

$\therefore$ Transfer function, $\dfrac{C(s)}{R(s)} = \dfrac{G_1G_2G_3}{1+G_2H - G_2G_3H + G_1 + G_1G_2H - G_1G_2G_3H}$

Problem 2.15. *Reduce the block diagram of the multi loop system as shown in Fig. 2.33.*

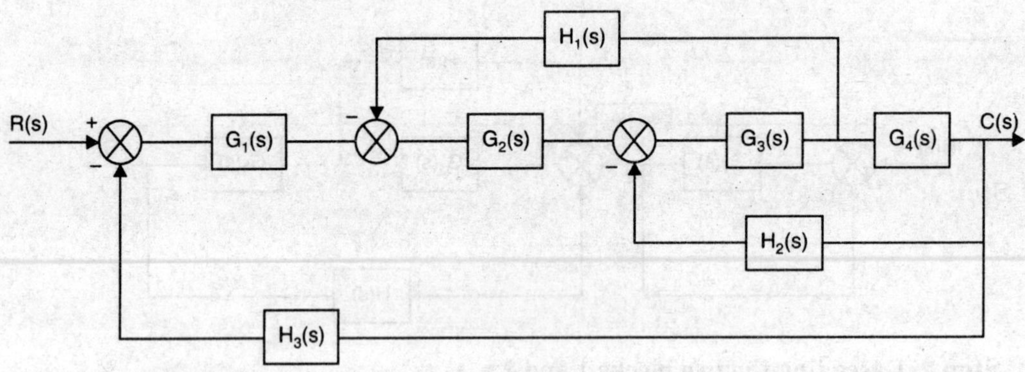

Fig. 2.33

Solution:

Step 1: Moving the take-off point after the block G_4

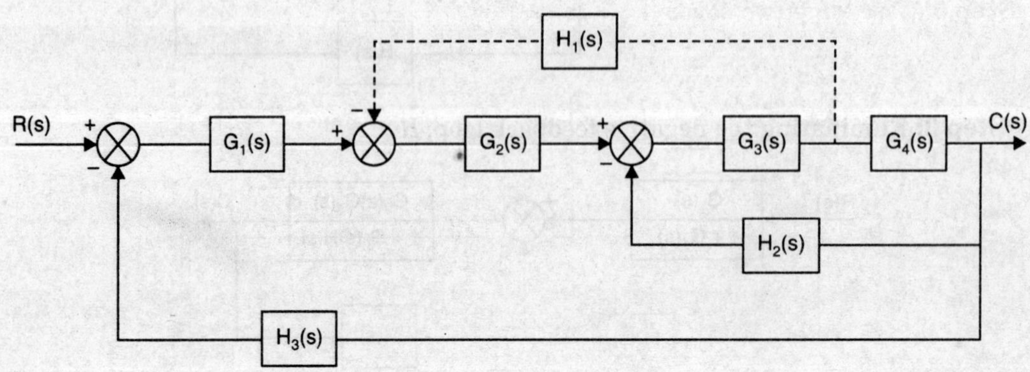

Step 2: Cascading the two blocks G_3 and G_4 and eliminating H_2

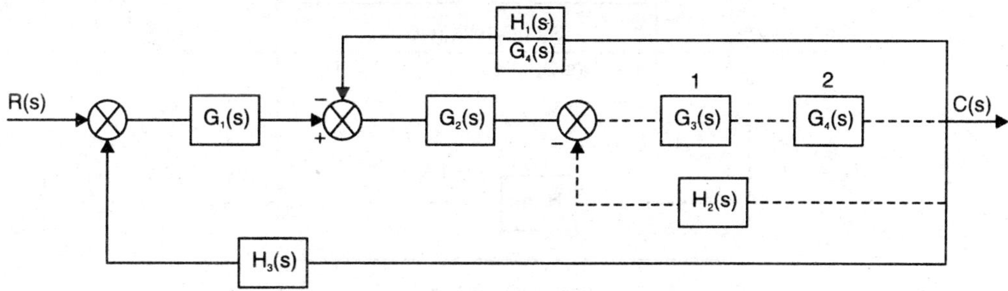

Step 3: Cascading the two blocks G_2 and $\frac{G_3G_4}{1+G_3G_4H_2}$

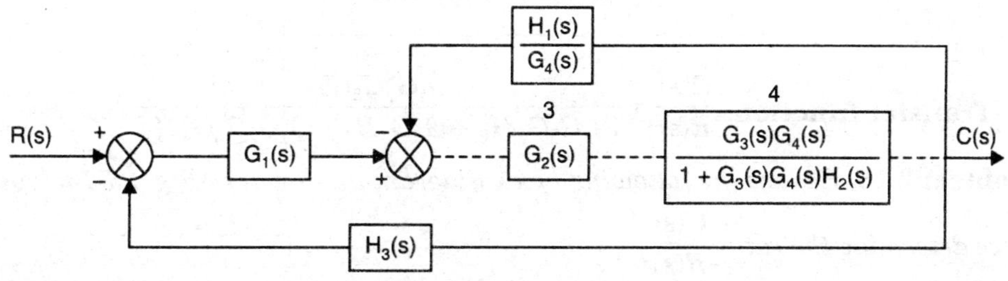

Step 4: Eliminating the loop $\frac{H_1}{G_4}$

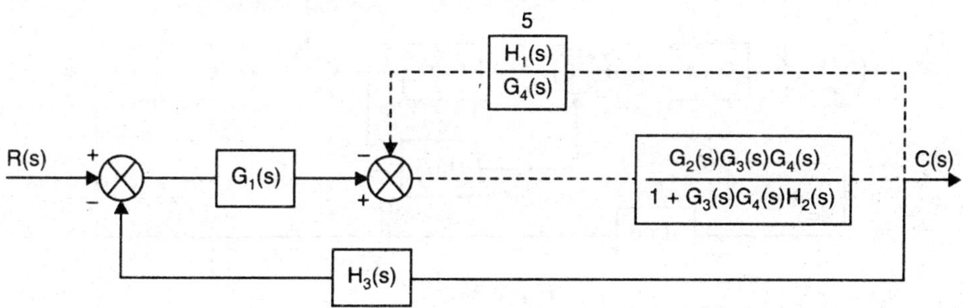

Step 5: Cascading two blocks

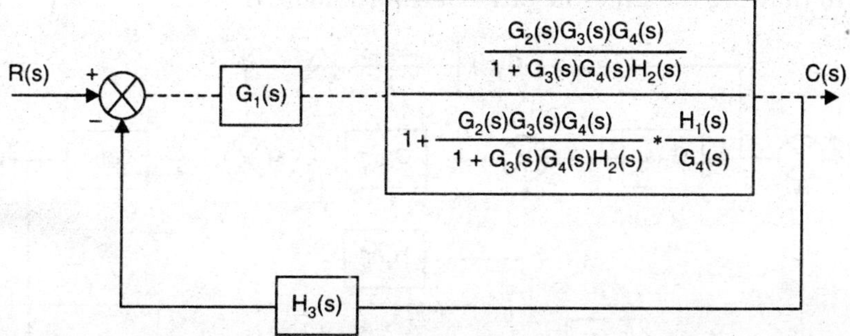

Step 6: Eliminating the negative feedback loop H_3

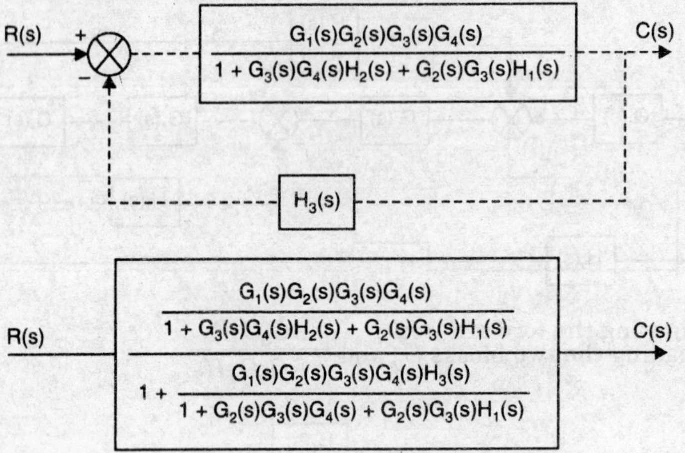

$\therefore$ **Transfer function,** $\dfrac{C(s)}{R(s)} = \dfrac{G_1G_2G_3G_4}{1 + G_3G_4H_2 + G_2G_3H_1 + G_1G_2G_3G_4H_3}$

Problem 2.16. *Reduce the following block diagram as shown in Fig. 2.34 to open loop form. Hence determine the ratio* $\dfrac{C(s)}{R(s)}$.

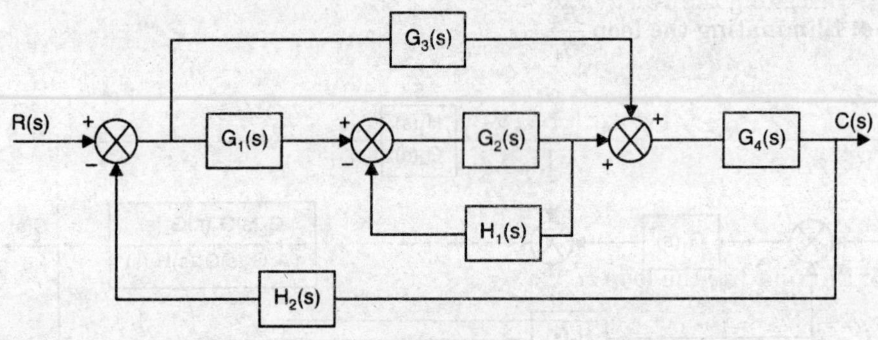

Fig. 2.34

Solution:

Step 1: Eliminating the inner negative feedback loop, H_1

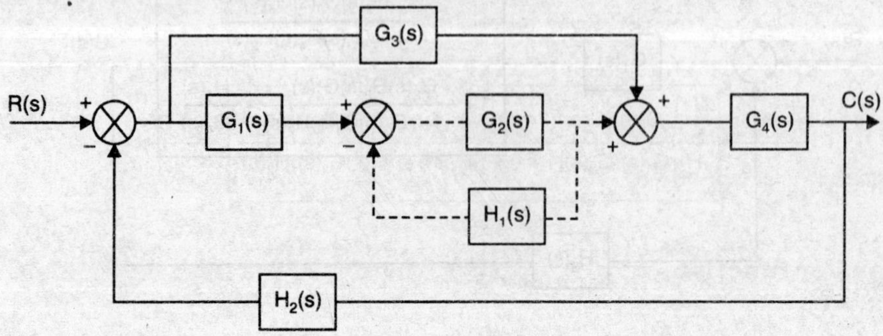

Step 2: Cascading the two blocks 1 and 2

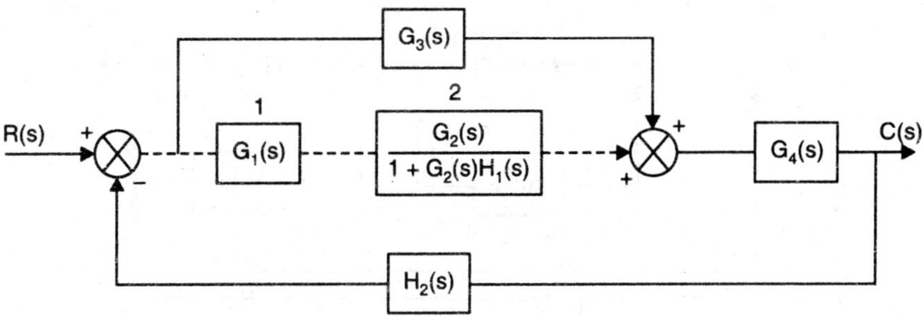

Step 3: Eliminating the forward loop 3

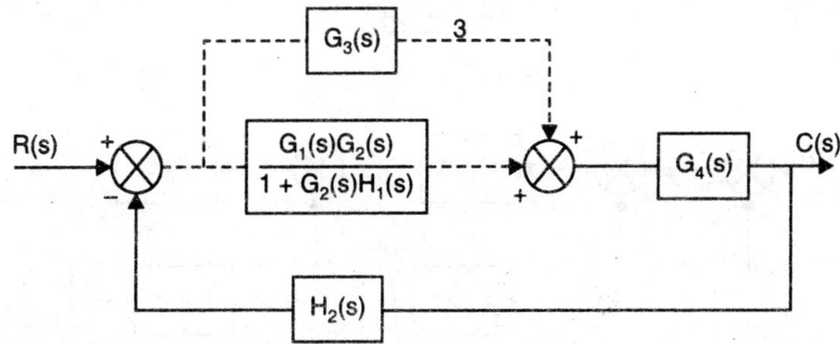

Step 4: Cascading the two blocks

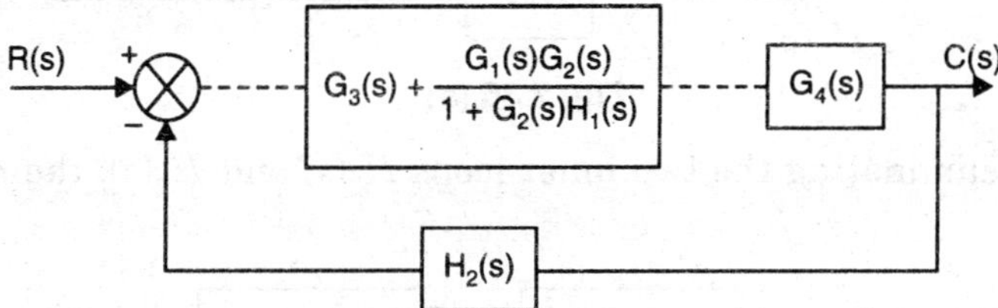

Step 5: Eliminating the loop H_2

R(s) + – $G_4(s)\left[G_3(s) + \frac{G_1(s)G_2(s)}{1 + G_2(s)H_1(s)}\right]$ C(s)

$H_2(s)$

R(s)

$$\frac{G_4(s)(G_3(s)(1 + G_2(s)H_1(s)) + G_1(s)G_2(s))}{1 + G_4(s)(G_3(s)(1 + G_2(s)H_1(s)) + G_1(s)G_2(s))H_2(s)}$$

C(s)

$\therefore$ **Transfer function,** $$\frac{C(s)}{R(s)} = \frac{G_4(G_3(1+G_2H_1)+G_1G_2)}{1+G_4H_2(G_3(1+G_2H_1)+G_1G_2)}$$

Problem 2.17. *Reduce the block diagram as shown in Fig. 2.35 to open loop form.*

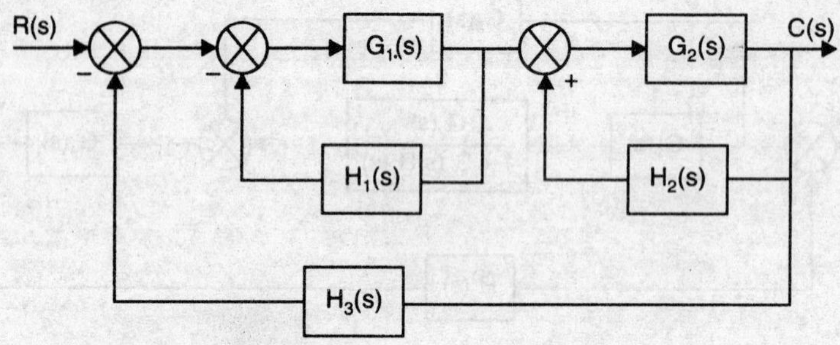

Fig. 2.35

Solution:

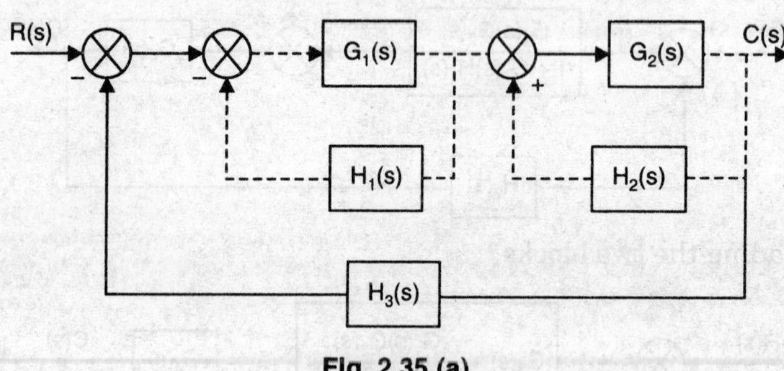

Fig. 2.35 (a)

Step 1: After eliminating the two inner loops $H_1(s)$ and $H_2(s)$, the reduced block diagram is

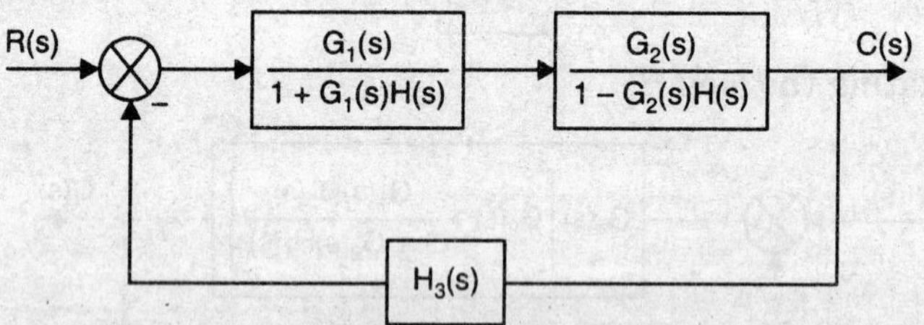

Step 2: Cascading the two blocks 1 and 2, the reduced block diagram is

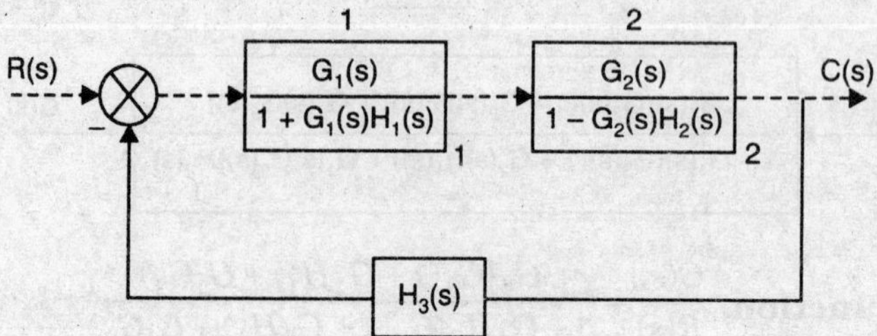

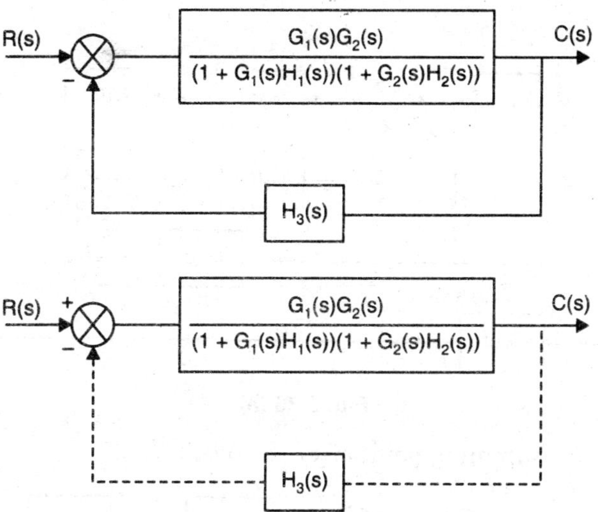

Step 3: Eliminating the feedback $H_3(s)$ in the above block diagram, the reduced block diagram is

R(s) →

$$\frac{\dfrac{G_1(s)G_2(s)}{(1+G_1(s)H_1(s))(1+G_2(s)H_2(s)}}{1+\dfrac{G_1(s)G_2(s)H_3(s)}{(1+G_1(s)H_1(s))(1+G_2(s)H_2(s)}}$$

→ C(s)

Simplifying the block

R(s) →

$$\frac{G_1(s)G_2(s)}{(1+G_1(s)H_1(s))(1-G_2(s)H_2(s)+G_1(s)G_2(s)H_3(s))}$$

→ C(s)

Problem 2.18. *Reduce the following block diagram as shown in Fig. 2.36 to open loop form and find the transfer function* $\dfrac{C(s)}{R(s)}$.

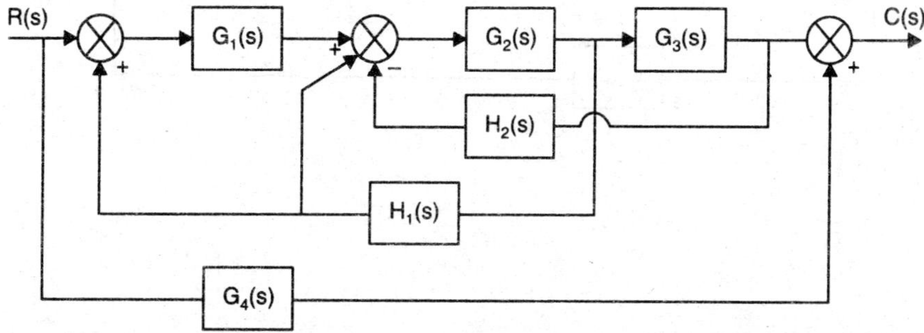

Fig. 2.36

Solution:

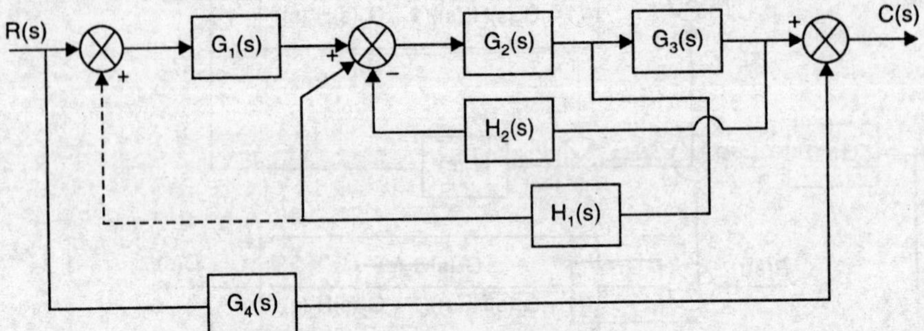

Fig. 2.36 (a)

Step 1: Moving the summing point after the block $G_1(s)$

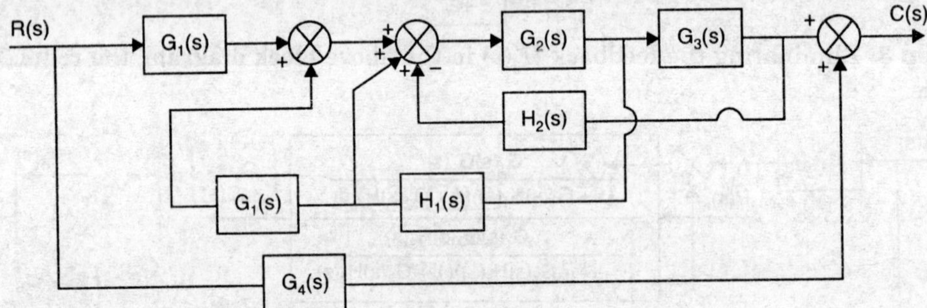

Step 2: Eliminating the $H_1(s)$ loop

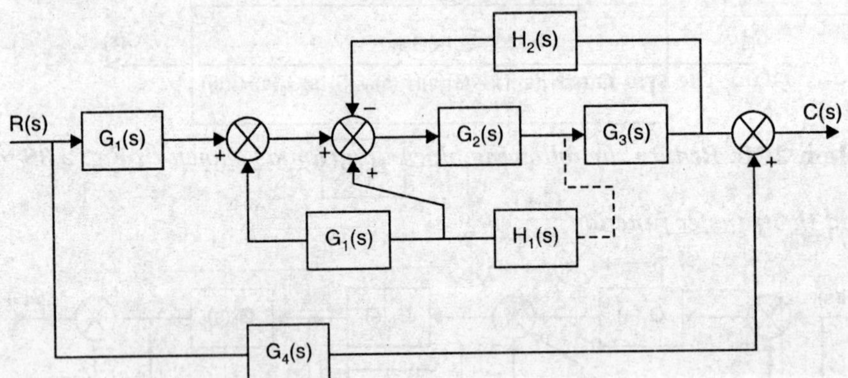

Step 3: Moving a take-off point 1 after $G_3(s)$

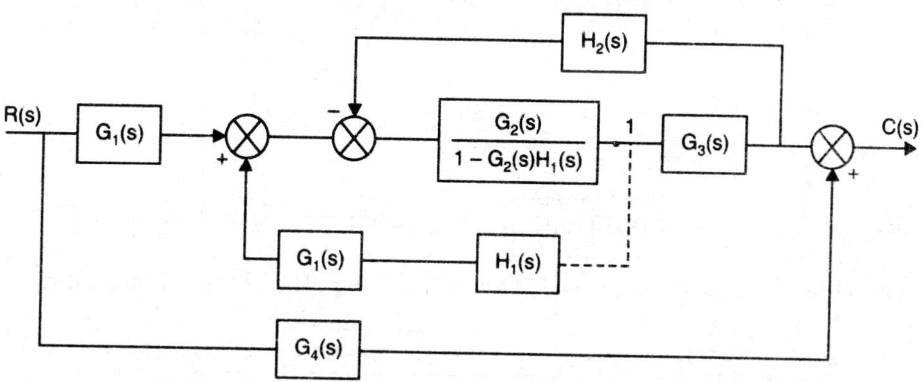

Step 4: Cascading the two blocks 1 and 2

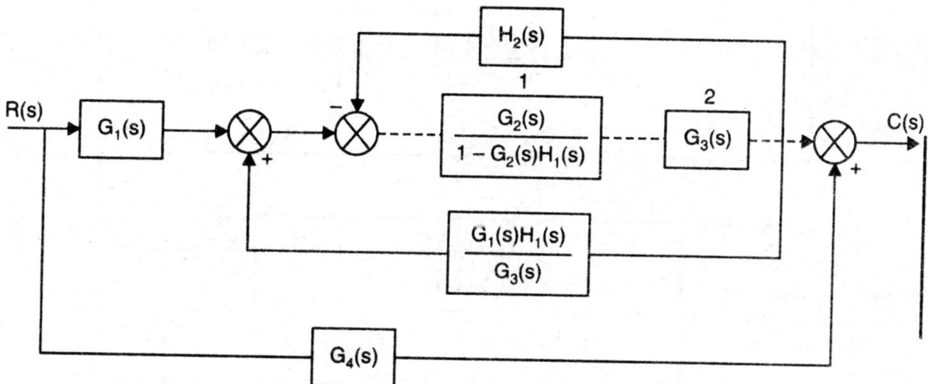

Step 5: Writing the two loops 3 and 4 in a single loop

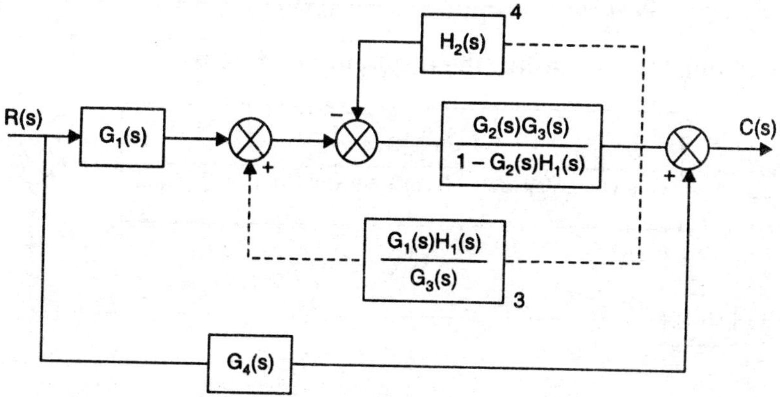

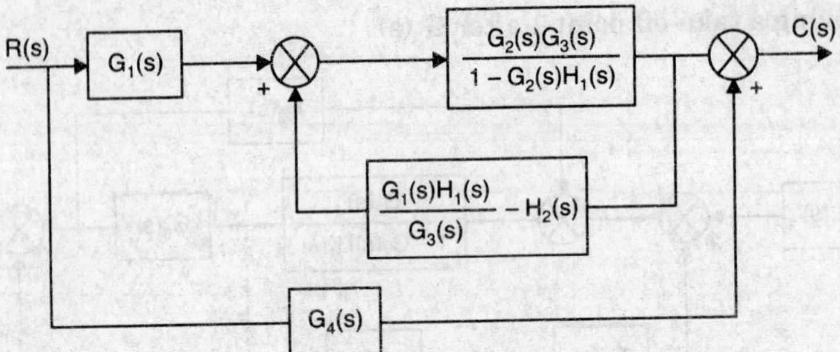

Step 6: Eliminating the inner positive feedback loop, the obtained block diagram is

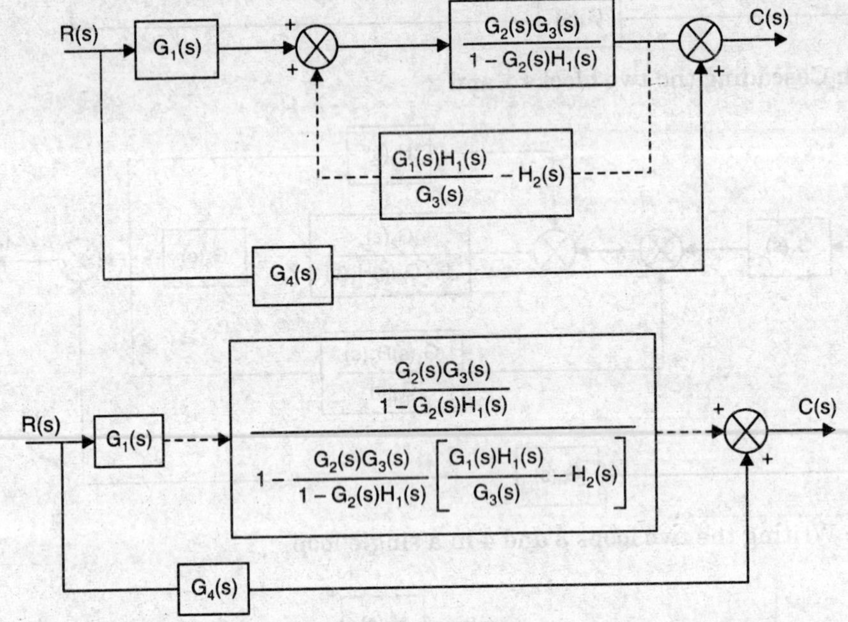

Step 7: Cascading the two blocks, the resultant diagram is

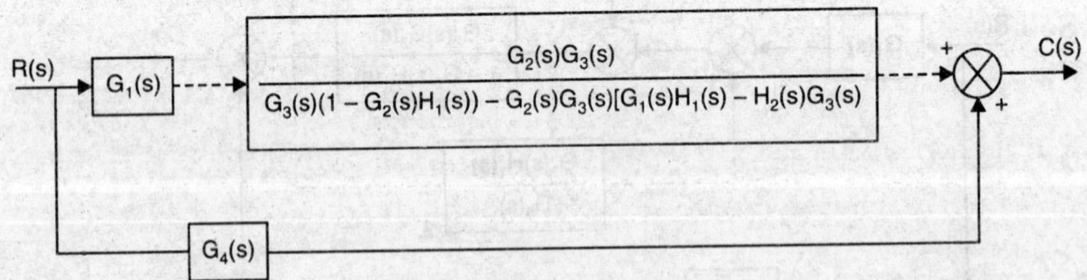

Step 8: Eliminating the forward loop

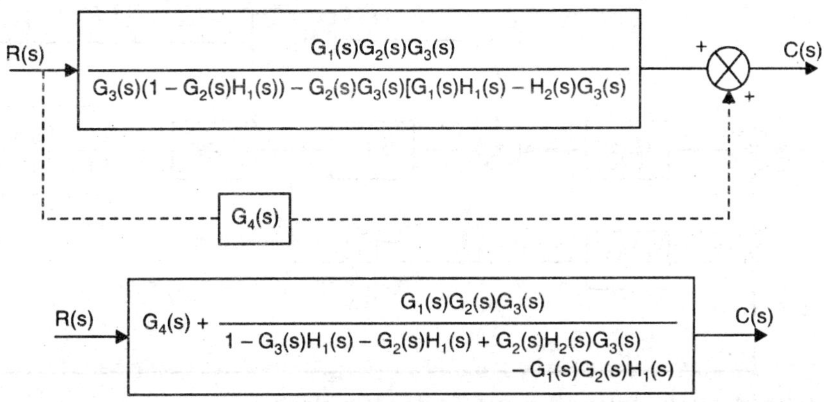

$\therefore$ Transfer function, $\dfrac{C(s)}{R(s)} = \dfrac{G_4(1 - G_3H_1 - G_2H_1 + G_2G_3H_2 - G_1G_2H_1) + G_1G_2G_3}{1 - G_3H_1 - G_2H_1 + G_2G_3H_2 - G_1G_2H_1}$

Problem 2.19. *Using block diagram reduction technique, find the transfer function for the system as shown in Fig. 2.37.*

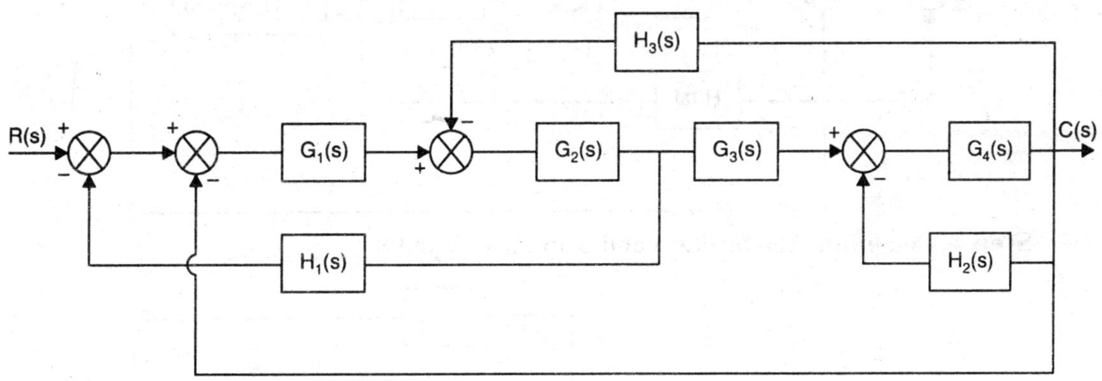

Fig. 2.37

Solution:

Step 1: Solving minor feedback loop of G_4 and H_2, the obtained block diagram is

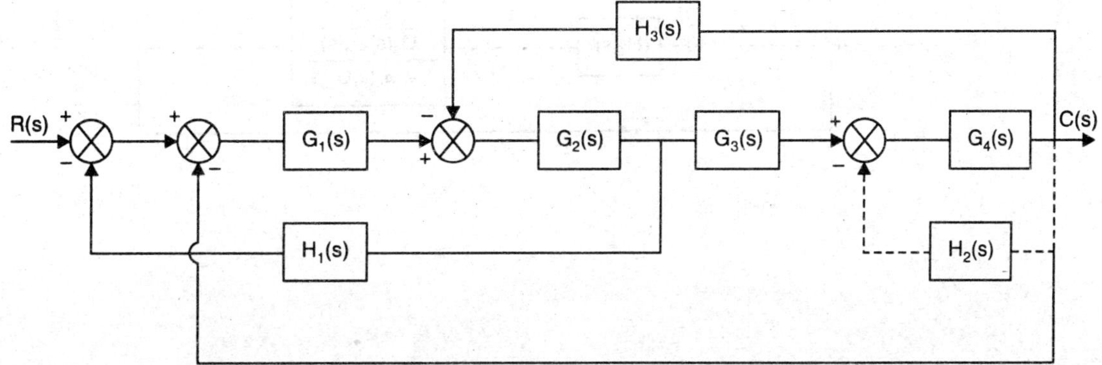

Step 2: Cascading the two blocks 1 and 2

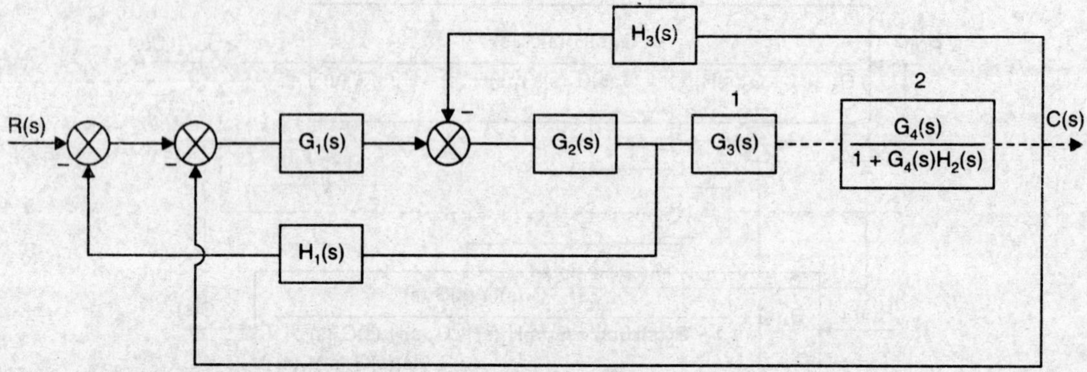

Step 3: Shifting the take-off point 3 after the block 4

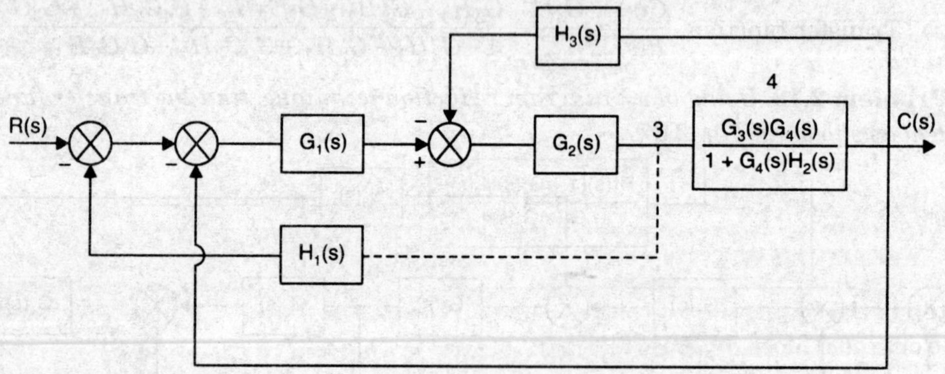

Step 4: Combining the blocks 5 and 6 and also 7 and 8

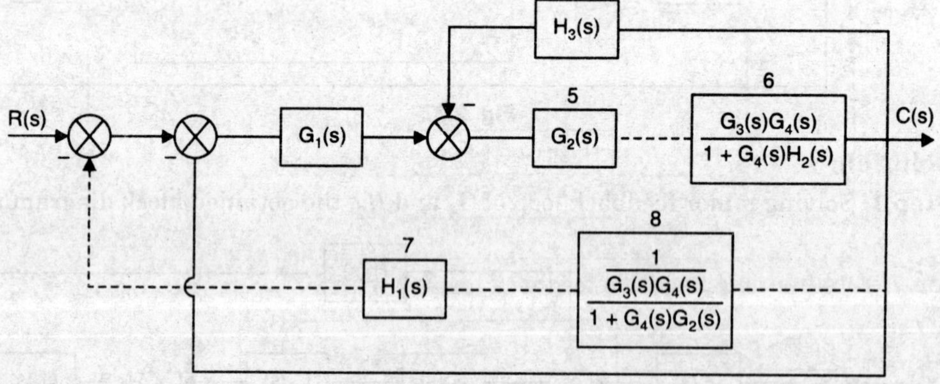

Step 5: Reduce the H_3 feedback loop, the obtained block diagram is

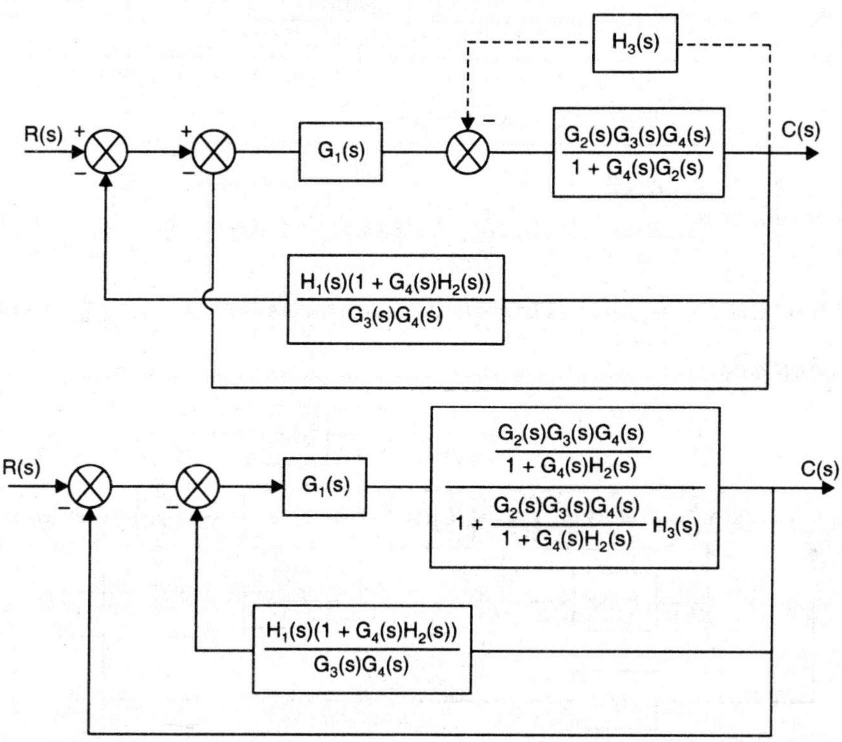

Step 6: Cascading the two blocks 9 and 10 and solving the two parallel blocks into unit block, the obtained block diagram is

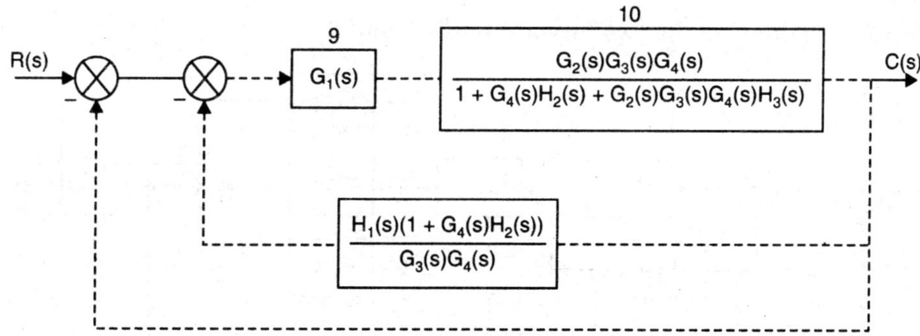

Step 7: Eliminationg negative feedback loop

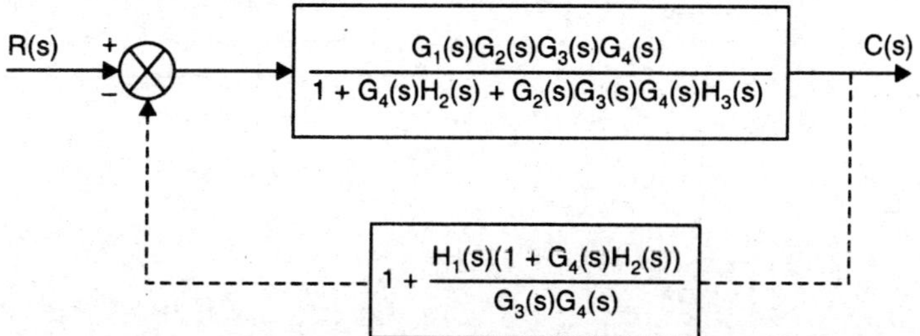

R(s) → C(s)

$$\frac{\dfrac{G_1(s)G_2(s)G_3(s)G_4(s)}{1 + G_4(s)H_2(s) + G_2(s)G_3(s)G_4(s)H_3(s)}}{1 + \dfrac{G_1(s)G_2(s)G_3(s)G_4(s)}{(1 + G_4(s)H_2(s) + G_2(s)G_3(s)G_4(s)H_3(s))} \times \dfrac{G_3(s)G_4(s) + H_1(s)(1 + G_4(s)H_2(s))}{G_3(s)G_4(s)}}$$

Transfer function, $\dfrac{C(s)}{R(s)} = \dfrac{G_1G_2G_3G_4}{1 + G_4H_2 + G_2G_3G_4H_3 + G_1G_2(G_3G_4 + H_1(1 + G_4H_2))}$

Problem 2.20. *Using method of reduction, determine the ratio* $\dfrac{C(s)}{R(s)}$ *in the block diagram as shown in Fig. 2.38.*

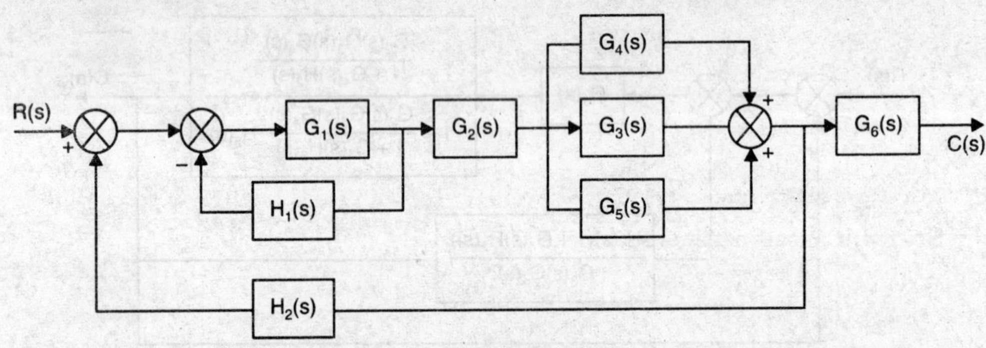

Fig. 2.38

Solution:

Step 1: Eliminating the two forward paths 1 and 2

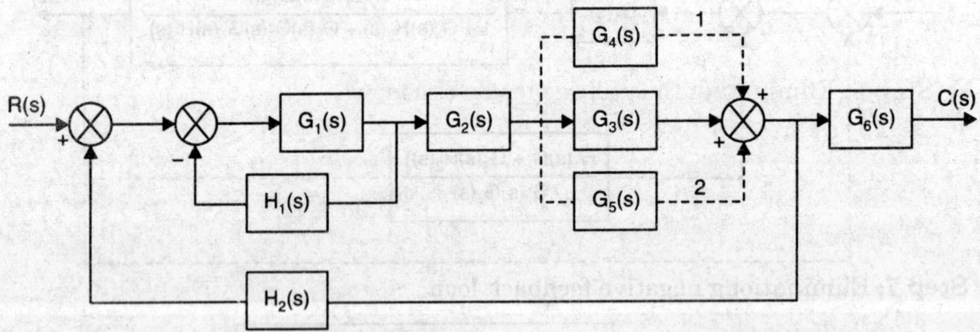

Step 2: Cascading the two blocks 3 and 4

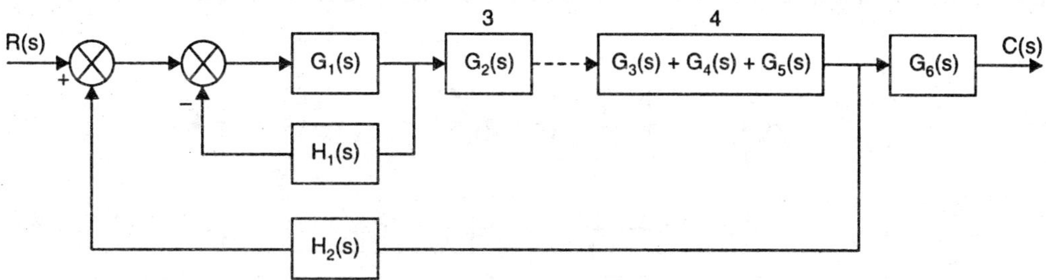

Step 3: Eliminating the negative feedback loop H_1

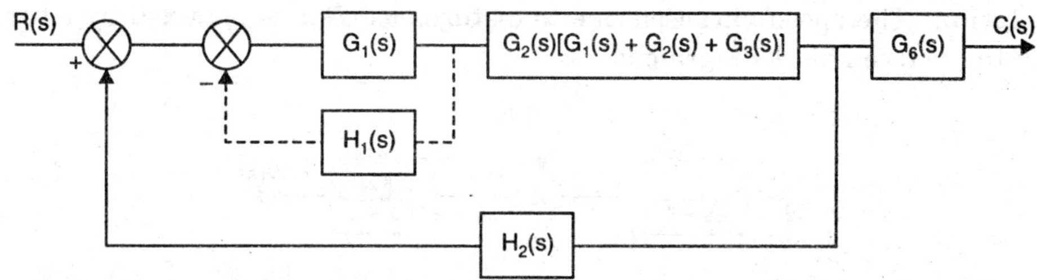

Step 4: Cascading the two blocks 5 and 6

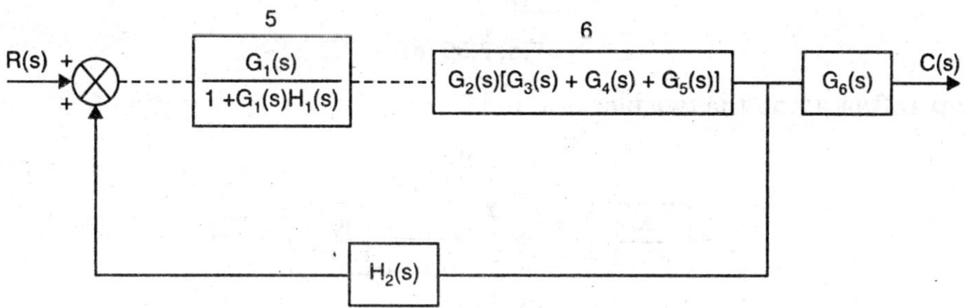

Step 5: Eliminating the positive feedback loop H_2

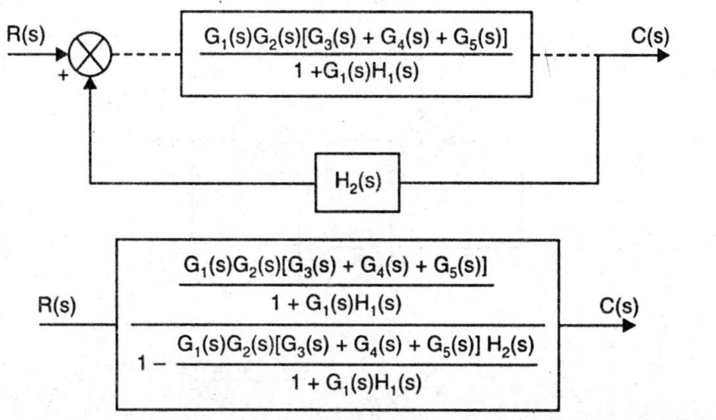

$$\therefore \quad \textbf{Transfer function, } \frac{C(s)}{R(s)} = \frac{G_1G_2G_3 + G_1G_2G_4 + G_1G_2G_5}{1 + G_1H_1 - G_1G_2G_3H_2 - G_1G_2G_4H_2 - G_1G_2G_5H_2}$$

Problem 2.21. *The block diagram as shown in Fig. 2.39, find the transfer function, the speed variation when a load torque disturbance D(s) is present in the system.*

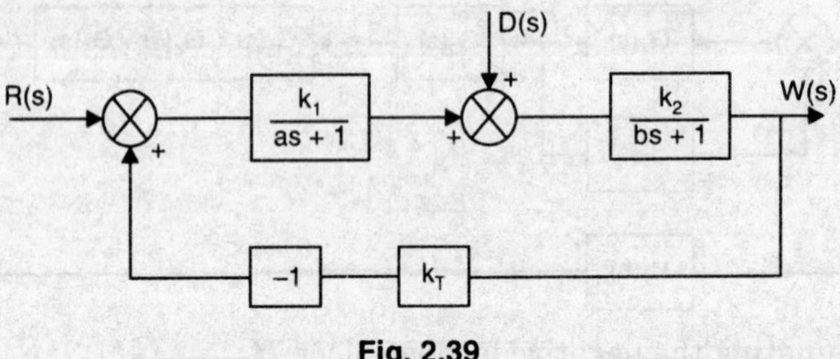

Fig. 2.39

Solution: The speed variation due to disturbance $D(s)$ is obtained by letting $R = 0$. Therefore the reduced block diagram is

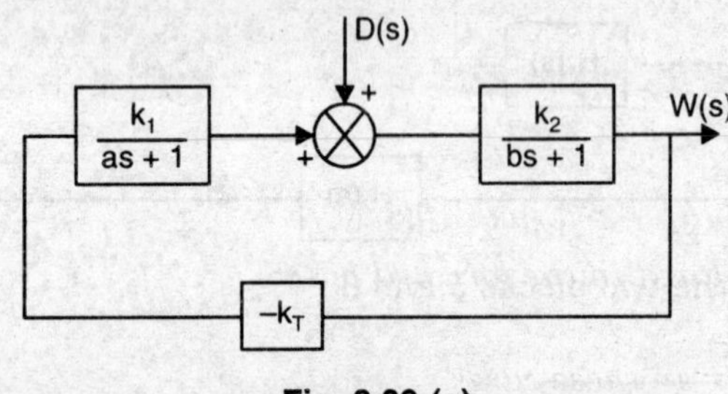

Fig. 2.39 (a)

Step 1: Cascading the two blocks 1 and 2

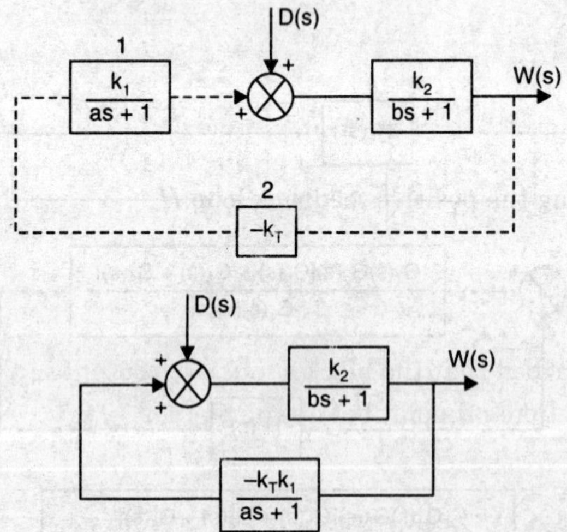

Step 2: Eliminating the positive feedback loop

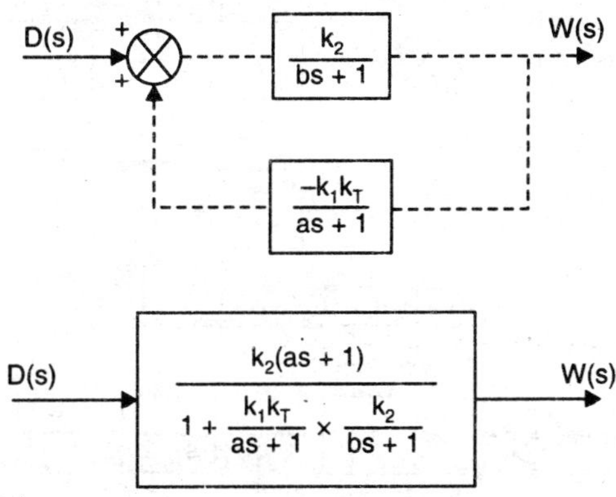

$$\therefore \quad \textbf{Transfer function, } \frac{W(s)}{D(s)} = \frac{K_2(as+1)}{(as+1)(bs+1) + K_1K_2K_T}$$

Problem 2.22. *Reduce the block diagram as shown in Fig. 2.40 to canonic forms when*

(i) Block k is isolated in the forward path

(ii) Block k is not isolated in the forward path.

Hence determine $\frac{C(s)}{R(s)}$ *in these cases.*

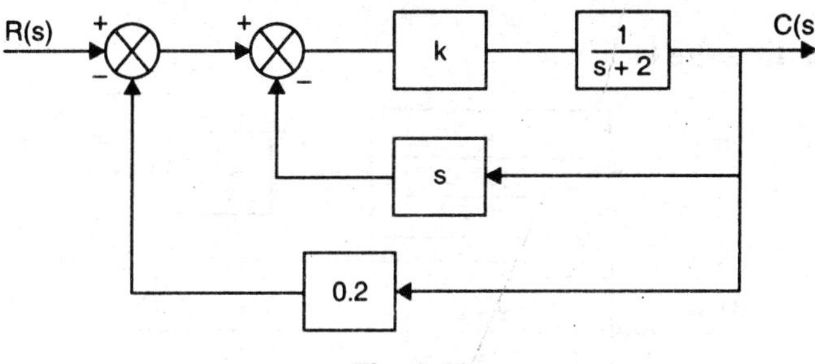

Fig. 2.40

Solution:

Case (*i*): It is prescribed that the block *k* must be present in the final block diagram.

Step 1: Moving the take-off point before the block 1

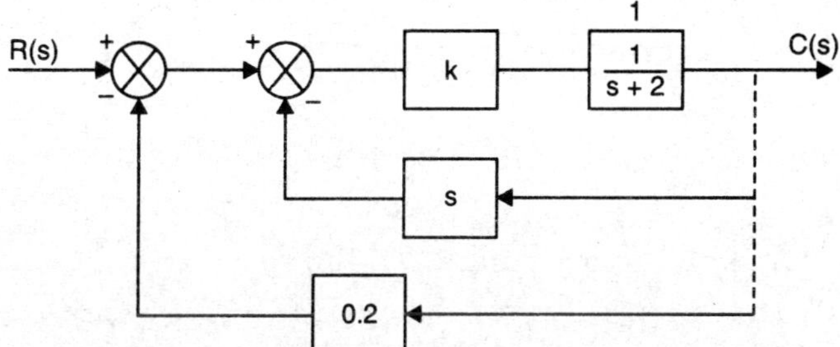

Step 2: Writing two summing points equal to one summing point w.r.t. their polarities.

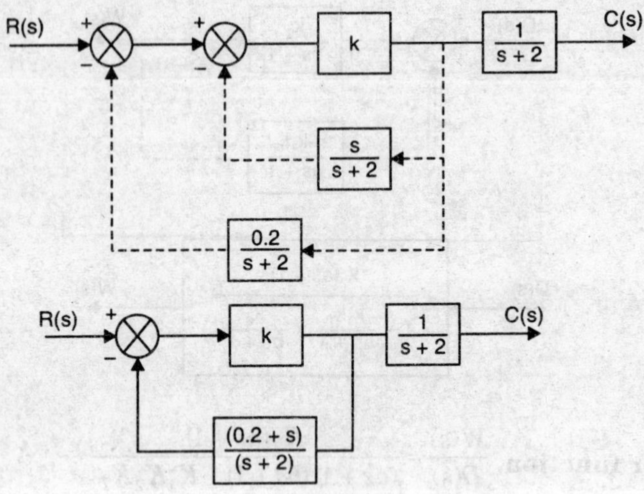

Step 3: Eliminating the negative feedback loop

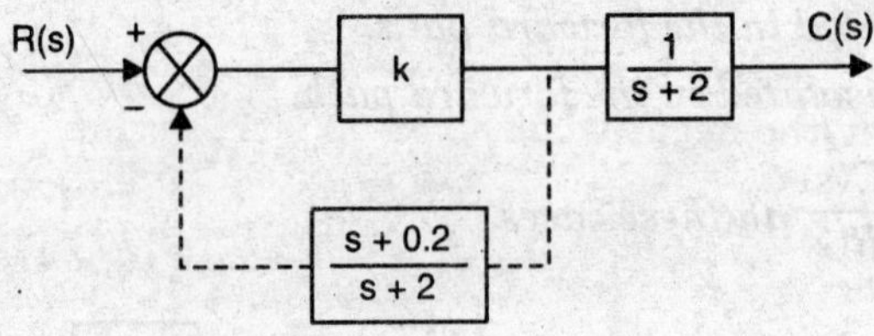

Step 4: Cascading the two blocks.

R(s) $\rightarrow \dfrac{k}{1+\dfrac{k(s+0.2)}{s+2}} \rightarrow \dfrac{1}{s+2} \rightarrow$ C(s)

R(s) $\rightarrow \dfrac{k(s+2)}{s+2+k(s+0.2)} \times \dfrac{1}{(s+2)} \rightarrow$ C(s)

$\therefore$ **Transfer function,** $\dfrac{C(s)}{R(s)} = \dfrac{k}{(s+2)+k(s+0.2)}$

Case (ii): Now it is prescribed that the block K should not be present in the forward path.

Step 1: Cascading the two blocks 1 and 2

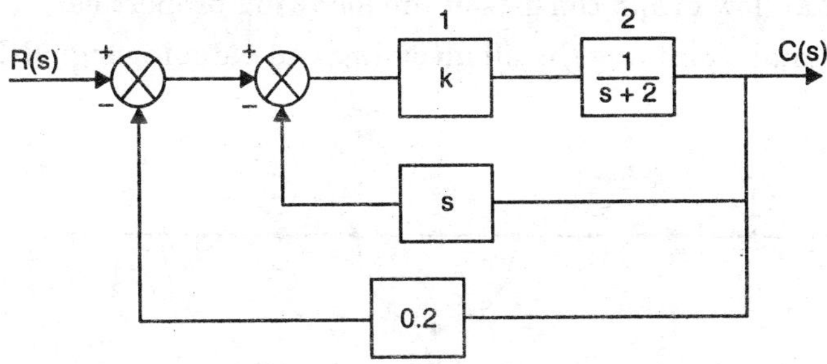

Step 2: Combining the parallel blocks 2 and 3

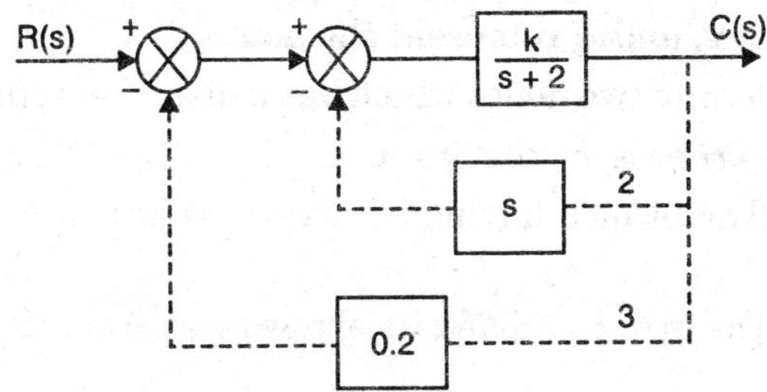

Step 3: Eliminating the negative feedback loop 4

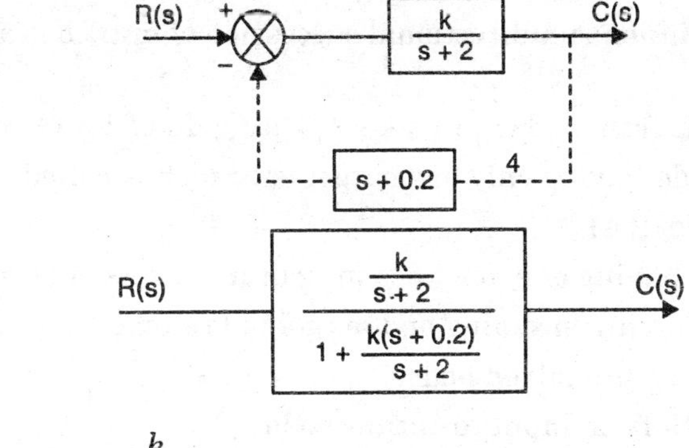

$$\frac{C(s)}{R(s)} = \frac{k}{(s+2) + k(s+0.2)}$$

2.9 SIGNAL FLOW GRAPH

Block diagram reduction technique is used to find the transfer function for large systems. But it has some following drawbacks.

- Block diagram reduction technique is a time consuming process.
- If the system is very large, block diagram reduction is somewhat complex.

Due to the above drawbacks present in the block diagram reduction technique, another alternative method is introduced to find the transfer function for large systems, which is known as signal flow graph approach.

Signal flow graph approach is a graphical method to find the transfer function of the system. Thus the signal flow graph consists of the following properties.

Node: It is a signal, summing the all incoming and outgoing signals.

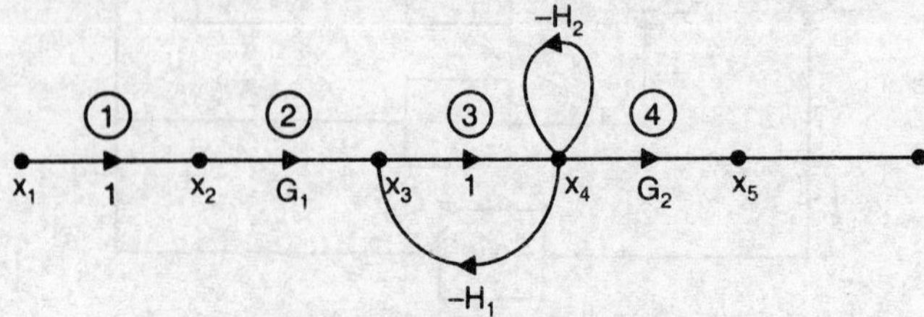

Fig. 2.41

In Fig. 2.41, x_1, x_2, x_3, x_4 and x_5 represent the nodes.

Branch: It is a line joining two nodes which has a direction represented by an arrow. In Fig. 2.41, 1, 2, 3 and 4 represent the branches.

Incoming branch: The branch having its arrow towards a node. At node x_3 incoming branch is G_1

Outgoing branch: The branch having its arrow away from the node. At x_3 node outgoing branch is one.

Transmittance: A branch which has magnitude is called transmittance, branch gain (or) transfer function. Gain of the branch 2 is G_1.

Path: This is the continuous unidirectional succession of branches along which no node is passed more than once.

In Fig 2.41, the path is from node x_1 to node x_5 with gain of $1 \times G_1 \times 1 \times G_2$.

Source node: The node having only one outgoing branch is called source node.

x_1 is source node in Fig. 2.41.

Sink node: The node having only one incoming branch, x_5 is sink node in Fig. 2.41.

Mixed node: A node having incoming and outgoing branches.

In Fig 2.41, x_2, x_3 and x_4 are mixed nodes.

Forward path: A path from input to output node.

In Fig. 2.41, forward path is $1 \times G_1 \times 1 \times G_2$.

Feedback path: A path which originates and terminates on same node.

Self loop: The loop consists of only one branch.

Path gain: Product of branch gains in going through a forward path.

In Fig. 2.41, path gain is $1 \times G_1 \times 1 \times G_2$.

Loop gain: The product of branch gains in a loop.

2.9.1 Rules Followed in Construction of Signal Flow Graph

(*i*) *The addition rules*: The value of variable represented by a node is equal to the sum of all signals entering at that node.

From Fig. 2.41 (*a*),

$$x_n = x_1\,G_1 + x_2\,G_2 \ldots\ldots x_i G_i$$

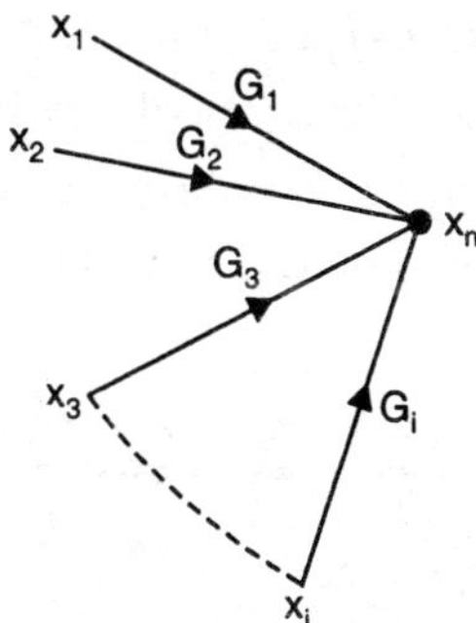

Fig. 2.41 (a)

(*ii*) *The transmission rule*: The value of the variable represented by a node, is transmitted on every branch which leaves that node. Hence the variable x represented by x_n node is transmitted into i node, is illustrated in Fig. 2.41 (*b*) and is represented by

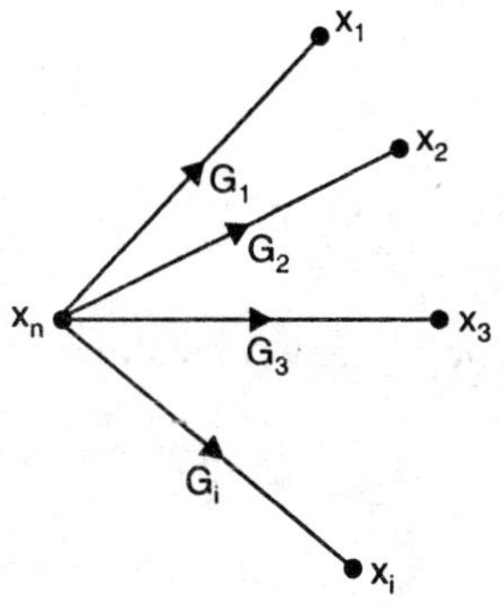

Fig. 2.41 (b)

(*iii*) *Multiplication rule*: A series connection of i branches with transmittances G_1, G_2 ...... G_i, shown in Fig 2.41 (*c*) can be replaced by a single branch with a new transmittance equal to the product of all branches of transmittance.

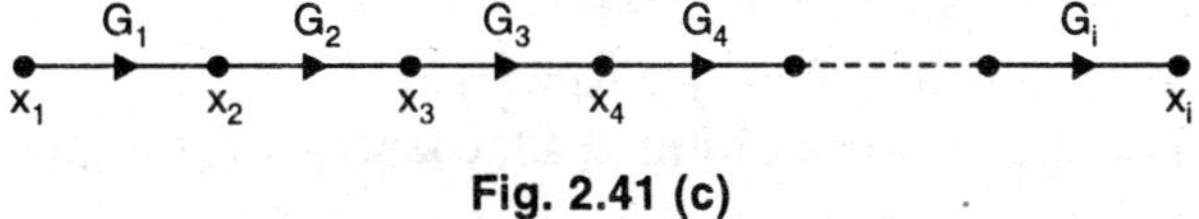

Fig. 2.41 (c)

2.9.2 Construction of Signal Flow Graph

The signal flow graph of a system is obtained from the equations which describes the system. Signal flow graph of systems can also be obtained from the block diagram of the system. The procedure used for construction of signal flow graph of a linear system is explained in the examples solved below.

Problem 2.23. *Draw the signal flow graph of the equations given below. Consider x_1 as input and x_6 as output nodes.*

$$x_2 = G_1 x_1 + H_1 x_3$$
$$x_3 = G_2 x_2$$

$$x_4 = G_3 x_3 + G_5 x_5$$
$$x_5 = H_2 x_4$$
$$x_6 = H_3 x_5$$

Solution: Using the above equations, the signal flow graph is constructed as shown in Fig. 2.42.

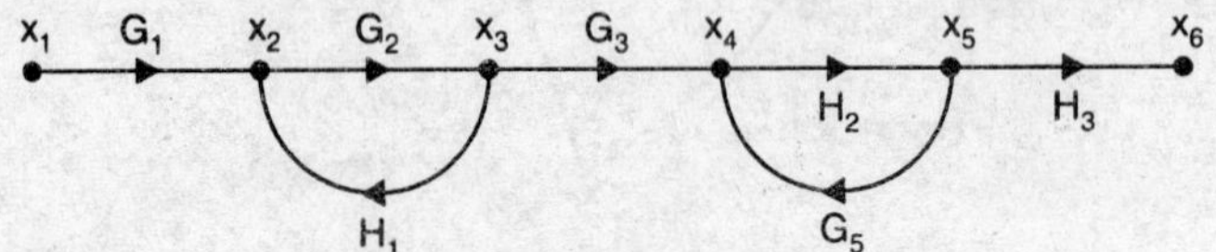

Fig. 2.42

Problem 2.24: *Draw the signal flow graph to represent the following equations and determine* $\frac{x_4}{x_1}$ *; how many forward paths are there and what is the transmittance of the graph?*

$$x_2 = G_1 x_1$$
$$x_3 = G_2 x_2 + G_3 x_2$$
$$x_4 = G_4 x_4$$

Solution: The signal flow graph is drawn as shown in Fig. 2.43.

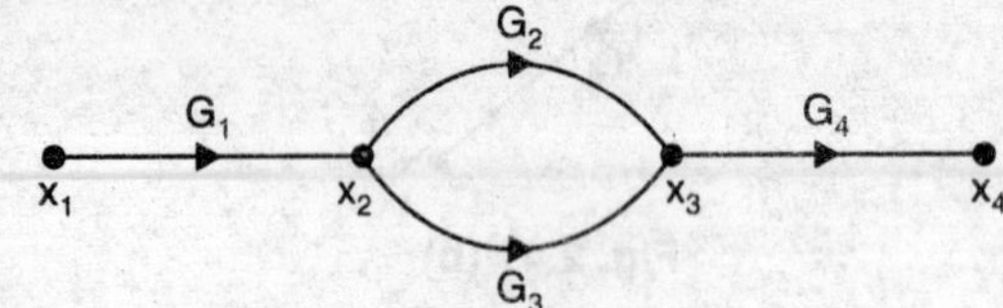

Fig. 2.43

Number of forward paths = 2,

$$P_1 = G_1 G_3 G_4, P_2 = G_1 G_2 G_4$$

Total transmittance of the graph = $P_1 + P_2 = G_1 G_4 (G_2 + G_3)$

Problems 2.25. For the network in Fig. 2.44, construct the signal flow graph.

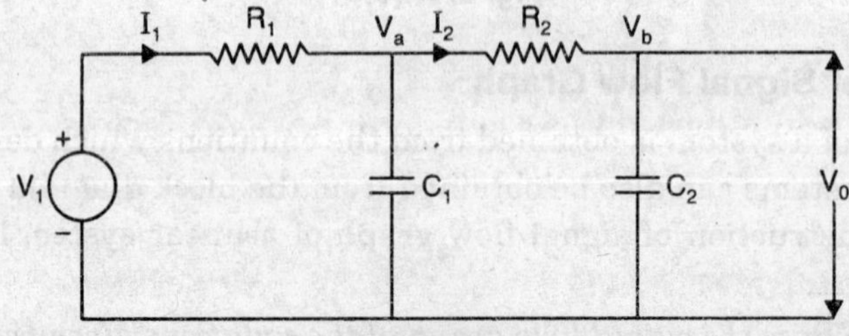

Fig. 2.44

Solution: For the given network the various signals are identified as

V_1 = input signal

I_1 = caused by V_1 in R_1

V_a = caused by I_1 through R_1

I_2 = current in R_2 caused by V_a

V_b = caused by I_2 in C_2

V_o = output signal.

The signals are inter related as follows:

$$I_1 = \left(\frac{V_1 - V_a}{R_1}\right) = \frac{V_1}{R_1} - \frac{V_a}{R_1}$$

$$V_a = (I_1 - I_2) \cdot \frac{1}{C_1 s} = \frac{I_1}{C_1 s} - \frac{I_2}{C_1 s}$$

$$I_2 = \frac{V_a - V_b}{R_2} = \frac{V_a}{R_2} - \frac{V_b}{R_2}$$

$$V_b = \frac{I_2}{C_2 s}$$

$$V_o = V_b = 1. V_b \tag{2.77}$$

These are six signals arranged at six nodes. Constructed signal flow graph is as shown in Fig. 2.45.

Edges are put between the respective nodes with their respective transmittances.

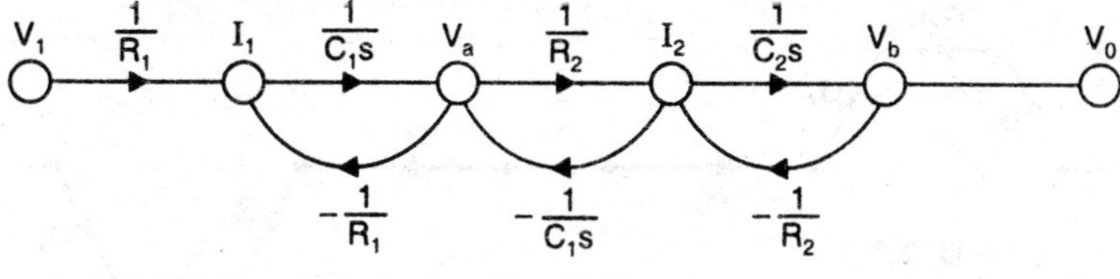

Fig. 2.45

2.9.3 Mason's Gain Formula

In signal flow graph approach, the overall gain or transfer function of the system is determined by using Mason's gain formula.

$$\text{Overall gain of the system, } T = \sum_{k=1}^{n} \frac{P_k \Delta_k}{\Delta}$$

The above relation is known as Mason's gain formula. The terms used in Mason's gain formula are explained below.

P_k = k^{th} forward path.

Δ = 1 – (sum of individual loop gains) + (sum of gain product of all possible combination of two non touching loops) – (sum of gain product of combination of three non touching loops) + ...

Δ_k = 1 – (sum of gains of non touching loops of kth forward path) + (sum of gain product of combination of two non touching loops for k^{th} forward path) – (sum of gain product of combination of three non touching loops for k^{th} forward path) + ...

Note: Δ_k is same as Δ but any loop not touching the kth forward path.

Problem 2.26. *Using Mason's gain formula, find the gain* $\frac{X_o}{X_i}$ *for the signal flow graph as shown in Fig. 2.46.*

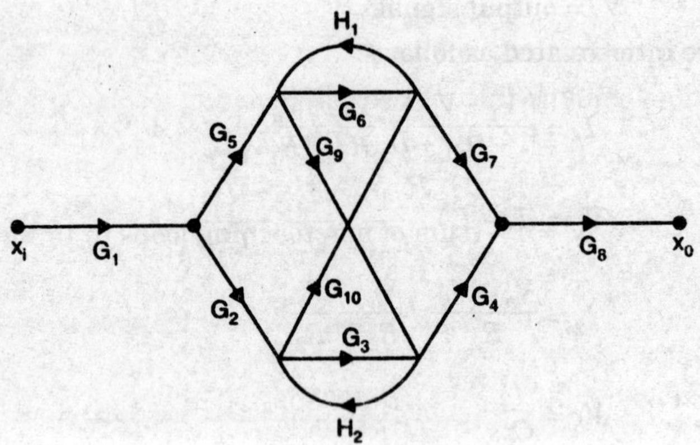

Fig. 2.46

Solution: There are six forward paths in the signal flow graph. Let the forward path gains be $P_1, P_2 \ldots\ldots P_6$.

Gain of forward paths,

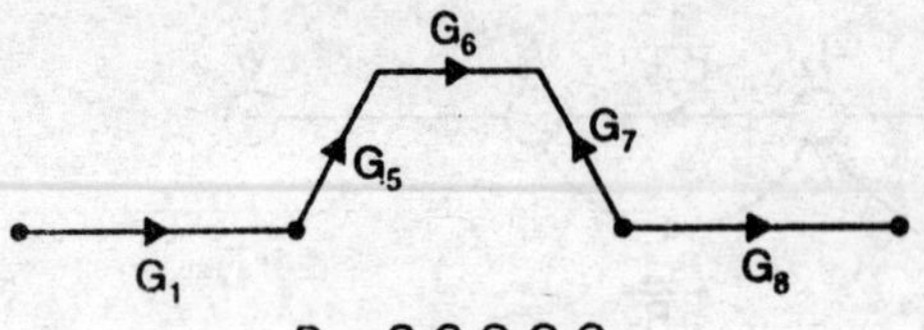

$$P_1 = G_1G_5G_6G_7G_8$$

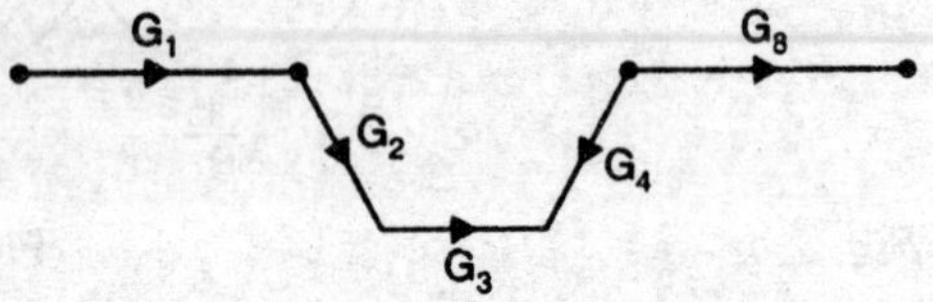

$$P_2 = G_1G_2G_3G_4G_8$$

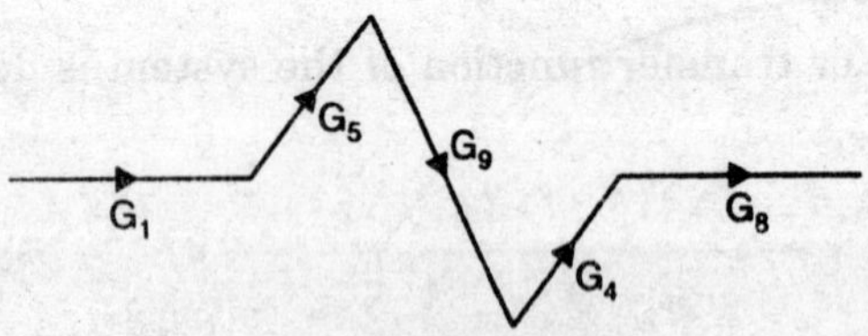

$$P_3 = G_1G_5G_9G_4G_8$$

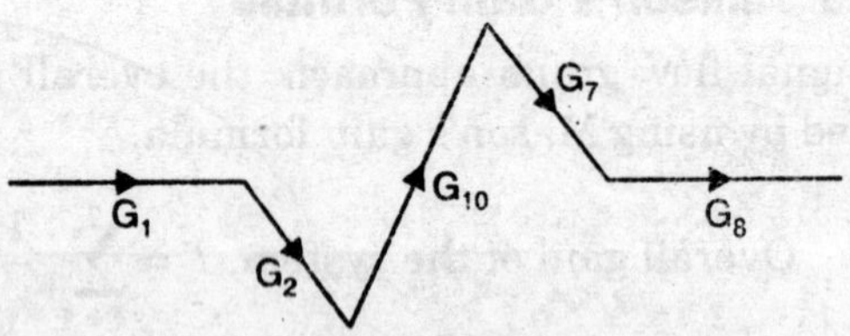

$$P_4 = G_1G_2G_{10}G_7G_8$$

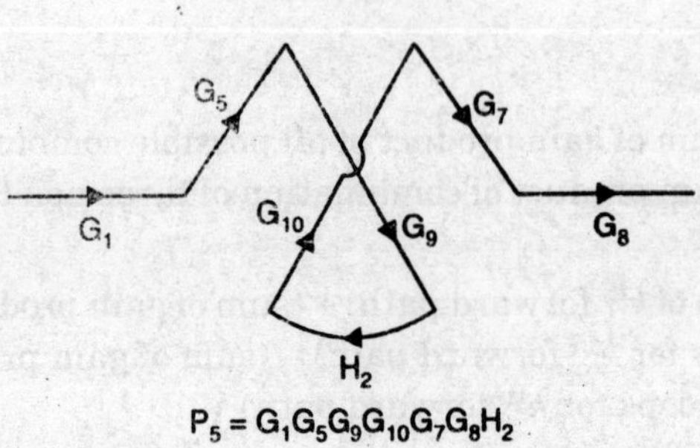

$$P_5 = G_1G_5G_9G_{10}G_7G_8H_2$$

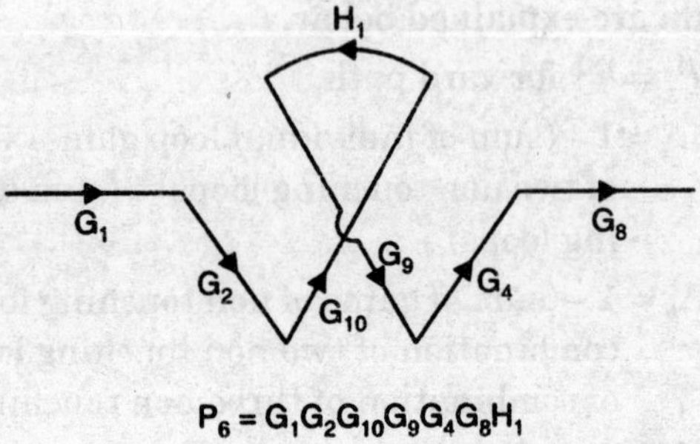

$$P_6 = G_1G_2G_{10}G_9G_4G_8H_1$$

There are three individual loops and their gains are

$$L_1 = G_6 H_1$$
$$L_2 = G_3 H_2$$
$$L_3 = G_9 H_2 G_{10} H_1$$

Non touching loops are L_1 and L_2

Combination of gain product of two non touching loops, $L_1 L_2 = G_3 G_6 H_1 H_2$

$$\Delta = 1 - (L_1 + L_2 + L_3) + L_1 L_2$$
$$= 1 - (G_6 H_1 + G_3 H_2 + G_9 G_{10} H_1 H_2) + G_3 G_6 H_1 H_2$$
$$\Delta_1 = 1 - (\text{sum of non touching loops to 1}^{\text{st}} \text{ forward path}) + \ldots\ldots$$
$$= 1 - (G_3 H_1)$$
$$\Delta_2 = 1 - (\text{sum of non touching loops to 2}^{\text{nd}} \text{ forward path}) + \ldots\ldots$$
$$= 1 - G_6 H_1$$
$$\Delta_3 = 1 - (\text{sum of non touching loops to 3}^{\text{rd}} \text{ forward path}) + \ldots\ldots$$
$$= 1 - (0) = 1$$

Similarly for Δ_4, Δ_5 and Δ_6

$$\Delta_4 = \Delta_5 = \Delta_6 = 1$$

Mason's gain formula, $\dfrac{X_o}{X_i} = \dfrac{P_1\Delta_1 + P_2\Delta_2 + P_3\Delta_3 + P_4\Delta_4 + P_5\Delta_5 + P_6\Delta_6}{\Delta}$

$$= \frac{\begin{array}{c} G_1G_5G_6G_7G_8(1 - G_3H_1) + G_1G_2G_3G_4G_8(1 - G_6H_1) + G_1G_5G_9G_4G_8 \\ + G_1G_2G_{10}G_7G_8 + G_1G_5G_9G_{10}G_7G_8H_2 + G_1G_2G_{10}G_9G_4G_8H_1 \end{array}}{1 - (G_6H_1 + G_3H_2 + G_9G_{10}H_1H_2) + G_3G_6H_1H_2}$$

Problem 2.27. *Using Mason's gain formula, find the overall gain of the system as shown in Fig. 2.46 (a).*

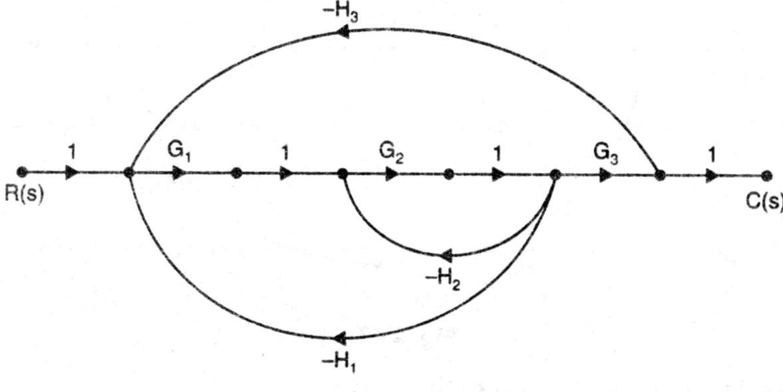

Fig. 2.46 (a)

Solution: There is one forward path

$P_1 = G_1G_2G_3$

There are three individual loops

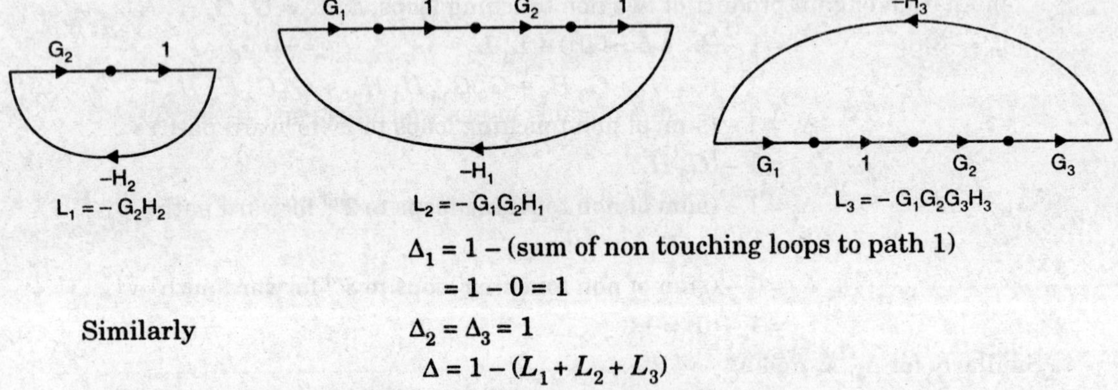

$$\Delta_1 = 1 - (\text{sum of non touching loops to path 1})$$

$$= 1 - 0 = 1$$

Similarly $\Delta_2 = \Delta_3 = 1$

$$\Delta = 1 - (L_1 + L_2 + L_3)$$

Since combination of two non touching loops is equal to zero.

Overall gain of the system

$$T = \frac{P_1\Delta_1}{\Delta}$$

$$T = \frac{G_1G_2G_3}{1 + G_2H_2 + G_1G_2H_1 + G_1G_2G_3H_3}$$

Problem 2.28. *Using Mason's gain formula, find overall gain of the system as shown in Fig. 2.46 (b).*

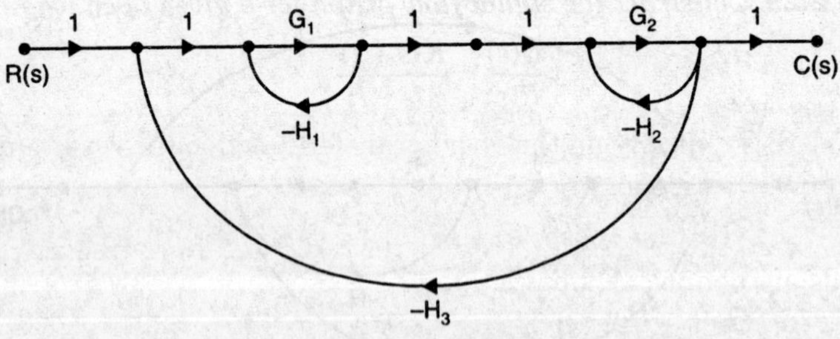

Fig. 2.46 (b)

Solution: There is one forward path

$P_1 = G_1G_2$

There are three individual loops

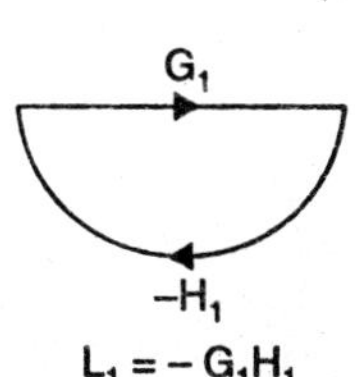

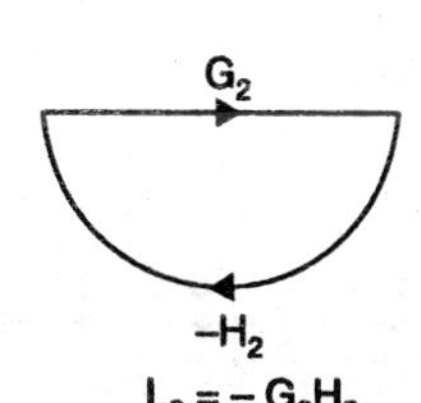

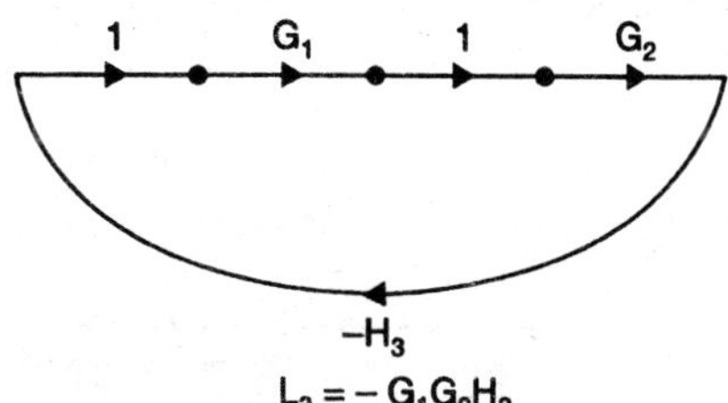

$$\Delta_1 = 1 - (\text{sum of gains of non touching loops to path 1})$$
$$= 1 - 0 = 1$$
$$\Delta = 1 - (\text{sum of gains of individual loops}) + (\text{sum of gain products of all possible combination of two non touching loops})$$
$$\Delta = 1 - (L_1 + L_2 + L_3) + L_1 L_2$$
$$\Delta = 1 - (G_1H_1 + G_2H_2 + G_1G_2H_3) + G_1G_2H_1H_2$$

Overall gain of the system

$$T = \frac{P_1\Delta_1}{\Delta}$$

$$T = \frac{G_1G_2}{1 + G_1H_1 + G_2H_2 + G_1G_2H_3 + G_1G_2H_1H_2}$$

2.9.4 Construction of Signal Flow Graph from a Transfer Function

Signal flow graphs can be constructed from the transfer function of the system by defining the signals with respect to the output and its time derivatives as well as the input and its time derivatives. The following examples illustrate the same.

Problem 2.29. *Construct the signal flow graph for a given open loop transfer function,*

$$\frac{C(s)}{R(s)} = \frac{K(s+1)}{s(s-1)}$$

Solution: Cross multiplying the above transfer function the resultant expression is

$$s(s-1)\,C(s) = K(s+1)\,R(s)$$

or
$$s^2C(s) = Ks\,R(s) + KR(s) + s\,C(s)$$

The signals are defined as follows:

$$x_1 = C(s),$$
$$x_2 = s\,C(s) = s\,x_1$$
$$x_3 = s^2\,C(s) = s\,x_2$$
$$y_1 = R(s)$$
$$y_2 = s\,R(s) = s\,y_1$$

i.e.,
$$x_1 = \frac{x_2}{s} \text{ and } x_2 = \frac{x_3}{s}$$

And $x_3 = Ky_2 + Ky_1 + x_2$

It is required to represent the given system equation in a signal flow graph as shown in Fig. 2.47.

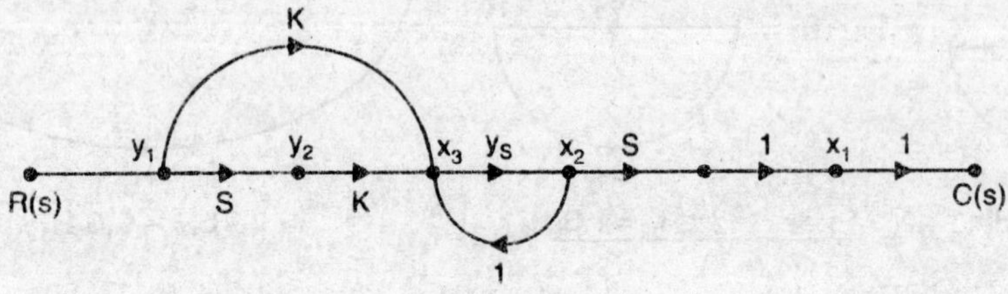

Fig. 2.47

Problem 2.30 *The signal flow graph is shown in Fig. 2.48. Using Mason's gain formula, determine its transfer function.*

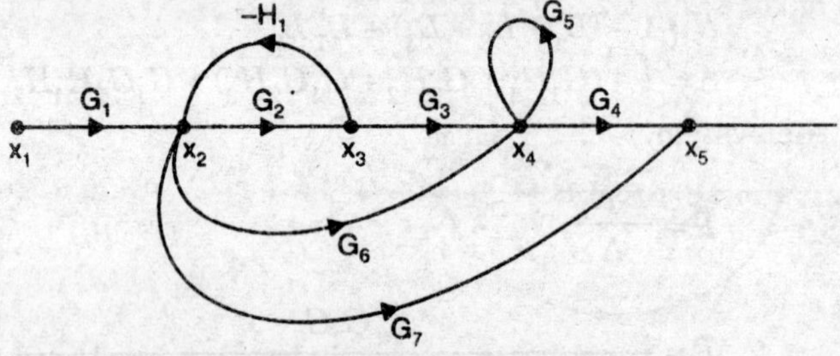

Fig. 2.48

Solution: There are three forward paths,

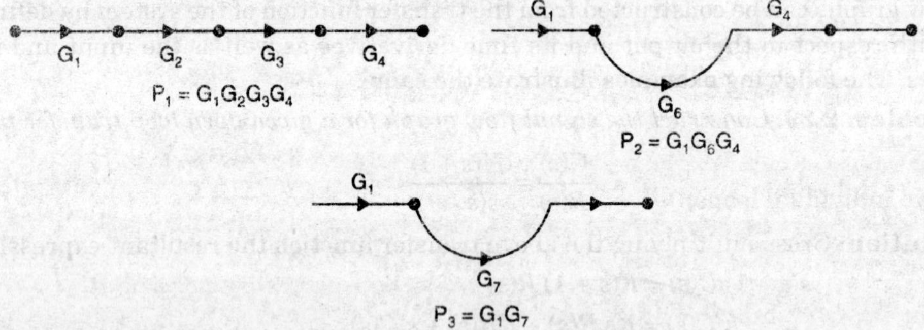

There are two individual loops,

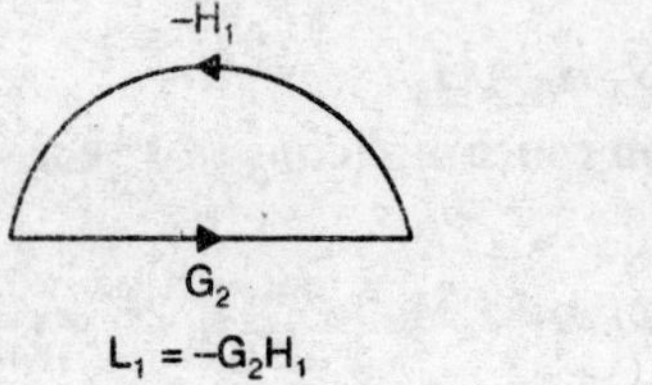

There is one combination of two non touching loops $L_1 L_2 = -G_2 G_5 H_1$

$\therefore \quad \Delta = 1 - (L_1 + L_2) + L_1 L_2 = 1 + G_2 H_1 - G_5 - G_2 G_5 H_1$

$\Delta_1 = 1 -$ (sum of gains of non touching loops to 1st forward path) + ...

$= 1 - (0) = 1$

$\Delta_2 = 1 - (0) = 1$

$\Delta_3 = 1 - G_5$

Overall gain,
$$T = \frac{P_1\Delta_1 + P_2\Delta_2 + P_3\Delta_3}{\Delta}$$

$$T = \frac{G_1G_2G_3G_4 + G_1G_6G_4 + G_1G_7\,(1 - G_5)}{1 + G_2H_1 - G_5 - G_2G_5H_1}$$

Problem 2.31. *Reduce the signal flow graph as shown in Fig. 2.49 and find the transmittance from node x_1 to output node x_5.*

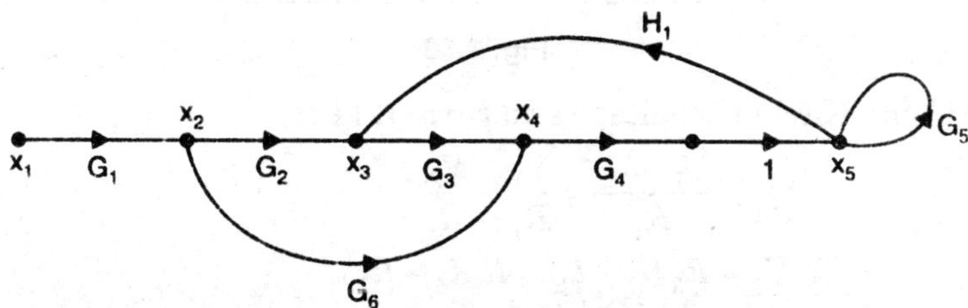

Fig. 2.49

Solution: There are two forward paths,

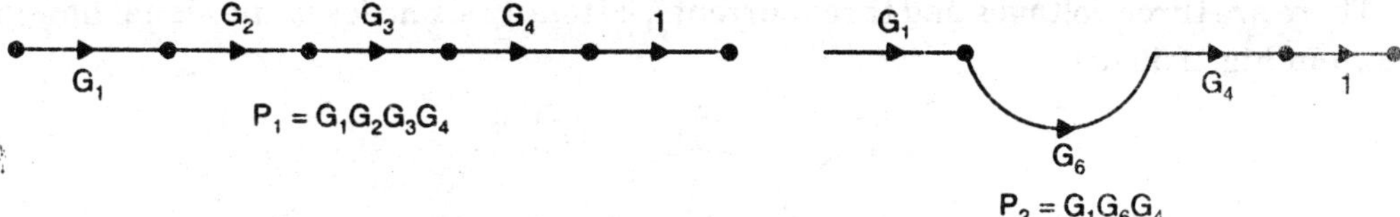

And two individual loops,

H₁

G₃ G₄

$L_1 = G_3G_4H_1$

G₅

$L_2 = G_5$

$\Delta = 1 - (L_1 + L_2) = 1 - G_3 G_4 H_1 - G_5$

$\Delta_1 = 1 -$ (sum of gains of non touching loops to 1st forward path) + ...

$= 1 - 0 = 1$

$\Delta_2 = 1 -$ (sum of gains of non touching loops to 2nd forward path)

$= 1 - 0 = 1$

Combination of any two non touching loops equals to zero.

Overall gain $$T = \frac{P_1\Delta_1 + P_2\Delta_2}{\Delta}$$

$$T = \frac{G_1G_2G_3G_4 + G_1G_6G_4}{1 - G_3G_4H_1 - G_5}$$

Problem 2.32. *Construct the signal flow graph for the network as shown in Fig. 2.50. Find also overall gain of the system.*

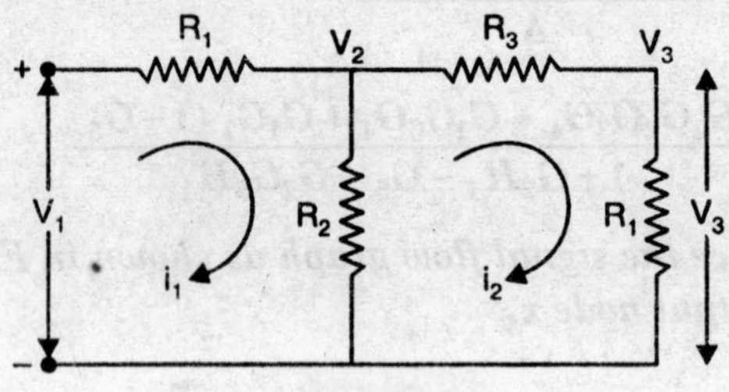

Fig. 2.50

Solution: Using Kirchoff's voltage and current laws.

$$i_1 = \frac{V_1 - V_2}{R_1} = \frac{V_1}{R_1} - \frac{V_2}{R_1}$$

$$V_2 = R_3 (i_1 - i_2) = R_3\, i_1 - R_3\, i_2$$

$$i_2 = \frac{V_2 - V_3}{R_2} = \frac{V_2}{R_2} - \frac{V_3}{R_2}$$

$$V_3 = R_4\, i_2$$

There are three voltages and three currents, so totally six nodes in the signal flow graph as shown in Fig. 2.50 (*a*).

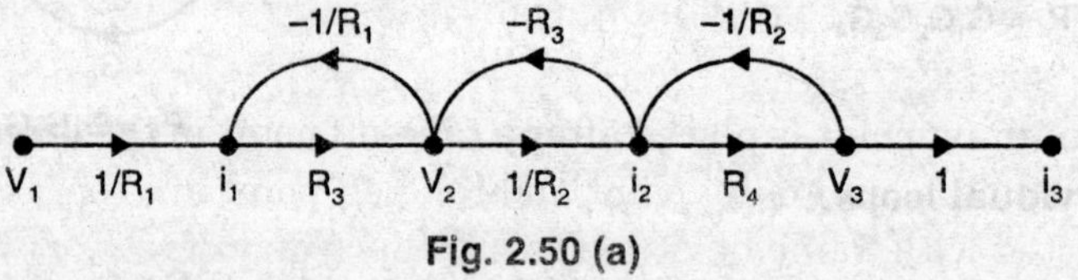

Fig. 2.50 (a)

There is only one forward path

$$P_1 = \frac{R_3R_4}{R_1R_2}$$

$\Delta_1 = 1 -$ (sum of gains of non touching loops to 1st forward path) = 1

Individual loops = There are three individual loops.

$$L_1 = \frac{-R_3}{R_1},\ L_2 = \frac{-R_3}{R_2},\ L_3 = \frac{-R_4}{R_2},\ L_1L_3 = \frac{R_3R_4}{R_2R_1}$$

There is no combination of three non touching loops.

Overall gain $T = \dfrac{P_1\Delta_1}{\Delta}$

$$T = \frac{\dfrac{R_3R_4}{R_1R_2}}{1+\dfrac{R_3}{R_1}+\dfrac{R_3}{R_2}+\dfrac{R_4}{R_2}+\dfrac{R_3R_4}{R_2R_1}}$$

$$T = \frac{\dfrac{R_3R_4}{R_1R_2}}{\dfrac{R_1R_2+R_3R_2+R_3R_1+R_4R_1+R_3R_4}{R_1R_2}}$$

$$T = \frac{R_3R_4}{R_1R_2+R_3R_2+R_3R_1+R_4R_1+R_3R_4}$$

2.9.5 Construction of Signal Flow Graph from Block Diagram

Block diagram of complex linear system can easily be analyzed by converting them into equivalent signal flow graphs and then applying direct Mason's gain formula. Hence generally, the signal flow graphs are obtained from the system block diagram before analyzing them. The correspondence between signal flow graph and block diagram is used to draw the equivalent signal flow graph of a system block diagram.

- Variable, summing points and take-off points are represented by nodes.
- If a summing point is placed before a take-off point in the direction of signal flow, then in such a case, summing point and take-off point are represented by a single node.
- If a summing point is placed after a take-off point in the direction of signal flow, then in such a case, summing point and take-off point are represented by separate nodes connected with a line having transmittance of unity.
- Transfer function is represented by line which is directed in the direction of signal flow.

Problem 2.33. *Using Mason's gain formula, find overall gain of the system as shown in Fig. 2.51 (a).*

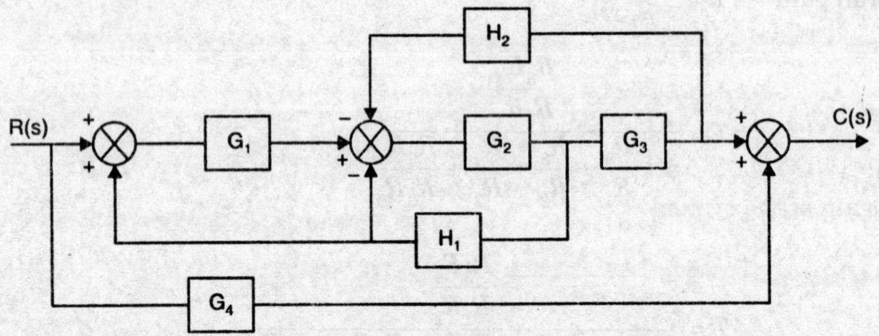

Fig. 2.51 (a)

Solution: For the above block diagram, signal flow graph is

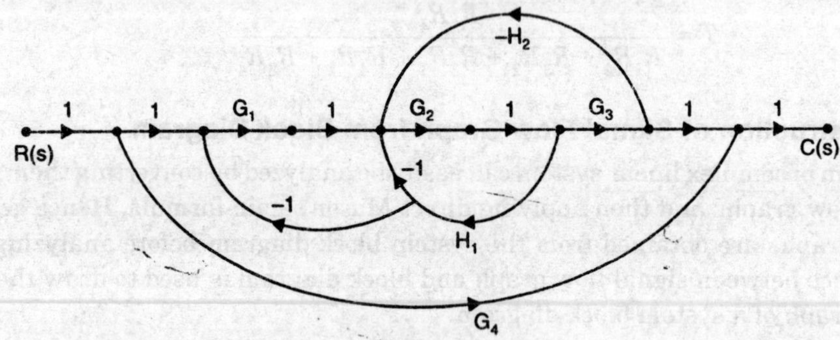

Fig. 2.51 (b)

There are two forward paths P_1 and P_2

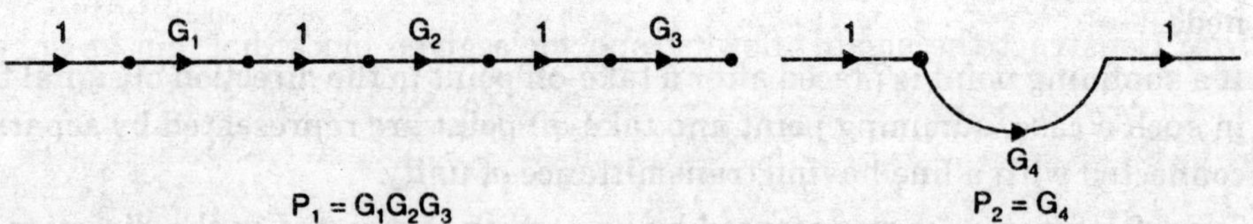

There are three individual loops

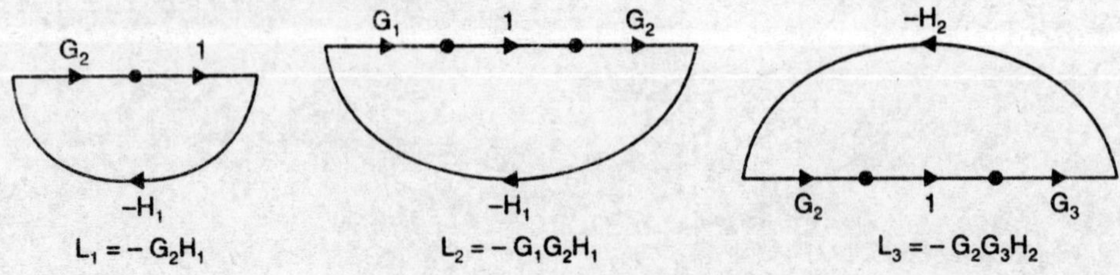

There is no combination of two non touching loops.

$$\Delta_1 = 1 - (\text{sum of gains of non touching loops to path 1})$$
$$= 1 - 0 = 1$$

Similarly
$$\Delta_2 = 1 - (L_1 + L_2 + L_3)$$
$$= 1 + G_2 H_1 + G_1 G_2 H_1 + G_2 G_3 H_2$$

Also,
$$\Delta = 1 - (L_1 + L_2 + L_3)$$
$$\Delta = 1 + G_1 H_1 + G_1 G_2 H_1 + G_2 G_3 H_2$$

Overall gain of the system

$$T = \frac{P_1\Delta_1 + P_2\Delta_2}{\Delta}$$

$$T = \frac{G_1G_2G_3 + G_4(1 + G_2G_3H_2 + G_1G_2H_1 + G_2H_1)}{1 + G_3G_2H_2 + G_1G_2H_1 + G_2H_1}$$

Problem 2.34: *Find the gain of the system using signal flow graph approach for a given block diagram as shown in Fig. 2.52.*

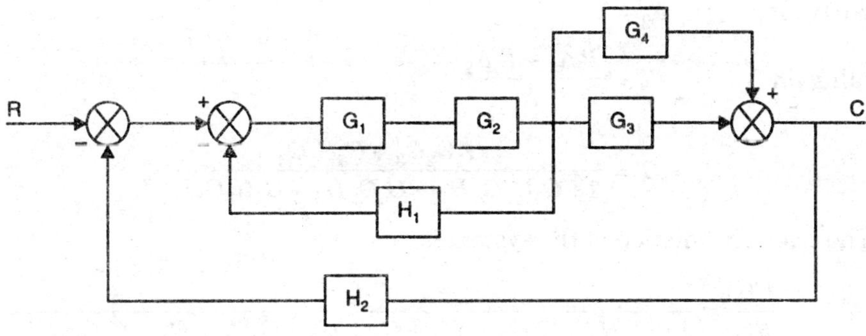

Fig. 2.52

Solution: Construct the signal flow graph for a given block diagram as shown in Fig. 2.53.

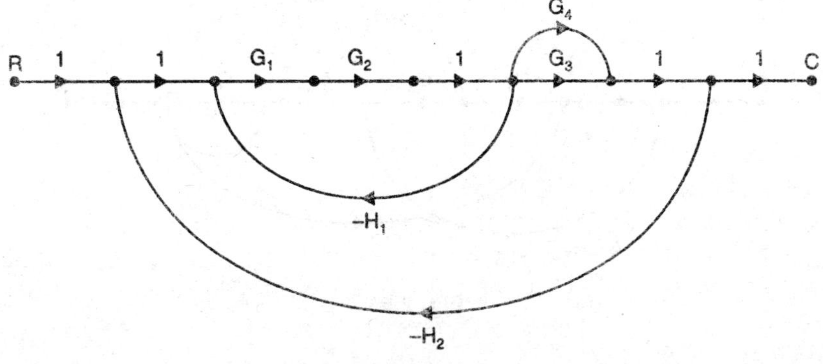

Fig. 2.53

There are two forward paths P_1 and P_2

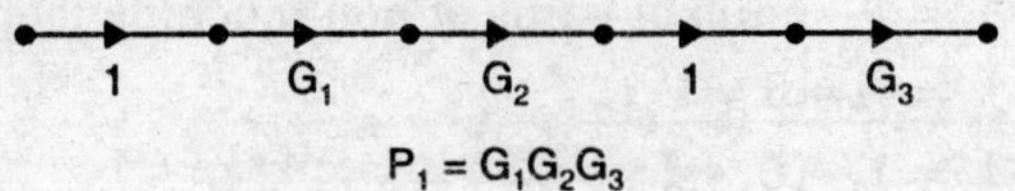

$P_1 = G_1G_2G_3$

G4

1 G1 G2

$P_2 = G_1G_2G_4$

There are three individual loops

$$L_1 = -G_1\,G_2\,G_4\,H_2 \qquad L_2 = -G_1\,G_2\,H_1 \qquad L_3 = -G_1\,G_2\,G_3\,H_2$$

Combination of two non touching loops is zero.

$$\therefore \quad \Delta = 1 - (L_1 + L_2 + L_3)$$

$$\Delta_1 = 1 - (\text{sum of non touching loops to 1}^{\text{st}} \text{ forward path})$$

$$= 1 - (0) = 1$$

Similarly for $\Delta_2 = 1$

Overall gain $$T = \frac{P_1\Delta_1 + P_2\Delta_2}{\Delta}$$

$$T = \frac{G_1G_2G_3 + G_1G_2G_4}{1 + G_1G_2G_4H_2 + G_1G_2H_1 + G_1G_2G_3H_2}$$

$\therefore$ The transfer function of the system is given by

$$\frac{C(s)}{R(s)} = \frac{G_1G_2G_3G_4}{1 + G_4H_2 + G_2G_3G_4H_3 + G_1G_2H_1 + G_1G_2G_3G_4 + G_1G_2G_4H_1H_2}$$

Problem 2.35. *A signal flow graph of the system is shown in Fig. 2.54. Find its transfer function by using Mason's gain formula.*

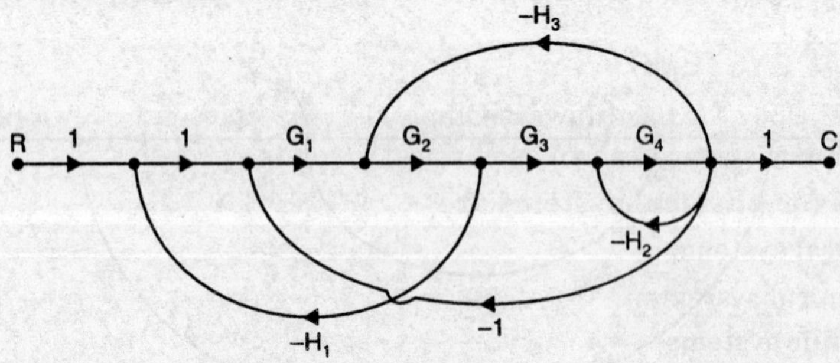

Fig. 2.54

Solution: There is only one forward path P_1

1 G_1 G_2 G_3 G_4 1

$P_1 = G_1G_2G_3G_4$

There are four individual loops

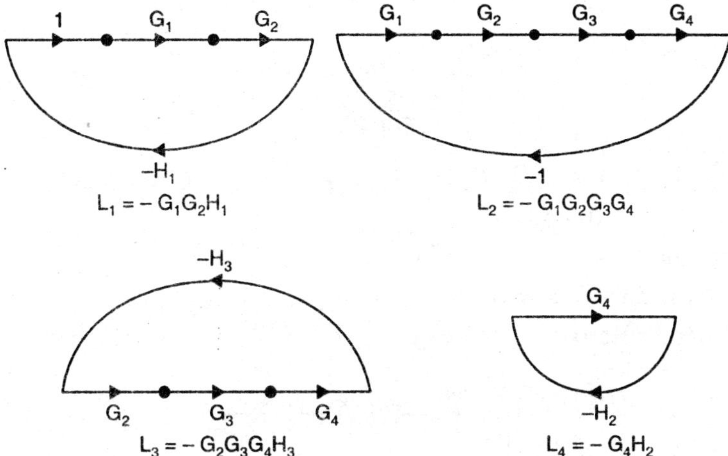

There is one combination of two non touching loops 1 and 4.

$$L_1 \times L_4 = G_4\, G_1\, G_2\, H_1\, H_2$$

$$\therefore \quad \Delta = 1 - (L_1 + L_2 + L_3 + L_4) + L_1 \times L_4$$

$$= 1 + (G_1G_2H_1 + G_1G_2G_3G_4 + G_2G_3G_4H_3 + G_4H_2) + G_1G_2G_4H_1H_2$$

As all the loops L_1, L_2, L_3 and L_4 touching the forward path P_1

$$\Delta_1 = 1 - 0 = 1$$

By using Mason's gain formula, the transfer function is

$$\frac{C}{R} = \frac{P_1\Delta_1}{\Delta} = \frac{G_1G_2G_3G_4}{1 + G_1G_2H_1 + G_1G_2G_3G_4 + G_2G_3G_4H_3 + G_4H_2 + G_1G_2G_4H_1H_2}$$

2.10 PHYSICAL SYSTEMS

The systems which obey the basic laws of Ohm's law, Kirchoff's law, Newton's law, Coulomb's law, D Alembert's law etc. are known as physical systems.

Examples for physical systems are

(*a*) Electrical systems

(*b*) Mechanical systems

(*c*) Hydraulic systems

(*d*) Pneumatic systems

(*e*) Thermal systems

(*a*) Electrical Systems

The systems which obey Ohm's law and Kirchoff's laws are known as electrical systems.

Kirchhoff's voltage and current laws are used in writing the electrical network equations. Before deriving the circuit equations, the voltage-current relationship of the three basic elements *i.e.*, resistance (R), inductance (L) and capacitance (C) are given below.

(*i*) Resistance R (Ohm)

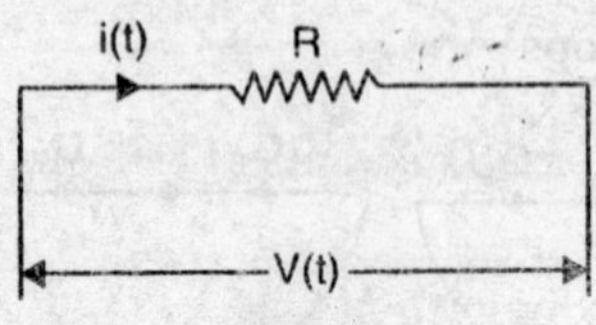

Fig. 2.55

$$V(t) = R\,i(t) \tag{2.78 (a)}$$

or

$$i(t) = G\,V(t) \tag{2.78 (b)}$$

where, G = conductance.

In Laplace transform $V(s) = R\,I(s)$

(*ii*) Inductance L (Henry)

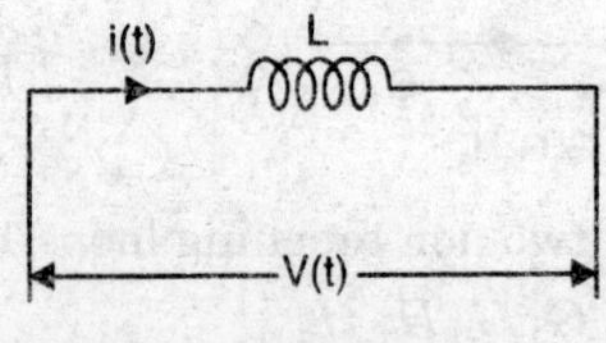

Fig. 2.56

$$V(t) = L\frac{di}{dt} \tag{2.79 (a)}$$

or

$$i(t) = \frac{1}{L}\int_0^t V(t)\,dt \tag{2.79 (b)}$$

In Laplace transform

$$V(s) = L[s\,I(s) - I_o]$$

where I_o = Initial current.

(*iii*) Capacitance C (farad)

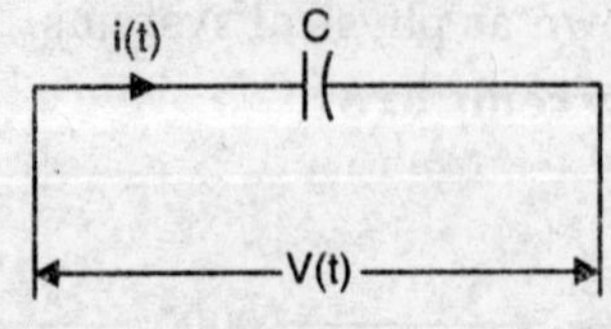

Fig. 2.57

$$V(t) = \frac{1}{C}\int_0^t i(t)\,dt \tag{2.80 (a)}$$

or

$$i(t) = \frac{CdV}{dt} \tag{2.80 (b)}$$

In Laplace transform

$$V(s) = \frac{1}{C}\frac{I(s)}{s} + \frac{V_o}{s}$$

where V_0 = initial voltage

(*b*) Mechanical Systems

The systems which obey the D Alembert's laws, Coulumb's law etc., such systems are known as mechanical systems.

Mechanical systems may be either translational system or rotational system. Components of mechanical translational system are mass, elasticity idealized by a helical spring and friction idealized by a damper. The components of rotational mechanical system are inertia, tortional spring and friction.

1. Translational Mechanical System

Mass:

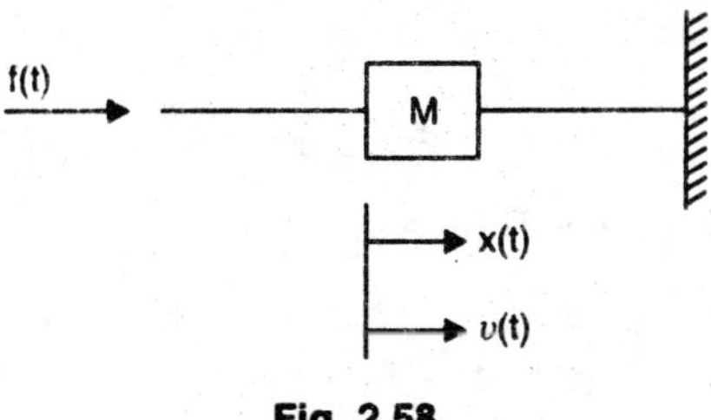

Fig. 2.58

M = Mass of the body

$f(t)$ = Force applied on mass of the body

$x(t)$ = Linear displacement

$v(t)$ = Linear velocity

According to Newton's second law

$$F(t) = Ma = \frac{Mdv}{dt} = M\frac{d^2x}{dt^2} \qquad \left[\because v = \frac{dx}{dt}\right] \qquad (2.81)$$

where a = acceleration due to gravity.

Spring:

In elasticity idealized as a spring with spring constant K force through the variable (input) and displacement x be the output.

Fig. 2.59

$$f(t) = Kx = K(x_1 - x_2)$$

$$= K \int_0^t (x_1 - x_2)\, dt$$

$$= K \int_0^t v\, dt \qquad (2.82)$$

Damper:

The input is the force F and the output is acceleration or rate of change of velocity.

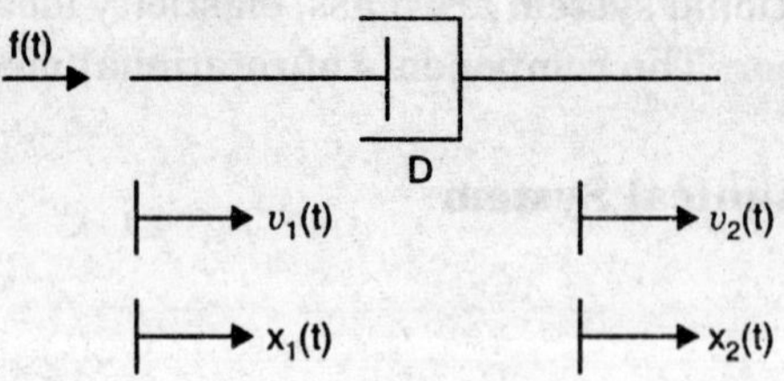

Fig. 2.60

$$F = Dv = D(v_1 - v_2)$$

$$= D \frac{d}{dt}(x_1 - x_2) \qquad (2.83)$$

Friction is idealized as damper (also called dash pot or shock absorber) where input is force 'F' and output is velocity 'v'. When a compression force is applied to a dash pot, it is translated into a velocity of the piston or rod. On removal of force, the piston does not return or returns gradually unlike a spring.

2. ***Rotational Mechanical Systems***

Inertial element:

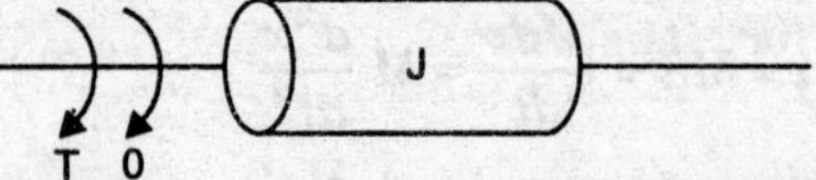

Fig. 2.61

Torque $$T = J\frac{d\omega}{dt} = J\frac{d^2\theta}{dt^2} \qquad \left[\because \omega = \frac{d\theta}{dt}\right] \qquad (2.84)$$

J = Moment of inertia

ω = Angular velocity

θ = Angular displacement

T = Torque on inertial body

Tortional spring

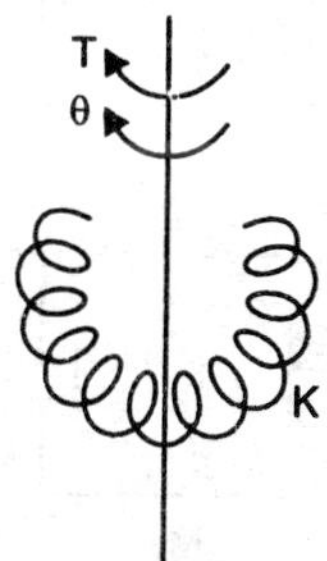

Fig. 2.62

$$T = K\theta = K\int_0^{\infty} \omega\, dt \tag{2.85}$$

Friction

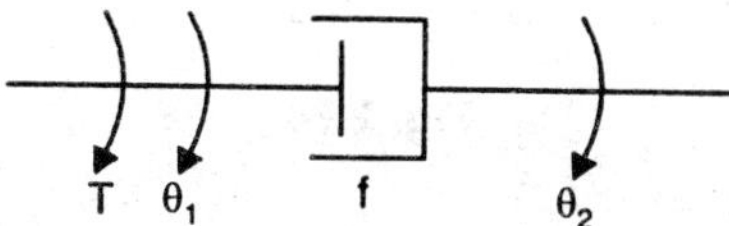

Fig. 2.63

$$T = f\omega = f\left[\frac{d\theta_1}{dt} - \frac{d\theta_2}{dt}\right] \tag{2.86}$$

The units are

M = newtons per m/sec^2

K = newtons per metre

$D = f$ = newtons per m/sec

J = kg-m^2

Procedure: The equations for translational mechanical systems are obtained by using the procedure described below.

1. Draw the physical system into an equivalent configuration with respect to a reference and define various co-ordinates with their positive directions.
2. Take the displacement at nodes. Therefore number of displacement represents number of nodes or nodal equations.
3. Write equations of equilibrium for all forces acting on elements using D'Alembert's principle, for transnational systems.
4. Write force-geometry relations for individual elements.

   Combine relations of steps (3) and (4) to obtain the system equation.

Problem 2.36. *Derive the transfer function* $\frac{x_1(s)}{F(s)}$ *for the mechanical system as shown in Fig. 2.64.*

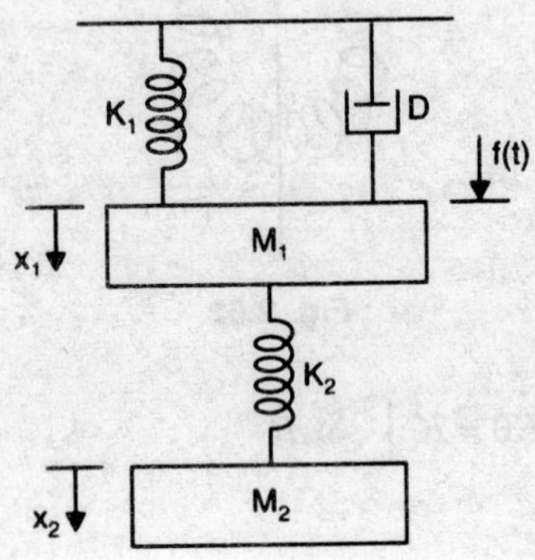

Fig. 2.64

Solution:

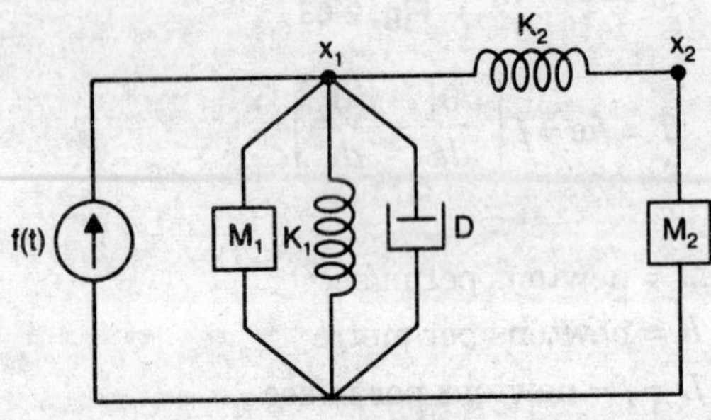

Fig. 2.65

Equivalent diagram of Fig. 2.64 with nodes x_1 and x_2 are shown in Fig. 2.65. According to D' Alembert's principle,

First nodal equation, $F(t) = M_1 \ddot{x}_1 + K_1 x_1 + D\dot{x}_1 + K_2 (x_1 - x_2)$

Second nodal equation, $0 = K_2 (x_2 - x_1) + M_2 \ddot{x}_2$

Note (*i*) The one dot over x indicates the first order derivative.

(*ii*) The two dots over x indicate the second order derivative.

Taking Laplace transformation of the above two equations.

$$F(s) = (M_1 s_2 + K_1 + Ds + K_2)\, x_1 (s) - K_2 x_2 (s) \tag{2.87}$$

$$0 = -K_2 x_1 (s) + (M_2 s^2 + K_2)\, x_2 (s) \tag{2.88}$$

From eqn. (2.88), writing $x_2(s)$ in terms of $x_1(s)$

$$x_2(s) = \frac{K_2}{M_2s^2 + K_2} x_1(s) \quad (2.89)$$

Substituting $x_2(s)$ from eqn. (2.89) in eqn. (2.87)

$$F(s) = (M_1s^2 + Ds + (K_1 + K_2)\, x_1(s) - \frac{K_2^{\,2}}{(M_2s^2 + K_2)} x_1(s)$$

$$F(s) = \left[\frac{(M_1s^2 + Ds + K_1 + K_2)(M_2s^2 + K_2) - K_2^{\,2}}{(M_2s^2 + K_2)}\right] x_1(s)$$

$$\frac{x_1(s)}{F(s)} = \frac{(M_2s^2 + K_2)}{(M_2s^2 + K_2)(M_1s^2 + Ds + K_1 + K_2) - K_2^{\,2}}$$

Problem 2.37. *A mechanical system is shown in Fig. 2.66. Derive its transfer function.*

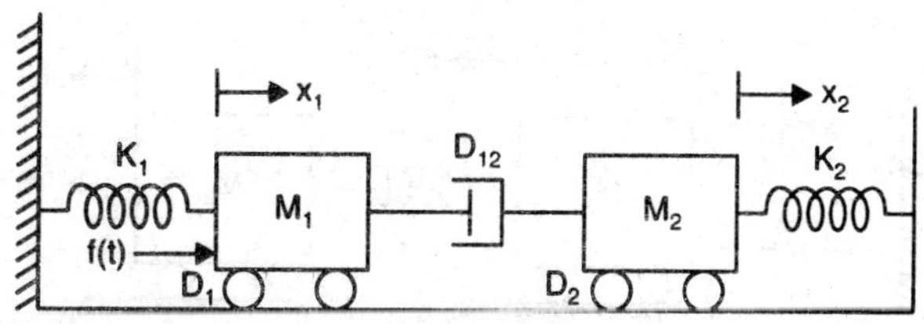

Fig. 2.66

Solution:

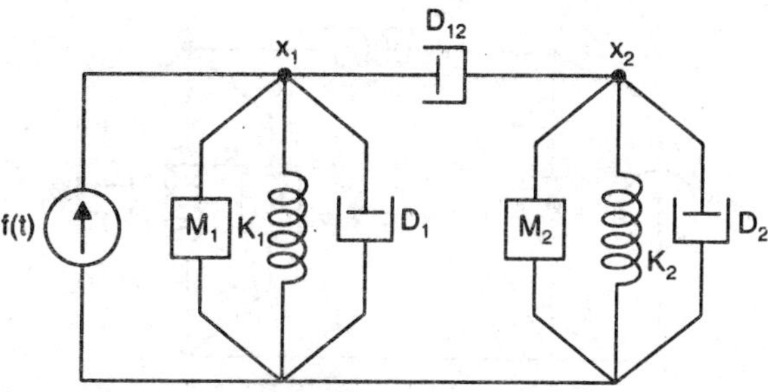

Fig. 2.67

Equivalent diagram for a given mechanical translational system is shown in Fig. 2.67, there are two nodes (displacements), so we will get two nodal equations.

1st nodal equation, $F(t) = M_1\ddot{x}_1 + K_1 x_1 + D_1 \dot{x}_1 + D_{12}(\dot{x}_1 - \dot{x}_2)$ (2.90)

2nd nodal equation, $0 = D_{12}(\dot{x}_2 - \dot{x}_1) + M_2\ddot{x}_2 + K_2 x_2 + D_2\ddot{x}_2$ (2.91)

Taking Laplace transformation of eqns. (2.90) and (2.91)

$$F(s) = (M_1s^2 + K_1 + D_1s + D_{12}s)\, x_1(s) - D_{12}\, s\, x_2(s) \quad (2.92)$$

$$0 = -D_{12}\, s\, x_1(s) + (M_2 s^2 + K_2 + D_2s + D_{12}s)\, x_2(s) \quad (2.93)$$

Writing the eqn. (2.93), $x_1(s)$ in terms of $x_2(s)$

$$x_1(s) = \left[\frac{M_2s^2 + K_2 + D_2s + D_{12}s}{sD_{12}}\right] x_2(s) \tag{2.94}$$

Substituting eqn. (2.94) in eqn. (2.92)

$$F(s) = \left\{[M_1s^2 + K_1 + s(D_1 + D_{12})]\frac{[M_2s^2 + K_2 + s(D_2 + D_{12})]}{sD_{12}} - sD_{12}\right\} x_2(s)$$

∴ Transfer function is

$$\frac{x_2(s)}{F(s)} = \frac{sD_{12}}{[M_1s^2 + K_1 + s[D_1 + D_{12}]]\,[M_2s^2 + K_2 + s[D_2 + D_{12}] - (sD_{12})^2}$$

Problem 2.38. *Find the transfer function* $\frac{x_2(s)}{F(s)}$ *of the system given in Fig. 2.68.*

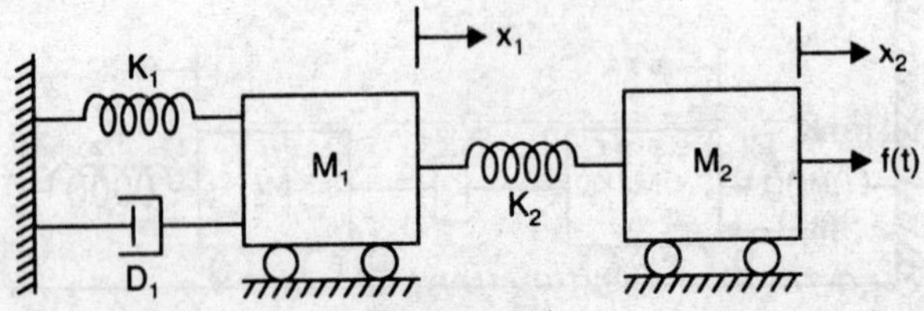

Fig. 2.68

Solution:

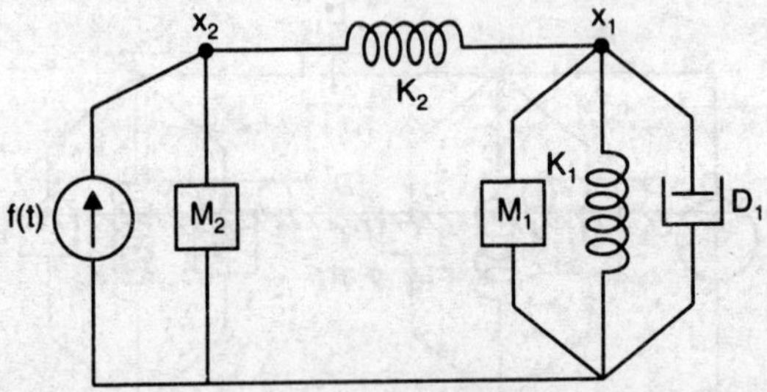

Fig. 2.69

Equivalent diagram for a given mechanical translational system consists of two nodes x_1 and x_2, as shown in Fig. 2.69.

First nodal equation according to D' Alembert's law is

$$F(t) = M_2\ddot{x}_2 + K_2(x_2 - x_1) \tag{2.95}$$

Second nodal equation

$$0 = K_2(x_1 - x_2) + M_1\ddot{x}_1 + K_1x_1 + D_1\dot{x}_1 \tag{2.96}$$

Taking Laplace transfer function on both sides of eqns. (2.95) and (2.96)

$$F(s) = (M_2s^2 + K_2)\, x_2(s) - K_2\, x_1(s) \tag{2.97}$$

$$0 = [K_2 + M_1 s^2 + K_1 + D_1 s]\, x_1(s) - K_2\, x_2(s) \tag{2.98}$$

Writing the eqn. (2.98), x_1 in terms of x_2

$$x_1(s)\,[K_2 + M_1s^2 + K_1 + D_1s] = K_2\, x_2(s)$$

$$\therefore \qquad x_1(s) = \frac{K_2 x_2(s)}{[M_1s^2 + D_1s + K_1 + K_2]} \tag{2.99}$$

Substituting the eqn. (2.99) in eqn. (2.97)

$$F(s) = [M_2s^2 + K_2]\, x_2(s) - \frac{K_2^{\,2}}{[M_1s^2 + D_1s + K_1 + K_2]}\, x_2(s)$$

$$\frac{x_2(s)}{F(s)} = \frac{[M_2s^2 + D_1s + K_1 + K_2]}{[M_1s^2 + D_1s + K_1 + K_2]\,[M_2s^2 + K_2] - K_2^{\,2}}$$

$$\therefore \quad \text{Transfer function, } \frac{x_2(s)}{F(s)} = \frac{[M_2s^2 + D_1s + K_1 + K_2]}{[M_1s^2 + D_1s + K_1 + K_2]\,[M_2s^2 + K_2] - K_2^{\,2}}$$

Problem 2.39: *Find the transfer function* $\frac{x_1(s)}{F(s)}$ *for a given mechanical translational system as shown in the Fig. 2.70.*

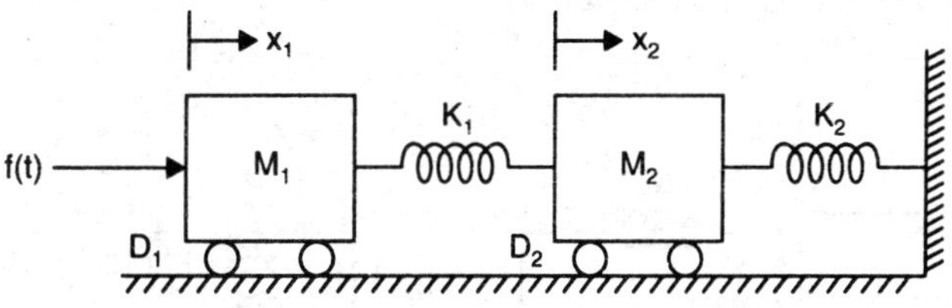

Fig. 2.70

Solution:

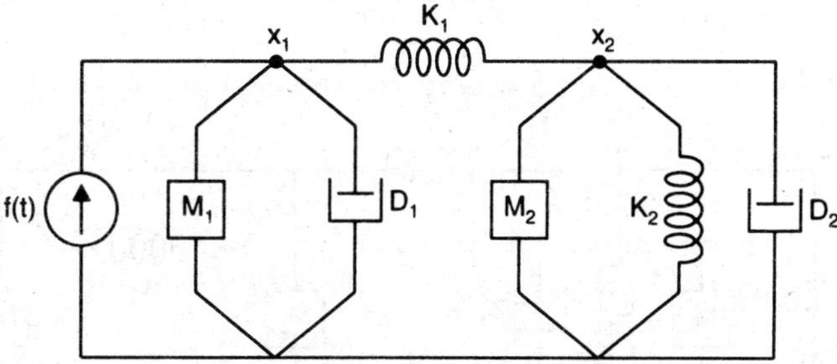

Fig. 2.71

The equivalent circuit diagram for the given mechanical system is shown in Fig. 2.71.

Number of nodal equations equals to number of displacements. First nodal equation is

$$f(t) = M_1\ddot{x}_1 + D_1\dot{x}_1 + K_1(x_1 - x_2) \quad (2.100)$$

Second equation $$0 = K_1(x_2 - x_1) + M_2\ddot{x}_2 + K_2x_2 + D_2\dot{x}_2 \quad (2.101)$$

Taking Laplace transformation on both sides of eqn. (2.100) and eqn. (2.101)

$$F(s) = [M_1s^2 + D_1s + K_1]\,x_1(s) - K_1\,x_2(s) \quad (2.102)$$

$$0 = -K_1\,x_1(s) + [M_2s^2 + D_2s + K_1 + K_2]\,x_2(s) \quad (2.103)$$

Writing eqn. (2.103), x_2 in terms of x_1

$$x_2(s) = \frac{K_1}{[M_2s^2 + D_2s + K_1 + K_2]}\,x_1(s) \quad (2.104)$$

Substituting the eqn.(2.104) in eqn. (2.102)

$$F(s) = [M_1s^2 + D_1s + K_1]\,x_1(s) - \frac{K_1^2}{[M_2s^2 + D_2s + K_1 + K_2]}\,x_1(s)$$

$$F(s) = x_1(s)\,\frac{[M_1s^2 + D_1s + K_1]\,[M_2s^2 + D_2s + K_1 + K_2 - K_1^2]}{[M_2s^2 + D_2s + K_1 + K_2]}$$

$$\therefore \quad \text{Transfer function, } \frac{x_1(s)}{F(s)} = \frac{M_2s^2 + D_2s + K_1 + K_2}{[M_1s^2 + D_1s + K_1]\,[M_2s^2 + D_2s + K_1 + K_2] - K_1^2}.$$

Problem 2.40. *Write the differential equations governing the mechanical system as shown in Fig. 2.72. Find also the transfer function* $\frac{x_1(s)}{F(s)}$.

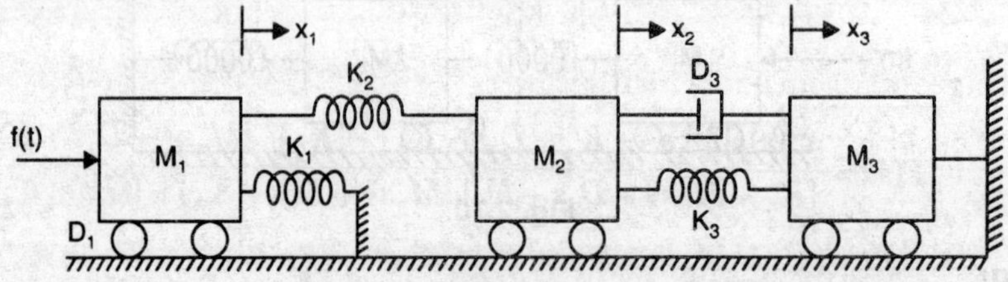

Fig. 2.72

Solution:

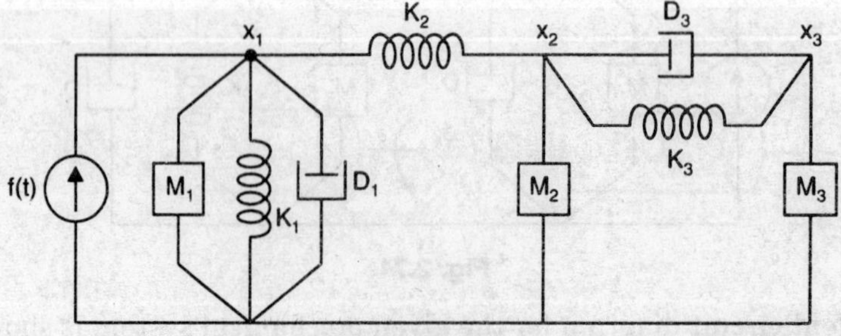

Fig. 2.73

In equivalent circuit, the number of nodes is equal to number of displacements. So, the first nodal equation for the given equivalent circuit is

$$F(t) = M_1\ddot{x}_1 + K_1 x_1 + D\dot{x}_1 + K_2(x_1 - x_2) \quad (2.105)$$

Note: (*i*) The dot over x indicates the first order derivative

(*ii*) The two dots over x indicate the second order derivative.

Second nodal equation is

$$0 = K_2(x_2 - x_1) + M_1\ddot{x}_2 + D_3(\dot{x}_2 - \dot{x}_3) + K_3(x_2 - x_3) \quad (2.106)$$

Third nodal equation is

$$0 = M_3\ddot{x}_3 + D_3(\dot{x}_3 - \dot{x}_2) + K_3(x_3 - x_2) \quad (2.107)$$

Taking Laplace transformation on both side of equations (2.105), (2.106) and (2.107)

$$F(s) = x_1(s)[M_1 s^2 + K_1 + D_1 s + K_2] - K_2 x_2(s) \quad (2.108)$$

$$0 = -K_2 x_1(s) + x_2(s)[K_2 + M_2 s^2 + D_3 s + K_3] - x_3(s)[D_3 s + K_3] \quad (2.109)$$

and

$$0 = -x_2(s)[D_3 s + K_3] + x_3(s)[M_3 s^2 + D_3 s + K_3] \quad (2.110)$$

From the eqn, (2.110)

$$x_3(s) = \frac{(D_3 s + K_3)}{[M_3 s^2 + D_3 s + K_3]} x_2(s) \quad (2.111)$$

Substituting eqn. (2.111) in eqn. (2.109)

$$x_1(s) = \frac{x_2(s)}{K_2}[K_2 + M_2 s^2 + D_3 s + K_3] - \frac{[D_3 s + K_3]^2}{[M_3 s^2 + D_3 s + K_3]} \quad (2.112)$$

$$x_2(s) = \frac{K_2[M_3 s^2 + D_3 s + K_3]\, x_1(s)}{[K_2 + M_2 s^2 + D_3 s + K_3][M_3 s^2 + D_3 s + K_3] - [D_3 s + K_3]^2} \quad (2.113)$$

Substituting $x_2(s)$ from eqn. (2.113) in eqn. (2.108)

$$F(s) = \frac{x_1(s)([M_1 s^2 + K_1 + D_1 s + K_2] - [K_2]^2[M_3 s^2 + D_3 s + K_3])}{[K_2 + M_2 s^2 + D_3 s + K_3][M_3 s^2 + D_3 s + K_3] - [D_3 s + K_3]^2}$$

$$\frac{x_1(s)}{F(s)} = \frac{[K_2 + M_2 s^2 + D_3 s + K_3][M_3 s^2 + D_3 s + K_3] - [D_3 s + K_3]^2}{[M_1 s^2 + K_1 + D_1 s + K_2]\, K_2^{\,2}\, [M_3 s^2 + D_3 s + K_3]}$$

Problem 2.41. *Find the transfer function* $\frac{\theta_2(s)}{T(s)}$, *for a given rotational mechanical system as shown in Fig. 2.74.*

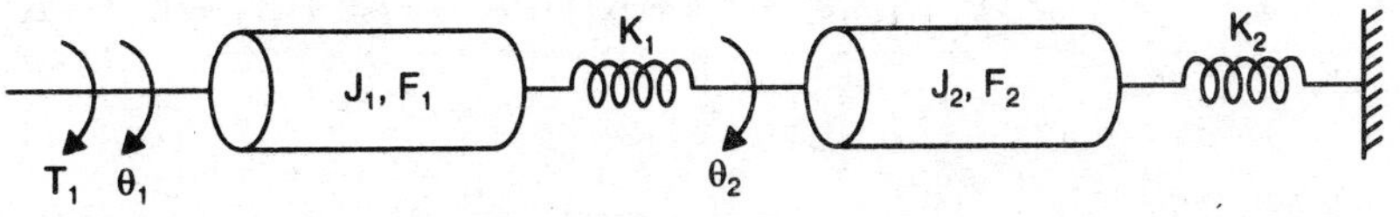

T₁ θ₁ J₁, F₁ K₁ θ₂ J₂, F₂ K₂

Fig. 2.74

Solution:

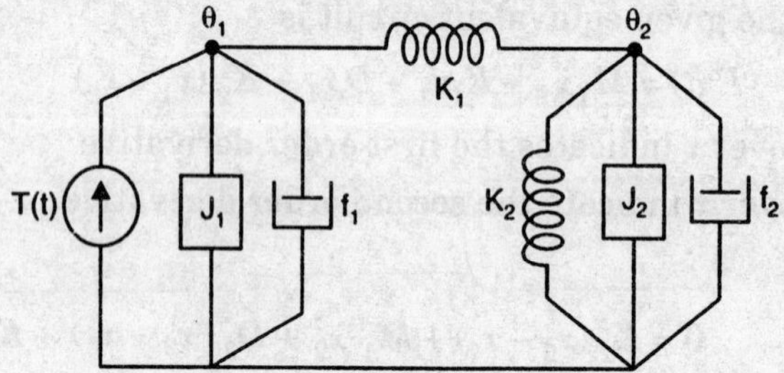

Fig. 2.75

Equivalent diagram for a given rotational mechanical system is shown in Fig. 2.75 and consists of two nodes which represent the angular displacements.

Elements J_1, f_1 are moment of inertia and friction coefficients of inertial body 1 are connected to node 1, in parallel, elements of inertial body 2 are connected in parallel to node 2, the element K_1 is connected between two nodes, and is common to θ_1 and θ_2. From the Fig. 2.75, first nodal equation is $T(t)$ be the torque applied on the system.

$$T(t) = J_1\ddot{\theta}_1 + f_1\dot{\theta}_1 + K_1(\theta_1 - \theta_2) \tag{2.114}$$

$$0 = K_1(\theta_2 - \theta_1) + J_2\ddot{\theta}_2 + K_2\theta_2 + f_2\dot{\theta}_2 \tag{2.115}$$

Taking Laplace transformation on both sides of eqns. (2.114) and (2.115)

$$T(s) = [J_1s^2 + f_1s + K_1]\,\theta_1(s) - K_1\,\theta_2(s) \tag{2.116}$$

$$0 = -K_1\,\theta_1(s) + [J_2s^2 + f_2s + K_1 + K_2]\,\theta_2(s) \tag{2.117}$$

Writing the eqn. (2.117), θ_1 in terms of θ_2

$$\theta_1(s) = \frac{[J_2s^2 + f_2s + K_1 + K_2]}{K_1}\,\theta_2(s) \tag{2.118}$$

Substituting the eqn. (2.118) in eqn. (2.116)

$$T(s) = \left[\frac{[J_1s^2 + f_1s + K_1]\,[J_2s^2 + f_2s + K_1 + K_2]}{K_1} - K_1\right]\theta_2(s)$$

$$\frac{\theta_2(s)}{T(s)} = \frac{K_1}{[J_1s^2 + f_1s + K_1]\,[J_2s^2 + f_2s + K_1 + K_2] - K_1^2}$$

$$\therefore \quad \text{Transfer function, } \frac{\theta_2(s)}{T(s)} = \frac{K_1}{[J_1s^2 + f_1s + K_1]\,[J_2s^2 + f_2s + K_1 + K_2] - K_1^2}$$

Problem 2.44: *Find the transfer function* $\dfrac{\theta_1(s)}{T(s)}$, *for the given rotational mechanical system as shown in the Fig. 2.76.*

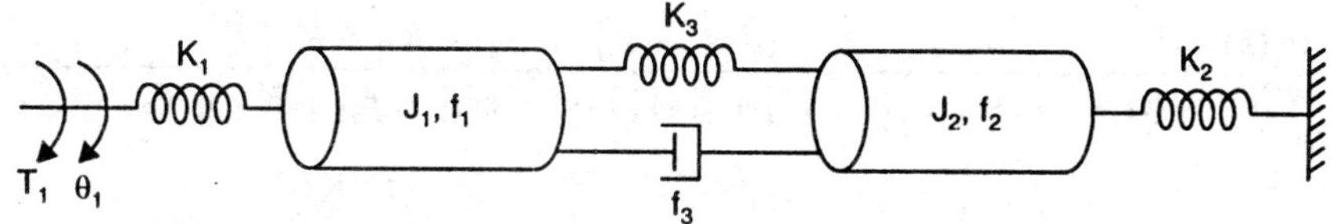

Fig. 2.76

Solution: Equivalent diagram for a given rotational mechanical system is shown in Fig. 2.77.

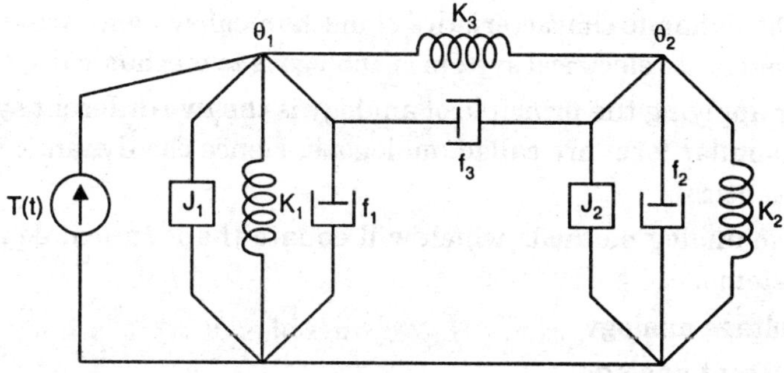

Fig. 2.77

First nodal equation

$$T(t) = J_1\ddot{\theta}_1 + K_1\theta_1 + f_1\dot{\theta}_1 + K_3(\theta_1 - \theta_2) + f_3(\dot{\theta}_1 - \dot{\theta}_2) \quad (2.119)$$

Second nodal equation

$$0 = K_3(\theta_2 - \theta_1) + f_3(\dot{\theta}_2 - \dot{\theta}_1) + J_2\ddot{\theta}_2 + f_2\dot{\theta}_2 + K_2\theta_2 \quad (2.120)$$

Taking Laplace transformation in eqns. (2.119) and (2.120)

$$T(s) = J_1 s^2\theta_1(s) + K_1\theta_1(s) + f_1 s\theta_1(s) + K_3(\theta_1(s) - \theta_2(s)) + f_3 s[\theta_1(s) - \theta_2(s)]$$

$$= [J_1 s^2 + K_1 + f_1 s + K_3 + f_3 s]\,\theta_1(s) - [K_3 + f_3 s]\,\theta_2(s) \quad (2.121)$$

$$0 = -(K_3 + f_3 s)\,\theta_1(s) + [K_3 + f_3 s + J_2 s^2 + f_2 s + K_2]\,\theta_2(s) \quad (2.122)$$

Writing the eqn. (2.122), $\theta_2(s)$ in terms of $\theta_1(s)$

$$\theta_2(s)[K_3 + f_3 s + J_2 s^2 + f_2 s + K_2] = [K_3 + f_3 s]\,\theta_1(s)$$

$$\theta_2(s) = \frac{[K_3 + f_3 s]}{[K_3 + f_3 s + J_2 s^2 + f_2 s + K_2]}\,\theta_1(s) \quad (2.123)$$

Substituting the eqn. (2.123) in eqn. (2.121)

$$T(s) = \left\{[J_1 s^2 + f_1 s + K_1 + K_3 + f_3 s] - \frac{[K_3 + f_3 s]^2}{[J_2 s^2 + f_2 s + K_2 + f_3 s + K_3]}\right\}\theta_1(s)$$

$$\frac{\theta_1(s)}{T(s)} = \frac{[J_2s^2 + s(f_2 + f_3) + K_2 + K_3]}{[J_1s^2 + s(f_1 + f_3) + K_1 + K_3]\,[J_2s^2 + s(f_2 + f_3) + K_2 + K_3] - [K_3 + f_3s]^2}$$

$\therefore$ Transfer function,

$$\frac{\theta_1(s)}{T(s)} = \frac{[J_2s^2 + s(f_2 + f_3) + K_2 + K_3]}{[J_1s^2 + s(f_1 + f_3) + K_1 + K_3]\,[J_2s^2 + s(f_2 + f_3) + K_2 + K_3] - [K_3 + f_3s]^2}.$$

2.11 ANALOGOUS SYSTEMS

When the dynamic equations of two systems are identical, such systems are said to be analogous systems.

The dynamic characteristics of electrical system are similar to characteristics of mechanical system. The dynamic characteristics of mechanical systems are similar to characteristics of electrical system. So electrical system is analogous to mechanical system and vice-versa.

The basis for applying the principle of analogy is the two different systems represented by equation of the similar form are called analogous. Hence the dynamic equations of analogous systems are identical.

There are two analog methods which will equate the identical equations of electrical and mechanical systems.

(*i*) Force–voltage analogy

(*ii*) Force–current analogy

(i) Force–voltage analog method

In force-voltage analog method, force (torque in rotational systems) is taken as reference in mechanical translational system and voltage in electrical system.

Table 2.2 (*a*)

Electrical system	*Translational mechanical system*	*Rotational mechanical system*
1. Voltage across an inductor	Force on mass M	Torque on inertial body
$V = L\dfrac{di}{dt}$	$F(t) = M\dfrac{dv}{dt}$	$T(t) = J\dfrac{d\omega}{dt}$
V is analogous to →	$F(t)$ is analogous to →	$T(t)$
L is analogous to →	M is analogous to →	J
i is analogous to →	v is analogous to →	ω
Current i is rate of change of charge 'q'		
that is, $i = \dfrac{dq}{dt}$		
$V = L\dfrac{d^2q}{dt^2}$	$F(t) = M\dfrac{d^2x}{dt^2}$	$T(t) = J\dfrac{d^2\theta}{dt^2}$
q is analogous to →	x is analogous to →	θ

2. Voltage across resistor $V = iR$ R is analogous to →	force on damper $F(t) = Dv$, D is analogous to →	Torque on damper $T(t) = f\omega$ f
3. Voltage across capacitor $V = \frac{1}{C}\int i dt$ $\frac{1}{C}$ is analogous to →	Force on spring $F(t) = Kx = K\int v dt$, K is analogous to →	Torque on spring $T(t) = K\theta = K\int \omega\, dt$, K

(ii) Force–current analog method

In force-current analog method, force (torque in rotational systems) is taken as reference in mechanical translational system and current in electrical system.

Table 2.2 (*b*)

Electrical system	*Translational mechanical system*	*Rotational mechanical system*
1. Current passing through a capacitor $i = C\frac{dV}{dt}$ i is analogous to → C is analogous to → V is analogous to →	Force on mass M $F(t) = M\frac{dv}{dt}$ $F(t)$ is analogous to → M is analogous to → v is analogous to →	Torque on inertial body $T(t) = J\frac{d\omega}{dt}$ $T(t)$ J ω
According to Faraday's law voltage is rate of change of flux linkages $V = \frac{d\psi}{dt}$ $V = L\frac{d^2\psi}{dt^2}$ ψ is analogous to →	$F(t) = M\frac{d^2x}{dt^2}$ x is analogous to →	$T(t) = J\frac{d^2\theta}{dt^2}$ θ
2. Current passing through the resistor $i = \frac{V}{R}$ $\frac{1}{R}$ is analogous to →	Force on damper $F(t) = Dv$, D is analogous to →	Torque on damper $T(t) = f\omega$ f

3 Current passing through an inductor $i = \frac{1}{L} \int V\,dt$	Force on spring $F(t) = Kx = K \int v\,dt$,	Torque on spring $T(t) = K\theta = K \int \omega\,dt$,
$\frac{1}{L}$ is analogous to →	K is analogous to →	K

Table 2.2 Force-Voltage and Force-Current Analog Method

Electrical system	*Force-voltage analog*		*Force-current analog*	
	Translational mechanical system	*Rotational mechanical system*	*Translational mechanical system*	*Rotational mechanical system*
Voltage V	$F(t)$	$T(t)$	ν	ω
Current I	v	ω	$f(t)$	$T(t)$
Resistance R	D	F	$1/D$	$1/f$
Inductance L	M	J	$1/K$	$1/K$
Capacitance C	$1/K$	$1/K$	M	J
Charge q	x	θ	–	–
Flux linkages	–	ψ	x	θ

Problem 2.43. *For problem no. 2.36, draw the electrical analog circuits in (i) Force-voltage analog method. (ii) Force-current analog method.*

Solution: From Fig. 2.64, nodal equations are

$$F(t) = M_1 \ddot{x}_1 + K_1 x_1 + D \dot{x}_1 + K_2 (x_1 - x_2) \tag{2.124}$$

$$0 = K_2 (x_2 - x_1) + M_2 \ddot{x}_2 \tag{2.125}$$

(*i*) Force-voltage analog method.

$f(t)$ is analogous to voltage $V(t)$

M is analogous to L

K is analogous to $\frac{1}{C}$

D is analogous to R

x is analogous to q

Analog equation for eqns. (2.124) and (2.125)

$$V(t) = L_1 \frac{d^2 q_1}{dt^2} + \frac{1}{C_1} q_1 + R \frac{dq_1}{dt} + \frac{1}{C_2} (q_1 - q_2) \tag{2.126}$$

$$0 = \frac{1}{C_2} (q_2 - q_1) + L_2 \frac{d^2 q_2}{dt^2} \tag{2.127}$$

Current i is rate of change of current

$$i = \frac{dq}{dt}$$

$$\Rightarrow \qquad dq = idt \qquad (2.128\ (a))$$

Taking integration on both sides of above expression

$$q = \int i\, dt \qquad (2.128\ (b))$$

Substituting eqns. (2.128 (a)) and (2.128 (b)) in eqns. (2.126) and (2.127), we get the equations in terms of current (i)

$$V(t) = L_1 \frac{di_1}{dt} + \frac{1}{C_1} \int i_1 dt + R\, i_1 + \frac{1}{C_2} \int (i_1 - i_2)\, dt \qquad (2.129)$$

$$0 = \frac{1}{C_2} \int (i_2 - i_1)\, dt + L_2 \frac{di_2}{dt} \qquad (2.130)$$

Equations (2.129) and (2.130) represent two loop equations. There are two loop currents i_1 and i_2. Draw the electrical analog circuit for eqns. (2.129) and (2.130) as shown in fig. 2.78.

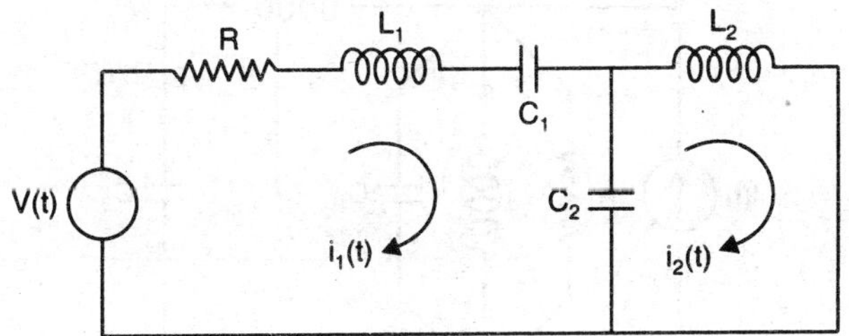

Fig. 2.78

(ii) Force-current analog method

$F(t)$ is analogous to $i(t)$

M is analogous to C

K is analogous to $\frac{1}{L}$

D is analogous to $\frac{1}{R}$

x is analogous to ψ

In force-current analog equations for eqns. (2.124) and (2.125) are

$$i(t) = C_1 \frac{d^2\psi_1}{dt^2} + \frac{1}{L_1}\psi_1 + \frac{1}{R}\frac{d\psi_1}{dt} + \frac{1}{L_2}\psi_1 - \psi_2 \qquad (2.131)$$

$$0 = \frac{1}{L_2}(\psi_2 - \psi_1) + C_2 \frac{d^2\psi_2}{dt^2} \qquad (2.132)$$

According to Faraday's law of electromagnetic induction, voltage equals to rate of change of flux linkages.

$$V = \frac{d\psi}{dt}, \quad d\psi = V\,dt \tag{2.133 (a)}$$

Taking integration on both sides of above equation

$$\psi = \int V\,dt \tag{2.133 (b)}$$

Substituting the eqns. (2.133 (*a*)) and (2.133 (*b*)) in eqns. (2.131) and (2.132), we get

$$i(t) = C_1\frac{dV_1}{dt} + \frac{1}{L_1}\int V_1 dt + \frac{V_1}{R} + \frac{1}{L_2}\int (V_1 - V_2)\,dt \tag{2.134}$$

$$0 = \frac{1}{L_2}\int (V_2 - V_1)\,dt + C_2\frac{dV_2}{dt} \tag{2.135}$$

There are two nodal equations with two node voltages V_1 and V_2. Draw the electrical nodal circuit for the eqns. (2.134) and (2.135) as shown in Fig. 2.79.

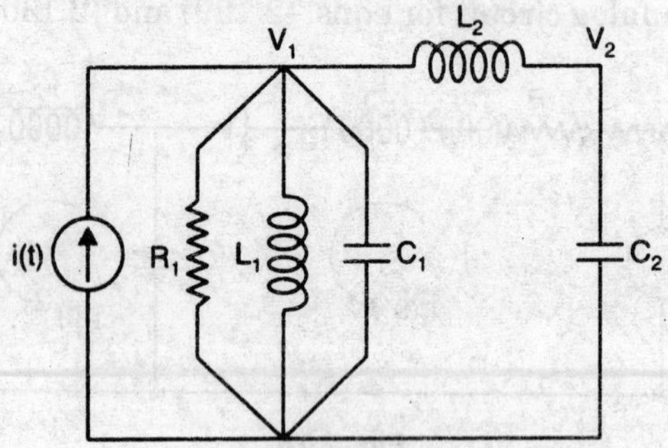

Fig. 2.79

Problem 2.44. *For problem no. 2.37, draw the electrical analog circuits in (i) Force-voltage analog method. (ii) Force-current analog method.*

Solution: From Fig. 2.66 nodal equations are

$$F(t) = M_1\ddot{x}_1 + D_1\dot{x}_1 + D_{12}(\dot{x}_1 - \dot{x}_2) + K_1x_1 \tag{2.136}$$

$$0 = D_{12}(\dot{x}_2 - \dot{x}_1) + M_2\ddot{x}_2 + D_2\dot{x}_2 + K_2x_2 \tag{2.137}$$

Force-voltage analog method:

$$F(t) \leftrightarrow V(t)$$

$$M \leftrightarrow L$$

$$K \leftrightarrow \frac{1}{C}$$

$$D \leftrightarrow R$$

$$x \leftrightarrow q$$

Writing the analog equations for above two eqns. (2.136) and (2.137), we get

$$V(t) = L_1\ddot{q}_1 + R_1\dot{q}_1 + R_{12}(\dot{q}_1 - \dot{q}_2) + \frac{1}{C_1}q_1 \quad (2.138)$$

$$0 = R_{12}(\dot{q}_2 - \dot{q}_1) + L_2\ddot{q}_2 + R_2\dot{q}_2 + \frac{1}{C_2}q_2 \quad (2.139)$$

From eqn. (2.128 (*b*)), $q = \int i dt$

Rewriting the eqns. (2.138) and (2.139) in terms of current i

$$V(t) = L_1\frac{di_1}{dt} + R_1 i_1 + R_{12}(i_1 - i_2) + \frac{1}{C_1}\int i_1\, dt \quad (2.140)$$

$$0 = R_{12}(i_2 - i_1) + L_2\frac{di_2}{dt} + R_2 i_2 + \frac{1}{C_2}\int i_2\, dt \quad (2.141)$$

There are two loop equations, consisting of two loop currents. From eqns. (2.140) and (2.141) electrical circuit in force-voltage analog method is shown in Fig. 2.80.

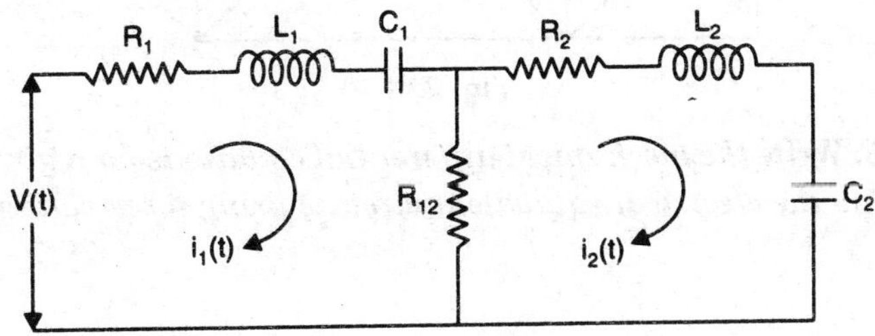

Fig. 2.80

(*ii*) Force-current analog method

$F(t) \leftrightarrow i(t)$

$M \leftrightarrow C$

$K \leftrightarrow \frac{1}{L}$

$D \leftrightarrow 1/R$

$\theta \leftrightarrow \psi$

Writing the analog equations for above two eqns. (2.136) and (2.137)

$$i(t) = C_1\ddot{\psi}_1 + \frac{1}{R_1}\dot{\psi}_1 + \frac{1}{L_1}(\psi_1 - \psi_2) \quad (2.142)$$

$$0 = \frac{1}{L_1}(\psi_2 - \psi_1) + C_2\ddot{\psi}_2 \frac{1}{L_2}\psi_2 + \frac{1}{R_2}\dot{\psi}_2 \quad (2.143)$$

where $\psi = \int V dt$, writing the eqns. (2.142) and (2.143) in terms of voltage.

There are two nodal equations, consisting of two node voltages V_1 and V_2

$$i(t) = C_1 \frac{dV_1}{dt} + \frac{V_1}{R_1} + \frac{1}{L_1} \int (V_1 - V_2)dt$$

$$0 = \frac{1}{L_1} \int (V_2 - V_1)dt + C_2 \frac{dv_2}{dt} + \frac{1}{L_2} \int V_2\, dt + \frac{V_2}{R_2}$$

There are two nodal equations with two nodal voltages V_1 and V_2. Electrical analog circuit in force-current analog method is as shown in Fig. 2.81.

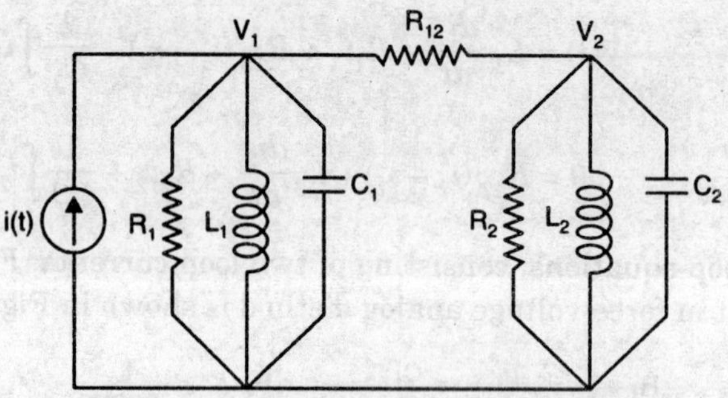

Fig. 2.81

Problem 2.45. *Write the mechanical differential equations for a given system as shown in Fig. 2.82 and draw the electrical equivalent circuits using force-voltage and force-current analog methods.*

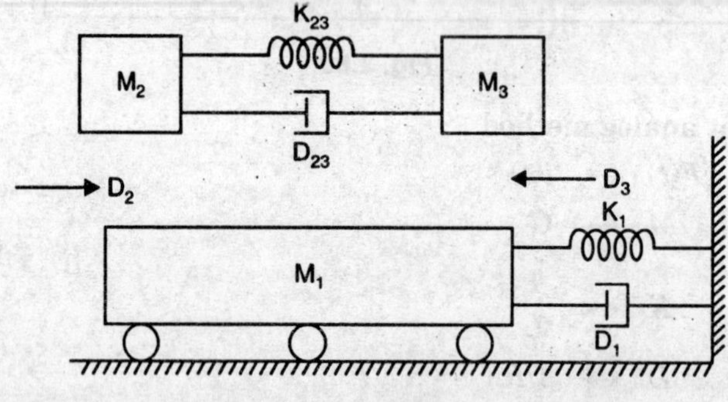

Fig. 2.82

Solution:

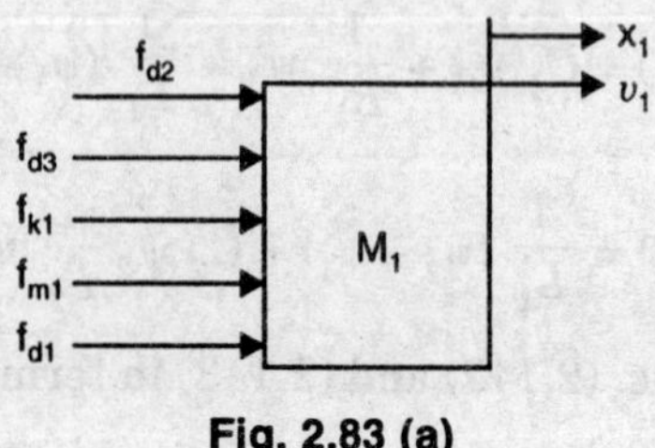

Fig. 2.83 (a)

For Fig. 2.83 (*a*), first nodal equation is

$$M_1\ddot{x}_1 + D_1\dot{x}_1 + k_1x_1 + D_2(\dot{x}_1 - \dot{x}_2) + D_3(\dot{x}_1 - \dot{x}_3) = 0 \quad (2.144)$$

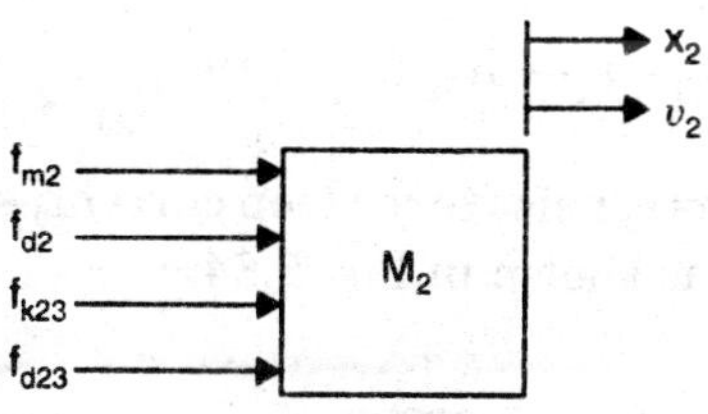

Fig. 2.83 (b)

For Fig. 2.83 (*b*), second nodal equation is

$$M_2\ddot{x}_2 + D_2(\dot{x}_2 - \dot{x}_1) + D_{23}(\dot{x}_2 - \dot{x}_3) + k_{23}(x_2 - x_3) = 0 \quad (2.145)$$

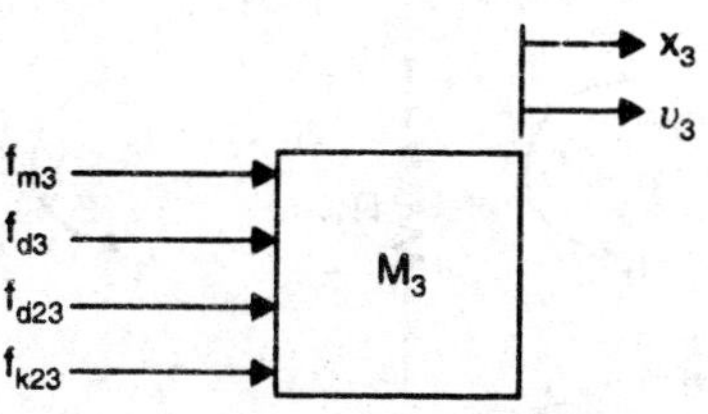

Fig. 2.83 (c)

For Fig. 2.83 (*c*), third nodal equation is

$$M_3\ddot{x}_3 + D_3(\dot{x}_3 - \dot{x}_1) + D_{23}(\dot{x}_3 - \dot{x}_2) + K_{23}(x_3 - x_2) = 0 \quad (2.146)$$

Replacing the displacement with velocity $x = \int v\, dt$

Writing the eqns. (2.144), (2.145) and (2.146) as

$$M_1\frac{dv_1}{dt} + D_1v_1 + k_1\int vdt + D_2(v_1 - v_2) + D_3(v_1 - v_3) = 0 \quad (2.147)$$

$$M_2\frac{dv_2}{dt} + D_2(v_2 - v_1) + D_{23}(v_2 - v_3) + K_{23}\int (v_2 - v_3)\, dt = 0 \quad (2.148)$$

$$M_3\frac{dv_3}{dt} + D_3(v_3 - v_1) + D_{23}(v_3 - v_2) + K_{23}\int (v_3 - v_2)\, dt = 0 \quad (2.149)$$

Force-voltage analogous equations for eqns. (2.147), (2.148) and (2.149) are

M is analogous to *L*

D is analogous to *R*

K is analogous to 1/*C*

V is analogous to *i*

$$L_1\frac{di_1}{dt} + R_1i_1 + \frac{1}{C_1}\int i_1dt + R_2(i_1 - i_2) + R_3(i_1 - i_3) = 0 \quad (2.150)$$

$$L_2 \frac{di_2}{dt} + R_2 (i_2 - i_1) + R_{23} (i_2 - i_3) + \frac{1}{C_{23}} \int (i_2 - i_3)\, dt = 0 \quad (2.151)$$

$$L_3 \frac{di_3}{dt} + R_3 (i_3 - i_1) + R_{23} (i_3 - i_2) + \frac{1}{C_{23}} \int (i_3 - i_2)\, dt = 0 \quad (2.152)$$

The above three equations consist three loop current, so force-voltage analogous circuit for the above three equation is as shown in Fig. 2.84.

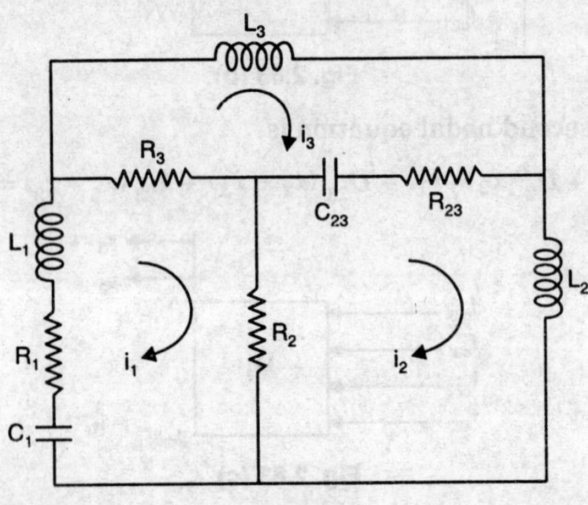

Fig. 2.84

(*ii*) Force current analogous method

M is analogous to C

D is analogous to $\frac{1}{R}$

K is analogous to $\frac{1}{L}$

v is analogous to V

∴ Electrical analog equations of eqns. (2.147), (2.148) and (2.149)

$$C_1 \frac{dV_1}{dt} + \frac{V_1}{R_1} + \frac{1}{L_1} \int V_1\, dt + \frac{1}{R_2} (V_1 - V_2) + \frac{1}{R_3} (V_1 - V_3) = 0 \quad (2.153)$$

$$C_2 \frac{dV_2}{dt} + \frac{1}{R_2} (V_2 - V_1) + \frac{1}{R_{23}} (V_2 - V_3) + \frac{1}{L_{23}} \int (V_2 - V_3)\, dt = 0 \quad (2.154)$$

$$C_3 \frac{dV_3}{dt} + \frac{1}{R_3} (V_3 - V_1) + \frac{1}{R_{23}} (V_3 - V_2) + \frac{1}{L_{23}} \int (V_3 - V_2)\, dt = 0 \quad (2.155)$$

Electrical analog circuit for eqns. (2.153), (2.154) and (2.155) is as shown in Fig. 2.85.

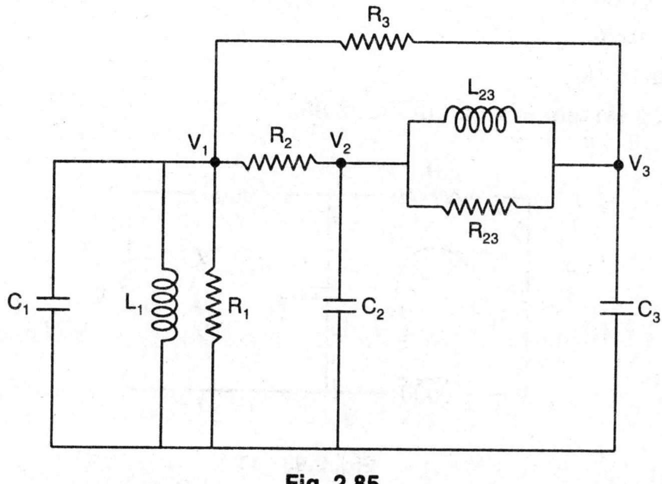

Fig. 2.85

Problem2.46. *Without writing analog equations draw the electrical circuit for the rotational mechanical system of Fig. 2.86 in the force-voltage analog system.*

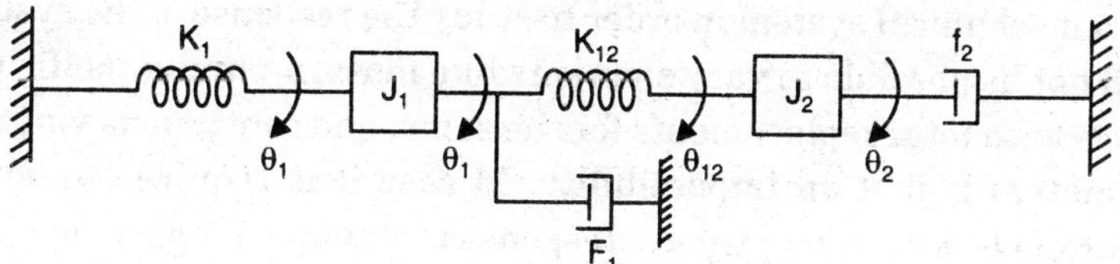

Fig. 2.86

Solution:

Equivalent circuit for Fig. 2.86 is as shown in Fig. 2.87.

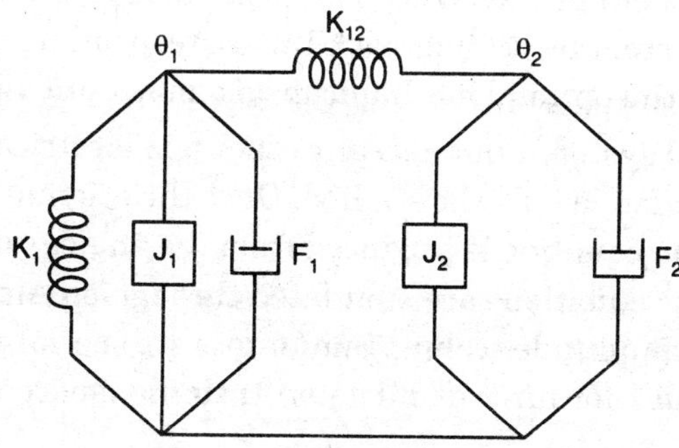

Fig. 2.87

In the mechanical system, J_1, K_1, and f_1 have the same angular displacement (θ_1) and are in series. J_2, f_2 are also in series. K_{12} is connected in parallel across the series combination of J_1, K_1 and f_1.

In force–voltage analog system

J is analogous to L

f is analogous to R

k is analogous to $1/C$

$\therefore$ The analog system is given in Fig. 2.88.

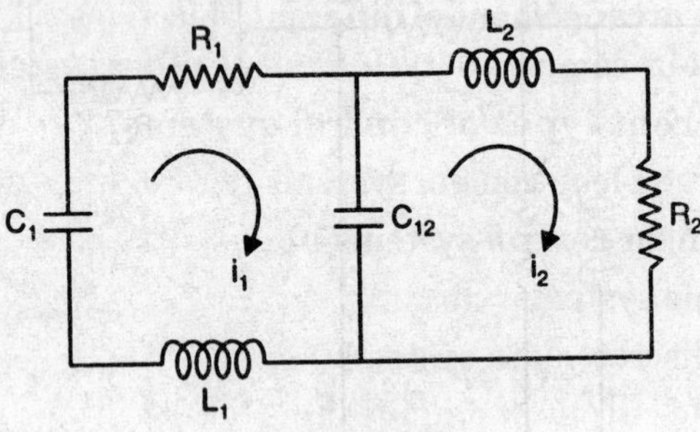

Fig. 2.88

2.12 IMPORTANCE OF ELECTRICAL ANALOG SYSTEMS

While designing a mechanical system in order to study the response of the system to parameter variations, it will not be possible to vary elements like mass, damping coefficients of dashpots, stiffness of springs with total replacements for the same, and continuous variation of the above mentioned parameters is just an impossibility. In case it is required to obtain the optimal value for those parameters so that optimum response is obtained for specified inputs, it becomes a very difficult situation. It will not be economically feasible even if it is sometimes physically viable. In such cases it is advantageous to construct an electrical analog system wherein varying the resistors, inductors, capacitors, voltages and currents is not at all a problem. Analog scales are derived while constructing such analog systems through which information on the system behavior can be transferred either way. Once the optimal value of the parameters are obtained on the electrical analog system by scaling back the corresponding values for the mechanical system are computed and the optimal mechanical system is built with success.

Not only mechanical systems, but also any other non electrical system such as thermal, pneumatic, hydraulic systems can also be studied. Once the original system is mathematically modeled, the analog system equation is produced from the analogy of the elements. On analog computers system dynamic equations are simulated through op Ams and potentiometers and solved in time scaling and amplitude scaling. Simulators are special purpose analog computers. We have simulators popular for aircraft pilots on training, space study applications, power system models, etc.

SHORT QUESTIONS AND ANSWERS

1. **What is a system?**

   A system is collection of objectives connected in a design manner to perform the specified task.

2. **What is meant by control system?**

   Control system is an arrangement of different objectives connected in such a manner so as to regulate, direct or command itself or some other system.

3. **What are the different types of control systems?**

   - Open loop and closed loop control systems
   - Linear and non linear control systems
   - Static and dynamic systems
   - Continuous and discrete data systems
   - Single input-single output and multi input-multi output systems.

4. **What is the basis of obtaining a mathematical model for the system?**

   The system is considered to be in a state of dynamic equilibrium.

5. **What is the test for the linearity of a system?**

   Superposition theorem.

6. **What is the salient feature of a control system over general class of systems?**

   Power advantage—low power control signal handles large power outputs at the load end.

7. **What is the problem caused by providing positive feedback in control systems?**

   It causes instability.

8. **What is the limitation of a transfer function?**

   It exists only for linear time invariant systems.

9. **Define the open loop control system.**

   When control action is independent of input it is known as open loop control system.

10. **What are the advantages and disadvantages of open loop control system?**

    Advantages of open loop system

    - Simplicity
    - Low cost
    - Easy to construct
    - Generally, open loop systems are stable.

    Disadvantages of open loop system

    - Lack of reliability and accuracy
    - Change in output due to change in disturbances is not corrected automatically.

11. **Define the closed loop control system.**

    Control action which somewhat depends upon input is known as closed loop control system.

12. What are the advantages and disadvantages of closed loop control system?

Advantages of closed loop system

- More accurate
- It compensates the disturbance
- Desired response can be obtained by appropriate design.

Disadvantages of closed loop system

- More complex and expensive
- Reduce the gain of the system.

13. What are the basic elements of thermal system?

The basic elements of thermal system are heat exchangers, and heat storage devices.

14. Define the transfer function.

It is defined as the ratio of Laplace transform of output to Laplace transform of input under all zero initial conditions.

15. What is a block diagram?

It is a pictorial representation of the functions performed by each component of the system.

16. What is signal flow graph?

Signal flow graphs can be constructed from the transfer function of the system by defining the signals with respect to the output and its time derivatives as well as the input and its time derivatives.

17. State the Mason's gain formula.

Mason's gain formula states that the overall gain or transfer function of the system is

$$T = \sum_{k=1}^{n} \frac{P_k \, \Delta_k}{\Delta}$$

18. What are the basic components required for modeling mechanical translational and rotational systems?

The basic components of mechanical translational system are mass, elasticity idealized by a helical spring and friction idealized by a damper.

The basic components of rotational mechanical system are inertia, tortional spring and friction.

19. What are the types of electrical analogous for mechanical system?

Two types of analog for the mechanical systems are

(*i*) Force–voltage analogy

(*ii*) Force–current analogy

OBJECTIVE TYPE QUESTIONS

1. A man with closed eyes is an example of
   (a) *Open loop control system* (b) Closed control loop system
   (c) Biological system (d) Linear control system
2. In closed loop control system
   (a) control action is independent of input
   (b) control action is independent of output
   (c) *control action somewhat depends upon input*
   (d) Control action somewhat depends upon output
3. In open loop system
   (a) control action is independent of input
   (b) *control action is independent of output*
   (c) control action somewhat depends upon input
   (d) Control action somewhat depends upon output
4. RLC network is an example for
   (a) Open loop control system (b) Closed loop control system
   (c) Time variant system (d) *Time invariant system*
5. In time invariant systems, input and output are
   (a) Independent of time (b) *Dependent on time*
   (c) Parameters dependent on time (d) None of these
6. Linear control system should satisfy
   (a) Homogeneity (b) Linearity
   (c) *Both (a) and (b)* (d) Non linearity
7. Transfer function is the
   (a) Ratio of output and input (b) Ratio of input and output
   (c) Ratio of Laplace transforms of input and input under zero initial conditions
   (d) *Ratio of Laplace transforms of output and input under zero initial conditions.*
8. Transfer function for RLC series network is
   (a) $\dfrac{Cs}{LCs^2 + RCs + 1}$ (b) $\dfrac{1}{LCs^2 + RCs + 1}$
   (c) $\dfrac{1}{RCs^2 + LCs + 1}$ (d) $\dfrac{1}{RLs^2 + LCs + 1}$
9. A position control system is a/an
   (a) Automatic regulating system (b) *Process control system*
   (c) Servo mechanism (d) Stochastic control system

10. Eliminating the negative feedback loop of equivalent is

(*a*) $\frac{G(s)}{1+G(s)H(s)}$ (*b*) $\frac{G(s)}{1-G(s)H(s)}$

(*c*) $G(s)\,H(s)$ (*d*) $G(s) + H(s)$

11. Eliminating the forward path $H(s)$ equivalent is

(*a*) $\frac{G(s)}{1+G(s)H(s)}$ (*b*) $\frac{G(s)}{1-G(s)H(s)}$

(*c*) $G(s)\,H(s)$ (*d*) $G(s) + H(s)$

12. One of the disadvantages of block diagram reduction technique is

(*a*) Time consuming

(*b*) for large systems number of blocks will increase, so simplification is difficult

(*c*) *Both (a) and (b)*

(*d*) We can find transfer function for large system

13. Equivalent analogy for current in mechanical translational system is

(*a*) Man (*b*) Acceleration

(*c*) *Velocity* (*d*) Displacement

14. Equivalent analog for flux linkages in mechanical rotational system is

(*a*) Angular velocity (*b*) *Angular displacement*

(*c*) Torque (*d*) Moment of inertia

15. Equivalent analogy for torque in mechanical translational system is

(*a*) *Force*

(*b*) Man

(*c*) Acceleration

(*d*) Displacement

16. The block diagram contains

(*a*) system output variable

(*b*) System input variable

(*c*) the functional relations of the variables

(*d*) *All the above*

17. In a block diagram, when a take off point is moved ahead of a block, G_1

(*a*) *The block G1 will be added in parallel*

(*b*) The block G1 will be added in the feedback path.

(*c*) The block G1 will be added in series.

(*d*) The block will be added in the feedback path.

18. What is the gain of the system (output/input) given below?

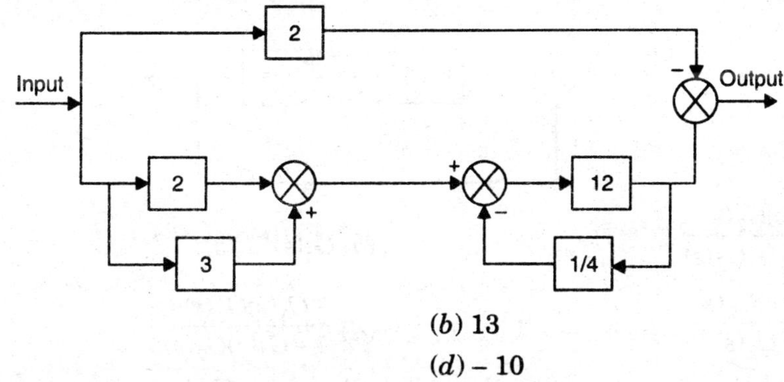

(*a*) 36 (*b*) 13

(*c*) 90 (*d*) – 10

19. The closed-loop gain of the system sketched below is

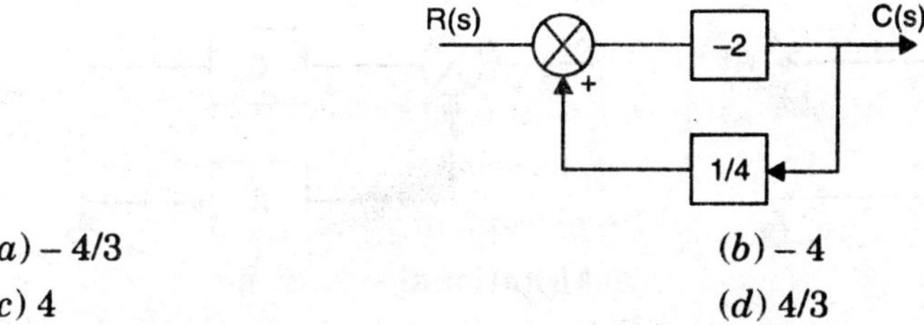

(*a*) – 4/3 (*b*) – 4

(*c*) 4 (*d*) 4/3

20. In the block diagram shown, the output θ(*s*) is equal to

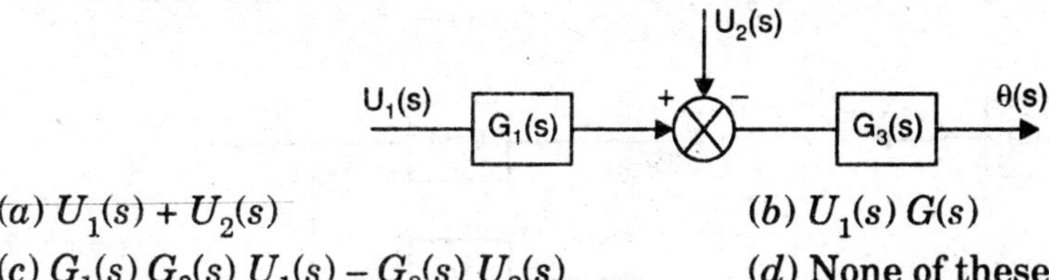

(*a*) $U_1(s) + U_2(s)$ (*b*) $U_1(s)\, G(s)$

(*c*) $G_1(s)\, G_3(s)\, U_1(s) - G_3(s)\, U_2(s)$ (*d*) None of these

21. The transfer function $E_0(s)/E_1(s)$ of the RC-network shown is given by

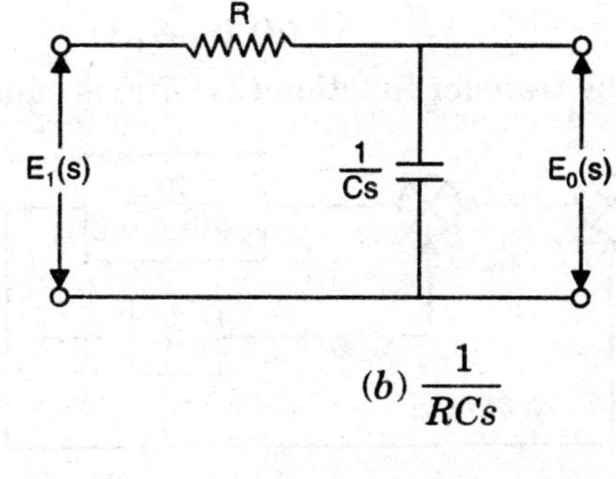

(*a*) $\dfrac{1}{RCs + 1}$ (*b*) $\dfrac{1}{RCs}$

(*c*) $\dfrac{RCs}{RCs + 1}$ (*d*) None of these

22. The block diagram of a certain system is shown below. The transfer function $Y(s)/U(s)$ is given by

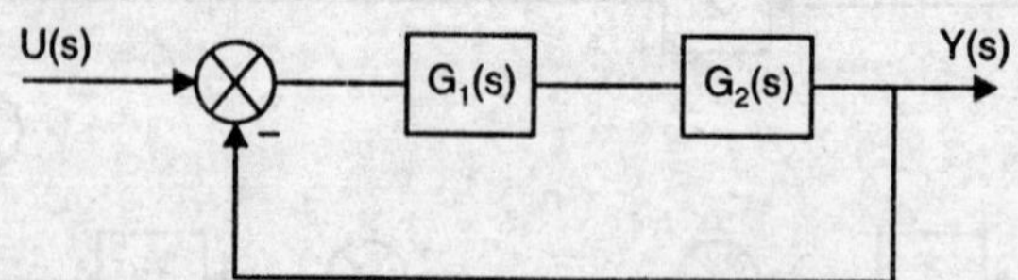

(a) $\dfrac{G_1(s)G_2(s)}{1-G_1(s)G_2(s)}$ (b) $G_1(s)\,G_2(s)$

(c) $\dfrac{1+G_1(s)G_2(s)}{G_1(s)G_2(s)}$ (d) $\dfrac{G_1(s)G_2(s)}{1+G_1(s)G_2(s)}$

23. The figure below gives two equivalent block diagrams. The values of transfer function of block marked '*X*' is given by

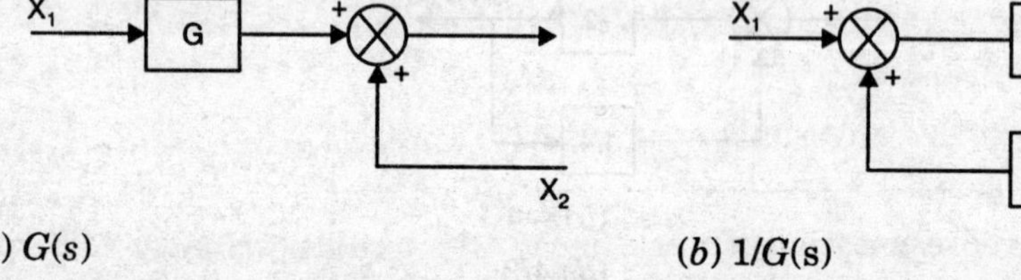

(a) $G(s)$ (b) $1/G(s)$

(c) 1 (d) $1+G(s)$

24. The figure shows two equivalent block diagrams. The transfer function of the block marked '*X*' is given by

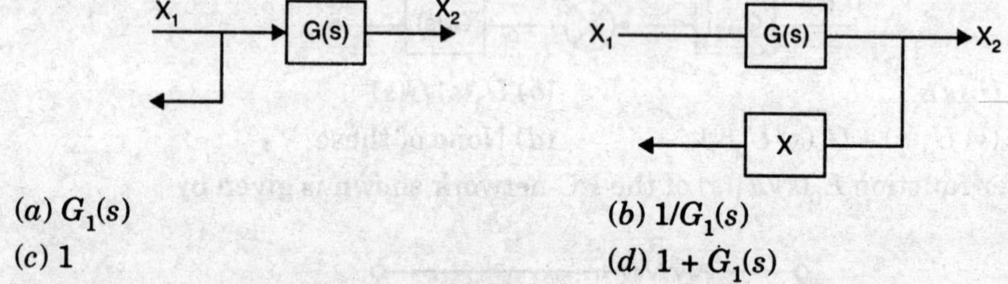

(a) $G_1(s)$ (b) $1/G_1(s)$

(c) 1 (d) $1 + G_1(s)$

25. For the system shown, the transfer function $C(s)/R(s)$ is equal to

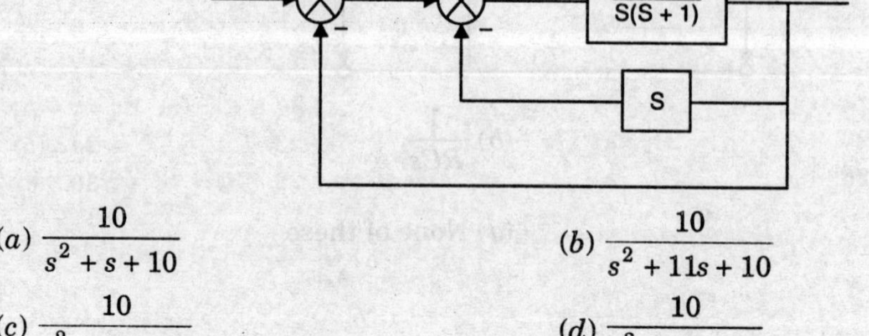

(a) $\dfrac{10}{s^2+s+10}$ (b) $\dfrac{10}{s^2+11s+10}$

(c) $\dfrac{10}{s^2+9s+10}$ (d) $\dfrac{10}{s^2+2s+10}$

26. In a signal flow graph, the nodes represent

(*a*) *The system variables* (*b*) The system gain

(*c*) The system parameters (*d*) All the above

27. The branch of a signal flow graph represents

(*a*) The system variables (*b*) The functional relations of the variables

(*c*) The system parameters (*d*) None of the above

28. By applying Mason's gain formula, it is possible to get

(*a*) the ratio of the output variable to input variable only

(*b*) the system functional relations between any two variables

(*c*) *the overall gain of the system*

(*d*) The ratio of any variable to input variable only

29. Two or more loops in a signal flow graph are said to be non-touching

(*a*) if they do not have any common branch

(*b*) if they do not have any common loop

(*c*) *if they have common node*

(*d*) if they do not have any common node

30. Signal flow graph is a

(*a*) Topological representation of a non-linear differential equation.

(*b*) Schematic graph

(*c*) *Special type of graph for analysis of modern control system.*

(*d*) Plot between frequency and magnitude in *dB*.

31. The signal flow graph shown, $X_2 = TX_1$ where T is equal to

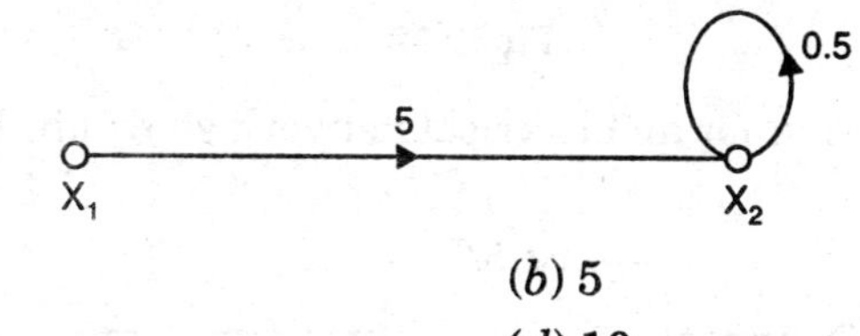

(*a*) 2.5 (*b*) 5

(*c*) 5.5 (*d*) 10

KEY

1. (*a*)	**2.** (*c*)	**3.** (*b*)	**4.** (*d*)	**5.** (*b*)	**6.** (*c*)
7. (*d*)	**8.** (*b*)	**9.** (*b*)	**10.** (*a*)	**11.** (*b*)	**12.** (*c*)
13. (*c*)	**14.** (*b*)	**15.** (*a*)	**16.** (*d*)	**17.** (*d*)	**18.** (*b*)
19. (*a*)	**20.** (*c*)	**21.** (*a*)	**22.** (*d*)	**23.** (*b*)	**24.** (*b*)
25. (*b*)	**26.** (*a*)	**27.** (*b*)	**28.** (*c*)	**29.** (*d*)	**30.** (*c*)
31. (*b*)					

EXERCISE

1. What are the differences between open loop control and closed loop control?

2. Describe the closed loop control system with an examples.

3. How do you represent a control system by the block diagram? Illustrate with an example.

4. What are the advantages and disadvantages of open loop control system?

5. What are the advantages and disadvantages of closed loop control system?

6. Explain the block diagram reduction technique and what are the disadvantages?

7. What is mean by signal flow graph and what are its advantages?

8. State and explain the Mason's gain formula.

9. Briefly discuss the different classification of control systems.

10. Explain the thermal and fluid systems with examples.

11. Define the transfer function.

12. Explain the need of electrical analog systems.

13. Derive the transfer function for an electrical network shown in Fig. 2.89.

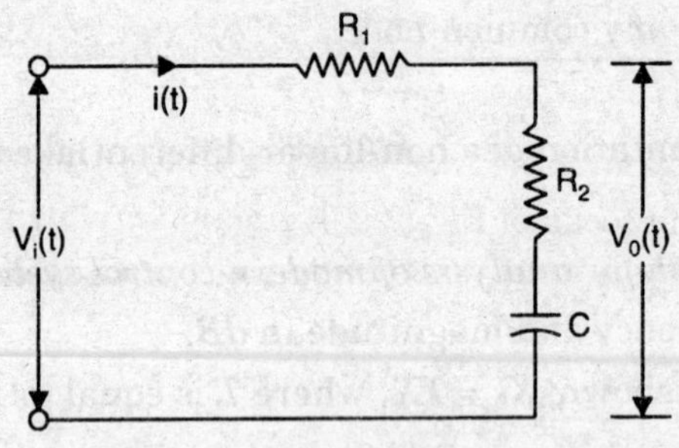

Fig. 2.89

14. Find the transfer function for an electrical network shown in Fig. 2.90.

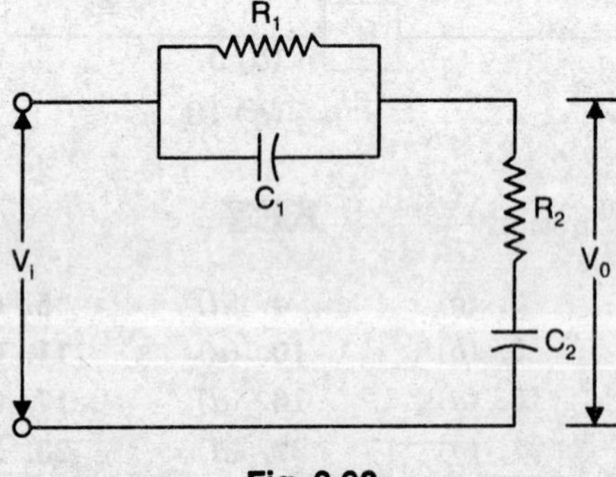

Fig. 2.90

15. Find the transfer function for a given differential equation $\frac{d^2y(t)}{dt^2} + 2\frac{dy(t)}{dt} + 5y(t) = u(t)$, if the input is unit step.

16. Transform the given network shown in Fig. 2.91 into s-domain and obtain the transfer function.

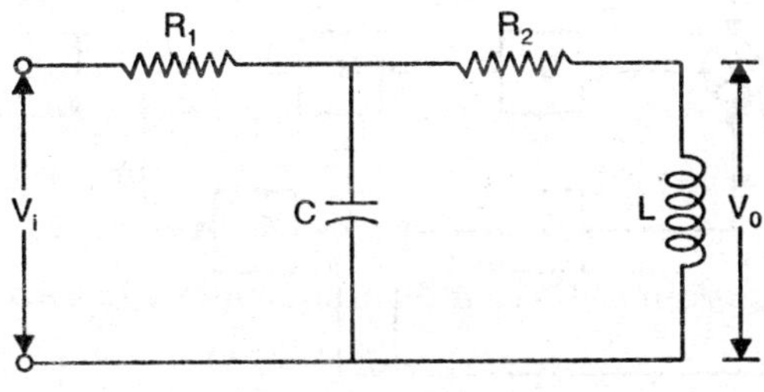

Fig. 2.91

17. Obtain the overall transfer function of the given block diagram shown in Fig. 2.92.

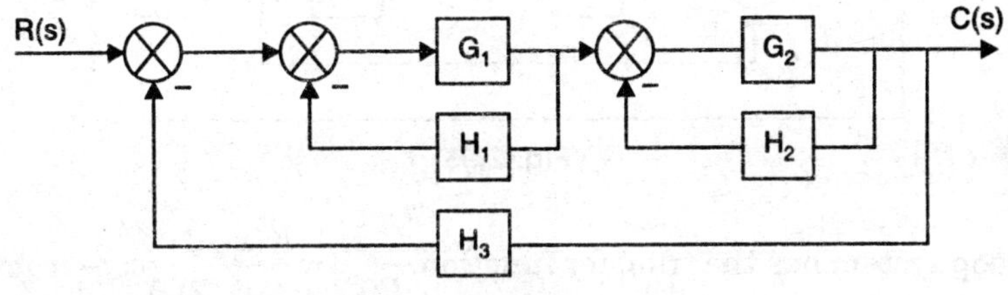

Fig. 2.92

18. Reduce the number of blocks of Fig. 2.93 into an equivalent one.

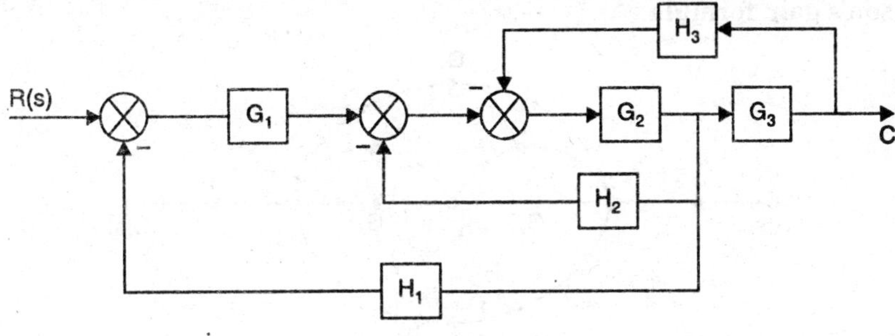

Fig. 2.93

19. Obtain the transfer function $\frac{C(s)}{R_1(s)}, \frac{C(s)}{R_2(s)}$ and $\frac{C(s)}{R_3(s)}$ for a given block diagram shown in Fig. 2.94.

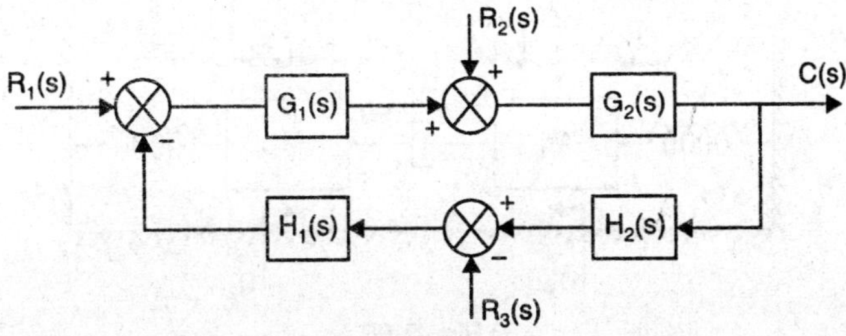

Fig. 2.94

20. Draw the signal flow graph for the given block diagram as shown in Fig. 2.95.

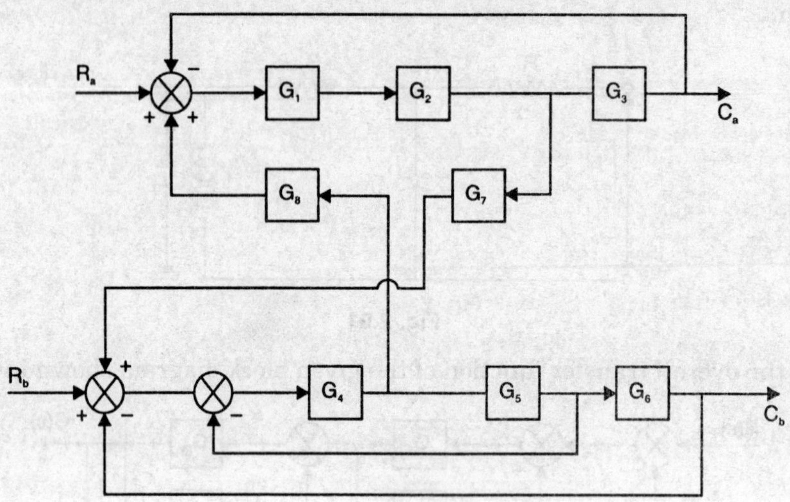

Fig. 2.95

21. A closed loop system has the transfer function $\frac{C(s)}{R(s)} = \frac{K(s+1)}{s(s+2)}$, construct the signal flow graph for the system.

22. Determine the transfer function for the signal flow graph as shown in Fig. 2.96, using the Mason's gain formula.

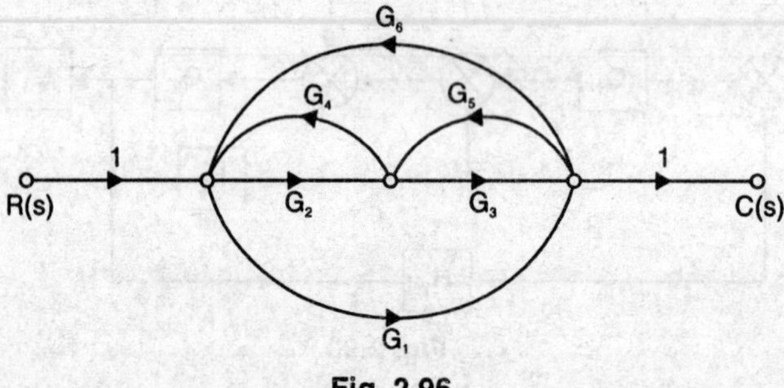

Fig. 2.96

23. Obtain the transfer function $\frac{X_2(s)}{F(s)}$ for a given mechanical translational system as shown in Fig. 2.97.

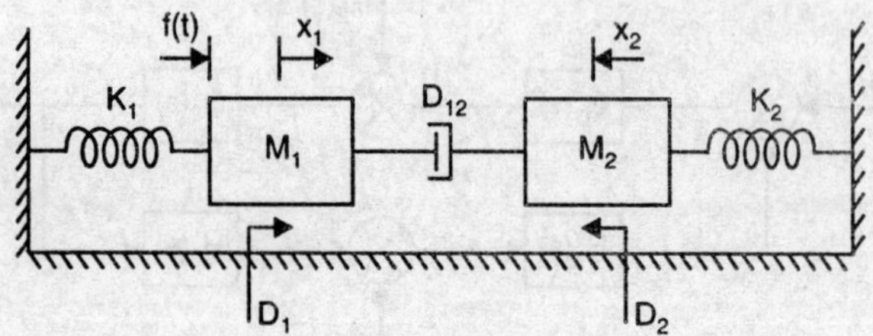

Fig. 2.97

24. For the mechanical system shown in Fig. 2.98, obtain the force-voltage analog system.

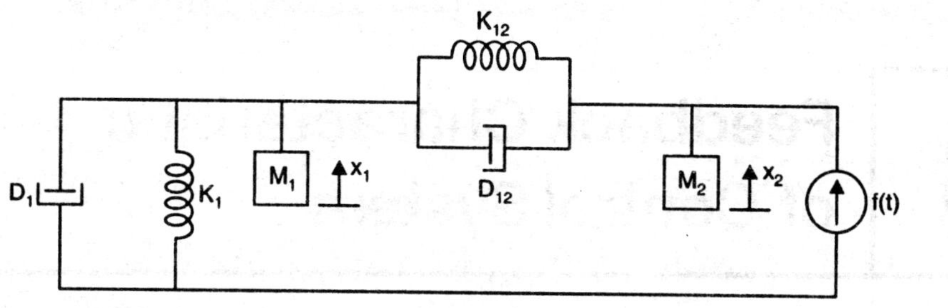

Fig. 2.98

3 Feedback Characteristic of Control System

3.1 INTRODUCTION

In the first unit, we discussed about the differences between open loop and closed loop systems. In this chapter, we shall deal with the effects of feedback on various aspects of closed loop performance.

What is a feedback?

A closed loop system can be represented by the general block diagram as shown in the Fig. 3.1. Such a system is composed of three basic elements like feedback element, controller and controlled system.

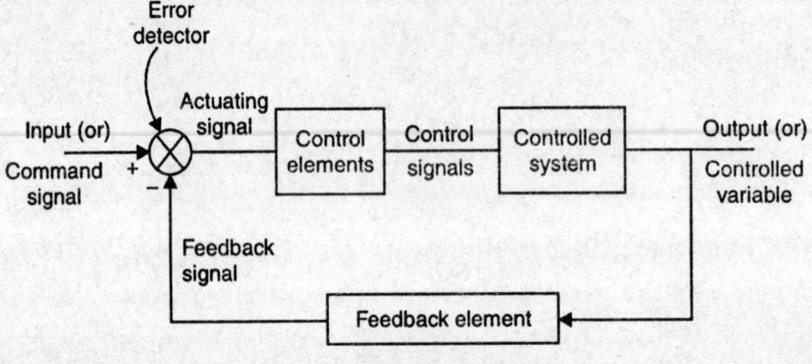

Fig. 3.1

The feedback element is a device which converts the output variable into another suitable variable which means the controller consists of an error detector and control elements. The error detector compares the feedback signal obtained from the plant output with the input signal (command) and the deviation known as the actuating signal. The actuating signal is usually at low power level. It is suitably manipulated by the control elements to produce a control signal. The manipulation may involve amplification, generation of a suitable function of the actuating signal and a low power stage. The power stage in control elements is essential so that the control signal can drive the controlled system to produce the desired output variable.

3.2 FEEDBACK CHARACTERISTICS OF CONTROL SYSTEMS

From the above discussion, feedback gives automatic regulation and control and also reduces the sensitivity of the system to parameter variations.

3.2.1 Reduction of Parameter Variation by Feedback

The parameters of a system may vary with age, with changing environment (*e.g.*, ambient temperature), etc. Suppose due to parameter variations, the open loop transfer function of the system changes to $[G(s) + \Delta G(s)]$, where $G(s) >> | \Delta G(s)|$.

$$\therefore \quad [C(s) + \Delta C(s)] = [G(s) + \Delta G(s)]\, R(s)$$

The output of the open loop system changes to

$$\Delta C(s) = \Delta G(s)\, R(s) \tag{3.1}$$

Similarly in the closed loop case, the output is

$$C(s) = \frac{G(s)}{1+G(s)\, H(s)} R(s)$$

$$\text{Output changes to } [C(s) + \Delta C(s)] = \frac{G(s) + \Delta G(s)}{1+G(s)\, H(s) + \Delta G(s)\, H(s)} R(s)$$

due to variation of $\Delta G(s)$ in $G(s)$.

Since $|G(s)| >> |\Delta G(s)|$, we have from the above, the variation in the output is,

$$\Delta C(s) \simeq \frac{\Delta G(s)}{1+G(s)H(s)} R(s) \tag{3.2}$$

From equations (3.1) and (3.2), it is seen that in comparison to the open loop system, the change in the output of closed loop system due to variation in $G(s)$ is reduced by a factor of $[1 + G(s)\, H(s)]$ which is much greater than one, in most practical cases.

3.2.2 Effect of Feedback on Sensitivity

Effect of parameters on the sensitivity of the system is derived mathematically by defining the sensitivity of the system.

The term sensitivity is used to describe the relative variation in the overall transformation, $T(s) = \dfrac{C(s)}{R(s)}$ due to variation in $G(s)$ and is defined below.

Let the variation in a system which is varying T, due to parameter variation K of the system. The sensitivity of the system parameter T to parameter K is expressed as

$$S = \frac{\text{Percentage change in } T}{\text{Percentage change in } K}$$

For small variation in $G(s)$, the sensitivity is written in quantitative form as,

$$S_G^T = \frac{\partial T/T}{\partial G/G}$$

where S_G^T denotes the sensitivity of T with respect to G.

For the simple open loop system as shown in Fig. 3.2.

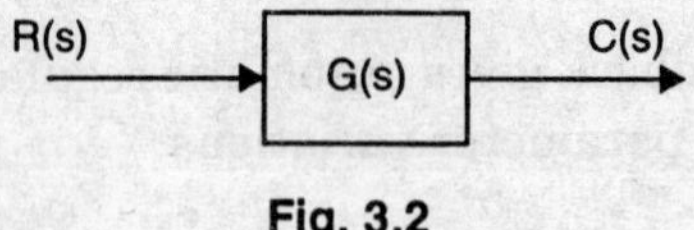

Fig. 3.2

$$S_G^T = \frac{\partial T/T}{\partial G/G} = 1 \text{ (in this case } T = G) \tag{3.3}$$

For the closed loop system as shown in Fig. 3.3.

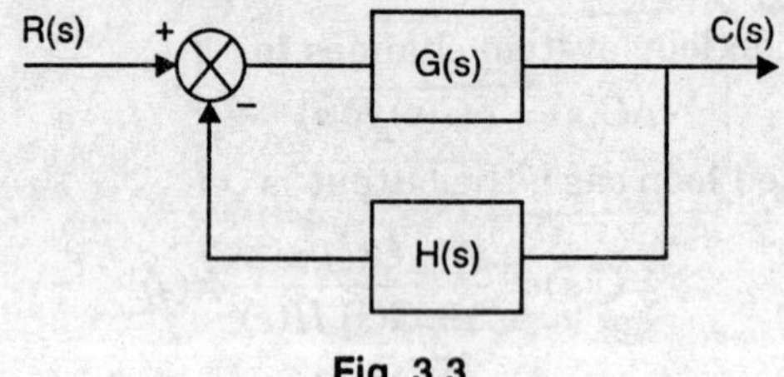

Fig. 3.3

S_G^T is given by

$$S_G^T = \frac{\partial T}{\partial G} \times \frac{G}{T} = \frac{1 + GH - GH}{(1 + GH)^2} \times \frac{\frac{G}{G}}{1 + GH} = \frac{1}{1 + GH} \tag{3.4}$$

From eqns. (3.3) and (3.4), the sensitivity of a closed loop system with respect to variation in G is reduced by a factor of $(1 + GH)$ as compared to that of an open loop system. The sensitivity of T with respect to H (the feedback sensor), is given as

$$S_H^T = \frac{\partial T}{\partial H} \times \frac{H}{T} = -\frac{[G(s)]^2}{[1 + G(s)\, H(s)]^2} \times \frac{\frac{H(s)}{G(s)}}{1 + G(s)\, H(s)} = \frac{G(s)H(s)}{1 + G(s)\, H(s)} \tag{3.5}$$

The eqn.(3.5) gives that for larger values of $G(s)\, H(s)$, the sensitivity of the feedback system with respect to H approaches unity. Thus we can see that the changes in H directly affect the system output. Therefore, it is important to use feedback elements which do not vary with environmental changes or can be maintained constant. Since $G(s)$ is made up of power elements and $H(s)$ is made up of measuring elements which operate at low power levels, the selection of accurate $H(s)$ is far less costly than that of $G(s)$ to meet the exact specification. The price for improvement in sensitivity by use of the feedback is paid in terms of loss of system gain. The open loop has a gain $G(s)$ while the gain of the closed loop system is $\dfrac{G(s)}{1 + G(s)\, H(s)}$.

Problem 3.1. *In the closed loop system shown in the Fig 3.4, find its sensitivities* $S_{G_1}^T$ *and* $S_{G_2}^T$.

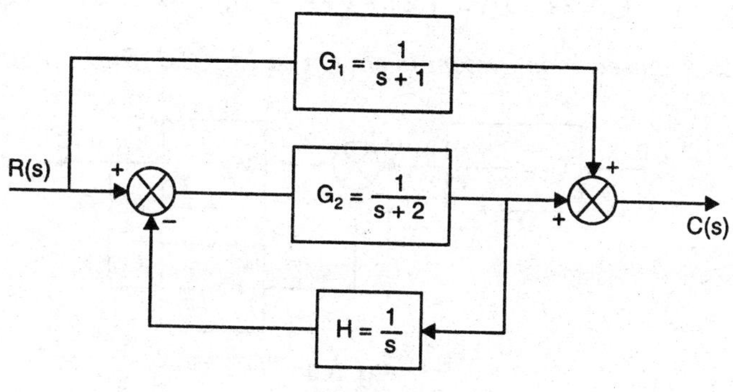

Fig. 3.4

Solution: The closed loop transfer function $T(s)$ of the Fig. 3.4 is given by

$$T(s) = G_1(s) + \frac{G_2(s)}{1+G_2(s)\,H(s)}$$

The sensitivity $S_{G_1}^T$ is calculated as

$$S_{G_1}^T = \frac{\partial T}{\partial G_1} \times \frac{G_1}{T} = 1 \times \frac{G_1}{G_1 + \dfrac{G_2}{1+G_2H}} = \frac{(1+G_2H)G_1}{G_1 + (1+G_2H) + G_2}$$

Substituting the given values in the above equation, we get

$$S_{G_1}^T = \frac{\dfrac{1}{s+1}\left(1+\dfrac{1}{(s+2)s}\right)}{\dfrac{1}{s+1}\left[1+\dfrac{1}{s(s+2)}\right]+\dfrac{1}{s+2}} = \frac{s(s+2)+1}{[s(s+2)+1]+s(s+1)} = \frac{s^2+2s+1}{2s^2+3s+1}$$

The sensitivity $S_{G_2}^T$ is then calculated as

$$S_{G_2}^T = \frac{\partial T}{\partial G_2} \times \frac{G_2}{T} = 1 \times \frac{(1+G_2H) - G_2H}{(1+G_2H)^2} \times \frac{G_2}{G_1 + \dfrac{G_2}{1+G_2H}}$$

$$= \frac{1}{(1+G_2H)^2} = \frac{G_2(1+G_2H)}{G_1 + G_1G_2H + G_2} = \frac{G_2}{(1+G_2H)(G_1 + G_1G_2H + G_2)}.$$

Substituting the given values in the above equation, we get

$$S_{G_2}^T = \frac{\dfrac{1}{(s+2)}}{\left[1+\dfrac{1}{s(s+2)}\right]\left[\dfrac{1}{s+1}+\dfrac{1}{s(s+1)(s+2)}\,\dfrac{1}{s+2}\right]}$$

$$= \frac{s^2(s+1)(s+2)}{(s(s+2)+1)\,[s(s+2)+1+s(s+1)]}$$

$$= \frac{s^2(s^2+3s+2)}{(s^2+2s+1)\,[s^2+2s+1+s^2+1)]} = \frac{s^4+3s^3+2s^2}{(s^2+2s+1)(2s^2+2s+2)}.$$

Problem 3.2. *A closed loop system is shown in Fig. 3.5, find sensitivities* $S_{G_1}^T$ *and* $S_{G_2}^T$.

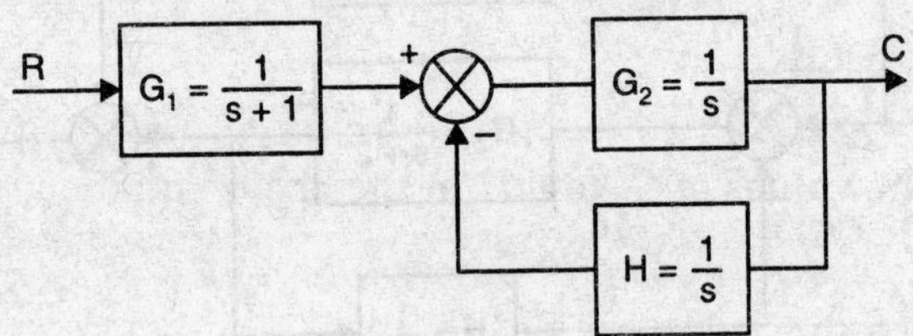

Fig. 3.5

Solution:

From Fig. 3.5, the closed loop transfer function is

$$T = \frac{C}{R} = \frac{G_1G_2}{1+G_2H}$$

Sensitivity

$$S_{G_1}^T = \frac{\dfrac{\partial T}{T}}{\dfrac{\partial G_1}{G_1}} = \frac{\partial T}{\partial G_1}\left(\frac{G_1}{T}\right)$$

$$\frac{\partial T}{\partial G_1} = \frac{G_2}{1+G_2H}$$

$$\frac{G_1}{T} = \frac{G_1}{\left(\dfrac{G_1G_2}{1+G_2H}\right)} = \frac{1+G_2H}{G_2}$$

$$\therefore \quad S_{G_1}^T = \left(\frac{\partial T}{\partial G_1}\right)\left(\frac{G_1}{T}\right) = \left(\frac{G_2}{1+G_2H}\right)\left(\frac{1+G_2H}{G_2}\right) = 1$$

$$S_{G_2}^T = \frac{\dfrac{\partial T}{T}}{\dfrac{\partial G_2}{G_2}} = \left(\frac{\partial T}{\partial G_2}\right)\left(\frac{G_2}{T}\right)$$

$$\frac{\partial T}{\partial G_2} = G_1\left(\frac{(1+G_2H)-G_2H}{(1+G_2H)^2}\right) = \frac{G_1}{(1+G_2H)^2}$$

$$\frac{G_2}{T} = \frac{G_2}{\left(\frac{G_1G_2}{1+G_2H}\right)} = \frac{(1+G_2H)}{G_1}$$

$$S_{G_2}^{T} = \left(\frac{\partial T}{\partial G_2}\right)\left(\frac{G_2}{T}\right) = \left(\frac{G_1}{(1+G_2H)^2}\right)\left(\frac{1+G_2H}{G_1}\right)$$

$$\therefore \qquad S_{G_2}^{T} = \frac{1}{(1+G_2H)}$$

By substituting the values of G_2 and H in the above equation, we get

$$S_{G_2}^{T} = \frac{1}{1+\left(\frac{1}{s}\right)\left(\frac{1}{s}\right)} = \frac{s^2}{1+s^2}$$

3.2.3 Control of System Dynamics by Feedback

Let us consider an elementary system as shown in Fig. 3.6 (*a*).

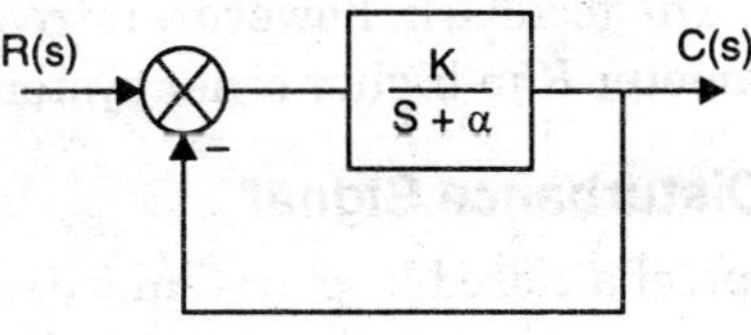

Fig. 3.6 (a)

The open loop transfer function of the system, $G(s) = \dfrac{K}{s+\alpha}$

which has a real pole at $s = -\alpha$

The output of the open loop system to a unit impulse *i.e.,* $R(s) = 1$, is given by

$$C(s) = \frac{K}{s+\alpha} \qquad (3.6)$$

And for the feedback system by

$$C(s) = \frac{K}{s+\alpha+K} \qquad (3.7)$$

Taking the inverse Laplace transform of the eqns. (3.6) and (3.7), we get

$C(t) = Ke^{-\alpha t}$ (for open loop system)

$= Ke^{-(K+\alpha)t}$ (for feedback system)

These responses are plotted in Fig. 3.6 (*b*).

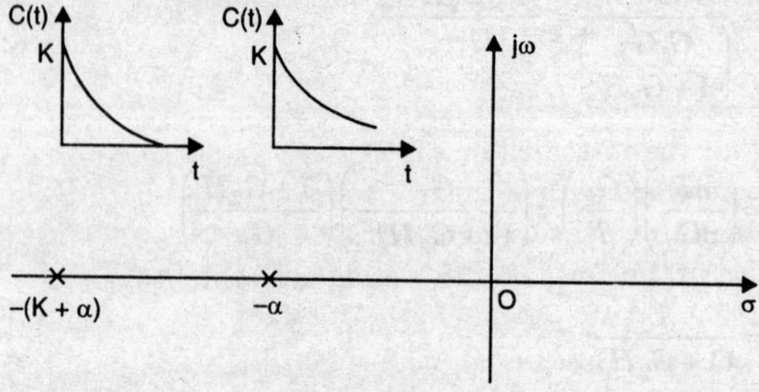

Fig. 3.6 (b)

For the non feedback system, the response is an exponential decay with a time constant of $T = \frac{1}{\alpha}$. For positive values of *K*, the effect of the feedback is to shift the pole negatively to $s = -(\alpha + K)$ and so the time constant reduces to $\frac{1}{K+\alpha}$. This implies, as *K* increases, the system dynamics continuously becomes faster. In second system, feedback can be used to control the damping and oscillations. The feedback, however, introduces the possibility of instability for large values of gain parameter *K* in higher order system.

3.2.4 Effects of Feedback on Disturbance Signal

Fig. 3.7 shows the signal flow graph of a closed loop system with the disturbance signal T_D in forward path.

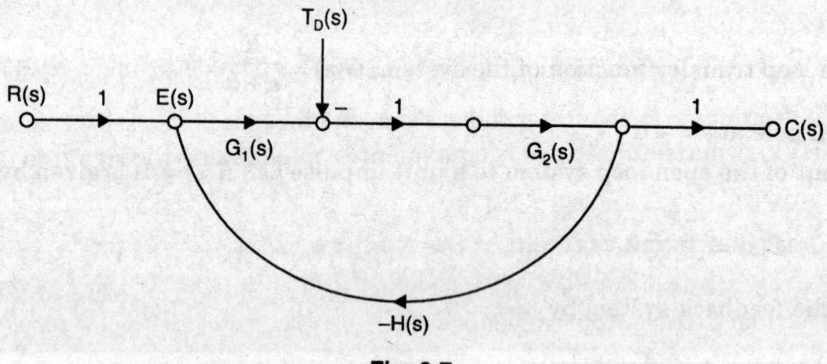

Fig. 3.7

The ratio of the output $C(s)$ to the disturbance signal $T_D(s)$ when $R(s) = 0$ is obtained by applying the Mason's gain formula to the graph of Fig. 3.7 and is given by

$$\frac{C(s)}{T_D(s)} = \frac{-G_2}{1 + G_1 G_2 H} \tag{3.8}$$

If $G_1G_2H >> 1$ over the working range of s, then from eqn. (3.8)

$$\frac{C(s)}{T_D(s)} = \frac{-1}{G_1H}$$

Thus it is seen that the introduction of feedback decreases the effects of disturbances and noise signals in the forward path of the feedback loop. Feedback is introduced by a set of additional elements, called the measurement sensor H which may itself generate from noise. Let us evaluate the effect of the noise on the system performance.

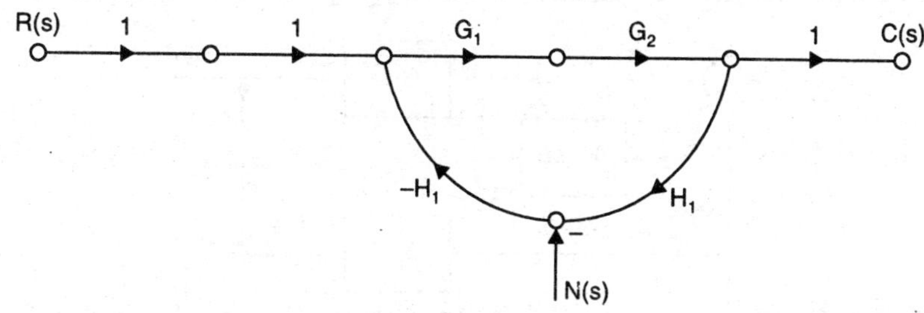

Fig. 3.8

Fig. 3.8 shows the signal flow graph of a system with the noise signal $N(s)$ in the feedback path. Using Mason's gain formula for the signal flow graph, the following result is obtained.

$$\left.\frac{C(s)}{N(s)}\right|_{R(s)=0} = \frac{-G_1G_2H_2}{1+G_1G_2H_1H_2}$$

For the large values of loop gain $|G_1 G_2 H_1 H_2| >> 1$ the above equation reduces to

$$\frac{C(s)}{N(s)} = -\frac{1}{H_1}$$

Thus for optimum performance of the system, the measurement sensor should be defined such that $H_1(s)$ is maximum which is equivalent to maximizing the signal to noise ratio of the sensor.

3.2.5 The Effect of Feedback on Gain of the System

Block diagram representation of regeneration feedback control system is shown in the Fig. 3.9. In regenerative feedback, the output is feedback with positive sign compared with the reference input produces the error signal $E(s)$.

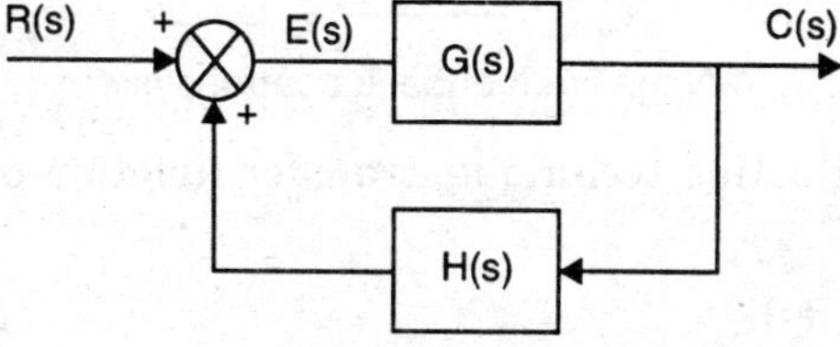

Fig. 3.9 A regenerative feedback control system

In this case, the transfer function is given by

$$\frac{C(s)}{R(s)} = \frac{G(s)}{1 - G(s)H(s)} \quad (3.9)$$

There is a negative sign in the denominator of eqn.(3.9) which indicates the possibility of denominator becoming equal to zero, thereby giving an infinite output for a finite input which is the condition of instability.

The regenerative feedback is sometimes used for increasing the loop gain of feedback systems. Fig. 3.10 shows a feedback system with inner loop having regenerative feedback.

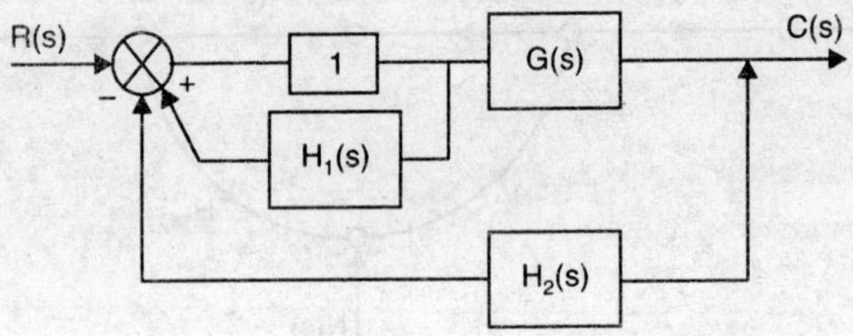

Fig. 3.10 Increasing loop gain by regenerative feedback

Using block diagram reduction technique, from Fig. 3.10 the transfer function is

$$\frac{C(s)}{R(s)} = \frac{\dfrac{G(s)}{1 - H_1(s)}}{1 + \dfrac{H_2 G(s)}{1 - H_1(s)}} \Rightarrow \frac{G(s)}{1 - H_1(s) + H_2 G(s)} \quad (3.10)$$

If $H_1(s)$ is selected to be nearly unity, the loop gain becomes very high and the closed loop transfer function approximates to

$$\frac{C(s)}{R(s)} = \frac{1}{H_2(s)} \quad (3.11)$$

Thus due to high loop gain provided by the inner regenerative feedback loop, the closed loop transfer function becomes insensitive to $G(s)$. The closed loop system with negative feedback is as shown in the Fig. 3.11.

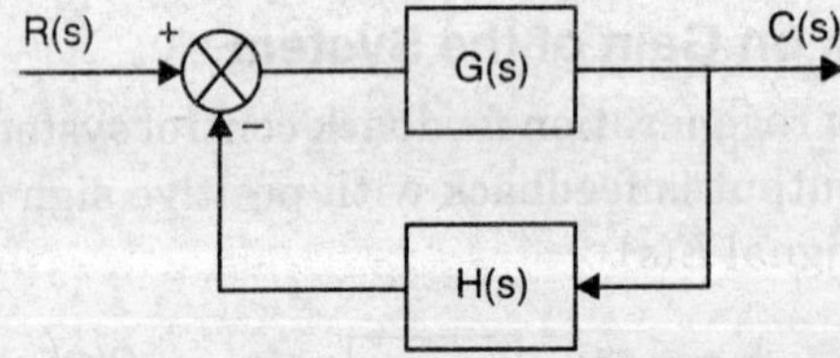

Fig. 3.11 Negative feedback control system

Using block diagram reduction technique, transfer function of the system as shown in Fig. 3.11, is

$$\frac{C(s)}{R(s)} = \frac{G(s)}{1 + G(s)\,H(s)} \quad (3.12)$$

From eqns. (3.9) and (3.12), it is observed that positive feedback reduces the system gain by the factor [1 – G(s) H(s)], whereas negative feedback system reduces the system gain by factor [1 + G(s) H(s)].

Hence reduction of gain is more for negative feedback as compared to the positive feedback.

3.3 EFFECTS OF FEEDBACK IN CONTROL SYSTEM

By providing feedback from the output signal to the error sensing device of a control system, the following effects are produced:

Advantages

- Reduces the sensitivity of the system.
- Improving the transient response and minimizing the effects of disturbance signals in control systems.
- The system can be made automatic.
- The system control can be made accurate.

Disadvantages

- The use of feedback increases the number of components of the system, thereby increasing its complexity further.
- It reduces the gain of the system and also introduces the possibility of instability.

3.4 CONTROL SYSTEM COMPONENTS

In this section, we will discuss different types of control system components.

3.4.1 Servomotors

The motors which are used in servo systems are called servomotors. Servomotors work on the principles of error signals. Such systems are generally automatic control systems in which the output is some mechanical variable like position, velocity, and acceleration. The error signals are amplified to drive the motors used in such systems. These motors are usually coupled to the output shaft i.e., load through gear train for motor power matching.

These motors are used to convert electrical signal applied into the angular velocity or movement of shaft.

Good servomotor requires

- Linear relationship between electrical control signal and the rotor speed.
- Its response should be as fast as possible.
- Inertia of rotor should be as low as possible.
- Operation should be stable without any disturbances.

Types of servomotor

Depending upon the nature of the supply, servomotors are classified into two types.

They are

1. DC servomotors and
2. AC servomotors

3.4.1.1 DC Servomotors

Basically DC servomotors are constructionally similar to DC motors. The difference between DC motor and DC servomotor is that all DC servomotors are separately excited types. This ensures the linear torque-speed characteristics.

The speed of DC servomotor can be controlled from field side or from armature side. Depending upon this, these are classified into armature controlled DC servomotor and field controlled DC servomotor.

(*i*) **Armature controlled DC Servomotor**

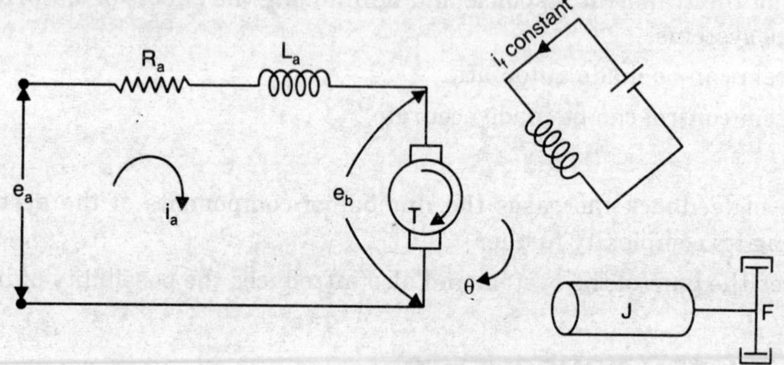

Fig. 3.12 (a) Armature controlled DC servomotor

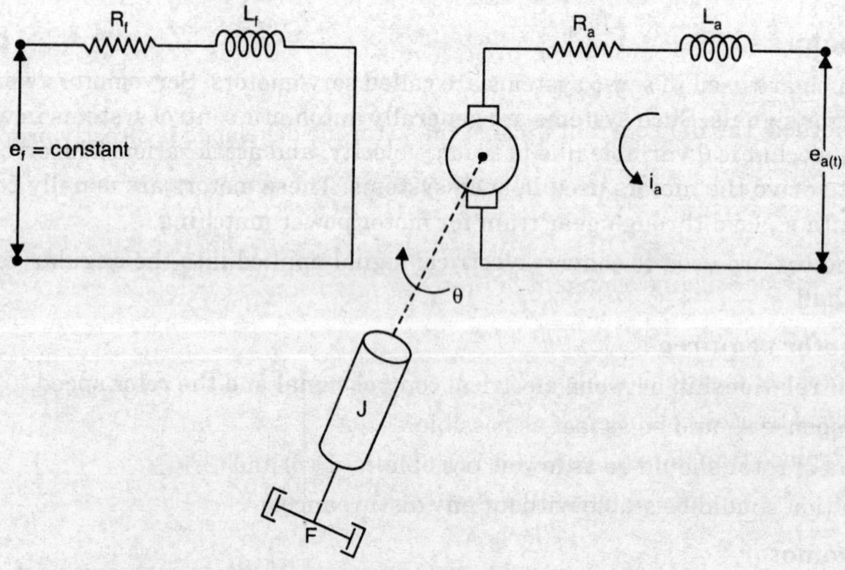

Fig. 3.12 (b)

Fig. 3.12 (*b*) is the equivalent diagram of armature controlled DC servomotor.

In this type of motor, input voltage e_a is applied to armature through resistance R_a and inductance L_a, field current is kept constant, and field is excited by permanent magnets. In such a case, no field coils are necessary.

When input voltage e_a is applied to the armature circuit, it allows the current i_a and produces back emf. This back emf will circulate the currents and these currents will run the motor against the moment of inertia J and viscous friction f.

Loop equation for electrical system of Fig. 3.12 (*b*) is

$$e_a(t) = R_a i_a(t) + L_a \frac{di_a}{dt}(t) + e_b(t) \tag{3.13}$$

where $e_a(t)$ = voltage applied to the armature circuit
$i_a(t)$ = armature current
R_a = armature resistance
L_a = armature inductance
$e_b(t)$ = back emf across the armature.

Back emf $\quad e_b(t) \propto \frac{d\theta}{dt}$

According to Faraday's law of electromagnetic induction, emf induced is directly proportional to the rate of change of flux linkages and these flux linkages are analogous to angular displacement.

$$\therefore \quad e_b(t) = K_t \dot{\theta} \tag{3.14}$$

where K_t = constant

The air gap flux produced is directly proportional to the field current.

i.e., $\quad \phi \propto i_f$

$$\therefore \quad \phi = K_f i_f \tag{3.15}$$

where K_f = constant

And the torque developed is directly proportional to the air gap flux and armature current.

i.e., $\quad T(t) \propto \phi\, i_a$

In this method, the field current is constant so torque developed is directly proportional to the armature current

i.e., $\quad T(t) \propto i_a(t)$

$$\therefore \quad T(t) = K_t i_a(t) \tag{3.16}$$

Torque developed against moment of inertia J and friction coefficient f is

$$T(t) = J\ddot{\theta} + f\dot{\theta} \tag{3.17}$$

where J = moment of inertia of shaft
f = viscous friction coefficient.

Taking Laplace transform of eqns. (3.13), (3.14), (3.16) and (3.17)

$$E_a(s) - E_b(s) = [R_a + L_a s]\, I_a(s) \tag{3.18}$$

$$E_b(s) = s\, K_b\, \theta(s) \tag{3.19}$$

$$T(s) = K_t I_a(s) \tag{3.20}$$

$$T(s) = [Js^2 + fs]\, \theta(s) \tag{3.21}$$

Substituting the eqn. (3.19) in eqn. (3.18)

$$E_a(s) - s\,K_b\,\theta(s) = [R_a + L_a s]\,I_a(s) \tag{3.22}$$

From eqn. (3.20),

$$I_a(s) = \frac{T(s)}{K_t}$$

Substituting $T(s)$ from eqn. (3.21) in above equation

$$\therefore \quad I_a(s) = \frac{[Js^2 + fs]}{K_t} = \theta(s) \tag{3.23}$$

Substituting eqn. (3.23) in eqn (3.22)

$$E_a(s) - sK_b\,\theta(s) = \frac{[R_a + L_a s][Js^2 + fs]\,\theta(s)}{K_t}$$

$$E_a(s) = \left[\frac{(R_a + L_a s)[Js^2 + fs]}{K_t} + sK_b\right]\theta(s)$$

$$\therefore \text{ Transfer function, } \quad \frac{\theta(s)}{E_a(s)} = \frac{K_t}{(R_a + L_a s)(Js^2 + fs) + K_b K_t s}$$

Neglecting the inductance of the armature circuit

$$\frac{\theta(s)}{E_a(1)} = \frac{K_t}{R_a\;(Js^2 + fs) + K_b K_t s}$$

Taking R_a and f common in denominator

$$\frac{\theta(s)}{E_a(s)} = \frac{K_t/(R_a f)}{\left[\frac{J}{f}s^2 + s\right] + K_b\,\frac{K_t}{R_a f}\,s} = \frac{K_m}{s(\tau_m s + 1) + K_b K_m s} \tag{3.24}$$

where K_m = Motor gain constant, $\dfrac{K_t}{R_a f}$

τ_m = Mechanical time constant, $\dfrac{J}{f}$

Unit block diagram representation of equations (3.18), (3.19), (3.20) and (3.21) are shown in Fig. 3.13 (*a*).

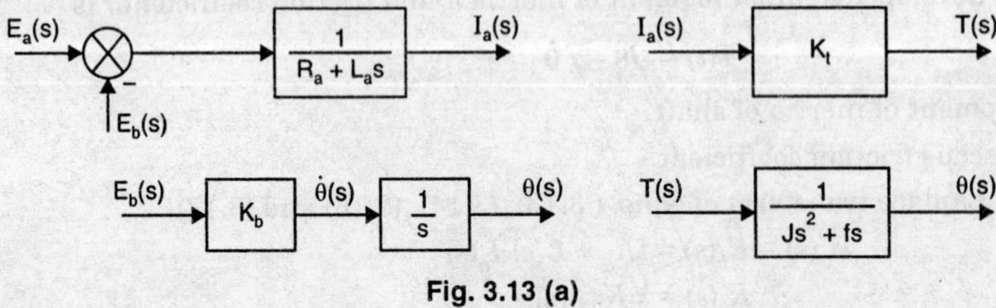

Fig. 3.13 (a)

Complete block diagram model representation of armature controlled DC servomotor is as shown in the Fig. 3.13 (*b*).

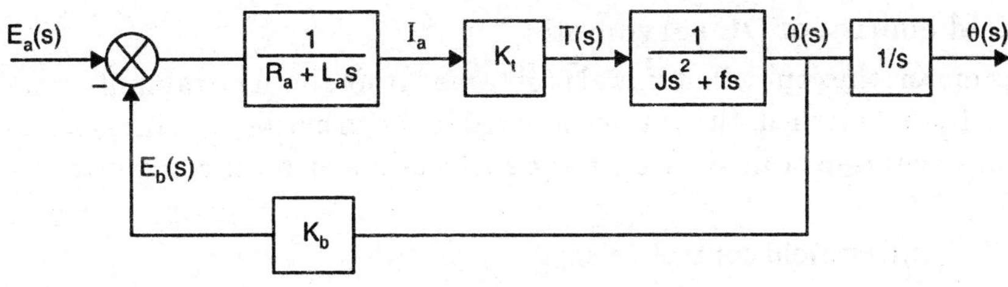

Fig. 3.13 (b)

Armature controlled DC servomotors are suitable for large rated motors. Due to its small time constant, its response is small, the efficiency and performance are better than field control system. It is an example of the closed loop system. Back emf provides the internal damping, which makes the system more stable.

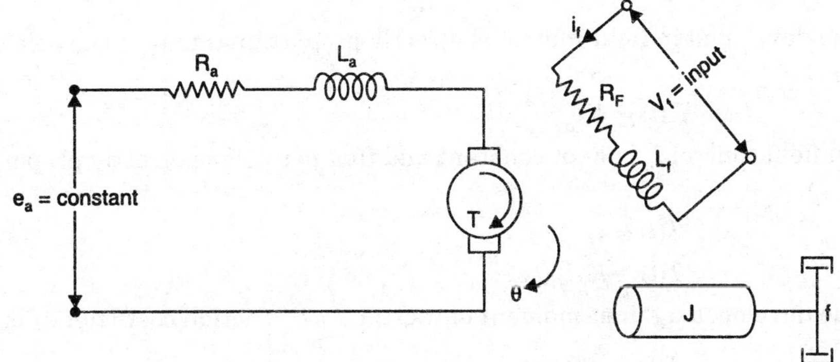

Fig. 3.14 (a) Field controlled DC servomtor

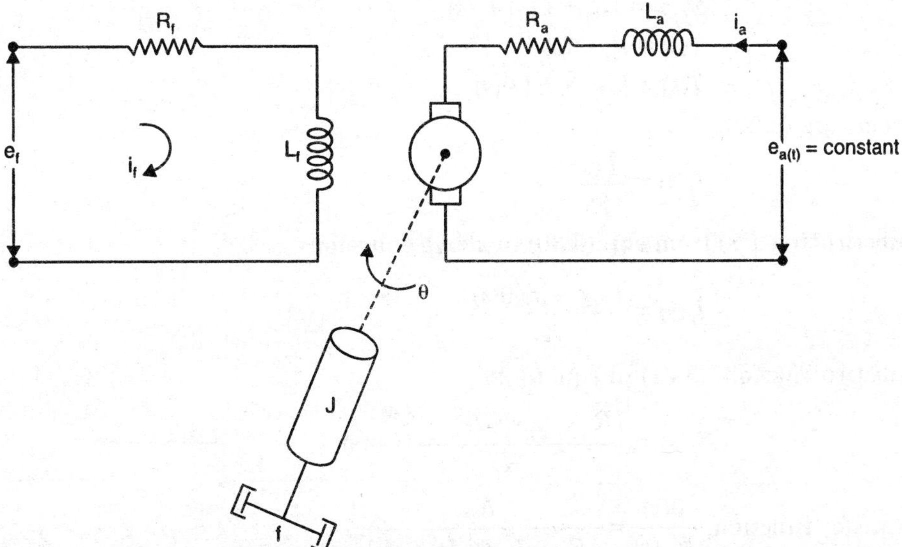

Fig. 3.14 (b) is the equivalent diagram of field controlled DC servomtor

(ii) **Field control of DC servomotor**

In this motor, the controlled signal is obtained from the servo amplifier and is applied to field winding. In this method, the voltage applied to the armature circuit is kept constant, and by varying the excitation or field current, the speed of the motor is controlled to above rated or base speed.

$e_f(t)$ = field control voltage
R_f = field resistance
L_f = field inductance
$i_f(t)$ = field current

Loop equation for electrical circuit of Fig. 3.14 (*b*) is

$$e_f(t) = R_f i_f(t) + L_f \frac{di_f(t)}{dt} \tag{3.25}$$

Torque developed in field control is directly proportional to flux per pole and armature current.

$$T(t) \propto \phi\, i_a$$

But in field control I_a is kept constant and flux per pole is directly proportional to field current ($\phi \propto i_f$).

$$\therefore \quad T(t) \propto i_f$$

$$T(t) = K_f i_f(t) \tag{3.26}$$

Torque developed against moment of inertia J and friction coefficient f is

$$T(t) = J\ddot{\theta} + f\dot{\theta} \tag{3.27}$$

Taking Laplace transform on both sides of eqns. (3.25), (3.26) and (3.27)

$$E_f(s) = [R_f + L_f s]\, I_f(s) \tag{3.28}$$

$$T(s) = K_t I_f(s) \tag{3.29}$$

$$T(s) = [Js^2 + fs]\, \theta(s) \tag{3.30}$$

From eqn. (3.29),

$$I_f(s) = \frac{T(s)}{K_t}$$

Substituting $T(s)$ from eqn.(3.30) in above equation

$$\therefore \quad I_f(s) = \frac{[Js^2 + fs]\theta(s)}{K_t} \tag{3.31}$$

Substituting eqn. (3.31) in eqn. (3.28)

$$E_f(s) = \frac{[R_f + L_f s]\,[Js^2 + fs]}{K_t}\, \theta(s)$$

Transfer function, $\dfrac{\theta(s)}{E_f(s)} = \dfrac{K_t}{[R_f + L_f s]\,[Js^2 + fs]}$

Taking R_f and f common in denominator

$$\frac{\theta(s)}{E_f(s)} = \frac{K_f/(R_f\, f)}{s\left[1+\dfrac{L_f}{R_f}s\right]\left[\dfrac{Js}{f}+1\right]} = \frac{K_m}{s(1+\tau_e s)(1+\tau_m s)} \tag{3.32}$$

where K_m = motor gain constant, $\dfrac{K_t}{R_f f}$

τ_e = electrical time constant, $\dfrac{L_f}{R_f}$

τ_m = mechanical time constant, $\dfrac{J}{f}$

Unit block diagram model representation of eqns. (3.28), (3.29) and (3.30) are shown in Fig. 3.14 (*c*).

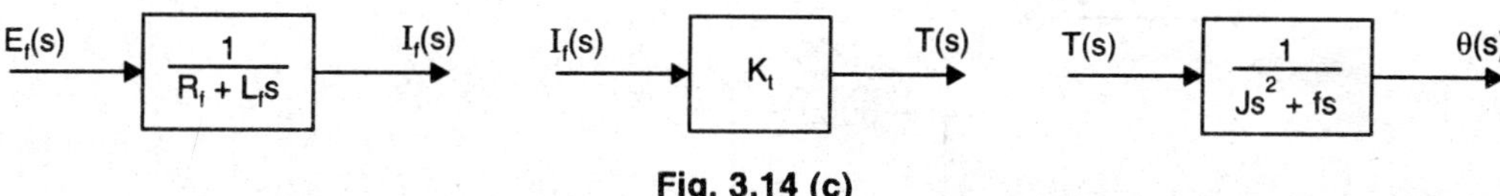

Fig. 3.14 (c)

Complete block diagram representation of field controlled DC servomotor is shown in Fig. 3.14 (*d*).

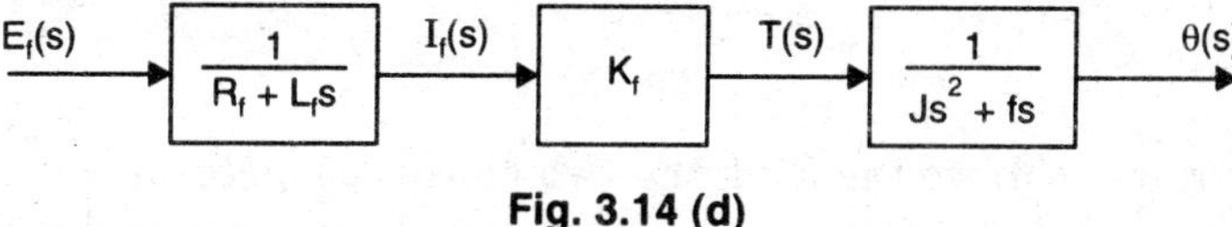

Fig. 3.14 (d)

In field controlled DC servomotors which are suitable for small rated motors, control circuit is simple.

3.4.1.2 AC Servomotors

Three phase squirrel cage induction motors run at nearly constant speed which make them the first choice for industrial drives. But in control applications, the basic requirement of a motor is a high starting torque over a wide range of speeds. AC motors available in the fractional horse power range upto about $\frac{1}{3}$ HP.

The principal advantages of AC motors are

- Absence of brushes or slip rings which introduce friction necessitating replacement of parts.
- Stalling without causing damage
- Almost instantaneous starting, stopping and reverse.
- Long life with high reliability
- Compatibility with AC signals of synchros, LVDCS and other AC transducers without the need for demodulation as in the case of DC motors.

An AC servomotor is basically a two phase induction motor. AC servomotor differs from a normal induction motor as follows.

The rotor of the servomotor is built with high resistance so that its X/R ratio is small and the torque-speed characteristic is nearly linear in contrast to the highly non-linear characteristic with large X/R ratio of induction motor.

It can be shown that if a conventional induction motor with X/R ratio is used for servo applications, then because of the positive slope for part of the characteristic, the system using such a motor becomes unstable. The rotor construction is usually squirrel cage or drag cup type. The diameter of the rotor is kept small in order to reduce inertia and thus to obtain good accelerating characteristics. Drag-cup construction is used for very low inertia applications.

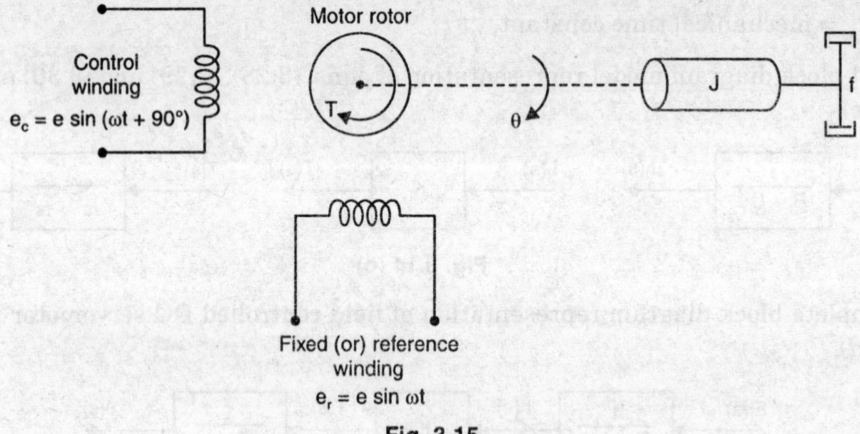

Fig. 3.15

Besides the rotor to drive the load, the two phase AC motor has two windings at right angles to each other and the two windings are supplied with voltages with 90° phase difference as through the amplifier. The rotor diameter is small but long to reduce inertia for faster motor response.

One winding called fixed or reference winding is supplied with a fixed voltage and frequency from a constant voltage source. The frequency is normally 50Hz or 400 Hz. The other winding is applied to the voltage whose amplitude can be varied but of the same frequency as shown in Fig. 3.15. The axes of the coils are in space with the quadrature and voltages in time quadrature. On account of this, there will be a resultant flux. This resultant flux will be rotating according to the phase difference between the two voltages at successive instants. The flux will be rotating at the frequency of supply at the synchronous speed, $\frac{120f}{p}$. This rotating flux cuts the conductors of rotor and causes a varying current. This varying current in turn produces a varying flux and reacts with the stator flux causing the rotor to rotate in a smooth manner. The speed-torque characteristic for normal induction motor and AC servomotor is as shown in Figs. 3.16 (*a*) and 3.16 (*b*) respectively.

The torque-speed characteristic of an AC motor are non-linear but may be considered linear for practical purposes.

From the speed-torque curve, the stall torque, $T_s = Ke_c$

where K = proportionality constant

e_c = control voltage

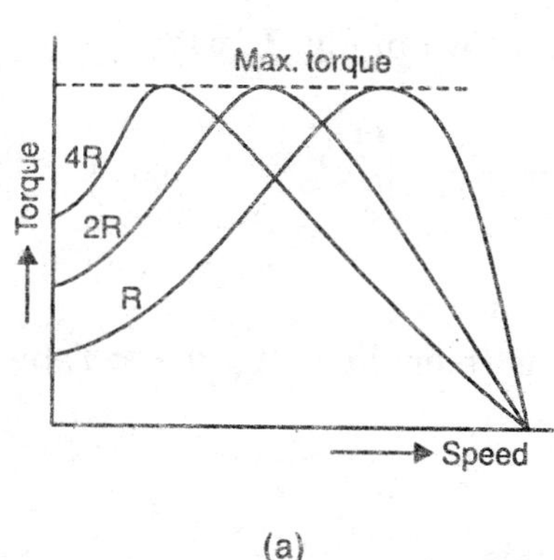

(a)

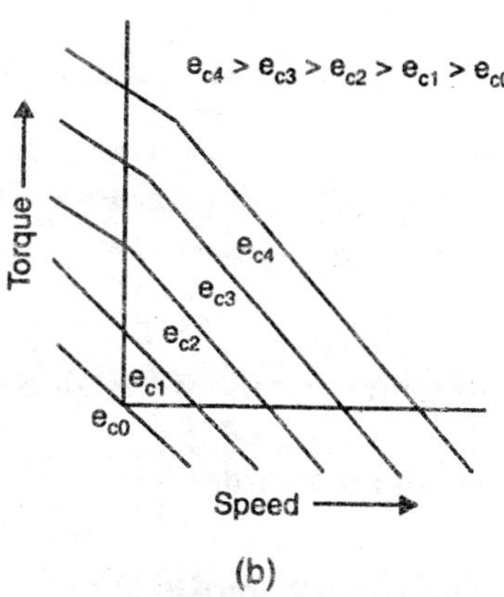

(b)

Fig. 3.16

No load speed, $$\omega_{nl} = \frac{Ke_c}{f_m} \quad (3.33)$$

K is different for different e_c.

As in the case of DC motors

$$T = Ke_c - f_m\,\omega \quad (3.34)$$

and $$T = J\frac{d^2\theta}{dt^2} + f_L\frac{d\theta}{dt} \quad (3.35)$$

where f_m = motor friction coefficient

f_L = load friction coefficient

From eqns. (2.34) and (2.35)

$$J\frac{d^2\theta}{dt_2} + f_L\frac{d\theta}{dt} = Ke_c - f_m\,\omega$$

$$J\frac{d^2\theta}{dt_2} + f\frac{d\theta}{dt} = Ke_c \quad (3.36)$$

Damping coefficient of motor and load, $f_m + f_L = f$

f is the viscous damping coefficient due to motor and load.

Taking Laplace transformation of eqn. (3.36)

Transfer function, $$G(s) = \frac{\theta(s)}{E_C(s)} = \frac{K}{Js^2 + fs} = \frac{K/f}{s(J/fs + 1)}$$

$$= \frac{K_m}{s(\tau_m s + 1)} \quad (3.37)$$

$$K_m = \text{Motor gain constant} = \frac{K}{f}$$

$$\tau_m = \text{Mechanical time constant} = \frac{J}{f}$$

The block diagram representation of eqn. (3.37) is shown in Fig. 3.16 (*c*).

$$E_c(s) \longrightarrow \boxed{\frac{K_m}{S(\tau_m S + 1)}} \longrightarrow \theta(s)$$

Fig. 3.16 (c)

The damping is high if the slope of the T-ω curve is steep. For $|\tau_m s| << 1$, the servo motor acts as an integrator $\left(\frac{1}{s}\right)$.

Applications of Servomotors

Servomotors are widely used in electromechanical actuators, process controllers, air craft control systems, robotics and machine tools etc.

Problem 3.3. *An AC servomotor has both windings excited with 115 V AC. It has a stall torque of 2 lb ft. Its coefficient of viscous friction is 0.2 lb ft. sec.*

(*a*) *Find its no load speed.*

(*b*) *It is connected to a constant load of 0.9 lb ft and coefficient of viscous friction of 0.05 lb ft sec through a gear pass with a ratio of 4. Find the speed at which the motor will run.*

Solution: (*a*) No load speed $\omega_{nl} = \frac{K_m V_m}{f_m}$

$$K_m V_m = \text{stall torque} = 2 \text{ lb ft}$$

$$\omega_{nl} = \frac{2}{0.2} = 10 \text{ rad/sec}$$

(*b*) Total viscous friction due to motor and load

$$f_{wt} = f_m + N^2 f_l$$

$$= 0.2 + 4^2 \times 0.05 = 1 \text{ lb. ft. s}$$

$$T_L = K_m V_m - F_{wt}\,\omega$$

$$0.9 = 2 - 1\,\omega$$

or $$\omega = 2 - 0.9 = 1.1 \text{ rad/sec.}$$

3.4.2 Synchros

Synchro is an electromechanical device. It produces an output voltage depending upon angular position of the rotor and not motor speed. In this way, it is different from a DC generator. The trade names for synchros are selsyn, antosyn, telesyn. There are four basic elements of synchros: such as transmitter, receiver, control transformer and differential transformer.

(*i*) Synchro transmitter

Constructional Details:

The constructional diagram and its equivalent electric circuits are shown in Figs. 3.17 (*a*) and 3.17 (*b*) respectively.

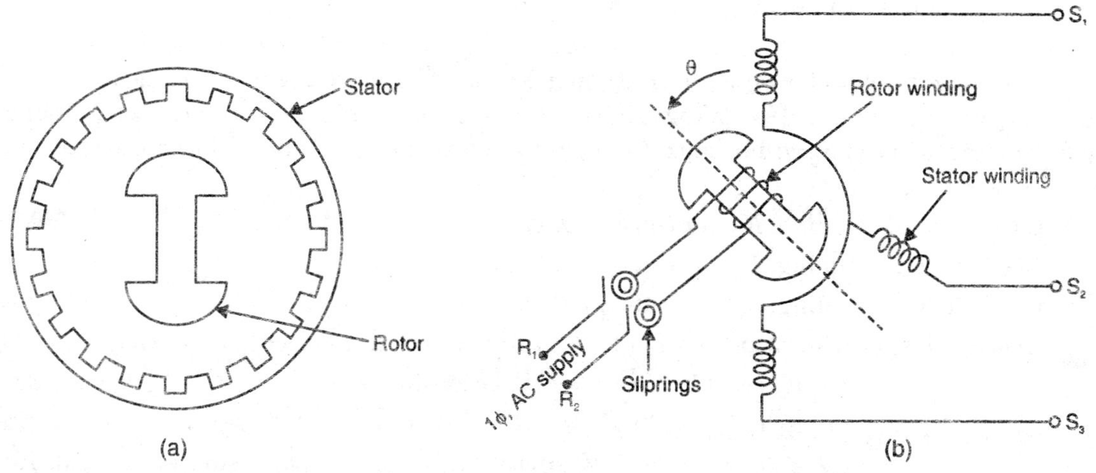

Fig. 3.17

The two main parts of synchrotransmitter are stator and rotor. The construction of stator of synchrotransmitter is very similar to that of stator of three phase alternator. The stator is made of laminated silicon steel and is slotted to accommodate balanced three phase winding of coils which are placed 120° apart. All three windings are stator.

The rotor is of dumpbell construction with a single concentric winding. A single phase A.C. supply voltage is given to the rotor winding through slip rings.

Operating principle:

When single phase A.C. voltage is applied to rotor winding, current flows through winding. Due to this current, magnetic field is produced around the rotor winding, an e.m.f. is induced in the stator coils by the rotor magnetic field just like in the transformer. The magnitude of e.m.f. induced in any stator coil will depend upon the angular position of coils axis with respect to the reference rotor axis.

Let V_r = instantaneous value of input A.C. voltage to rotor.

$V_{s_1}, V_{s_2}, V_{s_3}$ = instantaneous value of e.m.f induced in stator coils S_1, S_2 and S_3 respectively with respect to neutral.

V_r = maximum (peak) value of rotor input voltage.

ω = angular frequency rotor input voltage.

θ = angular displacement of rotor shaft with respect to reference axis

K_t = turn ratio of stator and rotor windings.

K_c = coefficient of coupling between stator and rotor windings.

The instantaneous value of input A.C. voltage to rotor, $V_r = E_r \sin \omega t$.

Let us assume that rotor is rotating in the anticlockwise direction, when the rotor rotates by an angle 'θ' with respect to reference axis, an e.m.f. is induced in stator windings. The frequency of induced e.m.f. in stator windings is equal to the rotor frequency. The magnitude of induced e.m.f.in any stator winding is proportional to the turn ratio and coefficient of coupling between the stator winding and rotor winding.

∴ Induced e.m.f in stator coil = $K_t K_c V_r \sin \omega t$

Let V_{s1} be reference phasor with reference to Fig 3.17 (*b*). When θ = 0, the flux linkages of coil S_1 is maximum and is equal to zero when θ = 90°. Therefore the flux linkages of coil S_1 are a function of cos θ. The flux linkages in coil S_3 will be maximum after rotation of 120° and the flux linkages in coil S_2 will be maximum after a rotation of 240° in anticlockwise direction from the reference axis.

Coefficient of coupling, K_{c1} for coil $S_1 = K \cos \theta$

Coefficient of coupling, K_{c2} for coil $S_2 = K \cos (\theta - 120°)$

Coefficient of coupling, K_{c3} for coil $S_3 = K \cos (\theta - 240°)$

Expressions for instantaneous e.m.f.s in stator coils with respect to neutral is

$$V_{s1} = K_t K_1 \cos \theta\, V_r \sin \omega\, t = K V_r \cos \theta \sin \omega\, t \quad (3.38)$$

$$V_{s2} = K_t K_1 \cos (\theta - 120°)\, V_r \sin \omega t = K V_r \cos (\theta - 120°) \sin \omega t \quad (3.39)$$

$$V_{s3} = K_t K_1 \cos (\theta - 240°)\, V_r \sin \omega t = K V_r \cos (\theta - 240°) \sin \omega t \quad (3.40)$$

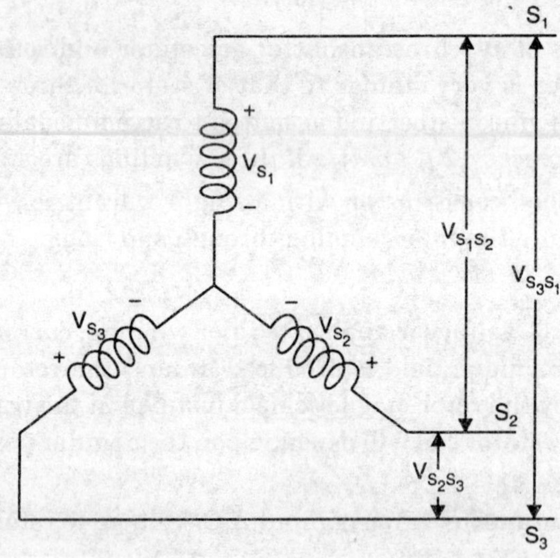

Fig. 3.17 (c) Induced e.m.f in stator coils

Expressions for coil instantaneous e.m.f.s

$$V_{s1s2} = V_{s1} - V_{s2} = KV_r \sin \omega t (\cos \theta - \cos (\theta - 120°))$$

$$= \sqrt{3}\, K V_r \sin (\theta + 120°) \sin \omega t \quad (3.41)$$

$$V_{s2s3} = V_{s2} - V_{s3} = KV_r \sin \omega t (\cos (\theta - 120°) - \cos (\theta - 240°))$$

$$= \sqrt{3} K V_r \sin \theta \sin \omega t \quad (3.42)$$

$$V_{s3s1} = V_{s3} - V_{s1} = KV_r \sin \omega t (\cos (\theta - 240°) - \cos \theta)$$
$$= \sqrt{3}\ K\,V_r \sin (\theta + 240°) \sin \omega t \tag{3.43}$$

When $\theta = 0$ (reference position) in eqn (3.38), induced e.m.f in stator coil S_1 is maximum. But from eqn. (3.42), it is observed that the voltagse V_{s2s3} is zero. This position of rotor is known as electrical zero position of the rotor.

Synchro transmitter will convert the angular position of rotor shaft (input) into a set of three stator coil-to-coil voltages. By measuring the set of three stator coil to coil voltages, the rotor position can be determined.

(*ii*) **Synchro control transformer:**

Construction Details:

The constructional diagram and its equivalent electric circuit are shown in Figs. 3.17 (*d*) and Fig. 3.17 (*e*) respectively.

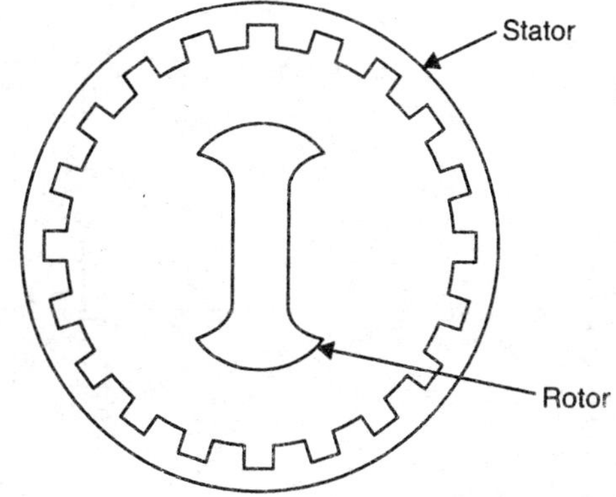

Fig. 3.17 (d)

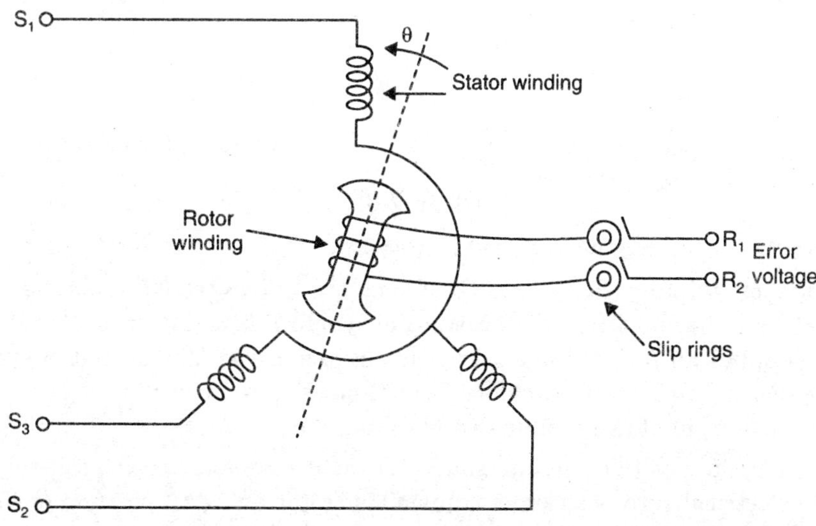

Fig. 3.17 (e)

The construction of synchro control transformer is very similar to that of synchro transmitter, except the shape of the rotor. The rotor of control transformer is in cylindrical shape, such that the air gap between stator and rotor is uniform. With this type of construction of rotor, rotor impedance is approximately constant with rotation of the shaft.

Operating Principle:

The induced e.m.f in the stator coils of synchro transmitter is applied to the stator coils of synchro control transformer. The rotor shaft is connected to the load side, whose position is to be controlled. The magnitude of induced e.m.f in the rotor winding of control transformer will depend upon the applied e.m.f to the stator and current position of the rotor. The induced e.m.f in the rotor is used to drive the motor (load) whose position can be controlled.

(iii) **Synchro pair as error detector:**

The synchro pair (synchro transmitter and synchro control transformer) will be used as error detector by interconnecting synchro transmitter and synchro control transformer as shown in the Fig. 3.17 (*f*).

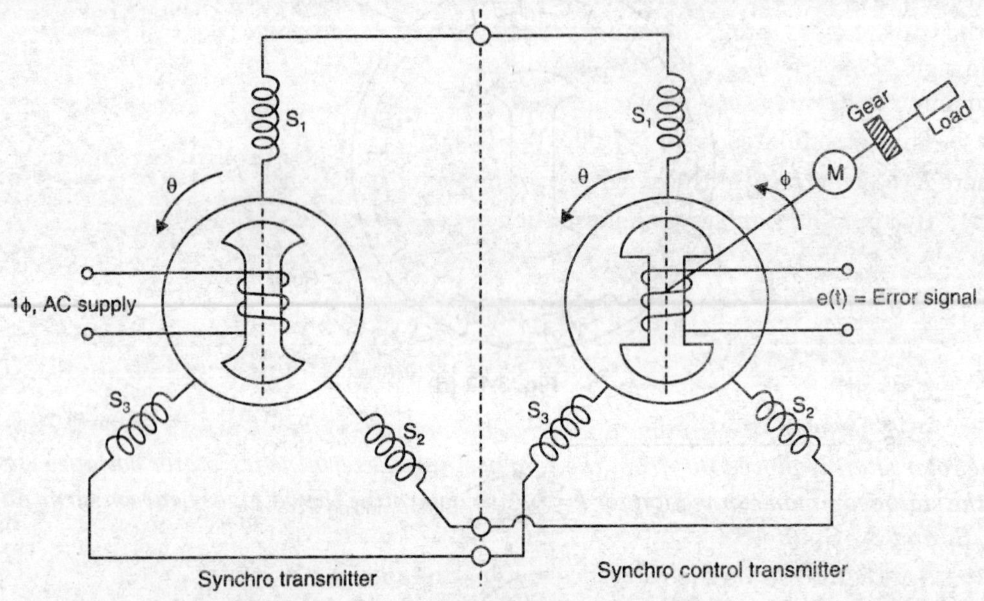

Fig. 3.17 (f)

synchro transmitter synchro control tramsitter

The loads of winding S_1, S_2 and S_3 of synchro control transformer are shown in the Fig. 3.17 (*f*). The angular position of rotor shaft of synchro transmitter is the reference input and this rotor supplies an A.C. voltage of angular frequency 'ω'. The output of synchro transmitter is given to synchro control transformer as input and the e.m.f. induced in rotor of control transformer is output of the system (error signal, etc.).

First, the angular positions of rotor shafts of synchro transmitter and control transformer are adjusted to electrical zero position of transmitter rotor and null position of control transformer respectively. If the rotor of synchro transmitter rotates through an angle 'θ' from its electrical zero position, then rotor of the control transformer will rotate through an angle 'ϕ' in the same direction. The net angular difference between the two rotors is equal to $\{90° + (\theta - \phi)\}$ and the e.m.f. induced in the rotor of control transformer is proportional to the cosine of the angular difference between two rotors.

The modulated error voltage signal across the rotor terminals of control transformer is given by

$$V = AV_r \cos (90 - \theta + \phi) \sin \omega t$$
$$= AV_r \cos (90 - (\theta - \phi) \sin \omega t$$
$$= AV_r \sin (\theta - \phi) \sin \omega t$$

where 'A' is the proportionality constant.

Let $\alpha(t) = (\theta - \phi)$

For small values of $(\theta - \phi)$

$$\therefore \quad V = AV_r\, \alpha\,(t) \sin \omega t \tag{3.44}$$

From the above equation, it is concluded that the output voltage of the synchro error detector is a modulated signal with carrier frequency ω which is equal to the supply frequencies of the transmitters rotor. The magnitude of the modulated carrier wave is proportional to $\alpha(t)$. The signal conditioning circuit demodulates the voltage signal and produces a demodulated and amplified error signal which drives the motor.

The demodulated error voltage, $V = K_s\, \alpha(t)$ (3.45)

where K_s = sensitivity of the synchro error detector in volt/deg.

By applying Laplace transformation to the above eqn. (3.45).

$$V(s) = K_s\, \alpha(s)$$

$$\frac{V(s)}{\alpha(s)} = K_s \tag{3.46}$$

Equation (3.46) represents the transfer function of synchro error detector.

Problem 3.4. *The voltage applied to the rotor of synchro control transmitter is 28V rms. The rotor shaft is moved 60° from the zero position. Determine the stator voltages with respect to the stator common connection for k= 1. Also determine the voltages between terminals S_1 and S_2, S_2 and S_3, S_3 and S_1.*

Solution: If S_1 is the reference stator winding

$$ES_2 = kE_R \cos \theta$$

where E_{S2} and E_R are r.m.s voltages.

$$E_{s_2} = 1 \times 20 \times \cos 60 = 14 \text{ V}$$
$$E_{s_1} = k\, E_R \cos (8 - 120) = 14 \text{ V}$$
$$E_{s_3} = k\, E_R \cos (\theta + 120) = -28 \text{ V}$$
$$E_{s_{31}} = \sqrt{3}\, k\, E_R \sin \theta = 42 \text{ V}$$
$$E_{s_{12}} = \sqrt{3}\, k\, E_R \sin (\theta - 120) = -42 \text{ V}$$
$$E_{s_{23}} = \sqrt{3}\, k\, E_R \sin (\theta + 120) = 0 \text{ V}$$

3.4.3 Potentiometers

A potentiometer is a variable resistance whose value varies according to angular position of the wiper contact. Slider contacts as in a rheostat are linear.

$$\frac{R}{R_T} = \frac{\theta_C}{\theta_T}$$

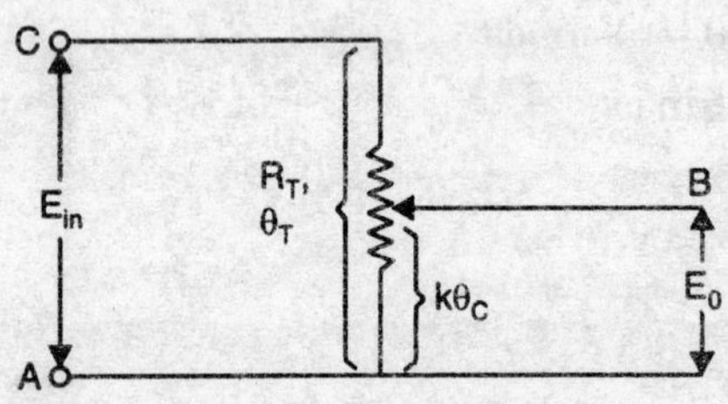

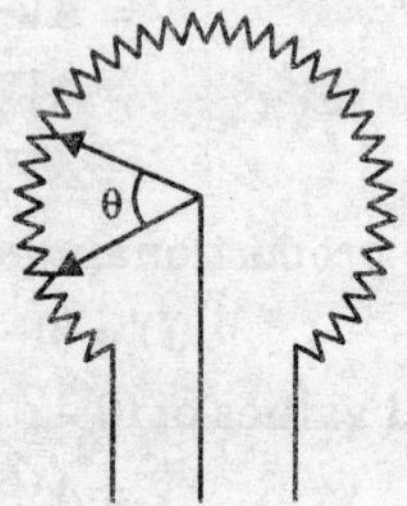

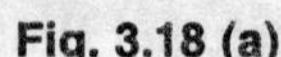

Fig. 3.18 (a)

The variation of resistance is directly proportional to the angular position. Also as the current flow through the resistance is uniform, the voltage is proportional to resistance.

$$\frac{R}{R_T} = \frac{\theta_C}{\theta_T} = \frac{E_0}{E_{in}}$$

$$E_0 = \frac{\theta_C}{\theta_T} \times E_{in} = \text{gain} \times \text{input} \tag{3.47}$$

So the shaft position is an indication of two output voltage and can be calibrated accordingly. The input is θ_c radians or degrees and gain $\frac{E_i}{\theta_T}$ = volts/radian so that the product output E_0 is volts.

The potentiometers are generally circular resistance element. One turn is 2π radians and for N turns $2\pi N$.

$\frac{E_i}{\theta_T}$ is the gain and can be varied for a fixed input E_i by varying θ_T ($\theta_T = 2\pi N$).

$$\text{Gain} = \frac{E_i}{2\pi N} = \text{volt/radian} \tag{3.48}$$

The ideal characteristic for a potentiometer is linear variation of resistance with angular position. This is best relied by having a large diameter, more number of turns and high resistance elements.

In wiping the turns the wiper makes simultaneous contact of adjacent turns to avoid discontinuity in output. Consequently the output is in the form of staircase steps. If there are 100 steps, each step will be 1% of total range. This is called **resolution.**

$$\therefore \quad \text{Percent resolution} = \frac{100}{\text{No. of turns of wire}}$$

The resolution of potentiometer is an important factor in the determination of minimum value of error signal. In addition, there will be loading effect on potentiometer, by connecting to the input of an amplifier.

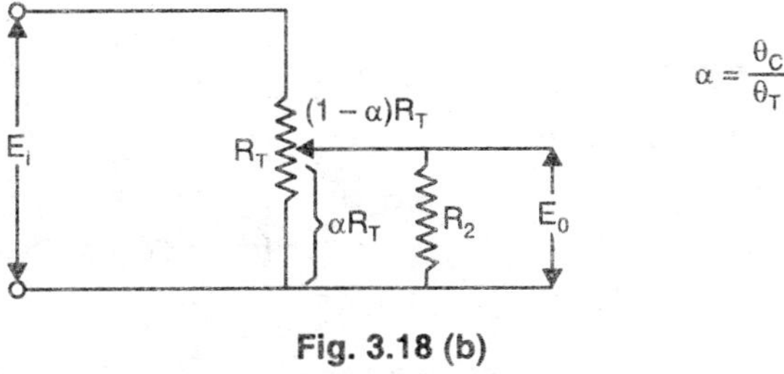

$$\alpha = \frac{\theta_C}{\theta_T}$$

Fig. 3.18 (b)

The resistance due to load at an angular position is

$$R = (1-\alpha)\,R_T + \frac{\alpha R_T R_L}{R_L + \alpha R_T} \tag{3.49}$$

The output voltage $E_0 = I\left(\dfrac{\alpha R_T R_L}{\alpha R_T + R_L}\right)$

$$= \frac{E_i}{R}\,\frac{\alpha R_T R_L}{\alpha R_T + R_L} \tag{3.50}$$

Substituting R from eqn. (3.49) in eqn. (3.50)

$$\therefore \qquad E_0 = \frac{\alpha E_i}{1+\alpha(1-\alpha)\dfrac{R_T}{R_L}}$$

Loading error

$$= \alpha\, E_i - \frac{\alpha E_i}{1+\alpha(1-\alpha)\dfrac{R_T}{R_L}}$$

$$\text{Error} = \left[\frac{\alpha^2\,(1-\alpha)}{\alpha(1-\alpha)\dfrac{R_L}{R_T}}\right] E_i \tag{3.51}$$

There are variations to the single ended potentiometer discussed so far. One variation is a double ended arrangement with the midpoint earthed. Voltage of opposite polarity is applied to the ends. The double ended potentiometer can be considered as two single ended potentiometers with their common point earthed.

$$E_0 = \frac{E_i}{\theta_T} \times (\theta_1 - \theta_2) \tag{3.52}$$

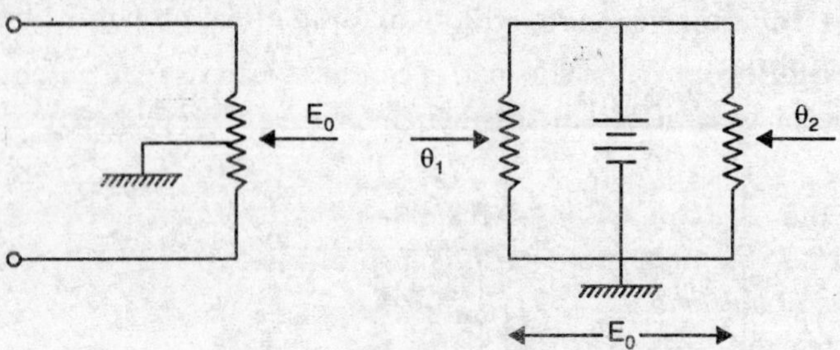

Fig. 3.18 (c)

These types of potentiometers are used in servomechanisms for error control.

The potentiometer also functions as an attenuator so that $E_0 \le E_i$. As an attenuator, gain constant or potentiometer constant is given by

$$K_P = \frac{E_i}{\theta_T} = \frac{\text{Input of excitation voltage}}{\text{Total angle of rotation}} \tag{3.53}$$

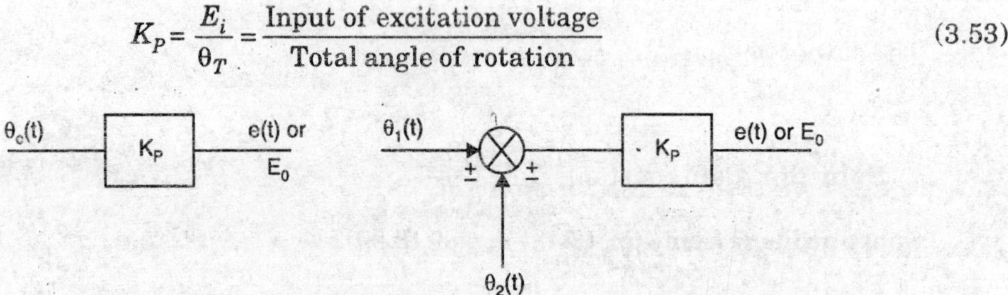

Fig. 3.18 (d) Block diagrams of single ended and double ended potentiometers

Performance of a potentiometer is specified by following characteristics:

(*i*) **Resolution:** It is the ratio of minimum change in output voltage to a total voltage applied to it.

$$\% \text{ Resolution} = \frac{\Delta E_0}{\Delta E_i} \times 100 \tag{3.54}$$

ΔE_0 = Change in output voltage, range from 0.5 to 0.02%

ΔE_i = Input voltage applied.

(*ii*) **Lift:** It is defined as maximum number of cycles of operation in which none of the electrical or mechanical characteristic depart from normal values by more than 50%.

(*iii*) **Linearity:** It is defined as the maximum deviation of actual curve from theoretical curve expressed as a percentage of applied voltage.

$$\% \text{ Linearity} = \frac{\Delta E_{max}}{\Delta E_i} \times 100, \tag{3.55}$$

Range from 0.5 to 0.02%

(*iv*) **Noise:** It indicates presence of various voltages. It is due to ripple caused by brush jump, vibration, stray capacitance's, etc.

(*v*) **Loading Error:** When the load is connected across the potentiometer, the resistance of the potentiometer gets affected, this causes error in its output which is called as loading error.

Problem 3.5. *A 10 K Ohm potentiometer is operated with a grounded centre tap as shown in Fig. 3.19. If the potentiometer has a rotation angle of 350°, what in the gain constant expressed in units of volt/rad.*

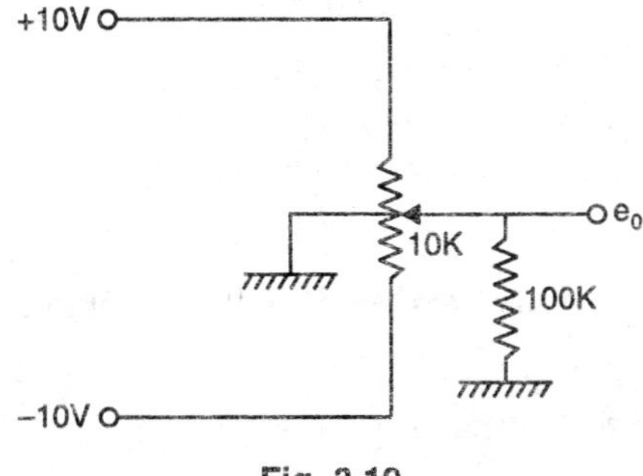

Fig. 3.19

Solution: Gain constant, $K_p = \dfrac{E_i}{\theta_{max}} = \dfrac{20}{350 \times \dfrac{2\pi}{360}} = 3.28$ V/rad

Problem 3.6. *Determine the number of turns of wire needed to provide a potentiometer with a resolution of 0.05 percent.*

Solution: Percent resolution = $\dfrac{100}{\text{Number of turns}}$

A multi turn potentiometer with many turns of wire will have a very good resolution.

$$0.05 = \frac{100}{N} \quad \Rightarrow \quad N = 2{,}000 \text{ turns.}$$

3.4.4 Tachometers (Tacho Generators)

Tachometer is an electromechanical unit which generates an electrical output proportional to the speed of the shaft. The tachometers are generally classified into two types. They are

1. DC tachometer and
2. AC tachometer.

3.4.4.1. DC tachometer

A tachometer is a miniature low voltage generator as shown in Fig. 3.20.

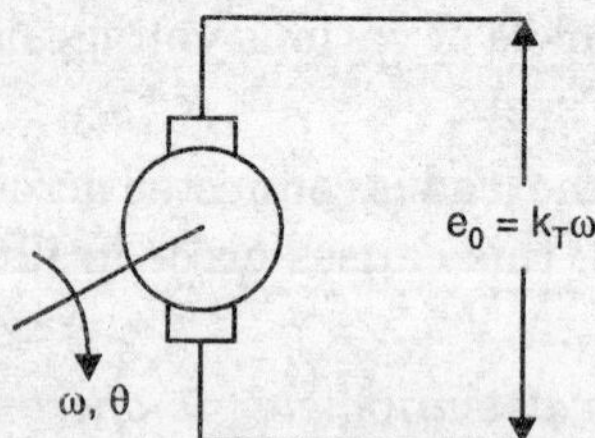

Fig. 3.20 Schematic diagram of DC tachometer

The output voltage of a generator is given by

$$E_g = K\,\phi\,\omega$$

where K = generator constant

ϕ = flux per pole

ω = armature speed

The output voltage generated is less by the drop in armature.

$$V_0 = E_g - I_a r_a \tag{3.56}$$

So the voltage output has a linear relationship with speed of rotation as the flux ϕ is a constant for the particular construction.

In a tachometer, the flux is produced by permanent magnets. As the output of the generator is AC, it is converted into DC by split commutator. But the DC output is derated by harmonics. To reduce it, more segments of commutator are used. Besides, the tachometer is operated at the highest possible speed which gives a larger output signal.

In servo systems, the tachometer is used as a transducer for obtaining signals proportional to speed. The **sensitivity or tachometer gain** is defined as output voltage divided by shaft speed in radians/sec.

$$K_T = \frac{E_0}{\omega}$$

$$E_0 = K_T\,\omega \tag{3.57}$$

where E_0 = output voltage

K_T = tachometer constant

ω = angular speed in rad/sec

3.4.4.2 AC tachometer

AC tachometers are called drag cup tachometers. It contains two stator windings at right angles to each other as shown in Fig. 3.21. One stator winding is excited by an AC reference voltage of fixed amplitude and frequency. The output is taken from the other stator winding. The rotor is in the form of a cup of non-magnetic material.

When the rotor cup is not rotating, there is no flux linkage in the stator windings as they are perpendicular. As the cup rotates, eddy currents are induced that generate a rotor flux. The rotor flux adds to the reference flux so that the direction of total flux is shifted. This results in an induced output voltage proportional to velocity. AC tachometers have good linearity and accuracy up to 0.1%.

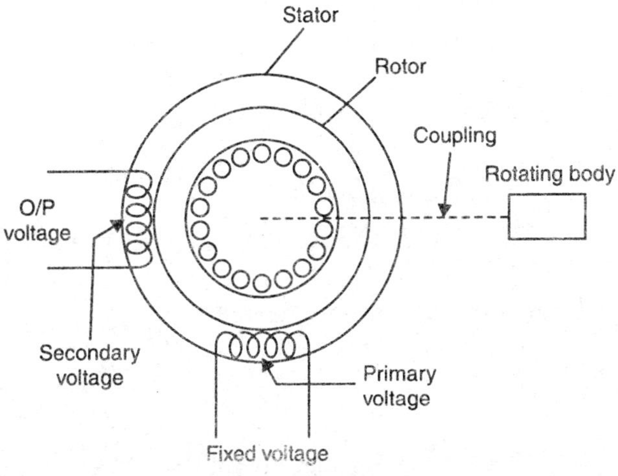

Fig. 3.21

The tachometer gain or sensitivity K_T is expressed in volts/angular velocity. In practice the output voltage is specified at a speed of 1000 rpm, values from 1 to 10 V are typical for DC tachometers.

$$E_0(t) = K_T \; \omega(t) = K_T \frac{d\theta}{dt}$$

Taking Laplace transform for the above equation

$$E_0(s) = K_T \, s \, \theta(s)$$

Transfer function, $G(s) = \dfrac{E_0(s)}{\theta(s)} = K_T \, s$ (3.58)

Problem 3.7. *A tachometer has a sensitivity of* $\frac{5}{1000}$ *volts per rpm. Express the gain constant of the tachometer in units of volts / radian / sec.*

Solution:

Gain constant $K_T = \dfrac{E_0}{\omega} = \dfrac{5}{1000} = \dfrac{5}{\frac{1000}{60}} = 0.3$ V/rps

$$= \frac{3}{10 \times 2\pi} = 0.048 \text{ V/rps}$$

Problem 3.8. *A tachometer has a gain of 0.05 V/(rad/s). Find the following:*

(*a*) *Output voltage when the shaft speed is 40 deg/sec.*

(*b*) *The output voltage when the shaft speed is 20 rad/sec.*

(*c*) *The shaft speed in radians/sec and deg/sec when the output voltage is 1.8 V.*

Solution: (*a*) $\frac{40}{180} \times \pi$ radians.

$$\text{Output voltage} = \frac{0.05 \times 40 \times \pi}{180} = 0.035$$

(*b*) Output voltage = 0.05 × 20 = 1 V

(*c*) Shaft speed rad/sec

$$\text{Gain constant} \quad K_g = \frac{E_0}{\omega}$$

$$\omega = \frac{E_0}{K_g} = \frac{18}{0.05} = 36 \text{ rad/sec}$$

3.4.5 Magnetic Amplifier

It has rugged construction and does not require heater supply unlike electronic valves. Good power amplification can be obtained. Disadvantages are pulsed output and drift due to temperature rise. These are generally used in the control of generator voltage, control of generator power.

It works on the principle of magnetic saturation. It has windings on a core just like transformer. Core is made up of high nickel alloy steels such as mumetal. The B–H curve of the mumetal is shown in Fig. 3.22.

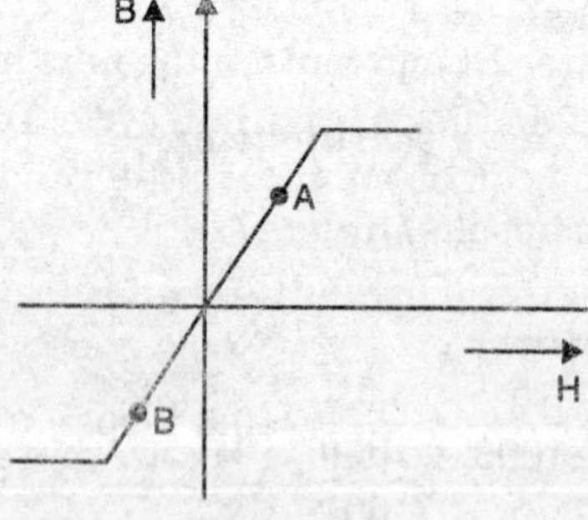

Fig. 3.22

The principle of operation is explained with the help of Fig. 3.23.

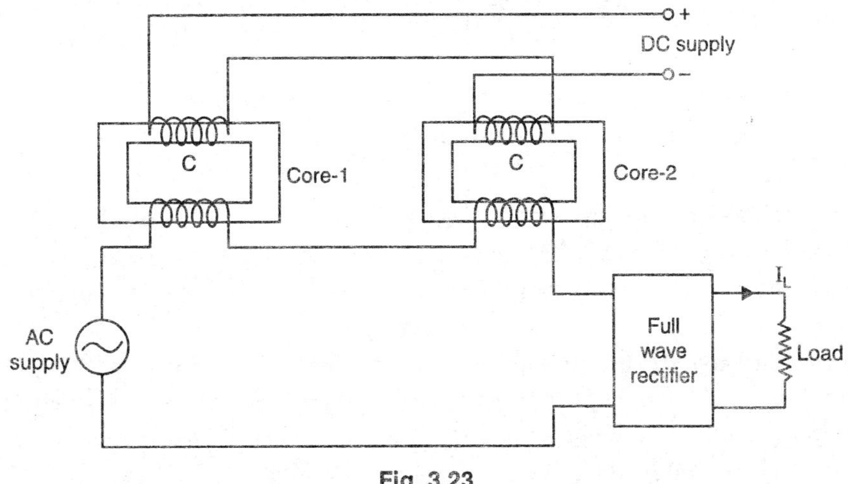

Fig. 3.23

There are two magnetic cores on each of which primary and secondary windings are wound. The two primary windings connected in series and excited by a DC supply, set up equal and opposite polarity fluxes in the two cores. Core 1 is at point '*A*' and core 2 at '*B*' in Fig. 3.22 indicate the operating points on the respective $B - H$ curves. The primary windings are called control windings. The secondary windings are load windings. They are connected in series and are supplied with AC supply. They are closed through full wave rectifier with DC load.

During the positive half cycle, the flux in the core 1 drives it to saturation (near the point A) and so there is no back emf induced in its secondary. At the same time, the flux in the core 2 can increase from the point B. Thus a back emf is induced in its secondary. Due to transformer action, an emf is also induced in its secondary. Due to transformer action an emf is induced in its primary winding and an alternating current I_{ac} is circulated through it. In Fig 3.24 some voltage will be dropped across its secondary and as a result, the net voltage across the rectifier increases causing a greater DC current. In the next half cycle core 2 is saturated, core 1 behaves the same way as core 2 in the positive half cycle. Due to the cross connection of the primary windings, the alternating current in the control winding is same. Thus there is a small increase in the average value (DC value) of the current through the control winding in both half cycles and a large value of direct current through the load.

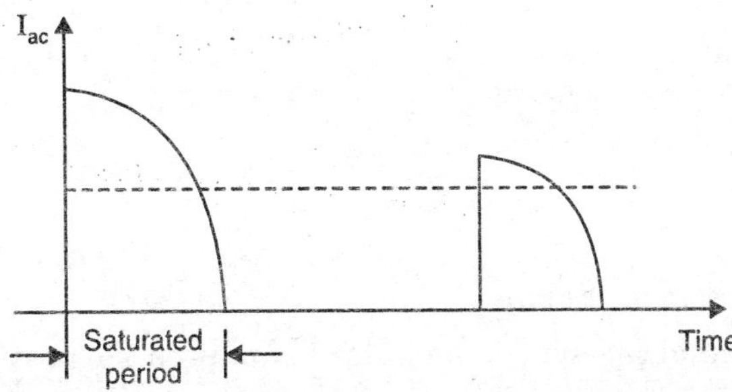

Fig. 3.24

Power amplification is given by $\left(\frac{\partial i_L}{\partial i_c}\right)^2 \frac{R_L}{R_c}$ (3.59)

where ∂i_L is the increase in the DC load current

∂i_c is the increase in the control winding current.

R_L is DC load resistance and

R_c is the control circuit resistance.

The direct current induced in the core reduces its magnetization bias and thus instant of saturation is reduced. This results in reduced value of the increase of load current. So the amplification is reduced.

In order to increase the amplification, positive feedback is provided to the DC winding from the secondary circuit through another rectifier. Thus is called 'Self excited magnetic amplifier'. In another kind of magnetic amplifier, the necessity of positive feedback is avoided by connecting the two secondaries in parallel instead of series.

SHORT QUESTIONS AND ANSWERS

1. What is mean by feedback element?

The feedback element is a device which converts the output variable into another suitable variable, which means the controller consists of an error detector and control elements.

2. What is the function of error detector?

The error detector compares the feedback signal obtained from the plant output with the input signal (command).

3. What is the need of feedback?

Feedback gives automatic regulation and control as well as reduces the sensitivity of the system to parameter variations.

4. What are the advantages and disadvantages of feedback in control system?

Advantages

- Reduces the sensitivity of the system.
- Improving the transient response and minimizing the effects of disturbance signals in control systems.
- The system can be made automatic.
- The system control can be made accurate.

Disadvantages

- The use of feedback increases the number of components of the system, thereby increasing its complexity further.
- It reduces the gain of the system and also introduces the possibility of instability.

5. **Why negative feedback invariably preferred in a closed loop system?**

The negative feedback results in better stability in steady state and rejects any disturbance signals. It also has low sensitivity to parameter variations. Hence, negative feedback is preferred in closed loop system.

6. **Why is positive feedback not useful in control system applications?**

An oscillatory response is not desirable. Positive feedback results in non minimization of error signal.

7. **What are the conditions on the loop gain to be satisfied for sustained oscillations to take place in the feedback loop?**

Magnitude of the loop gain is one and phase angle of the loop transfer function is zero.

8. **What is the effect of positive feedback on stability?**

The positive feedback increases the error signal and drives the output to instability.

9. **What are requirements of a good servomotor?**

Good servomotor requires

- Linear relationship between electrical control signal and the rotor speed.
- Its response should be as fast as possible
- Inertia of rotor should be as low as possible
- Operation should be stable without any disturbances.

10. **What are the types of servomotors?**

Depending upon the nature of thc supply, servomotors are classified into two types and they are

- DC servomotors and
- AC servomotors

11. **What is the difference between DC motor and DC servomotor?**

Ans. The difference between DC motor and DC servomotor is that all DC servomotors are separately excited types.

12. **What is the difference between AC servomotor and induction motor?**

The rotor of the AC servomotor is built with high resistance so that its $\frac{X}{R}$ ratio is small and the torque-speed characteristic is nearly linear in contrast to the highly non- linear characteristic with large $\frac{X}{R}$ ratio of induction motor.

13. **What is a synchro?**

It is an electromechanical transducer that produces an output voltage depending upon angular displacement.

14. **What are the basic elements of synchros?**

There are four basic elements of synchros such as transmitter, receiver, control transformer and differential transformer.

15. What is a potentiometer?

A potentiometer is a variable resistance whose value varies according to angular position of the wiper contact.

16. What is a tachometer?

Tachometer is an electromechanical unit which generates an electrical output proportional to the speed of the shaft.

17. How do you improve the resolution of a resistance potential divider used for position sensing?

By increasing the number of turns of the resistance wire so that the voltage per turn is decreased.

18. What is the advantage of using a tachogenerator over a tachometer?

A tachometer gives the velocity measurement on the spot. A tachogenerator gives the velocity measurement which can be transmitted to a distant place or can be used for controlling the velocity.

19. When do you use magnetic amplifier?

Magnetic amplifiers are preferred in rough environment conditions such as in missile guidance system.

20. How do you reverse the direction of rotation in AC servomotor?

By reversing the polarity of the voltage applied to the control winding.

OBJECTIVE TYPE QUESTIONS

1. In open loop system, sensitivity w.r.t. G is equal to

   (*a*) Zero (*b*) $\dfrac{G}{1+GH}$

   (*c*) Unity (*d*) $\dfrac{-1}{GH}$

2. In closed loop system, sensitivity w.r.t. G is equal to

   (*a*) Zero (*b*) $\dfrac{1}{1+GH}$

   (*c*) Unity (*d*) $\dfrac{-1}{GH}$

3. The effect of feedback reduces the

   (*a*) Sensitivity of the system (*b*) Non-linearity present in the system

   (*c*) Both (*a*) and (*b*) (*d*) Disturbance in the system

   (*e*) All of the above

4. The open loop transfer function $G(s) = \dfrac{K}{s+\alpha}$, by providing feedback, the time constant of the system is reduced by

(*a*) $-\frac{1}{T}$ (*b*) $-\frac{t}{\alpha}$

(*c*) $-\frac{t}{\alpha + K}$ (*d*) $\frac{-1}{t+\alpha}$

5. The effect of positive feedback
   (*a*) Reduces the gain (*b*) Increases the gain
   (*c*) Does not affect the gain (*d*) None of these
6. Synchro is basically a
   (*a*) Two phase induction motor (*b*) Three phase alternator
   (*c*) Three phase induction motor (*d*) Transformer
7. DC servomotor works on the principle of
   (*a*) Faraday's law of electromagnetic induction
   (*b*) Fleming right hand rule (*c*) Error signal
   (*d*) Kirchoff's law (*e*) None of these
8. DC servomotor is basically a
   (*a*) An AC motor (*b*) DC generator
   (*c*) Self excited DC motor (*d*) Separately excited DC motor
9. AC servomotor is basically a
   (*a*) 3ϕ induction motor (*b*) 2ϕ induction motor
   (*c*) 3ϕ alternator (*d*) Transformer
10. AC servomotor differs with normal induction motor in
    (*a*) small $\frac{X}{R}$ ratio (*b*) Large $\frac{x}{R}$ ratio
    (*c*) Linear speed-torque (*d*) Both (*a*) and (*c*)
11. A.C servo motor resembles-------
    (*a*) two phase induction motor (*b*) three phase induction motor
    (*c*) DC series motor (*d*) Universal motor
12. As a result of introduction of negative feedback which of the following will not decrease
    (*a*) band width (*b*) overall gain
    (*c*) distortion (*d*) instability
13. Regenerative feedback implies feedback implies feedback with
    (*a*) oscillations (*b*) step input
    (*c*) negative sign (*d*) positive sign
14. .... is an open loop control system
    (*a*) Ward Leonard control (*b*) field controlled DC motor
    (*c*) armature controlled DC motor (*d*) none of these

15. The band width, in a feedback system
    (*a*) remains unaffected
    (*b*) decreases by the same amount as the gain increases
    (*c*) increases by the same amount as the gain decreases
    (*d*) none
16. On which of the following factors does the sensitivity of a closed loop system to gain changes and load disturbances depend
    (*a*) frequency (*b*) loop gain (*c*) forward gain (*d*) all of the above
17. With the feed back --------increases
    (*a*) system stability (*b*) sensitivity
    (*c*) gain (*d*) effects of disturbing signals

KEY

1. (*c*)	2. (*b*)	3. (*e*)	4. (*c*)	5. (*a*)	6. (*b*)
7. (*c*)	8. (*d*)	9. (*b*)	10. (*d*)	11. (*a*)	12. (*a*)
13. (*d*)	14. (*b*)	15. (*c*)	16. (*d*)	17. (*a*)	

EXERCISE

1. What is meant by the feedback system?
2. Explain the reduction of parameter variation by feedback.
3. What do you mean by the sensitivity of the control system and discuss the effect of feedback on sensitivity?
4. Explain the effect of feedback on stability of the control system.
5. What are the merits and demerits of feedback control system?
6. What are the requirements of good servomotor?
7. Derive the transfer function and develop the block diagram of field controlled DC servo motor.
8. Derive the transfer function and develop the block diagram of armature controlled DC servomotor.
9. What are the advantages of AC servomotor over induction motor?
10. Explain the AC servomotor and draw its torque and speed characteristics.
11. Explain the construction and operation principle of synchro transmitter.
12. Describe the synchro pair as error detector with circuit diagram.
13. Explain the AC tachometer with circuit diagram.

4 Time Response Analysis

4.1 INTRODUCTION

Time domain analysis analyses the system response given by the system for the applied excitation which is the function of time. Analysis means how the system response relies w.r.t. time for a given input. For satisfactory performance of the system, output behaviour must be within the specified limits. Every system will have transient state before going to its steady state due to energy storage elements.

This chapter deals with the manner in which the first and second order control system respond to input signals during the period before steady state is reached.

In order to obtain the transient response or time domain response, we have to use mathematical model *i.e.*, the differential equation description of the system. The study of transient response of a control system is very important and it describes the dynamic behaviour of the system and the deviation between the response and the input.

Time Response

The response given by the system which is function of the time to applied excitation, is called time response.

The time response of a control system is usually divided into two parts.

i.e., $$C(t) = C_{tr}(t) + C_{ss}(t)$$

where $C_{tr}(t)$ = transient response

$C_{ss}(t)$ = steady state response

Transient Response $C_{tr}(t)$: The output variation during the time it takes to achieve its final value is called as transient response. The time that is required to achieve the final value is called transient period, or the transient response is defined as that part of the time response that goes to zero as time becomes large.

i.e., $$\underset{t \to \infty}{\text{Lt}}\; C_{tr}(t) = 0$$

Steady state Response: It is the part of the time response which remains after complete transient response has vanished from the system output.

The steady state response, when compared with the input or the desired response indicates the final accuracy of the system. The difference is the steady state error.

Steady state error: It is the difference between the desired output and actual output of the system at infinite time and is denoted by e_{ss}.

$$e_{ss} = \underset{t \to \infty}{\text{Lt}} \; (r(t) - c(t))$$

$r(t)$ = desired output

$c(t)$ = actual output

The system describing transient and steady state response of the system is shown in Fig. 4.1.

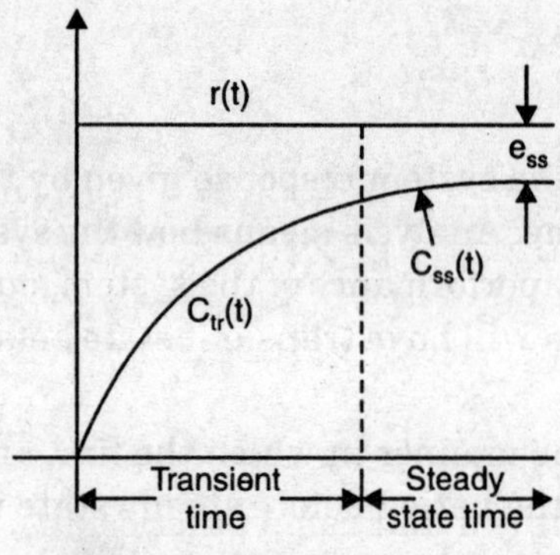

Fig. 4.1

4.2 TIME RESPONSE OF DIFFERENT INPUTS

Until now, we have developed a mathematical model of a system from its physical laws and represented by a block diagram. But we should relate theory to practice. But in practice many signals are available which are the functions of time and can be used as reference inputs, for various control systems. These signals are step, ramp, parabolic and impulse inputs etc.

These signals are most commonly used as reference inputs and are defined as standard test signals

(*i*) **Step input.** The most widely used input is the step input or function in testing dynamic behaviour of control system. Fig. 4.2 shows a step input.

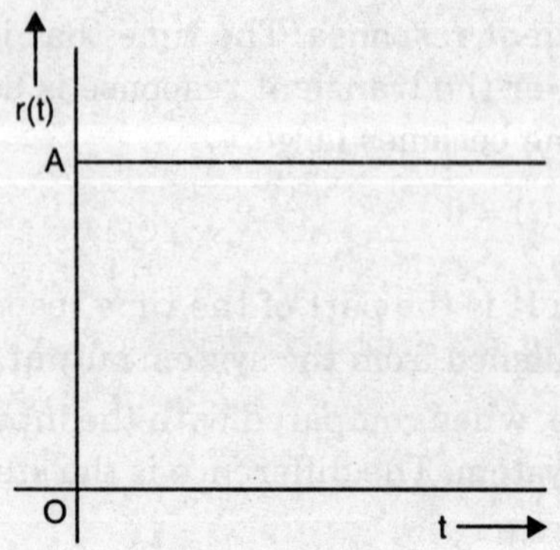

Fig. 4.2 Step input function

In a step input, all prior transients have completely died out at time $t = 0$, the input abruptly rises to a value of magnitude A and remains at that value until all transients caused by the response of the system to step of magnitude input have died out.

Mathematically, $r(t) = 0$, for $t < 0$

$r(t) = A$, for $t \geq 0$

It A is assigned a value of 1, it is a unit step function $r(t)$

i.e., $r(t) = A = 1$

Applying Laplace transform

$$L[r(t)] = L[1]$$

$$R(s) = \frac{1}{s}$$

(*ii*) **Ramp Input.** Prior to $t = 0$, $r(t) = 0$ and all transients have died out. From $t \geq 0$ the value of the function rises uniformly at a constant rate B, called the slope of the ramp, as shown in Fig. 4.3.

Mathematically, $r(t) = 0$, for $t < 0$

$r(t) = B\,t$, for $t \geq 0$

It B is assigned a value of 1, it is a unit ramp function $r(t)$

i.e., $r(t) = B\,t = t$ [since $B = 1$]

Applying Laplace transform

$$L[r(t)[= L\,[t]$$

$$R(s) = \frac{1}{s^2}$$

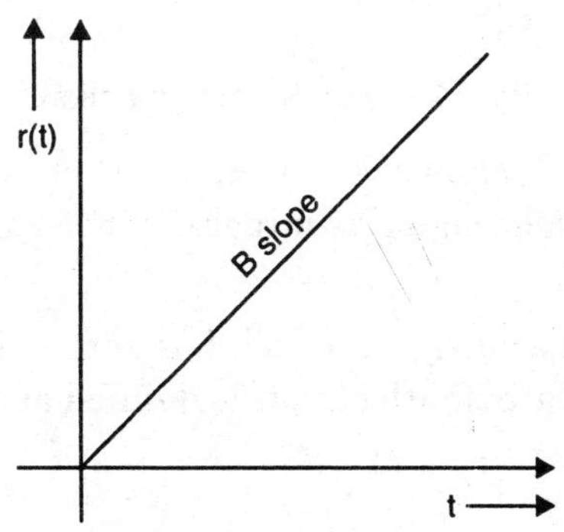

Fig. 4.3 Ramp input function

(*iii*) **Parabolic input.**

A unit Parabolic input is defined mathematically as

$$r(t) = 0, \text{ for } t < 0$$

$$= \frac{Ct^2}{2}, \text{ for } t \geq 0$$

It C is assigned a value of 1, it is a unit parabolic function $r(t)$

i.e., $$r(t) = \frac{t^2}{2}$$

Applying Laplace transform on both sides of above expression

$$L[r(t)] = L\left[\frac{t^2}{2}\right]$$

$$R(s) = \frac{1}{s^3}$$

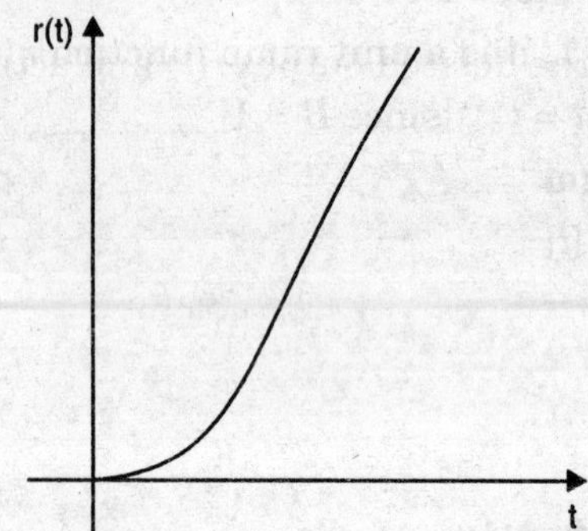

Fig.4.4 Parabolic input function

Graphical representation is shown in Fig. 4.4. It is observed that the ramp is the integral of a step signal and a parabolic signal is integral of the ramp signal.

(*iv*) **Impulse Input**

A unit impulse is defined as a signal which has zero value everywhere except at $t = 0$, where its magnitude is finite. Mathematically, it is defined as

$$\delta(t) = 1, \text{ for } t = 0$$

$$\delta(t) = 0, \text{ for } t \neq 0$$

Laplace transform of impulse input is

$$L[\delta(t)] = 1$$

Impulse function is the derivative of a step function

$$\delta(t) = \frac{dr(t)}{dt}$$

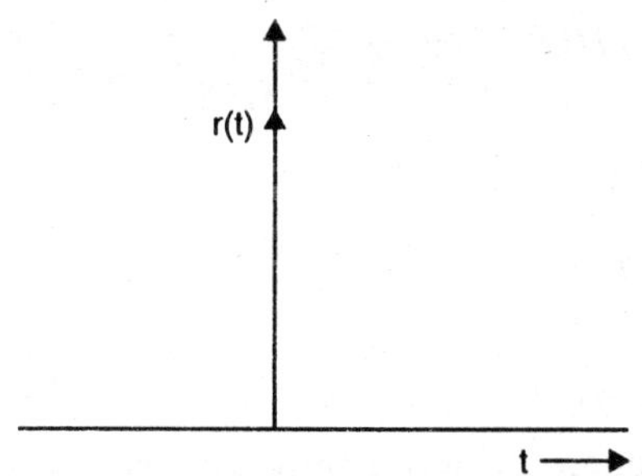

Fig. 4.5 Impulse input

Every system has two states *i.e.*, transient state and steady state due to energy storage elements. To get the desired output, the system must pass satisfactorily through transient period. Transient response must vanish after some time to get the final value closer to the desired value. Such systems in which transients die out after some time, are called stable systems.

4.3 STEADY STATE ERROR

It is the difference between the desired output $r(t)$ and actual output $c(t)$.

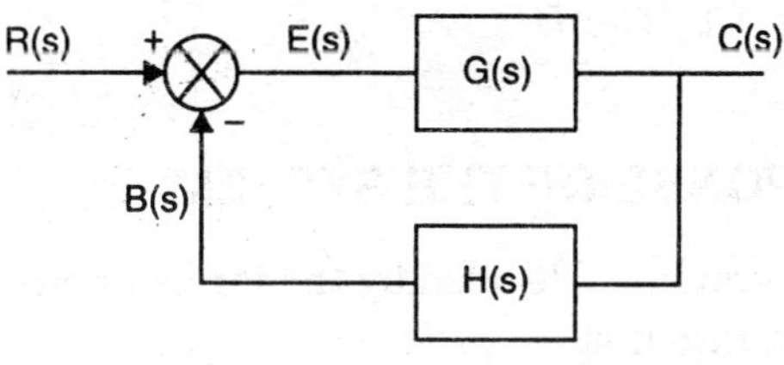

Fig. 4.6

Block diagram representation of closed loop system is shown in Fig. 4.6 with forward path transfer function $G(s)$ and feedback function $H(s)$. Now the reference input tells us the level of the desired output and actual output is fed back through the feedback element and compared with the reference input produces the actuating signal $E(s)$ which is input to the plant or process.

From the Fig. 4.6,

Error signal, $E(s) = R(s) - B(s)$ (4.1)

Feedback signal, $B(s) = C(s)\,H(s)$ (4.2)

Substituting eqn. (4.2) in eqn. (4.1)

$$E(s) = R(s) - C(s)\,H(s) \tag{4.3}$$

From the block diagram shown in Fig. 4.6

Output signal is

$$C(s) = G(s)\,E(s)$$

Substituting $C(s)$ in eqn. (4.3)

Error signal, $E(s) = R(s) - G(s)\,H(s)\,E(s)$

or $$E(s)\,[1 + G(s)\,H(s)] = R(s)$$

$$\therefore \quad E(s) = \frac{1}{1+G(s)H(s)}\,R(s) \tag{4.4}$$

For unit feedback system

$$E(s) = \frac{1}{1+G(s)}\,R(s) \qquad \text{(since } H(s) = 1\text{)}$$

From eqn. (4.4), it has been observed that the error signal is directly proportional to input signal and depends upon the type of the input, type and order of the system.

This $E(s)$ is the error in Laplace domain and is expressed in 's'. When we want to calculate the same error in time domain, corresponding error will be $e(t)$.

Steady state error, $e_{ss} = \underset{t \to \infty}{\text{Lt}}\ e(t)$

Using final value theorem, we can relate this by

$$e_{ss} = \underset{t \to \infty}{\text{Lt}}\ e(t) = \underset{s \to 0}{\text{Lt}}\ (sE(s))$$

Substituting the value of $E(s)$ from eqn.(4.4) in above expression

$$\therefore \quad e_{ss} = \underset{s \to 0}{\text{Lt}}\ \frac{s\,R(s)}{1+G(s)H(s)} \tag{4.5}$$

4.4 TRANSIENT RESPONSE OF THE SYSTEM

The order of a control system is decided by the largest power of 's' contained in it. From Fig. 4.5, the characteristic equation is

$$F(s) = 1 + G(s)\,H(s) \tag{4.6}$$

expressed as polynomial in 's'.

Already we know that steady state error depends upon the type of the input, and order of the system. In this section we will investigate the steady state error for different inputs and different orders of the system.

4.4.1 First Order System

A system which is described by a first order differential equation, is called a first order system. If we solve the equation the solution is an indication of the behavior of the system to an input function. A first order differential equation is of the form

$$f(t) = Ax + B\,\frac{dx}{dt}$$

where $f(t)$ is the known input function, x is the unknown output, and t is the independent variable, A and B are the coefficients.

The solution consists of two parts, one is called the forced or steady state response and the other part is natural or transient response. Both solutions together completely define the total response.

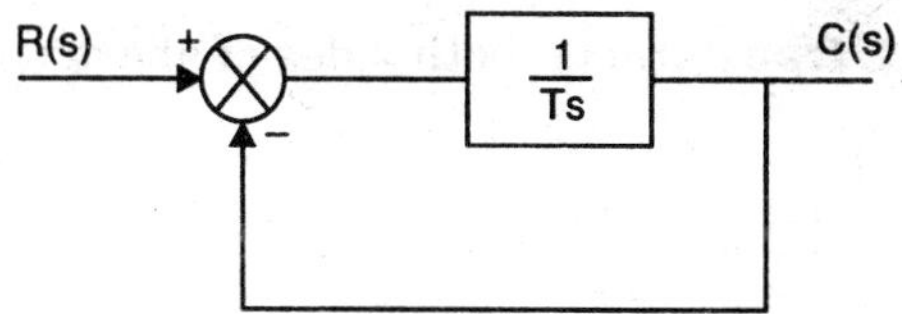

Fig. 4.7 Block diagram for first order system

Let us consider a first order system as shown in Fig. 4.7

where T = Time constant

$C(s)$ = Control output

$R(s)$ = Reference input

Closed loop transfer function after eliminating the unit negative feedback, is

$$\frac{C(s)}{R(s)} = \frac{1}{Ts+1}$$

or
$$C(s) = \frac{R(s)}{1+Ts}$$

We shall analyse the system response to different inputs such as step, ramp, parabolic and impulse.

4.4.1.1 Unit step input

The unit step function is given by

$$r(t) = 1, t > 0$$
$$= 0, t < 0$$

Laplace transform of unit step, $R(s) = \frac{1}{s}$

From the Fig. 4.7, the response of the first order system when it is subjected to unit step input, is

$$C(s) = \frac{1}{s(Ts + 1)}$$

Expanding it into partial fractions

$$C(s) = \frac{A}{s} + \frac{B}{Ts+1}$$

$$A(Ts + 1) + Bs = 1$$

For $s = 0$, $A = 1$

For $s = -\frac{1}{T}$, $B = -T$

$$\therefore \quad C(s) = \frac{1}{s} - \frac{T}{Ts+1} = \frac{1}{s} - \frac{1}{s+\frac{1}{T}}$$

Taking inverse Laplace transform on both sides of above equation

$$L^{-1}(C(s)) = L^{-1}\left[\frac{1}{s} - \frac{1}{s+\frac{1}{T}}\right]$$

$$= L^{-1}\left(\frac{1}{s}\right) - L^{-1}\left(\frac{1}{s+\frac{1}{T}}\right)$$

$$c(t) = 1 - e^{-t/T}, \ t \geq 0$$

Fig. 4.8

At $t = 0$, $c(t) = 0$ starts at origin, as $t \to \infty$, $c(t) = 1$ is the steady state response.

Steady state error, $e_{ss} = \underset{t \to \infty}{\text{Lt}} (r(t) - c(t))$

$$= \underset{t \to \infty}{\text{Lt}} \{1 - (1 - e^{-t/T})\}$$

$$= \underset{t \to \infty}{\text{Lt}} \ e^{-t/T} = 0$$

The time response in relation to above equation is shown in Fig. 4.8.

Thus we can observe from above equation that there is no steady state error for a unit step input applied to a first order system.

4.4.1.2 Unit ramp input

The unit ramp function is given by

$$r(t) = t, t \geq 0$$
$$= 0, t < 0$$

Laplace transform of unit ramp, $R(s) = 1/s^2$. From Fig. 4.7, the response of the first order system when it is subjected to unit ramp input, is

$$C(s) = \frac{1}{s^2 (Ts + 1)}$$

Expanding it into partial fractions

$$C(s) = \frac{A}{s^2} + \frac{B}{s} + \frac{C}{Ts+1}$$

$$A(Ts + 1) + Bs(Ts + 1) + Cs^2 = 1$$

For $s = 0$, $\qquad A = 1$

For $s = \frac{1}{T}$, $\qquad C = T^2$

Taking 's' terms, $AT + B = 0$

$$B = -T$$

$$\therefore \qquad C(s) = \frac{1}{s^2} - \frac{T}{s} + \frac{T^2}{Ts+1}$$

Taking inverse Laplace transform on both sides of above equation

$$L^{-1}[C(s)] = L^{-1}\left[\frac{1}{s^2} - \frac{T}{s} + \frac{T^2}{Ts+1}\right]$$

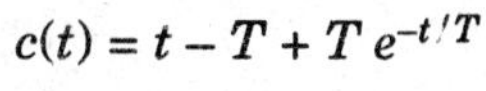

$$c(t) = t - T + T\,e^{-t/T}$$

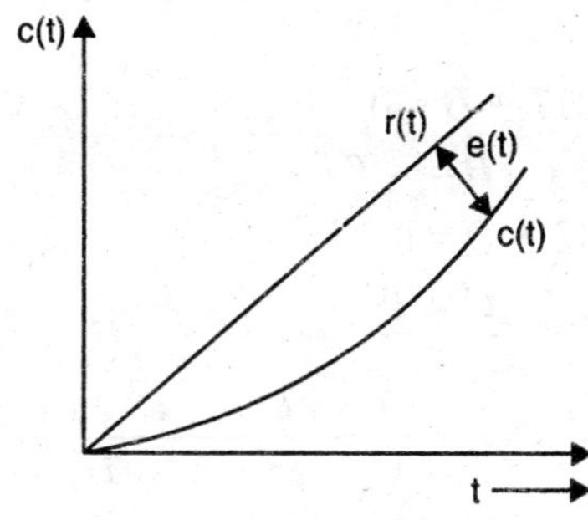

Fig. 4.9

At $t = 0$, $c(t) = 0$, response is zero

At $t = \infty$, $c(t) = \infty$, infinite response

Error signal $\;c(t) = (r(t) - c(t))$

$$= t - t + T - Te^{-t/T}$$

The steady state error, $e_{ss} = \underset{t \to \infty}{\text{Lt}} \{T[1 - e^{-t/T}]\} = T$

The time response in relation to above equation is shown in Fig. 4.9.

Hence steady state error of first order systems when it is subjected to unit ramp input depends upon the time constant of the system.

4.4.1.3 Unit parabolic input

The unit parabolic input function is given by

$$r(t) = \frac{t^2}{2},\ t \geq 0$$
$$= 0,\ t < 0$$

Laplace transform of unit parabolic, $R(s) = \frac{1}{s^3}$. From Fig. 4.7, the response of first order system when it is subjected to parabolic input, is

$$C(s) = \frac{1}{s^3(Ts+1)}$$

Expanding the above equation into partial fractions as

$$C(s) = \frac{A}{s^3} + \frac{B}{s^2} + \frac{C}{s} + \frac{D}{Ts+1}$$

$$1 = A(Ts+1) + Bs(Ts+1) + Cs^2(Ts+1) + Ds^3$$

For $s = 0$, $\quad A = 1$

$s = -\frac{1}{T}$, $\quad D = -T^3$

Comparing the 's' terms, $AT + B = 0$

$$B = -T$$

Comparing the s^3 terms, $CT + D = 0$

$$C = T^2$$

$$\therefore \quad C(s) = \frac{1}{s^3} - \frac{T}{s^2} - \frac{T^3}{1+sT}$$

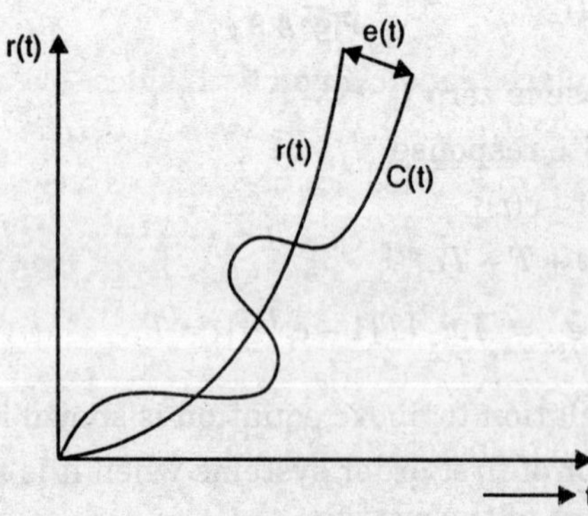

Fig. 4.10

Taking inverse Laplace transform on both sides of above equation

$$c(t) = \frac{t^2}{2} - Tt + T^2 - T^2 e^{-t/T}$$

$$= t\left(\frac{t}{2} - T\right) + T^2(1 - e^{-t/T})$$

For $t = 0$, $c(t) = 0$

For $t = \infty$, $c(t) = \infty$

Error signal $e(t)$ is the difference between described output and actual output.

$$e(t) = r(t) - C(t)$$

$$e(t) = \frac{t^2}{2} - \frac{t^2}{2} - Tt - T^2(1 - e^{-t/T})$$

$$= T\{t - T(1 - e^{-t/T})\}$$

Steady state error, $e_{ss} = \underset{t \to \infty}{\text{Lt}} \; T\{t - T(1 - e^{-t/T})\} = \infty$

The time response in relation to above equation is shown in Fig. 4.10.

The steady state error of first order system when it is subjected to parabolic input, is infinite.

4.4.1.4 Unit impulse input

An unit impulse function, $r(t) = \delta(t) = 1, t = 0$

$$= 0, t \neq 0$$

From Fig. 4.7, the response of first order system when it is subjected to unit impulse input, is

$$C(s) = \frac{1}{Ts + 1} = \frac{1}{T\left(s + \frac{1}{T}\right)}$$

Applying inverse Laplace transform on both sides

$$L^{-1} C(s) = L^{-1} \frac{1}{T\left(s + \frac{1}{T}\right)}$$

$$c(t) = \frac{1}{T} e^{-t/T}$$

For $t = 0$, $c(t) = \frac{1}{T}$, the response starts at $\frac{1}{T}$ exponentially decayed to zero.

At $t = \infty$ $c(t) = 0$, as shown in Fig. 4.11.

Error signal, $e(t) = r(t) - c(t)$

$$e(t) = 1 - \frac{1}{T} e^{-t/T}$$

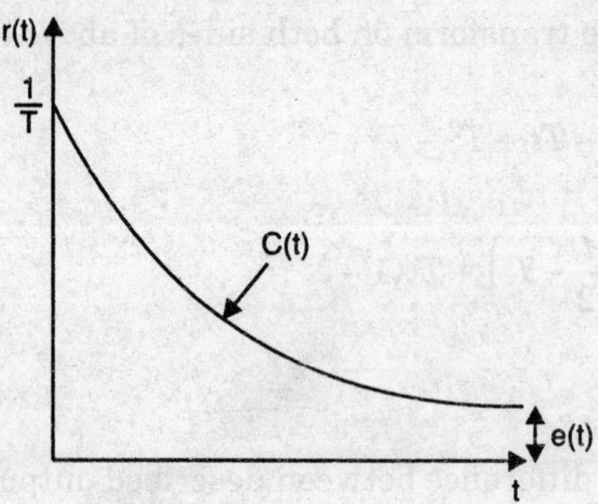

Fig. 4.11

Steady state error is

$$e_{ss} = \underset{t \to \infty}{\text{Lt}} \left(1 - \frac{1}{T} e^{-t/T}\right) = 1$$

For unit impulse input, the steady state error is unity.

4.4.1.5 Applications of first order systems

(*i*) **R-L Circuit**

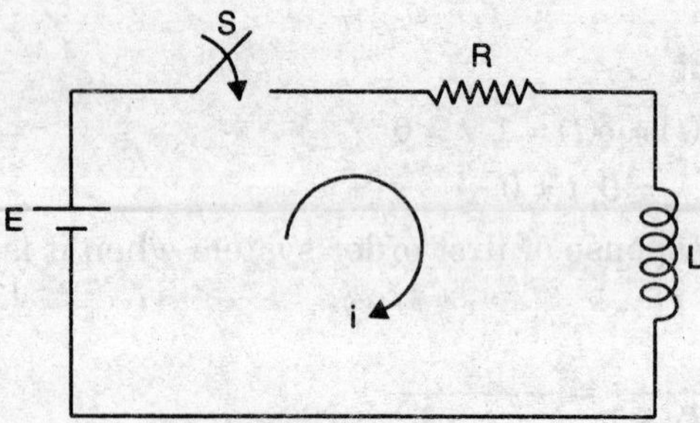

Fig. 4.12 Series RL circuit

Applying *KVL* for the system shown in Fig. 4.12, $e(t) = Ri + L \dfrac{di}{dt}$. This is the first order differential equation. Taking Laplace transform of the above equation on both sides

$$E(s) = I(s)\, [R + Ls]$$

Therefore transfer function to the network shown in Fig. 4.12, is

$$\frac{I(s)}{E(s)} = \frac{1}{R + Ls}$$

The response of the system, $I(s) = \dfrac{E(s)}{R + Ls}$

If the input voltage applied to the *RL* circuit is step voltage, $E(s) = \dfrac{E}{s}$.

$$\therefore \qquad I(s) = \frac{E}{s(R+Ls)} \tag{4.7}$$

Expanding the eqn. (4.7) into partial fractions

$$I(s) = E\left[\frac{A}{s} + \frac{B}{R+Ls}\right]$$

$$A(Ls + R) + Bs = 1$$

For $s = 0$, $\quad A = \frac{1}{R}$

$s = -\frac{R}{L}$, $\quad B = -\frac{R}{L}$

$$\therefore \qquad I(s) = \frac{E}{Rs} - \frac{L}{R}\frac{E}{L\left(\frac{R}{L}+s\right)} = \frac{E}{R}\left[\frac{1}{s} - \frac{1}{\frac{R}{L}+s}\right]$$

Applying inverse Laplace transform on both sides of above equation

$$i(t) = \frac{E}{R}\,[1 - e^{-t/T}]$$

where $T = \frac{L}{R}$ is the electrical time constant

$\frac{E}{R}$ is the steady state and

$\frac{-E}{R}\,e^{-Rt/L}$ is the transient currents.

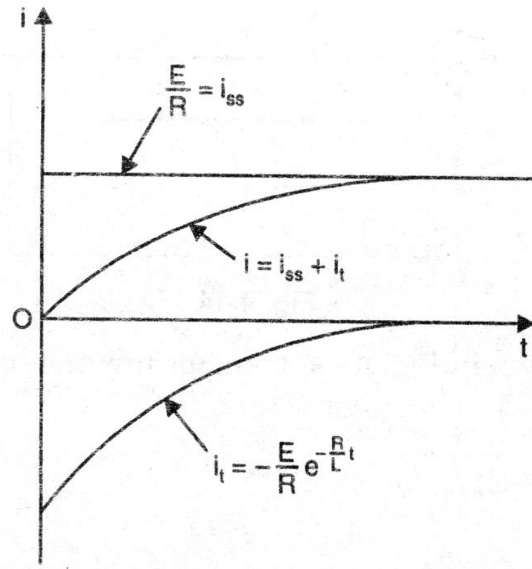

Fig. 4.13

It may be seen from the diagram shown in Fig. 4.13, at $t = 0$, the steady state value is equal to transient value but negative.

4.5 CHARACTERISTIC EQUATION OF CONTROL SYSTEMS

From the system as shown in Fig. 4.5, the transfer function is

$$\frac{C(s)}{R(s)} = \frac{G(s)}{1 + G(s)H(s)} \tag{4.8}$$

And that the characteristic equation obtained from eqn. (4.8)

i.e., $F(s) = 1 + G(s)\,H(s) = 0$

Equation (4.8) is usually a rational function with constant coefficient, so that its roots are either real numbers or pair of complex conjugate. The eqn. (4.8) can be written as

$$\frac{C(s)}{R(s)} = \frac{K \prod_{j=1}^{n} (s + z_j)}{\prod_{i=1}^{m} (s + P_i)} = \frac{K \prod_{j=1}^{n} (s + z_j)}{\prod_{i=1}^{m} (s + \sigma_i) \sum_{k=1}^{m} (s + (\alpha_k + jw_k)(s + (\alpha_k - jw_k)))}$$

where σ_i are the real roots and $\alpha_k \pm jw_k$ are the complex conjugate roots of the characteristic equation.

Time response of output $C(t)$ for any input $r(t) = L^{-1}[R(s)]$ is obtained by taking the inverse Laplace transform of $C(s)$.

4.6 TRANSIENT RESPONSE OF SECOND ORDER SYSTEM

A second order control system is one wherein the highest power of 's' in the denominator of its function equals two.

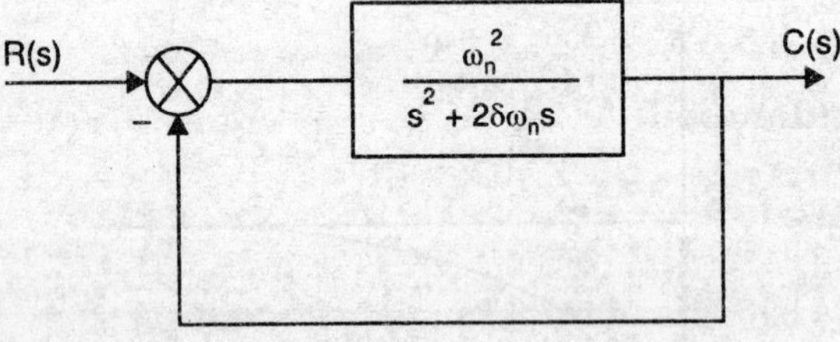

Fig. 4.14

For the system as shown in Fig. 4.14, transfer function is given by

$$\frac{C(s)}{R(s)} = \frac{\omega_n^2/(s^2 + 2\delta\omega_n s)}{1 + \omega_n^2/(s^2 + 2\delta\omega_n s)}$$

$$= \frac{\omega_n^2}{s^2 + 2\delta\omega_n s + \omega_n^2} \tag{4.9}$$

For unit step input, $R(s) = \frac{1}{s}$, and $C(s)$ is given by

$$C(s) = \frac{\omega_n^2}{s(s^2 + 2\delta\omega_n s + \omega_n^2)} \tag{4.10}$$

We have four situations depending on the value of δ, namely

$$\delta = 0$$
$$0 < \delta < 1$$
$$\delta = 1$$

and $$\delta > 1$$

Before we deal with the above cases, we express $C(s)$ of eqn. (4.10) into partial fraction expansion as

$$C(s) = \frac{A}{s} + \frac{Bs + C}{s^2 + 2\delta\omega_n s + \omega_n^2} \tag{4.11}$$

$$A(s^2 + 2\delta\omega_n s + \omega_n^2) + (Bs + C)s = \omega_n^2 \tag{4.12}$$

For $s = 0$, $A = 1$

Equating the s^2 terms of eqn. (4.12),

$$A + B = 1$$

∴ $$B = -1$$

Equating the 's' terms of eqn. (4.12),

$$2\delta\,\omega_n A + C = 1$$

∴ $$C = -2\delta\,\omega_n \tag{4.13}$$

Substituting the results of eqns. (4.12) and (4.13) in eqn.(4.11), we get

$$C(s) = \frac{1}{s} - \frac{s + 2\delta\omega_n}{s^2 + 2\delta\omega_n s + \omega_n^2} \tag{4.14}$$

Case I: δ = 0 (undamped)

$$C(s) = \frac{1}{s} - \frac{s}{s^2 + \omega_n^2} \tag{4.15}$$

$C(s)$ of eqn. (4.15) is of standard form and we can easily write from Laplace transform table

$$c(t) = 1 - \cos\omega_n t \tag{4.16}$$

From Fig. 4.15, we note that output oscillates about 1 between 0 and 2 with a constant frequency of ω_n radians/second. We also note that the overshoot is 100%. This sort of response is undesirable due to the undamped oscillations (*i.e.*, amplitude oscillations is constant at 1)

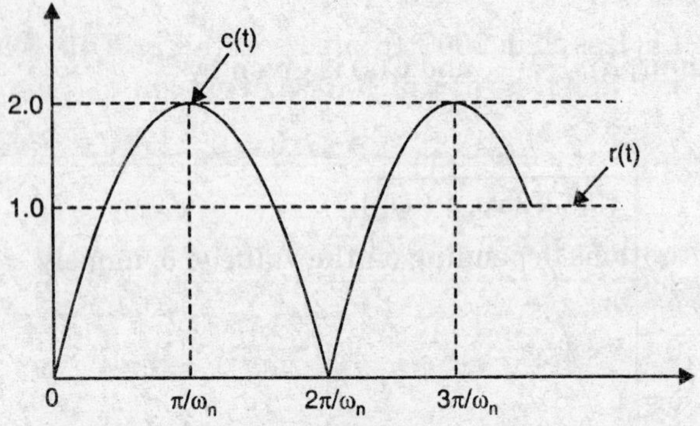

Fig. 4.15

Case II: δ < 1 (under damped oscillations)

Equation. (4.14) can be modified as follows

$$C(s) = \frac{1}{s} - \frac{s + \delta\omega_n}{(s + \delta\omega_n)^2 + \omega_n^2 (1 - \delta^2)} - \frac{\delta}{\sqrt{1-\delta^2}} \times \frac{\omega_n \sqrt{1-\delta^2}}{(s+\delta\omega_n)^2 + \omega_n^2 (1-\delta^2)} \tag{4.17}$$

Taking inverse Laplace transform on both sides of eqn. (4.17) for standard form as

$$c(t) = 1 - e^{-\delta\omega_n t} \cos (\omega_n \sqrt{1-\delta^2})t - \frac{\delta}{\sqrt{1-\delta^2}} e^{-\delta\omega_n t} \sin(\omega_n \sqrt{1-\delta^2})t$$

$$= 1 - \frac{e^{-\delta\omega_n t}}{\sqrt{1-\delta^2}} [\sqrt{1-\delta^2} \cos (\omega_n \sqrt{1-\delta^2})\, t + \delta \sin (\omega_n \sqrt{1-\delta^2})t]$$

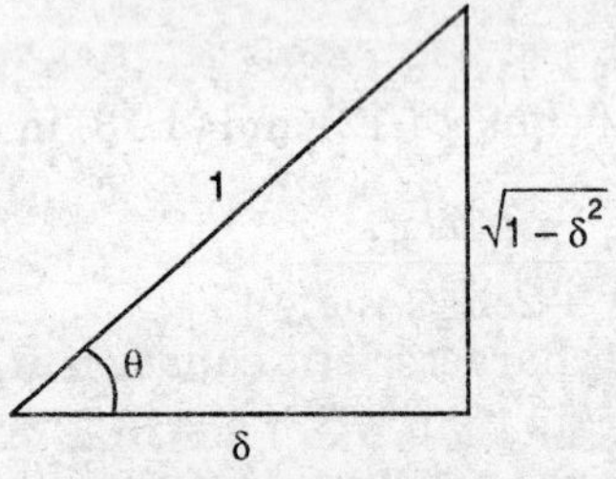

Fig. 4.16 (a)

From the triangle shown in Fig. 4.15 (a), the response of the second order system is

$$c(t) = 1 - \frac{e^{-\delta\omega_n t}}{\sqrt{1-\delta^2}} \sin \{\omega_n \sqrt{1-\delta^2})\, t + \theta\} \tag{4.18}$$

where $\theta = \tan^{-1} \dfrac{\sqrt{1-\delta^2}}{\delta}$

From Fig. 4.16 (b), we note that where δ < 1, the output oscillates about 1 with gradually diminishing amplitude. It means that the first positive peak represents the maximum over shoot and we note that it is less than 100% (compare with Fig. 4.15). We derive an expression for the maximum overshoot and time at which it occurs (t_p) in the next section.

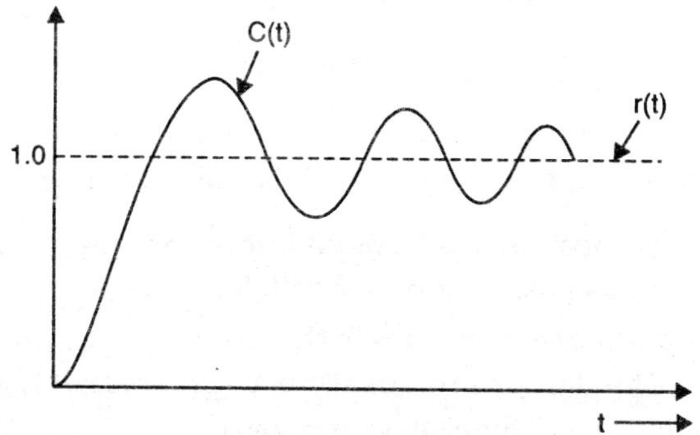

Fig. 4.16 (b) Under damped response of second order system to unit step input.

Case III: δ = 1 (critically damped)

In this case, we have two roots of the characteristic equation which are real and equal. The system is said to be critically damped, as $C(t)$ ceases to oscillate about 1. At this value of δ, eqn. (4.14) for $C(s)$ becomes

$$C(s) = \frac{1}{s} - \frac{s + 2\omega_n}{(s + \omega_n)^2}$$

$$= \frac{1}{s} - \frac{1}{s + \omega_n} - \frac{\omega_n}{(s + \omega_n)^2} \quad (4.19)$$

$$\therefore \quad C(t) = 1 - e^{-\omega_n t} - \omega_n t\, e^{-\omega_n t}$$

$$= 1 - e^{-\omega_n t}(1 + \omega_n t) \quad (4.20)$$

The output starts at $t = 0$ and gradually reaches unity as $t \to \infty$, without oscillations.

Case IV: δ > 1 (over damped)

In this case, two roots of the characteristic equation are real and unequal. As the output reaches to 1 at $t \to \infty$ at a slower rate than when δ = 1 (critically damped case), the system is said to be over damped when δ > 1. Fig. 4.17 shows diagrammatically the above cases.

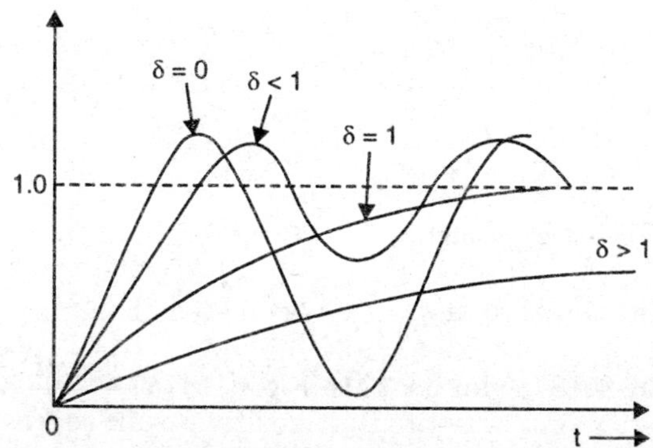

Fig. 4.17 Response of 2nd order systems to unit step input for different values of damping ratio δ.

4.6.1 Analysis

1. $\alpha = \delta\omega_n$ is called **damping co-efficient**. It is so called as it appears as an exponent of e in eqn. (4.18) and this controls the rate of decay or rise of time response.

2. T = Time constant is $\frac{1}{\alpha}$

3. ω_d **= damped frequency** $\omega_n\sqrt{1-\delta^2}$. This appears in the argument of the sine term. The damped frequency is always less than undamped frequency ω_n. This term requires further examination. The various possibilities are as follows.

(*a*) **$0 < \delta < 1$, undamped case:** when δ is 0, the damped frequency is equal to ω_n. As δ is increased from 0, the damped frequency gets progressively reduced. Poles (roots) are complex conjugates with negative real parts.

(*b*) **$\delta = 1$, critically damped case:** ω_d vanishes leaving only the undamped natural frequency ω_n. The poles are equal, negative and real.

(*c*) **$\delta > 1$, over damped case:** Both roots are negative, real and unequal. The response is slow and reaches steady state after a long time.

(*d*) **$\delta = 0$, undamped case:** As explained in a $\omega_d = \omega_n$, poles are complex conjugate $\delta = \pm j\omega_n$.

(*e*) **$\delta < 0$, negatively damped case:** poles are positive on s – plane.

4.6.2 Effect of Adding a Zero to a Second Order System

Let a zero at $s = -z$ be added to the transfer function of eqn. (4.9)

$$\frac{C(s)}{R(s)} = \frac{(s+z)\omega_n^2/z}{s^2 + 2\delta\omega_n s + \omega_n^2}$$

The numerator has been suitably multiplied by $\frac{1}{z}$ so as to make steady state value of $c(t)$ for unit-step input equal to one. Rewriting the above equation as

$$\frac{C(s)}{R(s)} = \frac{1}{z}\left(\frac{s\omega_n^2}{s^2 + 2\delta\omega_n s + \omega_n^2}\right) + \frac{\omega_n^2}{s^2 + 2\delta\omega_n s + \omega_n^2}$$

$$C_z(t) = L^{-1}\,[C(s)] \text{ for } R(s) = \frac{1}{s}$$

$$= \frac{1}{z}\,[sC(s)] + C(s)$$

$$= \frac{1}{z}\left[\frac{d}{dt}c(t)\right] + c(t)$$

where $c(t)$ is the output, when no zero at $s = -z$ was added. The graphical representation of $c_z(t)$ versus 't' is shown in Fig 4.18 (*a*) for $\delta < 1$. In Fig. 4.18 (*b*) $\frac{1}{z}\frac{dc(t)}{dt}$ versus 't' is shown.

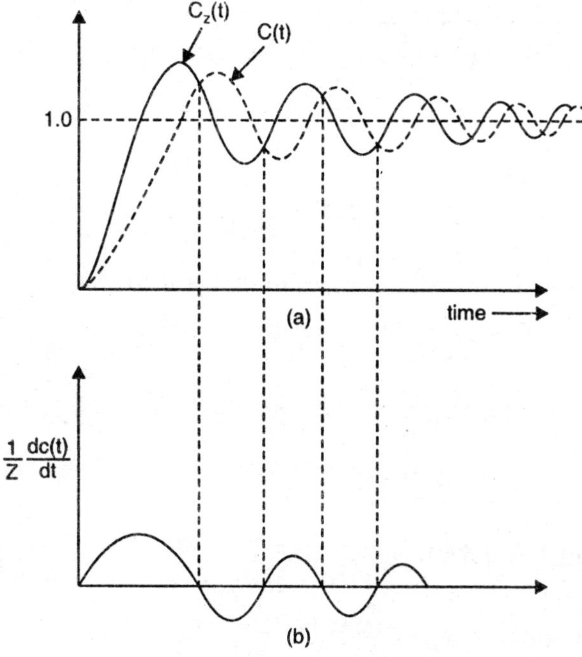

Fig. 4.18

It is seen that as z increases, the effect of adding the zero becomes very small and negligible if z is very large. As comparison between $c(t)$ and $c_z(t)$ will show that the overshoot increases with decreases of z. Thus the zeros on the real axis nearer to the origin are generally avoided in the design.

4.6.3 Application of Second Order System

1. **RLC circuit**

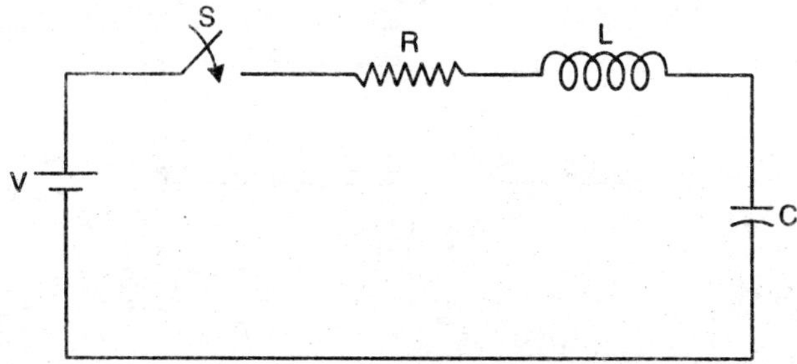

Fig. 4.19 Series RLC circuit

Applying *KVL* for the circuit shown in Fig. 4.19,

$$V = Ri + L\frac{di}{dt} + \frac{1}{C}\int idt$$

Taking Laplace transform on both sides of above equation

$$\frac{V(s)}{s} = \left[Ls + R + \frac{1}{Cs}\right] I(s)$$

Transfer function, $\dfrac{I(s)}{V(s)} = \dfrac{1/LC}{s^2 + \dfrac{R}{L}s + \dfrac{1}{LC}}$

If we equate the denominator to zero, the characteristic equation is a second order and its solution as per eqn. (4.14)

$$\therefore \qquad i(t) = \frac{V}{LC}\left[1 - \frac{e^{-\delta\omega_n t}}{\sqrt{1-\delta^2}} \sin(\omega_d t + \theta)\right]$$

$$\omega_n = \frac{1}{\sqrt{LC}};\ 2\delta\omega_n = R/L$$

Damping ratio, $\delta = \dfrac{R}{2}\sqrt{\dfrac{C}{L}}$

Gain, $K = \dfrac{1}{\sqrt{LC}}$

Damping coefficient, $\alpha = R/2L$

Damped natural frequency, $\omega_d = \dfrac{1}{\sqrt{LC}}\sqrt{1 - \dfrac{R^2C}{4L}}$

Time constant, $T = \dfrac{1}{\alpha} = \dfrac{2L}{R}$

2. For **parallel circuit** shown in Fig. 4.20.

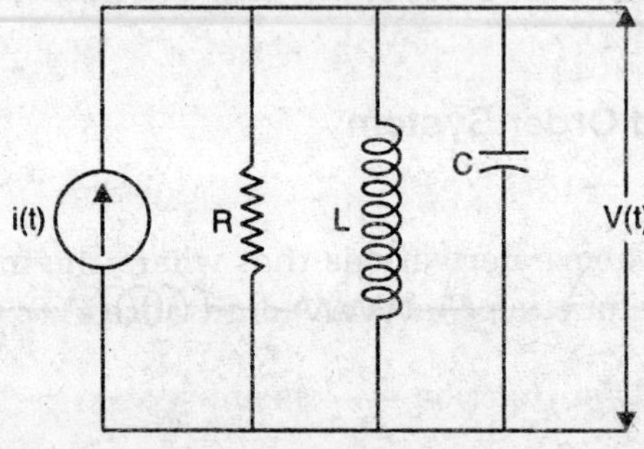

Fig. 4.20

The current supplied from the source is

$$i(t) = i_{(R)} + i_{(L)} + i_{(C)}$$

$$= \frac{V(t)}{R} + \frac{1}{L}\int_0^t V(t)\,dt + C\frac{dv}{dt}$$

Taking Laplace transform on both sides of above equation

$$I(s) = \frac{V(s)}{R} + \frac{V(s)}{Ls} + Cs\,V(s)$$

$$= V(s)\left[\frac{1}{R} + \frac{1}{Ls} + Cs\right]$$

$$= V(s)\left[\frac{s^2RCL + sL + R}{sLR}\right]$$

$$G(s) = \frac{V(s)}{I(s)} = Z(s)$$

$$= \frac{LRs}{s^2LCR + sL + R}$$

$$= \frac{sLR}{LCR\left(s^2 + s\frac{1}{RC} + \frac{1}{LC}\right)} = \frac{s\frac{1}{LC}}{\left(s^2 + s\frac{1}{RL} + \frac{1}{LC}\right)}$$

The natural frequency is the same as series circuit,

$$\omega_n = \frac{1}{\sqrt{LC}}$$

Damping ratio, $\delta = \frac{1}{2R}\sqrt{\frac{C}{L}}$,

Gain, $K = \frac{1}{LC}$

Damping coefficient, $\alpha = \delta\,\omega_n = \frac{1}{2RC}$

It may be noted in the above derivations that where the input is voltage source and output is current, the transfer function $G(s)$ is an admittance $Y(s)$; where the input is current and output is voltage, the transfer function $\frac{V(s)}{I(s)}$ is impedance $Z(s)$. It may not always be the same. The input and output both may be voltages $V_i(s)$ and $V_0(s)$.

For series circuit $= \frac{V_0(s)}{V_1(s)} = \frac{1}{LCs^2 + RCs + 1}$

Problem 4.1. *Find the transient response and steady state response of the given system*

$$c(t) = 1.2 - 1.5\,e^{-3t} + 2.5\,e^{-4t}$$

Solution: Steady state response will be obtained as $t \to \infty$.

As $t \to \infty$ both the exponential terms will approach to zero.

$\therefore$ $c(t) = 1.2$, is steady state response.

$c(t) = -1.5\,e^{-3t} + 2.5\,e^{-4t}$ is transient response.

Problem 4.2. *A unit feedback control system whose forward path transistor function,* $G(s) = \dfrac{8}{s(s+5)}$ *is subjected to unit step input. Find the steady state error according to final value theorem.*

Solution: Steady state error is

$$e_{ss} = \underset{s \to 0}{\text{Lt}}\ sE(s) = \underset{s \to 0}{\text{Lt}}\ s\,\frac{R(s)}{1 + G(s)\,H(s)}$$

Input signal is step signal, $R(s) = 1/s$

$$G(s) = \frac{8}{s(s+5)} \qquad H(s) = 1$$

$$\therefore \qquad e_{ss} = \underset{s \to 0}{\text{Lt}}\ \frac{s\frac{1}{s}}{1 + \frac{8}{s(s+5)} \times 1} = \underset{s \to 0}{\text{Lt}}\ \frac{s(s+5)}{s^2 + 5s + 8} = 0$$

Problem 4.3. *Determine the steady state voltage of the capacitor after the switch is closed as shown in Fig. 4.21.*

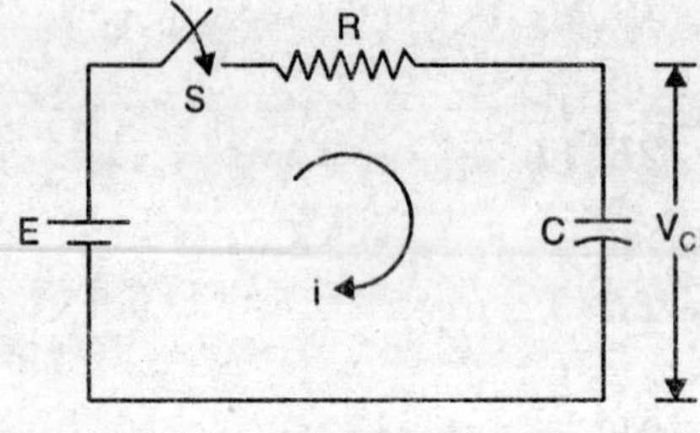

Fig. 4.21

Solution:

$$E = iR + V_c \ \text{ and } \ i = C\frac{dV_c}{dt}$$

$$E = RC\frac{dV_c}{dt} + V_c$$

Steady state voltage, $V_c = E$

4.7 TIME DOMAIN SPECIFICATIONS

We have discussed that the two characteristics of transient response of a system are natural frequency and damping ratio. There are a few more characteristics of a transient response arising from the nature of oscillation. They are, for a unit step function (*i*) delay time t_d, (*ii*) rise time t_r, (*iii*) peak time t_p, (*iv*) maximum overshoot M_p and (*v*) settling time t_s.

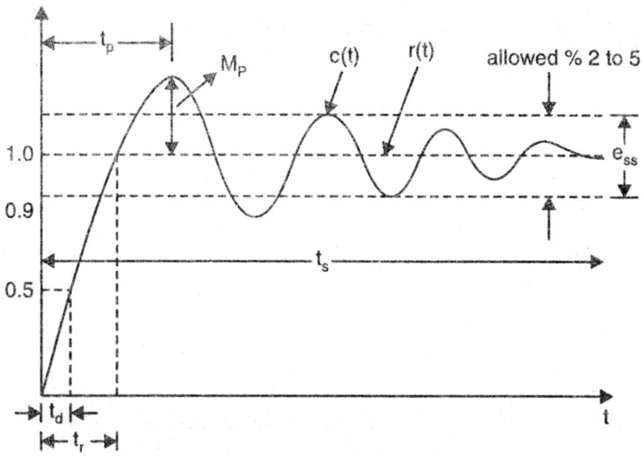

Fig. 4.22

All these are explained with respect to the curve shown in Fig. 4.22.

(*i*) **Delay time t_d.** It is the time for the response to reach half (50%) of its final value.

(*ii*) **Rise time t_r.** It is the time required by the step function response to reach from 10 to 90% or 0 to 100% of its final value (steady state). For undamped systems, 0 to 100% is used and for overdamped systems 10 to 90% rise time is used.

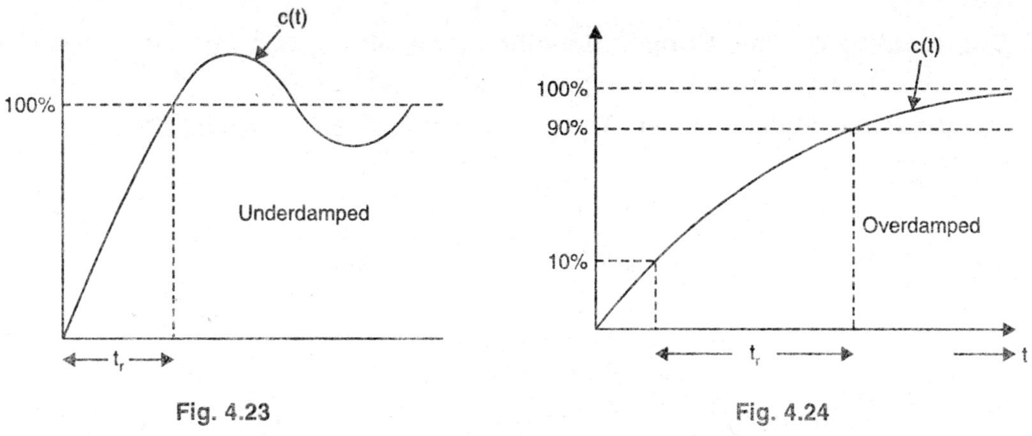

Fig. 4.23 Fig. 4.24

It may be noted from the Figs. 4.23 and 4.24, a fast responding system has less rise time, than slow responding system. We prefer a slow response to command by an elevator and a fast response for machine tool positioning.

(*iii*) **Peak time t_p.** It is the time taken by the controlled system to reach the peak of the first overshoot.

(*iv*) **Maximum overshoot M_p.** Maximum percent overshoot (M_p)

$$= \frac{\text{Maximum value of first overshoot} - \text{steady state value}}{\text{steady state value}} \times 100$$

M_p decides the relative stability of the system.

A 10% overshoot will have a damping ratio of 0.7.

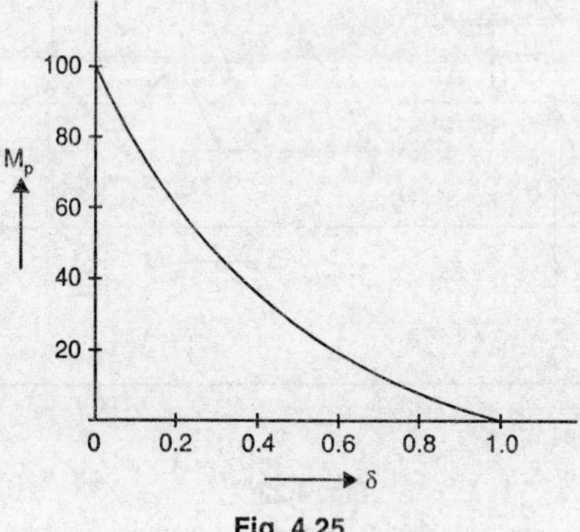

Fig. 4.25

(*v*) **Settling time t_s.** It is the time required by the system to reach and settle within a prescribed percentage of the final or steady value. Usually, it is 2 to 5% *i.e.*, 4 to 3 times of time constants $\left(T = \dfrac{1}{\delta\omega_n}\right)$ respectively.

4.7.1 Computation of Time Domain Specifications

In this section, expression for time domain specifications are derived.

(*i*) **Rise time.** Equate the eqn. (4.18) to 1 as the transient has to reach 1 or 100% in the shortest possible time.

$$C(t) = 1 - \frac{e^{-j\omega_n t_n}}{\sqrt{1-\delta^2}} \sin(\omega_d t_r + \theta) = 1$$

$$0 = \frac{e^{-j\omega_n t_r}}{\sqrt{1-\delta^2}} \sin(\omega_d t_r + \theta)$$

$e^{-\omega_n t_r}$ cannot be zero

So, $\quad \sin(\omega_d t_r + \theta) = 0$, when $\omega_d t_r = n\pi - \theta$ where $n = 0, 1, 2, 3$ .....

$$\therefore \qquad t_r = \frac{n\pi - \theta}{\omega_d} = \frac{n\pi - \theta}{\omega_n \sqrt{1-\delta^2}}$$

where $\theta = \tan^{-1} \dfrac{\sqrt{1-\delta^2}}{\delta}$

Since $0 < \delta < 1$ we have $0 < \theta < \dfrac{\pi}{2}$. Hence as δ is increased, δ increases and t_r is decreased. That is the system response become faster.

(*ii*) **Peak time t_p.** The peak time is obtained by differentiating the eqn. (4.18) w.r.t. t

$$\frac{dc(t)}{dt} = \frac{d}{dt}\left[1 - \frac{e^{-\delta\omega_n t_p}}{\sqrt{1-\delta^2}} \sin(\omega_d t_p + \theta)\right]$$

$$= -\left[\frac{e^{-\delta\omega_n t_p}}{\sqrt{1-\delta^2}} \omega_d \cos(\omega_d t_p + \theta) - \frac{\delta\omega_n}{\sqrt{1-\delta^2}} e^{-\delta\omega_n t_p} \sin(\omega_d t_p + \theta)\right]$$

$$= -\omega_n \frac{e^{-\delta\omega_n t}}{\sqrt{1-\delta^2}} [\cos(\omega_d t_p + \theta)\sqrt{1-\delta^2} - \delta \sin(\omega_d t_p + \theta)]$$

$$= -\omega_n \frac{e^{-\delta\omega_n t}}{\sqrt{1-\delta^2}} [\cos(\omega_d t_p + \theta)\cos\theta - \sin\theta \sin(\omega_d t_p + \theta)] \text{ [From Fig. 4.16 (a)]}$$

$$= -\omega_n \frac{e^{-\delta\omega_n t_p}}{\sqrt{1-\delta^2}} \cos(\omega_d t_p), \qquad t_p \geq 0$$

Equating the term to zero to obtain the maximum time

$$\cos \omega_n \sqrt{1-\delta^2}\, t_p = 0$$

$$\omega_n \sqrt{1-\delta^2}\, t_p = n\pi \qquad \text{where } n = 0, 1, 2, \ldots\ldots \text{ etc.}$$

$$\therefore \qquad t_p = \frac{n\pi}{\omega_n\sqrt{1-\delta^2}} = \frac{n\pi}{\omega_d} \tag{4.21}$$

The peak time t_p corresponds to one half cycle of frequency of damped oscillation. As δ decresed, t_p is also decreased.

(*iii*) **Maximum overshoot M_p.** Maximum overshoot occurs at $t_p = \frac{n\pi}{\omega_d}$, substituting t_p from eqn. (4.21) in eqn. (4.18)

$$C(t_p) = 1 - \frac{e^{-n\pi\delta/\sqrt{1-\delta^2}}}{\sqrt{1-\delta^2}} \sin(n\pi + \theta), \quad n = 1, 2, 3 \ldots\ldots$$

$$= 1 - \frac{e^{-n\pi\delta/\sqrt{1-\delta^2}}}{\sqrt{1-\delta^2}} (-\sin\theta)$$

From the triangle as shown in the Fig. 4.16 (*a*), $\sin\theta = \sqrt{1-\delta^2}$

$$C(t_p) = 1 + \frac{e^{-n\pi\delta/\sqrt{1-\delta^2}}}{\sqrt{1-\delta^2}} (\sqrt{1-\delta^2})$$

∴ Peak response of the second order system is

$$C(t_p) = 1 + e^{-n\pi\delta} / \sqrt{1-\delta^2} \quad (4.22)$$

The first overshoot is of maximum amplitude at substituting $n = 1$ in eqn. (4.22).

Maximum overshoot (M_p) is the difference in peak response to steady state response of the system

$$\therefore \quad M_p = 1 + e^{-\delta\pi/\sqrt{1-\delta^2}} - 1$$

$$= e^{-\delta\pi/\sqrt{1-\delta^2}}$$

% overshoot $\quad M_p = e^{-\delta\pi/\sqrt{1-\delta^2}} \times 100 \quad (4.23)$

For $\delta = 0$, M_p is normalised at 1.0 and when $\delta = 1.0$, M_p is monotonically decreases to zero. M_p droes not depend on the natural frequency ω_n. Thus M_p is very much used to represent the system damping indirectly. From Fig. 4.25 it is seen that when the % overshoot is not in acceptable limits, increase of δ is resorted to bring it to under check.

(*iv*) **Settling time t_s.** The settling time depends on the allowed tolerance say 2 to 5% of steady state value. For time constant 5, the transient response almost reaches steady value (0.993 of steady state value).

Settling time $t_s = \dfrac{4}{\delta\omega_n}$ or $\dfrac{3}{\delta\omega_n}$, for 2% or 5% tolerance respectively $\quad (4.24)$

Problem 4.4. *Differential equation of the spring mass system is* $\dfrac{d^2y}{dt^2} + 1.6\dfrac{dy}{dt} + 128y = 0$. *Determine the undamped natural frequency and damping ratio of the system.*

Solution: Differential equation of the given system is

$$\frac{d^2y}{dt^2} + 1.6\frac{dy}{dt} + 128y = 0.$$

Taking Laplace transformation of above equation with initial conditions are zero.

$$s^2 + 1.6s + 128 = 0$$

Comparing the above expression with standard second order characteristic equation,

Undamped natural frequency, $\omega_n^2 = 128$

$$\therefore \quad \omega_n = 11.3 \text{ rad/sec}$$

Damping ratio, $\delta = \dfrac{1.6}{2 \times 11.3} = 0.071$

Problem 4.5. *For the system shown below in Fig. 4.26, taking K = 10, find the values of 'a' and 'b' so that overshoot is 16% and time constant is 0.1 sec in its response to unit step input.*

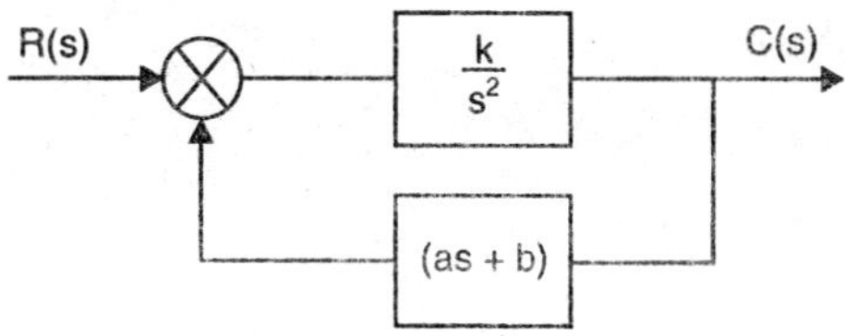

Fig. 4.26

Solution: Transfer function of Fig. 4.26,

$$\frac{C(s)}{R(s)} = \frac{K/s^2}{1+(K/s^2)(as+b)} = \frac{K}{s^2+aKs+bK}$$

$$= \frac{1}{b} \times \frac{bK}{s^2+aKs+bK}$$

$$\omega_n^2 = 10b, \qquad \omega_n = \sqrt{10b}$$

$$2\delta\omega_n = 10a$$

$\therefore$
$$\delta \frac{10a}{2\omega_n} = \frac{10a}{2\times\sqrt{10b}} = \frac{\sqrt{10}}{2} \times \frac{a}{\sqrt{b}} = \frac{1.5812a}{\sqrt{b}} \qquad (i)$$

Overshoot, $$M_p = e^{-\pi\delta/\sqrt{1-\delta^2}} = 0.16$$

$$\frac{-\pi\delta}{\sqrt{1-\delta^2}} = \ln 0.16$$

By solving the above equation, we get

$$\delta = 0.5037 \qquad (ii)$$

Equating the eqns (*i*) and (*ii*)

$$\frac{a}{\sqrt{b}} = \frac{0.5037}{1.5812} = 0.319 \qquad (iii)$$

Time constant, $$\frac{1}{\delta\omega_n} = \frac{1}{5a}$$

$\therefore$
$$\frac{1}{5a} = 0.1 \quad \text{or } a = 2$$

From eqn. (*iii*), $b = \dfrac{a^2}{(0.319)^2} = 39.31$

Problem 4.6. *For problem 4.5, $\delta = 0.115$ and $\omega_n = 92.41$ rad/sec, determine the rise time t_r, peak time t_p, maximum overshoot M_p and settling time t_s.*

Solution: (*i*) Rise time, $t_r = \dfrac{\pi-\theta}{\omega_d} = \dfrac{3.14-1.456}{91.8} = 0.018$ sec

(*ii*) Peak time, $t_p = \dfrac{\pi}{\omega_d} = \dfrac{\pi}{91.8} = 0.034$ sec (*iii*) Maximum overshoot,

$$M_p = e^{-\delta\pi/\sqrt{1-\delta^2}}$$

$$= e^{-0.115\pi/\sqrt{1-(0.115)^2}}$$

$$= e^{-0.364} = \frac{1}{1.44} = 0.695$$

Percentage overshoot = 0.695 × 100 = 69.5%

(*iv*) Settling time t_s for 2% criterion $= \dfrac{4}{\delta\omega_n} = \dfrac{4}{0.115 \times 92.41} = 0.376$ sec

For 5% criterion $= \dfrac{3}{10.63} = 0.282$ sec.

Problem 4.7. *The open loop transfer function of unit feedback system is given by G(s) =* $\dfrac{k}{s(s+2)}$. *The specification for the system transient response to a step input is as follows* t_p *: peak time 1sec per unit overshoot of 5%. In the above system, can these two specifications be met simultaneously?*

Solution: $$M_p = e^{-\delta\pi/\sqrt{1-\delta^2}} = \frac{5}{100} = 0.05$$

or $$\frac{\delta\pi}{\sqrt{1-\delta^2}} = 3,$$

∴ $$\delta = 0.69$$

$$G(s) = \frac{k}{s(s+2)};$$

Transfer function, $$T(s) = \frac{G(s)}{1+G(s)\,H(s)}$$

$$T(s) = \frac{k}{s^2 + 2s + k}$$

Compare the denominator with characteristic eqn. (4.28),

$$\omega_n = \frac{1}{0.69} = 1.45$$

Peak time, $$t_p = \frac{\pi}{\omega_d} = \frac{\pi}{\omega_n\sqrt{1-\delta^2}} = \frac{3.14}{1.45\sqrt{1-(0.69)^2}}$$

$$t_p = 2.992 \text{ sec.}$$

But peak time t_p given is 1 sec and hence these two specifications cannot be met.

Problem 4.8. *The open loop transfer function of unit feedback system is given by G(s)* $= \dfrac{K}{s(1+Ts)}$.

(i) By what factor should the amplifier gain K be multiplied so that the damping ratio is increased from 0.25 to 0.75.

(ii) By what factor should K be multiplied so that the maximum overshoot of step response is reduced from 80% to 20%.

Solution: The closed loop transfer function is given by

$$\frac{C(s)}{R(s)} = \frac{G(s)}{1+G(s)} = \frac{K}{s^2+s+k} = \frac{K/T}{s^2+\frac{1}{T}s+\frac{K}{T}}$$

The characteristic equation is $s^2 + \frac{1}{T}s + \frac{K}{T} = 0$

Comparing this equation with $s^2 + 2\delta\omega_n s + \omega_n^2 = 0$, we have

$$\omega_n = \sqrt{\frac{K}{T}},\ \delta = \frac{1}{2T\omega_n} = \frac{1}{2\sqrt{(TK)}}$$

Let K_1 and K_2 be the amplifier gains for damping ratio δ_1 and δ_2 respectively. where $\delta_1 = 0.25$ and $\delta_2 = 0.75$ (given)

$$\therefore \quad \frac{\delta_1}{\delta_2} = \sqrt{\frac{K_2}{K_1}}$$

$$\sqrt{\frac{K_2}{K_1}} = \frac{0.25}{0.75}$$

$$\frac{K_2}{K_1} = 0.111$$

$$K_2 = 0.111\,K_1$$

Therefore, the amplifier gain K should be multiplied by a factor of 0.111, so that damping ratio gets increased from 0.25 to 0.75.

The maximum overshoot is given by

$$M_p = e^{\frac{-\pi\delta}{\sqrt{1-\delta^2}}}$$

Therefore, we have

$$e^{\frac{-\pi\delta_1}{\sqrt{1-\delta_1^2}}} = 0.8 \quad \text{and} \quad e^{\frac{-\pi\delta_2}{\sqrt{1-\delta_2^2}}} = 0.2$$

From which $\dfrac{-\pi\delta_1}{\sqrt{1-\delta_1^2}} = 0.2231$

and $\dfrac{-\pi\delta_2}{\sqrt{1-\delta_1^2}} = 1.6094$

By solving the above equations, we get

$$\delta_1 = 0.0708$$

And $$\delta_2 = 0.4559$$

$$\delta_1 = \frac{1}{2\sqrt{(TK_1)}}$$

$$TK_1 = \frac{1}{4\delta_1^2} = \frac{1}{4(0.0708)^2} = 49.8739$$

Similarly, $$TK_2 = 1.2028$$

Therefore, we have, $$\frac{K_2}{K_1} = \frac{1.2028}{49.8739} = 0.0241$$

The factor by which K is to be multiplied so that maximum overshoot reduces from 80% to 20% is 0.0241.

Problem 4.9. *A unit feedback system has a forward path transfer function $G(s) = \frac{9}{s(s+1)}$. Find the value of damping ratio, undamped natural frequency of the system, percentage overshoot, peak time and settling time.*

Solution:

Given data, $G(s) = \frac{9}{s(s+1)}$ and the system is unit feedback system

So, $$\frac{C(s)}{R(s)} = \frac{G(s)}{1+G(s)}$$

$$= \frac{\frac{9}{s(s+1)}}{1+\frac{9}{s(s+1)}}$$

$$= \frac{9}{s^2+s+9}$$

Transfer function of second order system is

$$\frac{C(s)}{R(s)} = \frac{\omega_n^2}{s^2+2\delta\omega_n s+\omega_n^2}$$

On comparing, we get

$$\omega_n^2 = 9,$$

$$2\delta\omega_n = 1$$

or $$\delta = \frac{1}{2\omega_n}$$

$$\omega_n = 3,$$

$$\therefore \qquad \delta = \frac{1}{2(3)} = \frac{1}{6} = 0.16$$

$$\omega_d = \omega_n \sqrt{1-\delta^2} = 2.96 \text{ rad/sec}$$

Percentage overshoot

$$\%M_p = e^{-\delta\pi/\sqrt{1-\delta^2}} \times 100$$

$$= e^{\frac{-0.16 \times \pi}{\sqrt{1-0.16^2}}} \times 100 = 165\%$$

Peak time, $$t_p = \frac{\pi}{\omega_d} = \frac{3.14}{2.96} = 1.06 \text{ sec}$$

Time constant, $$T = \frac{1}{\delta\omega_n} = \frac{1}{0.48} = 2.08 \text{ sec}$$

$\therefore$ For 5% error, settling time is

$$t_s = 3T$$

$$= 3 \times 2.08 = 6.25 \text{ sec}$$

For 2% error, settling time, $t_s = 4T = 4 \times 2.08 = 8.32$ sec.

Problem 4.10. *A unit feedback control system has a loop transfer function, $G(s) = \frac{10}{s(s+2)}$. Find the rise time, percentage overshoot, peak time and settling time for a step input of 12 units.*

Solution: Given that open loop transfer function of the system,

$$G(s) = \frac{10}{s(s+2)}$$

Given, $H(s) = 1$ (unity feedback)

Closed loop transfer function,

$$\frac{C(s)}{R(s)} = \frac{G(s)}{1 + G(s)\,H(s)}$$

$$= \frac{\frac{10}{s(s+2)}}{1 + \frac{10}{s(s+2)}}$$

$$\frac{C(s)}{R(s)} = \frac{10}{s^2 + 2s + 10}$$

The characteristic equation of the system is

$$s^2 + 2s + 10 = 0$$

On comparing this characteristic equation with standard characteristic equation, we get

$$s^2 + 2\delta\omega_n s + \omega_n^2 = 0$$

$$\omega_n^2 = 10; \omega_n = \sqrt{10} = 3.16 \text{ rad/sec.}$$

$$2\delta(3.16) = 2$$

$$\delta = 0.316.$$

Rise time, $$t_r = \frac{\pi - \theta}{\omega_d}$$

where ω_d = damped frequency

$\theta = \cos^{-1}\delta$ in radians

$\theta = \cos^{-1} 0.316 = 1.248$ radians

$$\omega_d = \omega_n\sqrt{1-\delta^2}$$

$$= 3.16\sqrt{1-0.316} = 2.998$$

$$\therefore \quad t_r = \frac{3.14 - 1.248}{2.998} = 0.631 \text{ sec}$$

Percentage overshoot with 12 units step input

$$= 12\left(e^{\frac{-\pi\delta}{\sqrt{1-\delta^2}}}\right) \times 100$$

$$= 12\left(e^{\frac{-\pi(0.316)}{\sqrt{1-(0.316)^2}}}\right) \times 100 = 422.39\%$$

Peak time, $$t_p = \frac{\pi}{\omega_d} = \frac{\pi}{2.998} = 1.047 \text{ sec}$$

Settling time, $$t_s = \frac{4}{\delta\omega_n} = \frac{4}{0.316 \times 3.16} = 4 \text{ sec}$$

4.8 STEADY STATE RESPONSE

In order to obtain the transient response or time domain response, we require mathematical model *i.e*, the differential equation description of the system. The steady of transient response of a control system is very important as it describes the dynamic behaviour of the system and the deviation between the response and input.

The time response of a control system is usually divided into two parts.

$$C(t) = C_{tr} + C_{ss}(t)$$

$C_{tr}(t)$ = transient response

$C_{ss}(t)$ = steady state response

The steady state response for second order system when subjected to step input is unity.

Then the steady state response can be defined as that part of the response which remains after the transient has died out. The steady state response, when compared with the input or the desired response indicates the final accuracy of the system. The difference is the steady state error.

4.9 STEADY STATE ERROR AND ERROR CONSTANTS

From eqn. (4.5), using final value theorem

Steady state error, $e_{ss} = \underset{s \to 0}{\text{Lt}}\; sE(s) = \underset{t \to \infty}{\text{Lt}}\; e(t)$

$$e_{ss} = \underset{s \to 0}{\text{Lt}}\; s \times \frac{R(s)}{1 + G(s)H(s)}$$

where $E(s) = \dfrac{R(s)}{1 + G(s)H(s)}$

Unity feedback control systems are classified as type 0, type 1, type 2, and based on the nature of their forward path gain $G(s)$. The general form of $G(s)$ is as follows:

$$G(s) = \frac{K(1+sT_1)(1+sT_2)\ldots\ldots(1+sT_m)}{s^J\,(1+sT_a)(1+sT_a)(1+sT_C)\ldots\ldots(1+sT_n)} \qquad (4.25)$$

where K is a real positive number and the degree of denominator polynomial is greater than that of numerator polynomial. The system is said to be of type 'J', if $G(s)$ had a pole of order J'.

Hence for type 0 system, $J = 0$

Type 1 system, $J = 1$

Type 2 system, $J = 2$, and soon.

From eqn. (4.5) we know that error signal or steady state error depends upon the type of input, order and type of the system. Now we will see the steady state error of the system for different inputs, order and type of the system.

(*i*) **Step input:**

Considering input, $r(t) = Au(t)$

where $u(t)$ is unit step function

We have $R(s) = \dfrac{A}{s} = \dfrac{1}{s}$, for $A = 1$

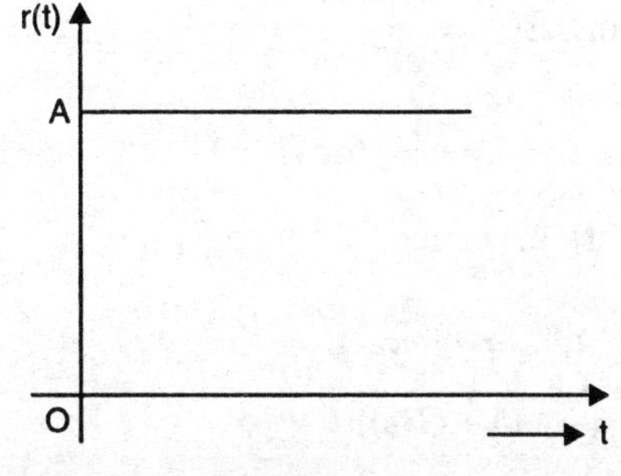

Fig. 4.27

From eqn. (4.5), the steady state error is

$$e_{ss} = \underset{s \to 0}{\text{Lt}} \frac{sR(s)}{1+G(s)H(s)}$$

For type '0' system, $G(s) = \dfrac{K(1+T_1 s)(1+T_2 s)......(1+sT_m)}{s^0 (1+T_a s)(1+T_b s)......(1+sT_n)}$

For unit feedback system

Steady state error, $e_{ss} = \underset{s \to 0}{\text{Lt}} \dfrac{s \times 1/s}{1+G(s)} = \dfrac{1}{1+\underset{S \to 0}{\text{Lt}} G(s)}$

where $\underset{s \to 0}{\text{Lt}} G(s) = \underset{s \to 0}{\text{Lt}} \dfrac{K(1+T_1 s)(1+T_2 s)......(1+T_m s)}{(1+T_a s)(T_{bs}\, s+1)......(1+T_n s)} = K = K_P$

$$\therefore \qquad e_{ss} = \frac{1}{1+K_P}$$

where $K_P = \underset{s \to 0}{\text{Lt}} G(s)$, is known as position error constant or proportional error constant. .

For type 1 system,

$$\underset{s \to 0}{\text{Lt}} G(s) = \underset{s \to 0}{\text{Lt}} \frac{K(1+T_1 s)(1+T_2 s)......(1+sT_m)}{s^1 (1+T_a s)(1+T_b s)......(1+sT_n)} = \frac{K}{s} = \infty$$

$$e_{ss} = \frac{1}{1+\underset{s \to 0}{\text{Lt}} G(s)} = 0$$

Type 2 system, $\underset{s \to 0}{\text{Lt}} G(s) = \underset{s \to 0}{\text{Lt}} \dfrac{K(1+T_1 s)(1+T_2 s)......(1+T_n s)}{s^2 (1+T_n s)(1+T_b s)......(1+T_n s)} = \dfrac{K}{s^2} = \infty$

$$e_{ss} = \frac{1}{1+\underset{s \to 0}{\text{Lt}} G(s)} = \frac{1}{1+\infty} = 0$$

Hence steady state error for the system subjected to step input signal more than type 0 system is zero.

(*ii*) **Ramp input:**

Consider $r(t) = Bt, t \geq 0$

Taking Laplace transform, $R(s) = \dfrac{B}{s^2}$

$$= \frac{1}{s^2}, \text{ for } B = 1$$

From eqn. (4.5), steady state error is

$$e_{ss} = \underset{s \to 0}{\text{Lt}} \frac{s \times \frac{1}{s^2}}{(1+G(s))}$$

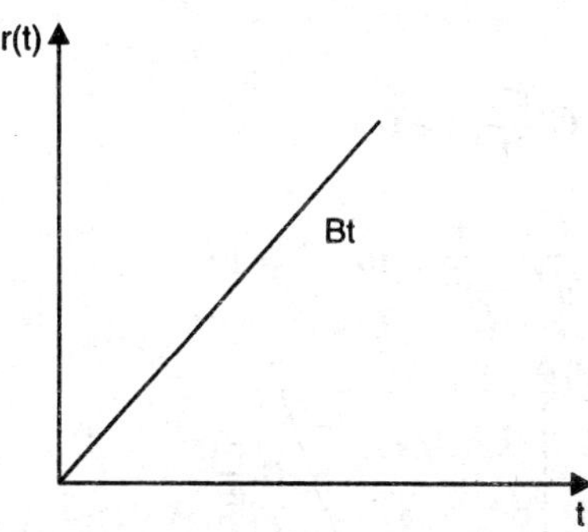

Fig. 4.28

$$\underset{s \to 0}{\text{Lt}} \frac{1}{s(1+G(s))} = \underset{s \to 0}{\text{Lt}} \frac{1}{s+sG(s)}$$

$$= \frac{1}{\underset{s \to 0}{\text{Lt}}\ s + \underset{s \to 0}{\text{Lt}}\ sG(s)} = \underset{s \to 0}{\text{Lt}} \frac{1}{sG(s)}$$

For type 0 system, the transfer function is

$$G(s) = \frac{K(1+T_1 s)(1+T_2 s)\ldots\ldots(T_m s+1)}{s^0 (1+T_a s)(1+T_b s)\ldots\ldots(1+T_n s)}$$

$$\underset{s \to 0}{\text{Lt}}\ sG(s) = 0$$

$$e_{ss} = \frac{1}{sG(s)} = \frac{1}{0} = \infty$$

For type 1 system, $G(s) = \dfrac{K(1+T_1 s)(1+T_2 s)\ldots\ldots(1+sT_m)}{s^1 (1+T_a s)(1+T_b s)\ldots\ldots(1+sT_n)}$

$$\underset{s \to 0}{\text{Lt}}\ sG(s) = K$$

$$e_{ss} = \frac{1}{\underset{s \to 0}{\text{Lt}}\ sG(s)} = \frac{1}{K} = \frac{1}{K_v}$$

where $K_v = \underset{s \to 0}{\text{Lt}}\ sG(s)$ is known as velocity error constant.

For type 2 system, $G(s) = \dfrac{K(1+T_1 s)(1+T_2 s)\ldots\ldots(1+sT_m)}{s^2 (1+T_a s)(1+T_b s)\ldots\ldots(1+sT_n)}$

$$\underset{s \to 0}{\text{Lt}}\ sG(s) = \infty$$

$$e_{ss} = \frac{1}{\underset{s \to 0}{\text{Lt}}\ sG(s)} = 0$$

Hence steady state error of the system subjected to ramp input for more than type 1 system is zero.

(*iii*) **Parabolic input :**

We take $r(t) = C\,\dfrac{t^2}{2},\ t \geq 0$

Taking *LT*, $R(s) = \dfrac{C}{s^3} = \dfrac{1}{s^3}$, for $C = 1$

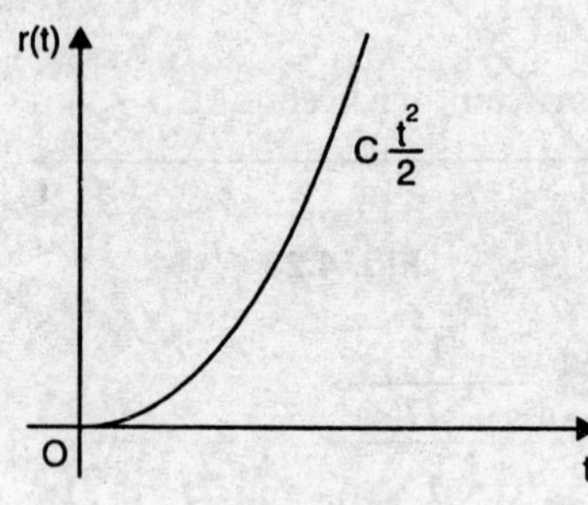

Fig. 4.29

From eqn. (4.5), steady state error is

$$e_{ss} = \underset{s \to 0}{\text{Lt}} \frac{s \times \dfrac{1}{s^3}}{1 + G(s)}$$

$$= \underset{s \to 0}{\text{Lt}} \frac{1}{s^2 + s^2\,G(s)} = \frac{1}{\underset{s \to 0}{\text{Lt}}\ s^2 + \underset{s \to 0}{\text{Lt}}\ s^2 G(s)}$$

$$e_{ss} = \frac{1}{\underset{s \to 0}{\text{Lt}}\ s^2\,G(s)}$$

For type 0 system, $G(s) = \dfrac{K(1+T_1 s)\,......\,(1+T_m s)}{s^0\,(1+T_a s)\,......\,(1+T_n s)}$

$$\underset{s \to 0}{\text{Lt}}\ s^2 G(s) = \underset{s \to 0}{\text{Lt}}\ s^2 K = 0$$

$$e_{ss} = \frac{1}{0} = \infty$$

Type 1 system, $G(s) = \dfrac{K(1+T_1 s)\,......\,(1+T_m s)}{s(1+T_a s)\,......\,(1+T_n s)}$

$$\underset{s \to 0}{\text{Lt}}\ s^2 G(s) = \underset{s \to 0}{\text{Lt}}\ sK = 0$$

$$e_{ss} = \frac{1}{0} = \infty$$

Type 2 system, $$G(s) = \frac{K(1+T_1s)......(1+T_ms)}{s^2(1+T_as)......(1+T_ns)}$$

$$\underset{s\to 0}{\text{Lt}}\ s^2G(s) = \underset{s\to 0}{\text{Lt}}\ K = K_a$$

$$e_{ss} = \frac{1}{K_a}$$

Where K_a is known as acceleration error constant.

For type 3 system, $$G(s) = \frac{K(1+T_1s)(1+T_2s)......(1+sT_m)}{s^3(1+T_as)(1+T_bs)......(1+sT_n)}$$

$$\underset{s\to 0}{\text{Lt}}\ sG(s) = \infty$$

$$e_{ss} = \frac{1}{\underset{s\to 0}{\text{Lt}}\ sG(s)} = 0 \qquad \left[\because \underset{s\to 0}{\text{Lt}}\ SG(s) = \infty\right]$$

Hence steady state error of the system subjected to parabolic input more than type 2 system is zero.

These three error constants *i.e.*, position, velocity and acceleration errors constant are known as **steady state error constants** (or) **static error constants.**

Steady state error for type 0, 1, 2 and 3 system is given in table 4.1.

Table 4.1

Type of system	K_p $\underset{s\to 0}{Lt}\ G(s)$	K_v $\underset{s\to 0}{Lt}\ sG(s)$	K_a $\underset{s\to 0}{Lt}\ s^2G(s)$	*Steady state error*		
				Step	*Ramp*	*Parabolic*
0	K	0	0	$\frac{1}{1+K}$	∞	∞
1	∞	K	0	0	$1/K$	∞
2	∞	∞	K	0	0	$1/K$
3	∞	∞	∞	0	0	0

The error constants describe the ability of a system to reduce or eliminate steady state error. As the type of system becomes higher, e_{ss} is eliminated. Systems higher than type 2, *i.e.*, with more than two integrations are not employed in practice because there is more difficulty to stabilize and dynamic system errors of such systems tend to be large, although their steady state performance is desirable.

The disadvantages of static error constants are

1. For any given type of system, only one of them will be finite, others being either zero or infinity.

2. e_{ss} evaluated with the above constants does not reveal its variation with time if any.

Note: K_p, K_v; and K_a can be used to evaluate e_{ss} when the input is a linear combination of step, velocity and acceleration. This is done by applying of superposition principle.

Problem 4.11. *Find the steady state error for the given system transient response* $C(t) = t - 1.25\, e^{-2t} + 2\, e^{+4t}$ *subjected to ramp input.*

Solution: $C(t) = t - 1.25\, e^{-2t} + 2\, e^{-4t}$

Input is ramp signal, $r(t) = t$

$\therefore$ Error signal $e(t)$ is the difference between desired output and actual output

$$e(t) = r(t) - c(t)$$

$$e(t) = t - t + 1.25\, e^{-2t} - 2e^{4t}$$

$$e(t) = 1.25\, e^{-2t} - 2e^{4t}$$

As $t \to \infty$, steady state error is

$$e_{ss} = \underset{t \to \infty}{\text{Lt}}\ e(t)$$

$$= \underset{t \to \infty}{\text{Lt}}\ 1.25\, e^{-2t} - 2e^{4t} = \infty$$

Problem 4.12. *Identify the type and order*

(a) $G(s) = \dfrac{k(1+2s)}{s^2}$ (b) $\dfrac{8}{s(s+5)}$ (c) $G = \dfrac{3}{(s^2+8)}$ *and* (d) $G = \dfrac{1}{s}$

Solution:

(a) $G(s) = \dfrac{k(1+2s)}{s^2}$, type 2 order 2

(b) $G(s) = \dfrac{8}{s(s+5)}$, type 1 order 2

(c) $G(1) = \dfrac{3}{(3s^2+8)}$, type 0 order 2

(d) $G = \dfrac{1}{s}$, type 1 order 1

Problem 4.13. *Determine the position, velocity and acceleration error constants for the unity feedback control system whose open loop transfer function is* $G(s) = \dfrac{K(s+25)(s+45)}{s^2(s^2+2s+8)}$

Solution: Position error constant is

$$K_p = \underset{s \to 0}{\text{Lt}}\ G(s)\, H(s)$$

$$= \underset{s \to 0}{\text{Lt}}\ \frac{k(s+25)(s+45)}{s^2(s^2+2s+8)} = \infty$$

Velocity error constant is

$$K_v = \underset{s \to 0}{\text{Lt}}\ sG(s)$$

$$= \underset{s \to 0}{\text{Lt}} \frac{k(s+25)(s+45)}{s^2(s^2+2s+8)} = \infty$$

Acceleration error, $K_a = \underset{s \to 0}{\text{Lt}}\ s^2 G(s)$

$$\underset{s \to 0}{\text{Lt}}\ s^2 \frac{k(s+25)(s+45)}{s^2(s^2+2s+8)} = \frac{1125K}{8}$$

Problem 4.14. *Determine the position, velocity and acceleration error constants for the unity feedback system whose open loop transfer function is* $G(s) = \dfrac{50}{(1+0.1s)(1+2s)}$

Solution: Position error constant, $K_p = \underset{s \to 0}{\text{Lt}}\ G(s)\,H(s)$

It may be noted that $H(s)$ is 1 as it is an open loop transfer function with unity feedback is $G(s)$

$$K_p = \underset{s \to 0}{\text{Lt}}\ G(s) = \underset{s \to 0}{\text{Lt}} \frac{50}{(1+0.1s)(1+2s)} = 50$$

Velocity error constant,

$$K_v = \underset{s \to 0}{\text{Lt}}\ sG(s)$$

$$= \underset{s \to 0}{\text{Lt}}\ s \frac{50}{(1+0.1s)(1+2s)} = 0$$

Acceleration error constant,

$$K_a = \underset{s \to 0}{\text{Lt}}\ s^2 G(s)$$

$$= \underset{s \to 0}{\text{Lt}}\ s^2 \frac{50}{(1+0.1s)(1+2s)} = 0$$

Problem 4.15. *Find the step, ramp and parabolic error coefficients and their corresponding steady state errors for unity feedback control system having the transfer function*

$$G(s) = \frac{14(s+3)}{s(s+5)(s^2+2s+5)}$$

Solution: The open loop transfer function of the system,

$$G(s) = \frac{14(s+3)}{s(s+5)(s^2+2s+5)},\ H(s) = 1$$

$$K_p = \underset{s \to 0}{\text{Lt}}\ G(s) = \underset{s \to 0}{\text{Lt}} \frac{14(s+3)}{s(s+5)(s^2+2s+5)} = \infty$$

Corresponding steady state error to

$$K_p = \frac{1}{1+K_p} = \frac{1}{\infty} = 0$$

$$K_v = \underset{s \to 0}{\text{Lt}}\ s \times \frac{14(s+3)}{s(s+5)(s^2+2s+5)} = \frac{14 \times 3}{5 \times 5} = \frac{42}{25}$$

The corresponding steady state error to K_v is

$$e_{ss} = \frac{1}{K_v} = \frac{1}{(42/25)} = 0.6$$

$$K_a = \underset{s \to 0}{\text{Lt}}\ s^2\, G(s) = \underset{s \to 0}{\text{Lt}}\ s^2 \times \frac{14(s+3)}{s(s+5)(s^3+2s+5)}$$

$$= \frac{0}{25} = 0$$

The corresponding steady state error to K_a is

$$e_{ss} = \frac{1}{K_a} = \frac{1}{0} = \infty \qquad [\because\ K_a = 0]$$

4.10 GENERALIZED ERROR COEFFICIENTS (OR) DYNAMIC ERROR COEFFICIENTS

The drawback in static error constants is that it does not show the variation of error with time and input should be a standard input. The generalized error coefficients give the steady state error as a function of time and also steady state error can be found for any type of input.

From eqn. (4.4), the error $E(s)$ of a feedback control system is

$$E(s) = \frac{R(s)}{1 + G(s)H(s)} \tag{4.26}$$

Rewriting the error eqn. (4.26) as product of two functions, we get

$$E(s) = W_e(s)\, R(s) \tag{4.27}$$

where
$$W_e(s) = \frac{1}{1 + G(s)H(s)} \tag{4.28}$$

Equation (4.28) is the error transfer function. Using convolution integral, we can get $e(t)$ from eqn. (4.27) as

$$e(t) = \int_{-\infty}^{t} W_e(\alpha)\, r(t-\alpha)\, d\alpha \tag{4.29}$$

If the n derivatives of $r(t)$ exist for all values of t, then expanding $r(t-\alpha)$ by using Taylor's expansion, we get

$$r(t-\alpha) = r(t) - \alpha \frac{dr(t)}{dt} + \frac{\alpha^2}{2!}\frac{d^2 r(t)}{dt^2} \tag{4.30}$$

Any standard test signal $r(t) = 0$ for $t < 0$

Substituting the eqn. (4.30) in eqn. (4.29)

$$\therefore \qquad e(t) = \int_0^t W_e(\alpha) \left[r(t) - \alpha \frac{dr(t)}{dt} + \frac{\alpha^2}{2!}\frac{d^2 r(t)}{dt^2} \ldots\ldots \right] d\alpha$$

$$e(t) = r(t) \int_0^t W_e(\alpha)\, d\alpha - \frac{dr(t)}{dt} \int_0^t \alpha W_e(\alpha)\, d\alpha + \frac{d^2 r(t)}{dt^2} \int_0^t \frac{\alpha^2}{2} W_e(\alpha)\, d\alpha \tag{4.31}$$

As before, steady state error is obtained from

$$e_{ss} = \underset{t \to \infty}{\text{Lt}}\; e(t)$$

i.e.,

$$e_{ss} = \int_0^\infty W_e(\alpha)\, r(t-\alpha)\, d\alpha$$

$$= r(t) \int_0^\infty W_e(\alpha)\, d\alpha - \frac{dr(t)}{dt} \int_0^\infty \alpha\, W_e(\alpha)\, d\alpha + \frac{d^2 r(t)}{dt^2} \int_0^\infty \frac{\alpha^2}{2} W_e(\alpha)\, d\alpha$$

Equation (4.31) is written as

$$e(t) = C_0 r(t) + C_1 \frac{dr(t)}{dt} + \frac{C_2}{2} \frac{d^2 r(t)}{dt^2}$$

where $C_0 = \int_0^\infty W_e(\alpha)\, d\alpha$

$$C_1 = - \int_0^\infty \alpha\, W_e(\alpha)\, d\alpha$$

$$C_2 = + \int_0^\infty \alpha^2\, W_e(\alpha)\, d\alpha$$

....

$$C_n = (-1)^n \int_0^\infty \alpha^n\, W_e(\alpha)\, d\alpha \tag{4.32}$$

Steady state error, $e_{ss} = \underset{t \to \infty}{\text{Lt}}\; e(t) = C_0 r(t) + C_1 \frac{dr(t)}{dt} + \frac{C_2}{2} \frac{d^2 r(t)}{dt^2} + \ldots$ (4.33)

Equation (4.33) is known as error series and the coefficients C_0, C_1, C_2 .... C_n are defined as the **generalized error coefficients** or simple error coefficients.

Since $W_e(\alpha)$ is the inverse Laplace transform of $W_e(s)$

i.e.,

$$W_e(s) = \int_0^\infty W_e(\alpha),\, e^{-s\alpha}\, d\alpha \tag{4.34}$$

Taking the limit on both sides of eqn. (4.34) as s approaches to zero, we have

$$\underset{s \to 0}{\text{Lt}}\; W_e(s) = \underset{s \to 0}{\text{Lt}} \int_0^\infty W_e(\alpha)\, d\alpha = C_0 \tag{4.35}$$

Hence the other coefficients can be derived as

$$C_1 = \underset{s \to 0}{\text{Lt}}\; \frac{dW_e(s)}{ds} \tag{4.36}$$

$$C_2 = \underset{s \to 0}{\text{Lt}}\; \frac{d^2 W_e(s)}{ds^2} \tag{4.37}$$

In general, $C_n = \underset{s \to 0}{\text{Lt}}\; \frac{d^n W_e(s)}{ds^n}$ (4.38)

Problem 4.16. *The open-loop transfer function of a control system with unity feedback is* $G(s) = \dfrac{150}{s(1+0.25s)}$

(i) Evaluate the error series for the system.

(ii) Determine the steady state error for an input $r(t) = (1 + t^2)\, u(t)$.

Solution:

$$W_e(s) = \frac{1}{1+G(s)H(s)} = \frac{1}{1+\dfrac{150}{s(1+0.25s)}}$$

$$= \frac{s(1+0.25s)}{0.25s^2+s+150}$$

Then $$C_0 = \underset{s\to 0}{\text{Lt}}\ W_e(s) = 0$$

$$C_1 = \underset{s\to 0}{\text{Lt}}\ \frac{dW_e}{ds}$$

$$= \underset{s\to 0}{\text{Lt}}\left[\frac{(0.25s^2+s+150)(0.5s+1) - s(1+0.25s)(0.5s+1)}{(0.25s^2+s+150)^2}\right]$$

$$= \frac{1}{150}$$

Similarly $C_2, C_3 \ldots\ldots C_n$, can be evaluated.

$$r(t) = (1 + t^2)\, u(t)$$

And $$\frac{dr(t)}{dt} = 2tu(t)$$

$$\frac{d^2r(t)}{dt^2} = 2u(t)$$

whereas higher powers of differentials is all zero.

Hence, $$e(t) = C_0 r(t) + C_1 \frac{dr(t)}{dt} + C_2 \frac{d^2r(t)}{dt^2} = 0 + \frac{1}{150}\, 2t + C_2 \,.\, 2$$

Then $$e_{ss} = \underset{t\to\infty}{\text{Lt}}\ e(t) = \infty$$

Problem 4.17. *For a negative feedback control system* $G(s) = \dfrac{10}{s(0.45+1)}$ *and* $H(s) = \dfrac{5}{s+4}$*, using generalized error series, determine the steady state error of the system when the input applied* $r(t)$ *is* $(1 + 3t + 4t^2)$.

Solution: The steady state error for non standard input can be obtained by using generalized error series as given below.

The error signal

$$e(t) = r(t)\,C_0 + \dot{r}(t)\,C_1 + \ddot{r}(t)\,\frac{C_2}{2!} + \ldots\ldots$$

Given $\qquad r(t) = 1 + 3t + 4t^2$

$$\dot{r}(t) = 3 + 8t$$

$$\ddot{r}(t) = 8$$

$$\dddot{r}(t) = 0$$

All derivatives of $r(t)$ are zero after second derivative. Hence we have to evaluate only three constants C_0, C_1 and C_2.

The generalized error constants are given by

$$C_0 = \underset{s\to 0}{\text{Lt}}\; W_e(s)\;;\; C_1 = \underset{s\to 0}{\text{Lt}}\; \frac{d}{ds}\,W_e(s)\;;\; C_2 = \underset{s\to 0}{\text{Lt}}\; \frac{d^2}{ds^2}\,W_e(s);$$

where $\qquad W_e(s) = \dfrac{1}{1+G(s)\,H(s)} = \dfrac{1}{1+\dfrac{50}{s(0.4s+1)\,(s+4)}}$

$$= \frac{0.4s^3 + 2.6s^2 + 4s}{0.4s^3 + 2.6s^2 + 4s + 50}$$

$$C_0 = \underset{s\to 0}{\text{Lt}}\; W_e(s) = \underset{s\to 0}{\text{Lt}}\; \frac{0.4s^3 + 2.6s^2 + 4s}{0.4s^3 + 2.6s^2 + 4s + 50} = 1/50 = 0.02$$

$$C_1 = \underset{s\to 0}{\text{Lt}}\; \frac{d}{ds}\,W_e(s) = \underset{s\to 0}{\text{Lt}}\; \frac{d}{ds}\left[\frac{0.4s^3 + 2.6s^2 + 4s}{0.4s^3 + 2.6s^2 + 4s + 50}\right]$$

$$= \underset{s\to 0}{\text{Lt}}\; \frac{(0.4s^3 + 2.6s^2 + 4s + 50)\,(1.2s^2 + 5.2s + 4) - (0.4s^3 + 2.6s^2 + 4s)\,(1.2s^2 + 5.2s + 4)}{(0.4s^3 + 2.6s^2 + 4s + 50)^2}$$

$$= \underset{s\to 0}{\text{Lt}}\left[\frac{60s^2 + 260s + 200}{0.4s^3 + 2.6s^2 + 4s + 50}\right] = 0.08$$

$$C_2 = \underset{s\to 0}{\text{Lt}}\; \frac{d^2}{ds^2}\,W_e(s) = \underset{s\to 0}{\text{Lt}}\; \frac{d}{ds}\left[\frac{dW_e}{ds}\right]$$

$$= \underset{s\to 0}{\text{Lt}}\; \frac{d}{ds}\left[\frac{60s^2 + 260s + 200}{0.4s^2 + 2.6s^2 + 4s + 50}\right]$$

$$= \frac{(0.4s^3 + 2.6s^2 + 4s + 50)\,2(120s + 260) - (60s^2 + 260s) + 200\,(0.4s^3 + 2.6s^2 + 4s + 50)\,(1.2s^2 + 5.2s + 4)}{(0.4s^3 + 2.6s^2 + 4s + 50)^4} = 0.09$$

$\therefore$ Error signal, $e(t) = r(t)\,C_0 + \dot{r}(t)\,C_1 + \ddot{r}(t)\,\dfrac{C_2}{2!}$

$$e(t) = \frac{(1 + 3t + 4t^2)}{50} + 0.08\,(3 + 8t) + 0.09 \times 8$$

$$= 0.02 + 0.06t + 0.2t^2 + 0.24 + 0.64t + 0.72$$

$$= 0.2t^2 + 0.7t + 0.96$$

Steady state error is obtained by letting t tend to ∞

$\therefore$ Steady state error, $e_{ss} = \underset{t \to \infty}{\text{Lt}}\; e(t)$

$$= \underset{t \to \infty}{\text{Lt}}\; [0.2t^2 + 0.7t + 0.96]$$

$\therefore$ Steady state error, $e_{ss} = \infty$

4.11 CONTROLLERS

A controller is a device which when introduced in the feedback or forward path of system, controls the steady state and transient response as per the requirement. Such device converts input to the controller to some other form of error which improves the transient and steady state response. In most of the practical systems, controller input is proportional to error generated. Such systems are called proportional to error mechanisms.

Consider a second order system as shown in Fig. 4.30.

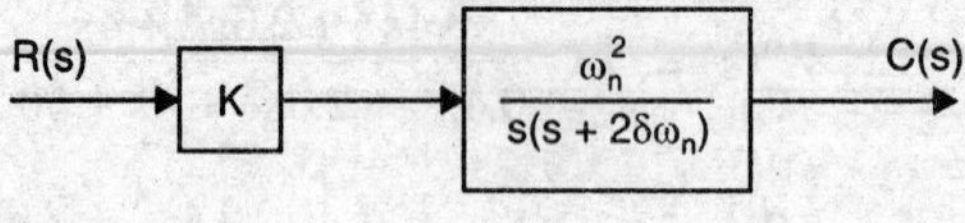

Fig. 4.30

where K is the proportionality constant.

Convert the given open loop system into closed loop by using negative unit feedback. Obtained block diagram is as shown in Fig. 4.31.

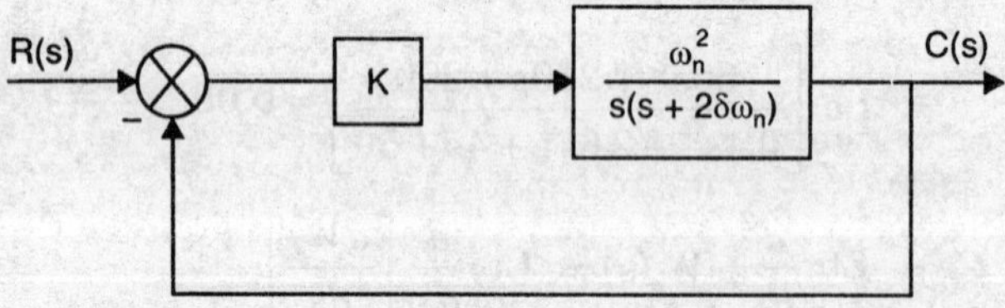

Fig. 4.31

where δ = damping ratio

ω_n = undamped natural frequency

For second order systems, velocity error constant $K_v = \frac{\omega_n}{2\delta}$ depends upon the damping ratio.

So, in order to improve the transient response of the system, we have to change the damping ratio. For good transient response less rise time, less overshoot, less settling time and small steady state error are needed.

Increasing the loop gain increases the velocity error and decreases steady state error. But due to high gain settling time and peak overshoot increases, this may cause system to be unstable. So compromise is made to keep steady state error and overshoot within the permissible limit by proving different types of controllers.

The types of controllers are

PD — Proportional plus derivative action.

PI — Proportional plus integral action.

PID — Proportional plus derivative plus integral action.

(*i*) **PD- Controller:** The proportional plus derivative controller produces an output signal consisting of two terms–one proportional to error signal and the other proportional to derivative of error signal.

Let the transfer function of PD controller with loop gain K_p and time constant T_d is $K_p(1 + T_d s)$.

where K_p = proportional gain

T_d = differential time

Block diagram representation of PD controller is shown in the Fig. 4.32.

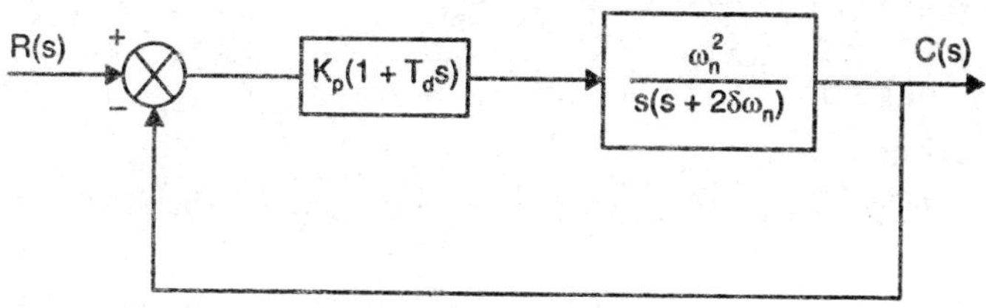

Fig. 4.32

Let the open loop transfer function $G(s)$ be second order system with transfer function $\frac{\omega_n^2}{s(s+2\delta\omega_n)}$. Now the closed loop transfer function is given by

$$\frac{C(s)}{R(s)} = \frac{K_p(1+T_d s)\dfrac{\omega_n^2}{s(s+2\delta\omega_n)}}{1+K_p(1+T_d s)\dfrac{\omega_n^2}{s(s+2\delta\omega_n)}}$$

$$= \frac{K_p(1+T_d s)\,\omega_n^2}{s(s+2\delta\omega_n)+K_p(1+T_d s)\,\omega_n^2}$$

$$= \frac{K_p\,(1+T_d s)\,\omega_n^{\,2}}{s^2+(2\delta\omega_n+K_d\omega_n^{\,2})\,s+K_p\,\omega_n^{\,2}}$$

where $K_d = K_p\,T_d$

$$= \frac{(K_p+K_d s)\,\omega_n^{\,2}}{s^2+(2\delta\omega_n+K_d\omega_n^{\,2})\,s+K_p\omega_n^{\,2}}$$

From closed loop transfer function, it is observed that the PD controller introduces a zero in the system and increases the damping ratio. The addition of zero may increase the peak overshoot and reduce the rise time. It improves the transient state without affecting the steady state.

(*ii*) **PI-Controller:** The proportional plus integral controller produces an output signal consisting of two terms–one proportional to error signal and other proportional to integral of error signal.

Block diagram representation of PI controller is shown in Fig. 4.33.

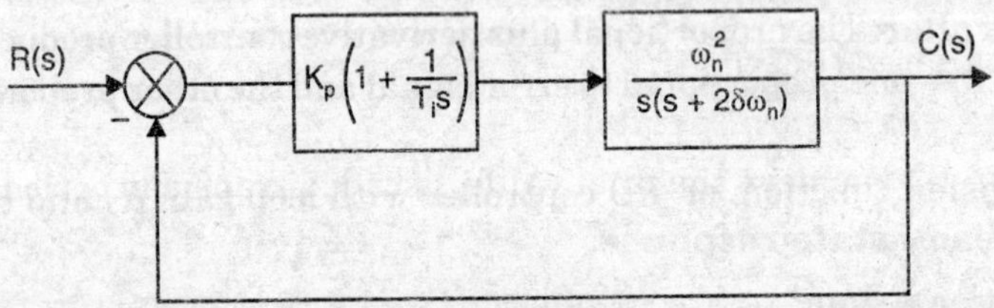

Fig. 4.33

Transfer function of PI controller: $K_p\left(1+\dfrac{1}{sT_i}\right)$

where K_p = Proportional gain

T_i = Integral time

If the transfer function of PI controller is connected in series with the second order system with unit transfer function $\dfrac{\omega_n^{\,2}}{s(s+2\delta\omega_n)}$, closed loop transfer function is given by

$$\frac{C(s)}{R(s)} = \frac{K_p\left(1+\dfrac{1}{T_i s}\right)\dfrac{\omega_n^{\,2}}{s(s+2\delta\omega_n)}}{1+K_p\left(1+\dfrac{1}{T_i s}\right)\dfrac{\omega_n^{\,2}}{s(s+2\delta\omega_n)}}$$

$$= \frac{K_p\omega_n^{\,2}(1+T_i s)}{s^3+2\delta\omega_n^{\,2}+K_i\omega_n^{\,2}(1+T_i s)}$$

where $K_i = \dfrac{K_p}{T_i}$. From closed loop transfer function, it is observed that PI controller introduces zero in the system and increases order by one. And also the type. Steady state error reduces tremendously for the same type of inputs. Design of K_i must be proper to maintain stability of system. So it makes the system relatively less stable because increase in order of the system results in less stability. Higher order systems are less stable than lower order systems. So, this *PI* controller improves steady state without affecting the transient state.

(*iii*) PID-Controller

As PD controller improves the transient state of the system and PI improves steady state of the system. Combination of these two may be used to improve overall time response of system. So suitable combination of the three basic modes proportional, integral and derivative can improve all aspects of the system performance.

The proportional controller stabilizes the gain but produces steady state error. Integral controller reduces or eliminates the steady state error. The derivative controller reduces the rate of change of error.

SHORT QUESTIONS AND ANSWERS

1. **Define the transient response.**

   It is defined as that part of the time response that goes to zero as time becomes large.

2. **What is mean by transient period?**

   The time required to achieve the final value, is called transient period.

3. **Define the steady state response.**

   It is the part of the time response which remains after complete transient response vanished from the system output.

4. **Define the steady state error.**

   It is the difference between the desired output and actual output of the system at infinite time called steady state error.

5. **What is the standard test signals used in time domain analysis?**

   The standard signals are step, ramp, parabolic and impulse signals.

6. **What is mean by first order system?**

   A system which is described by a first order differential equation is called a first order system.

7. **Define time constant of first order system.**

   It is the time taken for the output to change by 63.3% of its final value taken at the instant $t = 0$.

8. **What is the second order system?**

   A second order system is one wherein the highest power of '*s*' in the denominator of its function equals two.

9. **What are the time domain specifications?**

   The time domain specifications are : (*i*) delay time (*ii*) rise time (*iii*) peak time (*iv*) maximum overshoot and (*v*) settling time.

10. **Define the delay time.**

    It is the time for the response to reach half (50%) of its final value.

11. Define the rise time.

It is the time required by the step function response to reach from 10 to 90% or 0 to 100% of its final value.

12. Define the peak time.

It is the time taken by the controlled system to reach the peak of the first overshoot.

13. Define the peak overshoot.

It is defined as the difference between the peak value of step response and the steady output.

14. Define the settling time.

It is the time required by the system to reach and settle within a prescribed percentage of the final or steady value.

15. What is the difference between type and order of a system?

The type of the system indicates the number of poles at the origin whereas the order of the system gives the highest degree of numerator polynomial of the transfer function of the system.

16. What is the effect of the damping factor being less than one in a second order system subjected to step input?

The output overshoots and the steady state value oscillates before reaching the steady state. It also increases the speed of response as the damping factor is reduced.

17. What is meant by natural frequency of the system?

It is the frequency of the output of the system, if there is no damping and is also called free response to impulse input.

18. What is the nature of the output of a second order system of step input and having no damping?

It is oscillatory. The output oscillates with constant magnitude.

19. What is the relation between the peak overshoot and settling time in terms of changing the damping factor?

The peak overshoot and settling time are inversely proportional to damping factor.

20. What are the static error constants?

Position, velocity and acceleration errors constants are known as steady state error constants or static error constants.

21. What are the disadvantages of static error constants?

The disadvantages are

- For any given type of system, only one of them will be finite, others being either zero or infinity.
- e_{ss} evaluated with the above constants does not reveal its variation with time if any.

22. What are the advantages of generalized error constants?

The generalized error constants give the steady state error as a function of time and also steady state error can be found for any type of input.

23. What are the types of controllers?

The types of controllers are

- PD – Proportional plus derivative action
- PI – Proportional plus integral action
- PID – Proportional plus derivative plus integral action.

24. What is the need of PID controller?

As PD controller improves the transient state of the system and PI improves steady state of the system. Combination of the two may be used to improve overall time response of system.

25. What is the type and order of the transfer function is $G(s) = \frac{1}{s(s^2 + 3s + 10)}$.

Type-1 and Order-3

26. What is the effect of PI controller on system performance?

The PI controller increases the order of the system by one, which results in reducing the steady state error. But the system becomes less stable than the original system.

27. What is the effect of PD controller on system performance?

The effect of PD controller is to increase the damping ratio and so the peak overshoot is reduced.

OBJECTIVE TYPE QUESTIONS

1. In time domain analysis response of the system varies w.r.t.

(*a*) Time (*b*) Frequency
(*c*) Both time and frequency (*d*) Constant

2. Steady state error of first order control system subjected to ramp input is equal to

(*a*) Zero (*b*) Unity
(*c*) Time constant (*d*) Infinity

3. Undamped natural frequency for $s^2 + 2s + 1 = 0$ is

(*a*) Zero (*b*) One
(*c*) Two (*d*) Infinity

4. Damping ratio for problem 3 is equal to

(*a*) Zero (*b*) One
(*c*) 0.5 (*d*) Two

5. Order of the given open loop transfer function $G(s) = \frac{s+2}{s(s^2 + 2s + 2)}$

(*a*) One (*b*) Two
(*c*) Three (*d*) Zero

6. Type of the system given in problem 5 is equal to

(a) One (b) Two

(c) Three (d) Zero

7. Rise time for a given transfer function $G(s) = \dfrac{16}{s^2 + 1.6 + 16}$

(a) 0.35 sec (b) 0.45 sec

(c) 0.6 sec (d) 1 sec

8. Peak overshoot at which the response is equal to

(a) Zero (b) Infinity

(c) Unity (d) Two

9. Maximum overshoot is the difference between

(a) Peak response and steady state response

(b) Steady state response and peak response

(c) Output and input (d) Input and output

10. Position error constant K_p is equal to

(a) $\underset{s \to 0}{\text{Lt}}\ sG(s)$ (b) $\underset{s \to 0}{\text{Lt}}\ s^2G(s)$

(c) $\underset{s \to 0}{\text{Lt}}\ G(s)$ (d) $\underset{s \to 0}{\text{Lt}}\ s^3G(s)$

11. Velocity error constant K_v is equal to

(a) $\underset{s \to 0}{\text{Lt}}\ sG(s)$ (b) $\underset{s \to 0}{\text{Lt}}\ s^2G(s)$

(c) $\underset{s \to 0}{\text{Lt}}\ G(s)$ (d) $\underset{s \to 0}{\text{Lt}}\ s^3G(s)$

12. One of the disadvantages of static error coefficients

(a) Applicable only for specified inputs

(b) Will not give the internal state of the system

(c) Both a and b (d) Time consuming process

13. In generalized error constants, constants C_1 is equal to

(a) $\underset{s \to 0}{\text{Lt}}\ W_e(s)$ (b) $\underset{s \to 0}{\text{Lt}}\ sW_e(s)$

(c) $\underset{s \to 0}{\text{Lt}}\ s^2W_e(s)$ (d) $\underset{s \to 0}{\text{Lt}}\ s^3 sW_e(s)$

14. Steady state response of the system $C(t) = 1 + 1.2\,e^{-t} + 1.2\,e^{t}$ is equal to

(a) $1 + \dfrac{1.2}{s-1} + \dfrac{1.2}{s+1}$ (b) $\dfrac{1}{s} + \dfrac{1.2}{s+1} + \dfrac{1.2}{s-1}$

(c) $\dfrac{1}{s} - \dfrac{1.2}{s-1} + \dfrac{1.2}{s+1}$ (d) $\dfrac{1}{s} + \dfrac{1.2}{s-1} - \dfrac{1.2}{s+1}$

15. Effect of addition of pole, the stability of the system will

(*a*) Decrease (*b*) Increase

(*c*) Constant (*d*) Not effected

16. Effect of addition of pole increases the system

(*a*) Order (*b*) Type

(*c*) Order and type (*d*) Steady state error

17. Effect of addition of zero increases

(*a*) Order (*b*) Type

(*c*) Order and type (*d*) Steady state error

18. The acceleration lag error of type 2 system is

(*a*) Independent of gain constant.

(*b*) Directly proportional to 50% of gain constant.

(*c*) Inversely proportional to the inverse of gain constant.

(*d*) Inversely proportional to the gain constant.

19. The presence of non-linearities in a control system tends to introduce

(*a*) Transient error (*b*) Instability

(*c*) Static error (*d*) Steady-state error

20. The static acceleration constant of a type 2 system is

(*a*) Infinite (*b*) Zero

(*c*) Cannot be found out (*d*) Finite

21. The transfer function of the system which will have more steady state error for step input is

(*a*) $\frac{80}{(s+1)(s+2)(s+3)}$ (*b*) $\frac{120}{s(s+1)(s+15)}$

(*c*) $\frac{60}{(s+0.5)(s+3)(s+5.5)}$ (*d*) $\frac{120}{(s+1)(s+4)(s+15)}$

22. The presence or absence of steady-state error for any given system depends upon

(*a*) Presence or absence of pole at the infinity.

(*b*) Presence or absence of poles and zeros at the origin.

(*c*) Absence or presence of zeros at the origin.

(*d*) Absence or presence of pole at the origin.

23. When the gain 'K' of a system is increased, the steady-state error of the system

(*a*) Increases (*b*) Remains unchanged

(*c*) May increase or decrease (*d*) Decreases

24. The plant is represented by the transfer function $\frac{k}{s+a}$. The system is given a degenerative feedback. The effective of the feedback is to shift the pole

(*a*) Positively to $s = (a + k)$ and reduce the time constant to $\alpha + \left(\frac{1}{k}\right)$

(*b*) Negatively to $s = -(\alpha + k)$ and to increase the time constant to $\alpha + k$.

(*c*) Negatively to $s = -(\alpha + k)$ and to decrease the time constant $\left[\left(\frac{1}{\alpha + k}\right)\right]$.

(*d*) Negatively to $s = -(\alpha + k)$ and to reduce the time constant to $\alpha + \left(\frac{1}{k}\right)$.

25. In a system with input $R(s)$ and output $C(s)$, the transfer functions of the plant and the feedback system is given by $G(s)$ and $H(s)$ respectively. The system has got a negative feedback. Then the error signal is given by the expression.

(*a*) $E(s) = \frac{G(s)\,R(s)}{1+G(s)\,H(s)}$ (*b*) $E(s) = C(s)\,G(s)$

(*c*) $E(s) = \frac{1}{1+G(s)\,H(s)}$ (*d*) $E(s) = \frac{R(s)}{1+G(s)\,H(s)}$

26. The static error constants depends on

(*a*) The order of the system (*b*) The type of the system

(*c*) Both type and order of the system (*d*) None of the above

27. For which of the following input, the error series using dynamic error coefficients doesn't converge

(*a*) Step input (*b*) Ramp input

(*c*) Acceleration (parabolic) input (*d*) Sinusoidal input

28. The radial distance between a pole and the origin gives

(*a*) Damped frequency of oscillation (*b*) Undamped frequency of oscillation

(*c*) Time constant (*d*) Natural frequency of oscillation

29. For a type 1, second order control system, when there is an increase of 25% in its natural frequency, the steady-state error to unit ramp input is

(*a*) Increased by 20% of its value

(*b*) Equal to $\frac{2\delta}{\omega_n}$, where δ = damping factor

(*c*) Decreased by 21%

(*d*) Decreased effectively by 20%

30. In a type 1, second order system, first peak overshoot occurs at a time equal to

(a) $\frac{\pi\omega_n}{\sqrt{1-\delta^2}}$ (b) $\frac{\omega_n}{\pi\sqrt{1-\delta^2}}$

(c) $\frac{\pi\omega_n}{\sqrt{1+\delta^2}}$ (d) $\frac{\frac{\pi}{\omega_n}}{\pi\sqrt{1-\delta^2}}$

31. Type number of a system gets decreased if

(a) First an integrator and then a differentiator is included in the system

(b) An integrator is included in the forward path

(c) A differentiator is included in a parallel path

(d) A differentiator is included in the forward path

32. When the pole of a system is moved towards the imaginary axis, then

(a) Settling time decreases

(b) Settling time increases by 20% of initial value.

(c) Steady-state error is reduced to zero

(d) Settling time of the system increases

33. For changing time constant providing sufficient damping in the control system, the factor that is responsible, is

(a) first derivative of input signal

(b) 50% of the second derivative of the input signal

(c) The input signal

(d) Second derivative of the input signal

34. If the phase ϕ which is the angle between radial line connecting a pole and origin, is equal to 45%, then the peak overshoot is

(a) 0.5% (b) 1.2%

(c) 3% (d) 4.32%

35. The damping factor of a second order system whose response to unit step input is having sustained oscillations, is

(a) = 1 (b) > 1

(c) < 1 (d) = 0

36. The transient response of a system with feedback when compared to that without feedback

(a) Decays slowly (b) Rises slowly

(c) Rises more quickly (d) Decays more quickly

37. The settling time for the system $G(s) = \frac{25}{s^2 + 5s + 25}$ is ... seconds when the output settles within ± 2% for a unit step input.

(*a*) 0.8 (*b*) 1.2

(*c*) 2.0 (*d*) 1.6

38. The type of the system whose transfer function is given by $G(s) = \dfrac{(s+3)}{s^5 + s^4 + s^3 + 3s^2 + 2s}$ is

(*a*) 3 (*b*) 2

(*c*) 5 (*d*) 1

39. The transfer function of a plant is given by the expression $\dfrac{K}{s(s+1)}$ and the transfer function of the feedback block is $1 + K_h s$. The values of K and K_h such that peak overshoot is 20% and the time for peak overshoot is 1 sec, are

(*a*) 10.5, 2 (*b*) 16, 1

(*c*) 9, 0.5 (*d*) 12.5, 0.178

40. Physically the damping ratio represents the

(*a*) Energy available for transfer

(*b*) Energy available for exchange

(*c*) Ratio of energy available for exchange to that available for transfer

(*d*) Ratio of energy lost to the energy available for exchange

41. Four systems *A*, *B*, *C* and *D* are found to have the dominant poles at the points $-3 + 5j$, $-5 + 4j$, $-12 + 1j$, $-1 + 6j$ respectively. The choice of the system having maximum damping and minimum natural frequency would be

(*a*) *C* and *A* respectively (*b*) *B* and *C* respectively

(*c*) *C* and *D* respectively (*d*) *A* and *C* respectively

42. A system is required to have a peak overshoot of 4.32% and a natural frequency of 5 rad/sec. The required location of the dominant pole is

(*a*) $-3.535 + 5.535j$ (*b*) $-2.535 + 2.525j$

(*c*) $-2.535 + 3.535j$ (*d*) $-3.535 + 3.535j$

43. The time constant 'τ' for a system is the time taken by the exponential function e^{-at} to decay to

(*a*) 63% of its original value at $a = 0$ (*b*) 37% of its steady-state value

(*c*) 63% of its steady-state value (*d*) 37% of its original value at $t = 0$

44. The resonant and the damping frequency of a certain system was found to be 7.07 rad/sec and 8.666 rad/sec respectively, the real co-ordinate of the dominant pole is :

(*a*) -8.12 (*b*) -7

(*c*) 0.8 (*d*) 0.6

45. The resonant peak value of a system is found to be 1.042. The damping ratio δ is

(*a*) 0.7 (*b*) 0.5

(*c*) 0.8 (*d*) 0.6

46. The time domain specification which is dependent only on the damping factor is

(*a*) Rise time (*b*) Peak time

(*c*) Settling time (*d*) Peak overshoot

47. An integral controller is used to improve the transient response of a first-order system. If $G(s) = \frac{1}{1+s}$, and the system is operated in closed-loop with unity feedback, what is the value of T_i if the integral controller transfer function is $\frac{1}{T_i}$ to provide damping ratio of 0.5?

(*a*) 0.5 (*b*) 2

(*c*) 1 (*d*) 4

48. A step response of a system is given below.

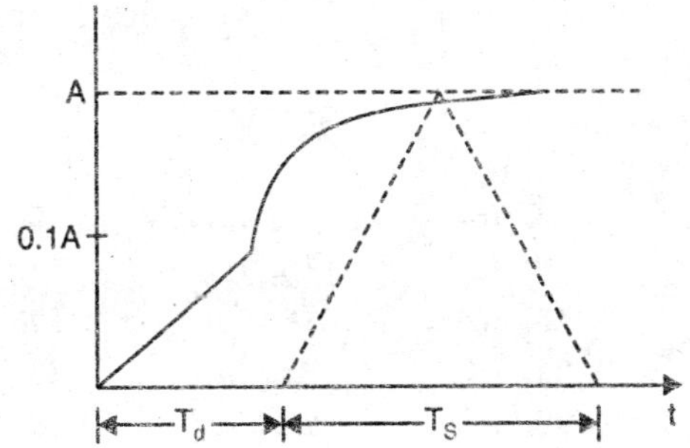

T_d represents the delay due to transportation lag and T_s is the rise time of the system. As a thumb rule, the system is easily controllable if $\frac{T_s}{T_d}$ is

(*a*) Less than 1 (*b*) Less than 3

(*c*) Greater than 10 (*d*) Equal to 6

49. A system has open-loop transfer given by $\frac{1}{(1+s)(1+0.5s)}$. The performance of this system is made faster with a controller of the form $\frac{K(1+T_1 s)}{(1+T_2 s)}$. The system with controller is operated in closed-loop with unity feedback. In order to increase the speed of response

(*a*) $T_1 = 1$ (*b*) $T_1 = 0.5$ and $T_2 = 1$

(*c*) $T_1 = 1$ and $T_2 = 1$ (*d*) $T_1 = 0.5$ and $T_2 < 0.5$

50. If stability error for step input and speed of response be the criteria for design, what controller would you recommend?

(*a*) P controller (*b*) PD controller

(*c*) PI controller (*d*) PID controller

51. An ON-OFF controller is a

(*a*) P controller (*b*) Integral controller

(*c*) Non-linear controller (*d*) PID controller

52. The term 'reset control' refers to

(*a*) Proportional control (*b*) Integral control

(*c*) Derivative control (*d*) PID controller

53. The pole-zero plot given below is that of a/an

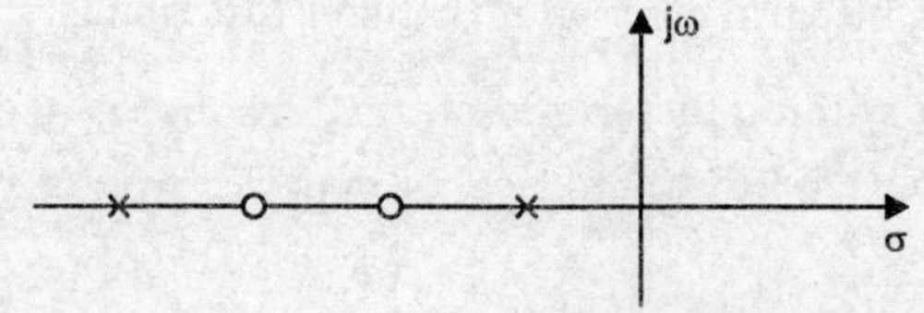

(*a*) Integrator (*b*) PD controller

(*c*) PID controller (*d*) Lag-lead compensating network

KEY

1. (*a*)	**2.** (*c*)	**3.** (*b*)	**4.** (*c*)	**5.** (*c*)	**6.** (*a*)
7. (*a*)	**8.** (*c*)	**9.** (*a*)	**10.** (*c*)	**11.** (*a*)	**12.** (*c*)
13. (*a*)	**14.** (*b*)	**15.** (*b*)	**16.** (*a*)	**17.** (*b*)	**18.** (*c*)
19. (*b*)	**20.** (*d*)	**21.** (*d*)	**22.** (*d*)	**23.** (*d*)	**24.** (*c*)
25. (*d*)	**26.** (*c*)	**27.** (*d*)	**28.** (*d*)	**29.** (*d*)	**30.** (*d*)
31. (*d*)	**32.** (*d*)	**33.** (*d*)	**34.** (*d*)	**35.** (*d*)	**36.** (*d*)
37. (*d*)	**38.** (*d*)	**39.** (*d*)	**40.** (*d*)	**41.** (*d*)	**42.** (*d*)
43. (*d*)	**44.** (*d*)	**45.** (*d*)	**46.** (*d*)	**47.** (*c*)	**48.** (*c*)
49. (*d*)	**50.** (*d*)	**51.** (*c*)	**52.** (*b*)	**53.** (*d*)	

EXERCISE

1. What is meant by step, ramp, parabolic and impulse inputs ?
2. Define time domain specifications.
3. Determine the response of second order system with unit step input.
4. Derive the time domain specifications of second order system with unit step input.
5. Explain the effect of adding a zero to a second order system response.
6. Define the steady state error and error constants of different types of inputs.
7. What are disadvantages of static error constants ?
8. Derive the generalized error constants.
9. Briefly discuss the different types of controllers.

10. A unity feedback system is characterized by an open loop transfer function $G(s) = \dfrac{K}{s(s+10)}$.

Determine the gain K so that the system will have a damping factor of 0.5. For this value of K determine the natural frequency of the system. It is subjected to a unity step input. Obtain the closed loop response of the system in time domain.

11. A unity feedback system is characterized by the open loop transfer function $G(s) = \dfrac{1}{s(1+0.5s)(1+0.2s)}$.

Determine the steady state error for unity step, unity ramp and unity acceleration inputs. Also determine the damping factor and natural frequency of dominant roots.

12. Given the open loop transfer function of a servo system with unity feed back is $G(s) = \dfrac{10}{s(1+0.1s)}$. Obtain the steady state error of the system when subjected to an input signal given by $r(t) = a_0 + a_1 t + \dfrac{a^2 t^2}{2}$

13. Damping factor and natural frequency of the system are 0.115 and 92.41 rad/sec respectively. Determine the rise time (t_r), peak time (t_p), maximum peak overshoot (m_p) and settling time (t_s).

14. Determine the position, velocity and acceleration error constants for the unity feedback control system whose open loop transfer function is $G(s) = \dfrac{K(1+2s)(1+4s)}{s^2(s^2+2s+8)}$

15. The open loop transfer function of a control system with unity feed back system is

$$G(s) = \frac{150}{s(1+0.25s)}$$

(*a*) Evaluate the error series for the system.

(*b*) Determine the steady state error for an input $\left(1+5t+\dfrac{3t^2}{2}\right)$.

5 Stability Analysis and Root Locus Technique

5.1 CONCEPT OF STABILITY

The first step in control system is to get a system, then we have to prepare a mathematical model by writing mathematical equations. These mathematical equations obtained are in time domain changes w.r.t. time. The time solution of the system is classified into two types: one in which the output tends to a finite value and second in which output approaches towards infinite value.

BIBO system for stability: The system is said to be bounded input–bounded output if

- The output is bounded for a given bounded input.
- The absence of input, output must be zero.

For unstable:

- For a bounded input, output is unbounded.
- In the absence of input, the output may not return to zero.

For example:

The stability of the system depends upon the position of poles on *s*-plane.

Case 1: If the roots of characteristic equation of system are lying on left side of *s*-plane, the system is said to be stable. The roots may be either negative, real or imaginary roots for a given BIBO.

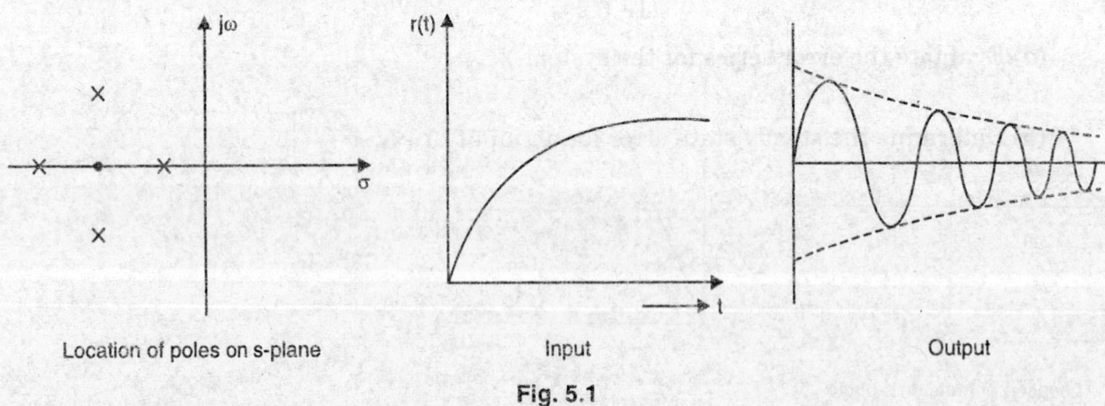

Fig. 5.1

Case 2: If the roots are located on imaginary axis, the output oscillates with constant frequency and amplitude for a bounded input. Such systems are said to be marginally or critically stable.

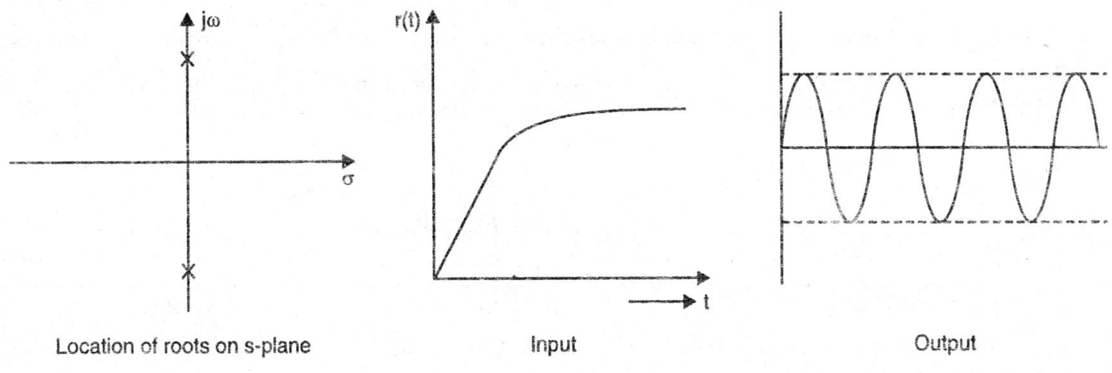

Fig. 5.2

Case 3: If the roots are repetitively located on imaginary axis of s-plane, the system is said to be unstable.

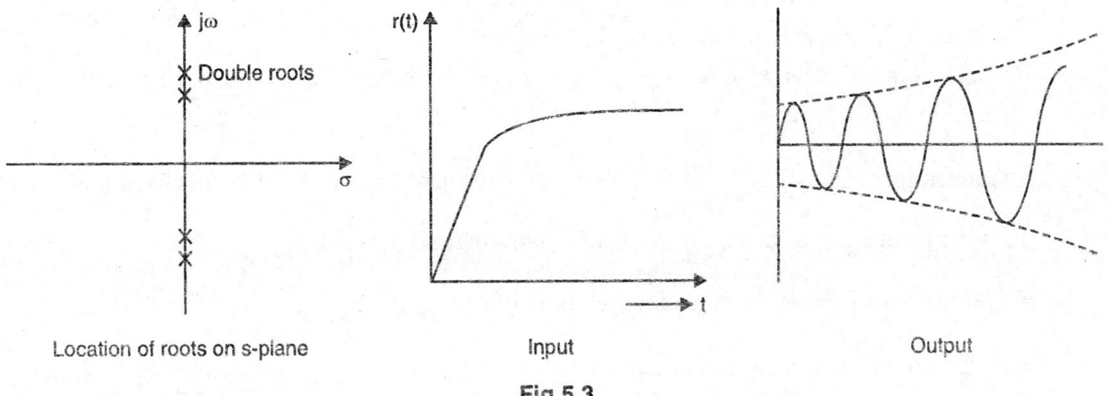

Fig 5.3

Case 4: If the roots are lying to right side of s-plane the system is said to be unstable.

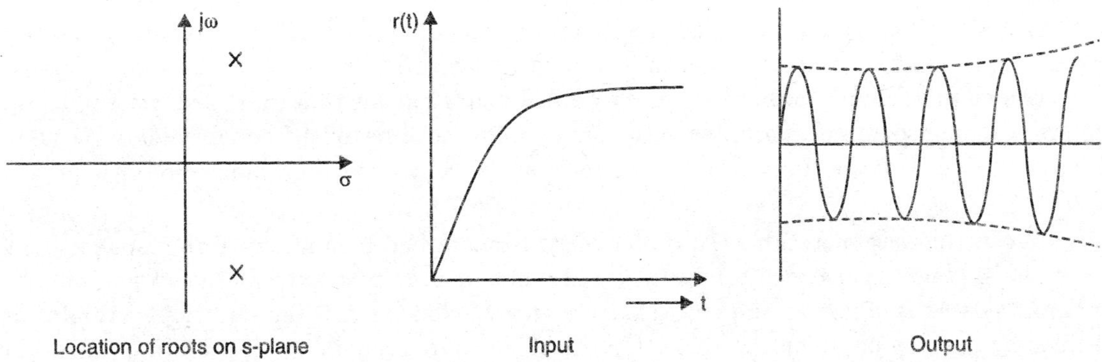

Fig. 5.4

So location of poles determines the stability of system, which are roots of characteristic equation.

Problem 5.1. *Determine if the systems and the following roots are stable, marginally stable and unstable.*

(a) – 3, – 6 (b) – 2 + 1

(c) – 1, 0 (d) – 1 + j, – 1 – j

(e) – 1, 1, – j, j (f) – 2 + j3, – 2 – j3, – 2

(g) an integrator (h) a step input

(i) $x(t) = \cos \omega t$ (j) $x(t) = e^{-t} \sin 4t$

Solution:

(a) Stable, negative real parts

(b) Unstable, one root is positive real part.

(c) Marginally stable as the roots are non positive

(d) Stable, the negative real part only counts

(e) Unstable, positive root 1

(f) Stable

(g) An integrator will be $\frac{F(s)}{s}$ or its characteristic equation is $s = 0$. As there is no root with positive real part, an integrator is marginally stable.

(h) A step input gives $\frac{F(s)}{s}$ and hence marginally stable.

(i) Unstable because there is no decay.

(j) Stable decaying exponentially.

5.2 ROUTH AND HURWITZ STABILITY CRITERIONS

In time domain analysis we studied the dynamic behavior of a feedback control system for specific inputs such as unit step, unit ramp and unit parabolic; there is a steady state response and a transient response to the system. We noted in our analysis that there are two important factors to response namely damping ratio (δ) and undamped natural frequency (ω_n). We shall now study about another important characteristic of a control system, the extent or degree of stability.

A control system is stable if it is able to avoid sustained oscillations. If in response to an input the system experiences to violent oscillations or driven to extreme it is unstable. An unstable system is of no use and only a system with feedback cause instability. So stability is decided by the way in which the characteristic of a system respond to different inputs rather than the nature of input.

There is also another definition of stability. A stable system is one which produces a bounded output to a bounded input (BIBO). A bounded signal is finite. A ramp input in not a bounded input.

5.2.1 Causes for Instability

The chief cause for instability is the inability of the system elements to respond quickly to input. In a feedback control system the quickness with which the system elements respond to error signal determines the ability to suppress unwanted oscillations. The delay to response is caused by energy storage elements such as inductance, capacitance, motors with magnetic field and rotating shaft. These elements store a large quantum of energy and it takes time for the energy to reach steady state value after oscillations to a reference input. If the oscillations die out the system is stable, if not it is unstable.

5.2.2 Absolute Stability and Relative Stability

There are two types of stability, absolute and relative stability. Absolute stability merely indicates whether a system is stable or not. But this information is insufficient from design point of view. We should also know the extent or degree of stability, that is, the extent to which the system can be operated before instability sets in. This is known as **relative stability**.

From the definitions of stability and causes for instability we find a close relation between relative stability and transient response of feedback control system which we studied in detail in the previous section. For a stable system the oscillations must die out and so there should be exponential decay. Thus for stability, the coefficient of the exponential terms in the transient response must be either negative real numbers or complex numbers with negative real parts. If the roots of characteristic equation lie on the left side of the s-plane, the system is stable; if they are on the imaginary axis or right half of s-plane the system is unstable. Thus the characteristic equation is a clue to the transient response of a system and hence its stability.

The characteristic equation can be related to the system's transfer function when it is known as characteristic transfer function of the system.

From eqn. (4.6),

$$\text{transfer function, } T(s) = \frac{C(s)}{R(s)} = \frac{G(s)}{1+G(s)H(s)}$$

or

$$C(s) = \frac{G(s)}{1+G(s)H(s)} R(s) \tag{5.1}$$

$G(s)\,R(s)$ is the excitation (or) driving function and stability does not depend on input or excitation. Only the steady state value depends on input function. So $G(s)\,R(s)$ may be set to zero without affecting the form of transient.

From eqn. (5.1), the characteristic equation is

$$1 + G(s)\,H(s) = 0 \tag{5.2}$$

This is the characteristic equation of a closed loop system and system stability can be evaluated from roots of $G(s)\,H(s)$. Being a product it can be factored in terms of its roots and can be written as a ratio of polynomials in s.

Open loop transfer function is

$$G(s)\,H(s) = \frac{N(s)}{D(s)} = \frac{\text{Numerator polynomial}}{\text{Denominator polynomial}} \tag{5.3}$$

The characteristic equation is

$$1 + G(s)\,H(s) = 1 + \frac{N(s)}{D(s)} = \frac{D(s) + N(s)}{D(s)} = 0$$

or $$D(s) + N(s) = 0 \tag{5.4}$$

The roots of the characteristic equation give an idea of the stability. This method though straight forward is cumbersome as it is difficult to factorize powers more than 2. An easier method is the Routh's and Nyquist criteria for stability. The other methods are Root locus plot, Bode diagram and Lyaupanov's stability criterion. The characteristic polynomial is given by

$$F(s) = a_0\, s^n + a_1 s^{n-1} + \ldots\ldots\, a_{n-1}\, s + a_n = 0 \tag{5.5}$$

where all the coefficients are real numbers.

5.2.3 Hurwitz Stability Criterion

Given the characteristic polynomial $F(s)$ of eqn. (5.5), n determinants are formed from the coefficients $a_{n1}\, a_{n-1}, \ldots\ldots a_1, a_0$, which must be positive, where the determinants are taken as the principal minors of the $(n \times n)$ Hurwitz determinant, given by

$$\begin{vmatrix} a_{n-1} & a_n & 0 & 0 & 0 & \cdot & \cdot & 0 & 0 \\ a_{n-3} & a_{n-2} & a_{n-1} & a_n & 0 & \cdot & \cdot & \cdot & 0 \\ a_{n-5} & a_{n-4} & a_{n-3} & a_{n-2} & a_{n-1} & a_n & 0 & \cdot & 0 \\ a_{n-7} & a_{n-6} & a_{n-5} & a_{n-4} & a_{n-3} & a_{n-2} & a_{n-1} & a_n & \cdot & 0 \\ \cdot & \cdot & \cdot & \cdot & \cdot & \cdot & \cdot & \cdot & \cdot \\ \cdot & \cdot & \cdot & \cdot & \cdot & \cdot & \cdot & \cdot & \cdot \\ a_{-n+1} & \cdot & \cdot & \cdot & \cdot & \cdot & a_2 & a_1 & a_0 \end{vmatrix}$$

where all the coefficients with larger than 'n' or negative are replaced by zeros. Thus the various determinants evaluated taken on the principal minors are

$$\Delta_1 = a_{n-1} > 0$$

$$\Delta_2 = \begin{vmatrix} a_{n-1} & a_n \\ a_{n-3} & a_{n-2} \end{vmatrix} > 0$$

$$\Delta_3 = \begin{vmatrix} a_{n-1} & a_n & 0 \\ a_{n-3} & a_{n-2} & a_{n-1} \\ a_{n-5} & a_{n-4} & a_{n-3} \end{vmatrix} > 0$$

$$\vdots \qquad \vdots$$

$$\Delta_n = \begin{vmatrix} a_{n-1} & a_n & 0 & 0 & 0 & \cdot & \cdot & 0 \\ a_{n-3} & a_{n-2} & \cdot & \cdot & \cdot & \cdot & \cdot & 0 \\ a_{n-5} & a_{n-4} & \cdot & \cdot & \cdot & \cdot & \cdot & 0 \\ \cdot & \cdot & \cdot & \cdot & \cdot & \cdot & \cdot & \cdot \\ \cdot & \cdot & \cdot & \cdot & \cdot & \cdot & \cdot & \cdot \\ a_{-n+1} & \cdot & \cdot & \cdot & \cdot & \cdot & a_1 & a_0 \end{vmatrix} > 0$$

When $\Delta_{n-1} = 0$, the system is critically stable.

The following examples illustrate the procedure.

Problem 5.2. *Investigate the stability of the system whose characteristic polynomial is*

$$F(s) = s^4 + 3s^3 + 2s^2 + s + 14 = 0$$

Solution: By inspection of characteristic polynomial all coefficients are positive and all powers of s are present. Thus the necessary conditions are satisfied. The (4 × 4) Hurwitz determinant is

$$\begin{vmatrix} 3 & 1 & 0 & 0 \\ 1 & 2 & 3 & 1 \\ 0 & 14 & 1 & 2 \\ 0 & 0 & 0 & 14 \end{vmatrix}$$

We have $\Delta_1 = 3 > 0$

$$\Delta_2 = \begin{vmatrix} 3 & 1 \\ 1 & 2 \end{vmatrix} = 5 > 0$$

$$\Delta_3 = \begin{vmatrix} 3 & 1 & 0 \\ 1 & 2 & 3 \\ 0 & 14 & 1 \end{vmatrix} = -121 < 0$$

$$\Delta_4 = \begin{vmatrix} 3 & 1 & 0 & 0 \\ 1 & 2 & 3 & 1 \\ 0 & 14 & 1 & 2 \\ 0 & 0 & 0 & 14 \end{vmatrix} = -1684 < 0$$

Since Δ_3 and Δ_4 are negative, the system is unstable.

Problem 5.3. *Investigate the stability of the system whose characteristic polynomial is,* $F(s) = s^4 + 2s^3 + 8s^2 + 4s + 3 = 0$.

Solution: By inspection $F(s)$ satisfies the necessary conditions of stability. For testing the sufficient conditions, we shall form the Hurwitz determinant.

$$\begin{vmatrix} 2 & 1 & 0 & 0 \\ 4 & 8 & 2 & 1 \\ 0 & 3 & 4 & 8 \\ 0 & 0 & 0 & 3 \end{vmatrix}$$

And we have from this $\Delta_1 = 2 > 0$

$$\Delta_2 = \begin{vmatrix} 2 & 1 \\ 4 & 8 \end{vmatrix} = 36 > 0$$

$$\Delta_4 = 3\Delta_3 = 108 > 0$$

Since all Δ's are positive, the system is stable.

5.2.4 Routh Stability Criterion

It is based on ordering the coefficients of the characteristic polynomial of eqn. (5.5) into an array, called the Routh array.

The necessary conditions for stability (that is, roots of the characteristic eqn. (5.5) with real parts are)

(*i*) All the coefficients $a_0, a_1, \ldots\ldots, a_n$ have the same sign.

(*ii*) None of the coefficients vanishes.

(*iii*) $a_0 > 0$

The sufficient condition for stability is that there should not be any sign changes in the first column of Routs array. The number of sign changes indicates the numbers of roots with positive real parts.

The Routh array for eqn. (5.5) is obtained as follows:

$$\begin{array}{c|cccccc} S^n & a_0 & a_2 & a_4 & \cdots & a_{n-2} & a_n \\ S^{n-1} & a_1 & a_3 & a_5 & \cdots & a_{n-1} & 0 \\ S^{n-2} & b_1 & b_3 & b_5 & \cdots & 0 & 0 \\ S^{n-3} & c_1 & c_3 & c_5 & \cdots 0 & 0 & 0 \\ . & . & . & . & . & . & . \\ . & . & . & . & . & . & . \\ S^0 & d_1 & & & & & \end{array} \tag{5.6}$$

where $b_1 = \dfrac{a_1a_2 - a_0a_3}{a_1}$; $b_3 = \dfrac{a_1a_4 - a_0a_5}{a_1}$ and so on.

$$c_1 = \frac{b_1a_3 - a_1b_3}{b_1} \; ; \; c_3 = \frac{b_1a_5 - a_1b_5}{b_1} \tag{5.7}$$

and so on till the last element of the row becomes zero. The procedure is continued till all the $(n + 1)$ rows are completed and the last row for s^0 has a single element.

Statement of stability:

For the system to be stable, it is necessary and sufficient that each term of the first column of the Routh array of eqn. (5.6) of the characteristic equation is positive if $a_n > 0$. If this condition is not satisfied, the system is unstable and the number of sign changes of the terms in the first column of the Routh array corresponds to the number of roots of the characteristic polynomial in the right half of the s-plane.

Problem 5.4. *Test the stability of the system whose characteristic equation is* $s^5 + 6s^4 + 3s^3 + 2s^2 + s + 1 = 0$.

Solution: In the given characteristic equation all coefficients are present and all coefficients have the same sign, so necessary condition is satisfied.

For sufficient condition, the Routh array is

$$\begin{array}{c|ccc} s^5 & 1 & 3 & 1 \\ s^4 & 6 & 2 & 1 \\ s^3 & 2.67 & 0.83 & 1 \\ s^2 & 0.135 & 1 & 0 \\ s & -18.95 & 0 & \\ s^0 & 1 & & \end{array}$$

There are two sign changes from s^2 row to s^1 row and from s^1 row to s^o row.

Hence the system is unstable and it has two roots with positive real parts *i.e.*, two roots on the right half of s-plane.

Problem 5.5. *Given the system characteristic equation as* $s^4 + 6s^3 + 21s^2 + 36s + 20 = 0$, *determine whether the system is stable or not by applying Routh-Hurwitz criterion.*

Solution: Routh array is

s^4	1	21	20
s^3	6	36	0
s^2	$\dfrac{6 \times 21 - 1 \times 36}{6}$	20	0
s	$\dfrac{1536 - 6 \times 20}{15}$	228	0
s^0	20		

No sign change in the first column and both the necessary and sufficient condition are satisfied, hence the system is stable

Problem 5.6. *The open loop transfer function of system is given by*

$G(s)\,H(s) = \dfrac{K}{s(s^2 + 2s + 4)}$ *where k is an adjustable loop gain. Find the values of k for which the system is stable.*

Solution: Characteristic equation is, $s(s^2 + 2s + 4) + k = 0$

$$s^2 + 2s^2 + 4s + k = 0$$

Routh array is

s^3	1	4
s^2	2	k
s^1	$\dfrac{8-k}{2}$	0
s^0	k	

For stability the value of k is $k > 0$

and $$\frac{8-k}{2} > 0$$

or $$8 - k > 0$$

or $$k < 8$$

For stability the value of k is, $0 < k < 8$

Problem 5.7. *The open loop transfer function of unity feedback system given by* $G(s) = \dfrac{k}{s(s^2 + 8s + T)}$. *Using Routh's criterion, determine the values of k and T which will correspond to a stable system.*

Solution: Given open loop transfer function,

$$G(s) = \frac{k}{s(s^2 + 8s + T)}$$

The characteristic equation of the system is

$$1 + G(s)\,H(s) = 0$$

$$\left[1 + \frac{k}{s(s^2 + 8s + T)}\right] = 0 \qquad [\because \; H(s) = 1 \text{ (unity feedback)}]$$

$$s(s^2 + 8s + T) + k = 0$$

$$s^3 + 8s^2 + Ts + k = 0$$

Constructing the *R-H* array as

$$\begin{array}{c|cc} s^3 & 1 & T \\ s^2 & 8 & k \\ s & \dfrac{8T - k}{8} & 0 \\ s^0 & k & \end{array}$$

For the system to be stable, all the elements in the first column of Routh array should be positive.

i.e., $$\frac{8T - k}{8} > 0$$

$\Rightarrow$ $k - 8T < 0$

$\Rightarrow$ $k < 8T$

and $k > 0$

$\therefore$ The range of k for the system to be stable is, $0 < k < 8T$.

5.2.5 Difficulties and Remedies

In applying the Routh criterion, there may be some special cases.

Difficulty 1: When the first element in any row of the Routh array is zero while rest of the row has at least one nonzero element.

Because of this element, the elements in the next row become infinite and Routh's test fails. The following remedies can be used to overcome this difficulty.

Remedy 1: The zero is replaced by a small positive number 'e' and the array continued. The sign changes can be ascertained by letting e approach to zero.

Problem 5.8. *Test the stability of the system with characteristic equation*

$$s^5 + s^4 + 2s^3 + 2s^2 + 3s + 5 = 0.$$

Solution:

The Routh array is

$$\begin{array}{c|ccc} s^5 & 1 & 2 & 3 \\ s^4 & 1 & 2 & 5 \\ s^3 & 0 & -2 & \end{array}$$

Replacing the zero in third row with positive value of 'e', s^3 row will become

$$\begin{array}{c|cc} s^3 & e & -2 \\ s^2 & \dfrac{2e + 2}{e} & 5 \\ s & -\dfrac{(4e + 4 + 5e^2)}{2e + 2} & \\ s^0 & 5 & \end{array}$$

Examine the signs of the terms in first column by letting e approach to zero. The first term of the fourth row has a positive sign *i.e.*, $\frac{2e+2}{e}$ is a positive quantity as $e \to 0$ and first term of fifth row $\frac{-4e-4-5e^2}{2e+2}$ tends to -2 as $e \to 0$. Thus there are two sign changes (*i.e.*, from ∞ to -2 : first sign change and -2 to 5 : second sign change) and hence the system is unstable having two poles lying on the right half of s-plane.

Remedy 2: Replacing the variable 's' with $\frac{1}{z}$, convert the 's' variable characteristic equation into 'z' variable characteristic equation. Evaluate the Routh array and examine elements of the first column of Routh array in z-plane.

Problem 5.9: *For problem 5.8, applying second remedy.*

Solution: The Routh array is

s^5	1	2	3
s^4	1	2	5
s^3	0	-2	

First element in s^3 row is zero so replacing $s = \frac{1}{z}$, convert the 's' variable characteristic equation into z variable characteristic equation.

i.e.,
$$\frac{1}{z^5} + \frac{1}{z^4} + 2\frac{1}{z^3} + 2\frac{1}{z^2} + 3\frac{1}{z} + 5 = 0$$

$$5z^5 + 3z^4 + 2z^3 + 2z^2 + z + 1 = 0$$

For z variable characteristic equation, Routh array is

z^5	5	2	1
z^4	3	2	1
z^3	-1.33	-0.66	0
z^2	1.95	1	
z	0.022		
z^0	1		

Examine the signs of first column of Routh array, there are two sign changes in the array *i.e.*, from 3 to -1.33 and from -1.33 to 1.95. So the system is unstable because two poles lie on the right of s-plane.

Difficulty 2: All terms in any one row may be zero.

This condition indicates there are symmetrically located roots in the s-plane. Because of a zero row in the array, the Routh's test fails. The following remedy can be used to overcome this difficulty.

Remedy: The zero row was replaced by writing auxiliary equation of the row just above the zero row and then first derivative of the auxiliary equation whose coefficient of 's' variable are the elements in the next row.

Routh array is

s^n	a_0	a_2	a_4
s^{n-1}	a_1	a_3	a_5
s^{n-2}	0	0	0
.			
.			
s^0			

$$\frac{d}{ds}(s^{n-2}) \text{ row} = 0$$

Replace the zero row with whose the elements are coefficients of *s* variable in equation $\frac{d}{ds}(s^{n-1})$.

Problem 5.10. *Test the stability of the system with the following characteristic equation by Routh's test:*

$$s^6 + 2s^5 + 8s^4 + 20s^2 + 16s + 16 = 0.$$

Solution:

The Routh array is

s^6	1	8	20	16
s^5	2	12	16	
	1	6	8	
s^4	2	12	16	
	1	6	8	
s^3	0	0		

Since the terms in s^3 row are all zero, the Routh test breaks down. Next the auxiliary polynomial is formed from the coefficients of the s^4-row which is given by

$$\Delta(s) = s^4 + 6s^2 + 8$$

Its derivative $$\frac{d\Delta(s)}{ds} = 4s^2 + 12s$$

The zero of s^3 row is now replaced by the coefficient 4 and 12. The Routh array is completed as follows:

s^6	1	8	20	16
s^5	1	6	8	
s^4	1	6	8	
s^3	4	12		
	1	3		
s^2	3	8		
s	1/3			
s^0	8			

We see that there is no change of sign in first column of the new array. The roots of the auxiliary polynomial namely, $s^4 + 6s^2 + 8 = 0$ are given by

$$s = \pm j\sqrt{2}, \quad s = \pm j2$$

The two pairs of roots are also the roots of original characteristic equation. But no root of the characteristic equation lies on the right half of s-plane. Hence the system under consideration is limitedly stable.

5.2.6 Relative Stability

If a system is unstable because of the closed loop poles lying in the right half of s-plane, one may be interested to know whether any of those pole(s) lie to the right of a line parallel to the $j\omega$-axis and at a distance of 'h' from it as in Fig. 5.5.

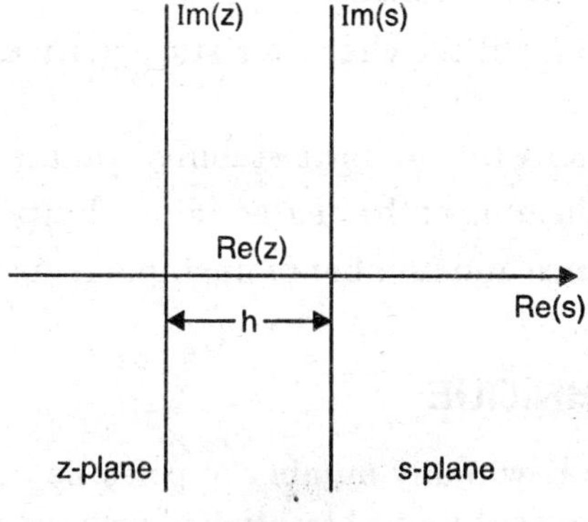

Fig. 5.5

For this we conceive another two dimensional z-plane such that $z = s - h$. The imaginary axis of the z-plane coincides with $s = h$ line. By putting $z = s - h$ in $F(s)$, it gets transformed into $F(z)$. Applying Routh–Hurwitz criterion to $F(z)$, the number of closed loop poles in the right half of z-plane, which is same as to the right of $s = h$ line of the s-plane is obtained.

Problem 5.11. *Determine the number of roots of a given polynomial with real parts between zero and – 1,* $8s^5 + 44s^4 + 126s^3 + 219s^2 + 258s + 85 = 0$.

Solution: Given the characteristic equation of the system as

$$8s^5 + 44s^4 + 126s^3 + 219s^2 + 258s + 85 = 0$$

Now, to get Routh's array with respect to the line $s = -1$, replace $s = z - 1$. Convert the s variable characteristic equation into z variable characteristic equation.

i.e., $$8(z-1)^5 + 44(z-1)^4 + 126(z-1)^3 + 219(z-1)^2 + 258(z-1) + 85 = 0$$

or $$8(z-1)^2(z-1)^3 + 44(z-1)(z-1)^3 + 126(z-1)^3 + 219(z-1)^2 + 258(z-1) + 85 = 0$$

or $$8(z^2 - 2z + 1)(z^3 - 1 - 3z^2 + 3z) + 44(z-1)(z^3 - 1 - 3z^2 + 3z)$$
$$+ 126(z^3 - 1 - 3z^2 + 3z) + 219(z^2 + 1 - 22) + 258(z-1) + 85 = 0$$

or $$8z^5 - 40z^4 + 80z^3 - 80z^2 + 40z - 8 + 44z^4 - 176z^3 + 264z^2 - 176z + 44$$
$$+ 126z^3 - 378z^2 + 378z - 126 + 219z^2 - 438z + 219 + 258z + 258 + 85 = 0$$

or $$8z^{-5} + 4z^4 + 30z^3 + 25z^2 + 62z - 44 = 0$$

Constructing *R-H* array as

z^5	8	30	62
z^4	4	25	– 44
z^3	– 20	150	0
z^2	55	– 44	
z	134	0	
z^0	– 44		

There are three sign changes in the elements of first column of *R-H* array. This indicates three roots will lie on the right side of $s = -1$, *i.e.*, three roots will lie between 0 and – 1 for the given characteristic equation of the system.

5.2.7 Disadvantages of Routh Criterion

- This method is valid only if the characteristic equation is algebraic and that all the coefficients are real.
- This gives an idea of only the absolute stability, that is stable or not.
- It does not give an indication of the degree of instability and the means of avoiding it.
- It indicates the presence and number of unstable roots but not their values.

5.3 ROOT LOCUS TECHNIQUE

Routh stability criterion gives how many number of poles lie on the right side of *s*-plane. It does not give the graphical representation of location of poles on *s*-plane. Due to the drawback in Routh criteria another alternative method is used to find stability of the system which is the Root Locus Technique.

In this section we are going to deal with the locus of the roots of characteristic equation in *s*-plane as the gain k is varied from 0 to ∞.

The advantages dealing with the roots and their location in *s*-plane are:

We get information about (*i*) transient response directly and (*ii*) to a certain extent, about the sinusoidal frequency behaviour indirectly.

Root locus method is a simple graphical procedure of evaluating the roots of the characteristic equation of a system for its poles and zeros. This method provides a very satisfactory way to carry out many analysis and design problems in time domain. Through frequency–response methods or root locus techniques could be applied for the study of system behaviour, we must understand that the methods complement each other and both should be applied to obtain optimum results.

5.3.1 Basic Concept

The root locus is a plot of the locations of the roots of an equation as some real-valuated parameter varies from $-\infty$ to $+\infty$. For example consider a polynomial equation

$$as^2 + bs + c = 0 \tag{5.8}$$

Suppose $a = 1$, $b = 2$ and we would like to plot the root locations as 'c' varies. The roots are

$$s = -1 \pm \sqrt{1-c} \tag{5.9}$$

It is clearly seen from the eqn. (5.9) that

(*i*) If $c < 1$, roots are real and distinct.

(*ii*) If $c = 1$, roots are real and repeated

(*iii*) If $c > 1$, roots are complex conjugate

The root locations for $0 \le c \le \infty$ are shown in Fig. 5.6.

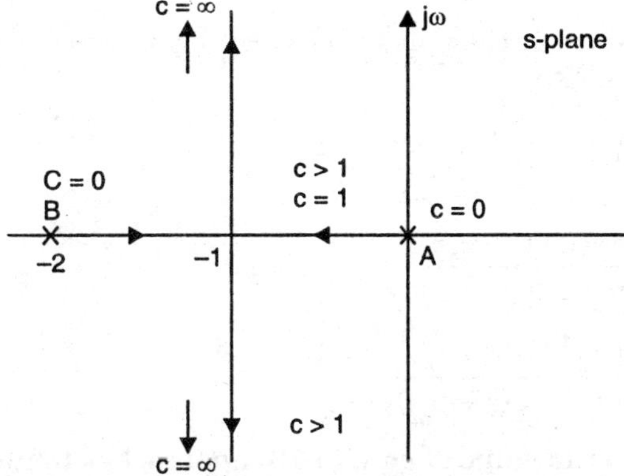

Fig. 5.6 Root locus for $s^2 + 2s + c = 0$, $c > 0$

The two root locations denoted by A and B are at $s = 0$ and $s = -2$ for $c = 0$. The arrows indicate the direction of root movement as c increases. As $c \to \infty$, the roots approach $s = 1 \pm j\infty$.

Consider a second order system transfer function

$$\frac{C(s)}{R(s)} = \frac{\omega_n^2}{s^2 + 2\delta\omega_n s + \omega_n^2}$$

The characteristic equation is

$$s^2 + 2\delta\, \omega_n\, s + \omega_n^2 = 0 \tag{5.10}$$

Changing the coefficients of characteristic equation will change the roots of characteristic equation as well as location of poles on s-plane. Routh criteria determine whether the system is stable or unstable. But it does not give the graphical representation of location of poles on s-plane. Root locus technique is a graphical method for obtaining the roots of characteristic equation on s-plane and thereby investigating the system performance. This technique clearly indicates the effects of the system performance in time and frequency domains.

5.4 THE GENERAL PROCEDURE FOR DETERMINING THE ROOT LOCI FROM LOOP TRANSFER FUNCTION

Step 1: The root locus plot is symmetric about the real axis as the complex roots occur in conjugate pairs. Hence, it is enough to consider upper half plane of the plot.

Step 2: The number of loci equals the number of poles of $G(s)\,H(s)$. Consider the open loop transfer function.

$$G(s)\,H(s) = \frac{k(s+z_1)(s+z_2)......(s+z_n)}{(s+p_1)(s+p_2)......(s+p_m)} = \frac{kN(s)}{D(s)},\; m > n \tag{5.11}$$

In general, the number of poles (n) is greater than the number of zeros (m).

$\therefore$ The characteristic equation $F(s) = D(s) + kN(s) = 0$ (5.12)

Since the order of $N(s)$ is less than or equal to the order of $D(s)$ which is the order of the eqn. (5.11), the number of zeros and poles of $G(s)\,H(s)$ is equal to the order of $N(s)$ and $D(s)$ respectively.

Step 3: (*i*) The loci start at the poles of $G(s)\,H(s)$ with $k = 0$

Proof: From eqn. (5.12)

$$\frac{N(s)}{D(s)} = \frac{-1}{k}$$

where $\dfrac{N(s)}{D(s)} = \dfrac{(s+z_1)(s+z_2)....(s+z_n)}{(s+p_1)(s+p_2)....(s+p_m)}$

$$\therefore \quad \frac{(s+z_1)(s+z_2)....(s+z_n)}{(s+p_1)(s+p_2)....(s+p_m)} = \frac{-1}{k} \tag{5.13}$$

As k approaches zero the value of eqn. (5.13) approaches to infinity. Again the value of eqn. (5.13) becomes infinity, at poles $s = -p_1, -p_2....., -p_m$. It means that the starting point (*i.e.*, $k = 0$) of each root loci is a pole (*i.e.*, $s = -p_1, -p_2, ..., -p_m$) of the open loop transfer function.

(*ii*) The loci terminate with $k = \infty$.

Proof:

As k approaches infinity the value of eqn. (5.12) approaches to zero. Again, the value of eqn. (5.12) becomes zero, at zeros of $s = -z_1, -z_2, ..., -z_n$. It means that the terminating point (*i.e.*, $k = \infty$) of each root loci is a zero (*i.e.*, $s = -z_1, -z_2....., -z_n$) of the open loop transfer function.

The poles terminate either at the zeros of $G(s)\,H(s)$ or at infinity. If 'z' is the number of zeros and 'p' is the number of poles, then the root loci segments going to infinity is $(p - z)$.

Step 4: A point on the real axis lies on root loci if number of poles and number of zeros on the real axis to the right of that point is odd.

Example 5.1. *The given open loop transfer function is* $G(s) = \dfrac{k}{s(s+1)(s+2)}$

Solution:

Number of poles = 3

Number of zeros = 0

Number of asymptotes = order of the system = 3

The points between 0 to – 1 and – 2 to – ∞ on the real axis lie on the root locus (represented by thick line as shown in Fig. 5.7(*a*)).

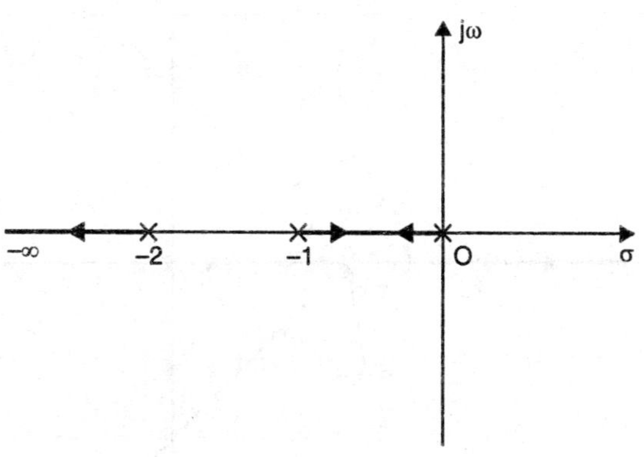

Fig. 5.7(a)

Step 5: Asymptotes of root loci as $s \to \infty$. The loci that do not terminate at a zero approach infinity along asymptotes. The angles that the asymptotes make with real axis are

$$\alpha_n = \frac{\pm n\pi}{p - z}$$

p = no. of open loop poles

z = no. of open loop zeros

n = odd integer 1, 3, 5 ...

Number of asymptotes equal to order of the system.

Step 6: The asymptotes intersect the real axis (for all values of k) at centroid

$$\therefore \quad \text{centroid,} \quad \sigma = \frac{\text{sum of real parts of poles} - \text{sum of real parts of zero}}{p - z}$$

$$\sigma = \frac{\sum p - \sum z}{p - z}$$

Example 5.2. *Consider example 5.1,*

Solution:

Number of poles = 3

Number of zeros = 0

Number of asymptotes = order of the system = 3

The points between 0 to − 1 and − 2 to − ∞ on the real axis lie on the root locus

$$\text{Centroid, } \sigma = \frac{\sum p - \sum z}{p - z}$$

$$= \frac{(0 - 1 - 2) - 0}{3 - 0} = -1$$

Centroid for a given system is − 1.

$$\text{Angle of asymptotes, } \alpha_n = \frac{\pm n\pi}{p - z}$$

$\alpha_1, \alpha_3, \alpha_5$ are equal to $\frac{\pm\pi}{3}, \pm\pi, \frac{\pm 5\pi}{3}$. These steps are shown in Fig. 5.7(*b*).

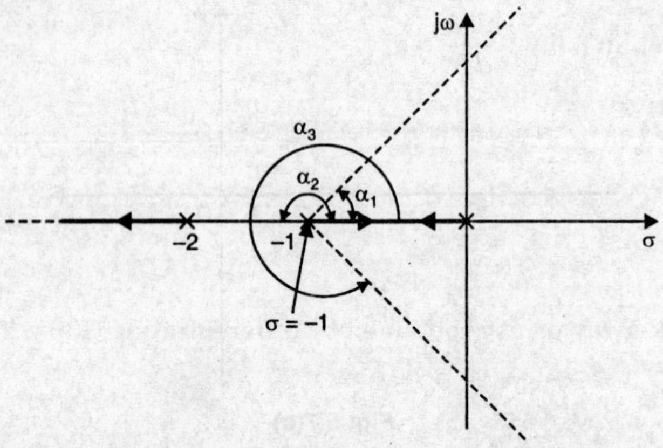

Fig. 5.7(b)

Step 7: The location of saddle points (break-away and break-in points) is found by determining where the parameter k attains a local maximum or minimum on the real axis. That is, the values of s at

$$\frac{d}{ds}(1 + G(s)\,H(s)) = \frac{d}{ds}[G(s)\,H(s)] = 0 \tag{5.14}$$

Proof: Differentiating the eqn.(5.12) w.r. t. s and equating to zero

$$\frac{dF(s)}{ds} = D'(s) + kN'(s) = 0 \tag{5.15}$$

where $D'(s) = \dfrac{dD(s)}{ds}$ and $N'(s) = \dfrac{dN(s)}{ds}$

The particular value of k which will yield multiple roots of the characteristic equation as obtained from eqn. (5.15).

$$\therefore \qquad k = \frac{-D'(s)}{N'(s)} \tag{5.16}$$

Substituting the eqn. (5.16) in eqn. (5.12)

$$F(s) = D(s) - \frac{-D'(s)}{N'(s)}N(s) = 0$$

or $$D(s)N'(s) - D'(s)\,N(s) = 0 \tag{5.17}$$

Again, from eqn. (5.12), $k = -\dfrac{D(s)}{N(s)}$

$$\text{Then } \frac{dk}{ds} = \frac{D'(s)N(s) - D(s)N'(s)}{N^2(s)} = 0 \tag{5.18}$$

The break-away points can be simply determined from the roots of eqn. (5.18).

Case 1: For break-away point, $\dfrac{d^2k}{ds^2} < 0$

Case 2: For break-in point, $\frac{d^2k}{ds^2} > 0$

Note: All the roots of $\frac{dk}{ds} = 0$ are not the break-away point.

Example 5.3. *Consider example 5.1 and determine the break-away point*

i.e., $$G(s) = \frac{k}{s(s+1)(s+2)}$$

Solution: Break-away point is obtained by differentiating G(s) w.r.t. s and equating it to zero.

$$\frac{dG(s)}{ds} = 0$$

$$\frac{d}{ds}\left(\frac{k}{s(s+1)(s+2)}\right) = 0$$

$$\frac{d}{ds}(s^3 + 3s^2 + 2s) = 0$$

$$3s^2 + 6s + 2 = 0$$

The roots are s_1 and $s_2 = \frac{-6 \pm \sqrt{36 - 4 \times 3 \times 2}}{6} = -1.57$ and -0.42

The valid break-away point is, $s = -0.42$, since root locus branch exists between 0 to -1.

Step 8: The points of which the loci cross the imaginary axis and the corresponding values of k can be determined by Routh-Hurwitz criterion.

Example 5.4. *For example 5.1, determine the real axis cutting point.*

Solution: Characteristic equation is

$$s^3 + 3s^2 + 2s + k = 0$$

Constructing the Routh table

s^3	1	2
s^2	3	k
s^1	$\frac{6-k}{3}$	0
s	k	

Condition for the stability: each entry in the first column of Routh array should be positive.

So $k > 0$ and $\frac{6-k}{3} > 0$

or $k < 6$

$\therefore$ The system is stable if k lies in between 0 and 6.

Take the maximum value of $k = 6$.

The auxiliary equation from the Routh table is

$$3s^2 + k = 0 \qquad \text{(or)} \quad 3s^2 + 6 = 0$$

$$s^2 = -2\,; \quad s = \pm j\,1.414$$

So the root locus intersects the imaginary axis at $\pm j1.414$

The complete root locus plot is shown in Fig. 5.7 (*c*).

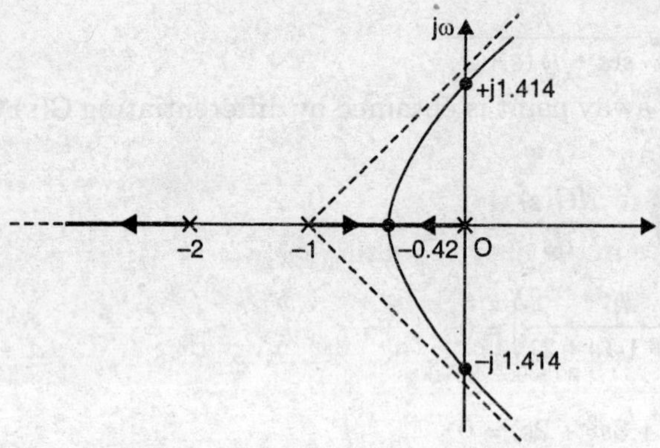

Fig. 5.7 (c) Root locus for $G(s) = \dfrac{k}{s(s+1)(s+2)}$

Step 9. Angles of departure from a pole or the angles of arrival at a zero of $G(s)\,H(s)$ are determined by choosing an arbitrary point infinitesimally close to the pole or zero respectively.

Angle of departure, $\theta_D = 180 - (\phi_p - \phi_z)$ (5.19)

Angle of arrival, $\theta_A = 180 - (\phi_z - \phi_p)$ (5.20)

where ϕ_P and ϕ_Z are algebraic sum of the angles of poles and zeros.

Case I: Angle of departure

There are four poles P_1, P_2, P_3, P_4 and a zero z_1. Angle of departure of pole 2 can be calculated by adding the angles θ_1, θ_2 and θ_3 from poles P_1, P_3, P_4 and θ_4 from zero z_1.

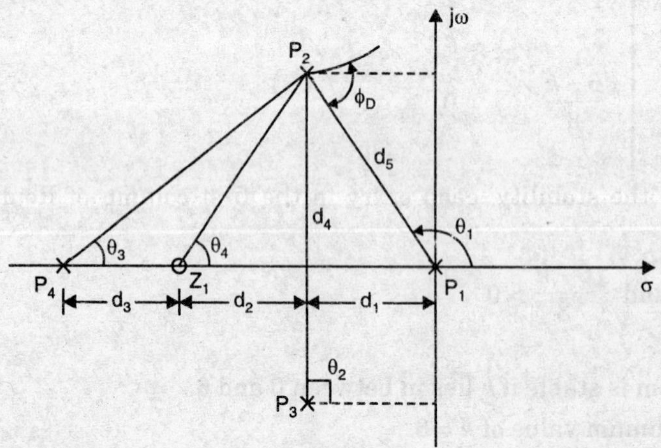

Fig. 5.8

Angle at P_1, $\theta_1 = 180 - \tan^{-1} \dfrac{d_4}{d_1}$

At P_3, $\theta_2 = 90°$ w.r.t. P_2

P_4, $\theta_3 = \tan^{-1} \dfrac{d_4}{d_2 + d_3}$

Z_1, $\theta_4 = \tan^{-1} \dfrac{d_4}{d_2}$

Angle of departure, $\phi_D = 180 - (\phi_p - \phi_z)$

$$\phi_p = \theta_1 + \theta_2 + \theta_3$$

$$\phi_z = \theta_4$$

Substituting angles in the above equation

Angle of departure $\phi_D = 180 - (\theta_1 + \theta_2 + \theta_3 - \theta_4)$

If ϕ_D is of negative value measure the angle in clockwise direction from the pole reference. For positive value measure the angle in anti clockwise direction from the pole reference.

Example 5.5: *For the given open loop transfer function,* $G(s) = \dfrac{k(s+2)}{s(s^2 + 2s + 2)(s+4)}$. *Calculate the angle of departure.*

Solution: Angle of departure at pole $(-1 + j1)$ can be calculated from Fig. 5.9.

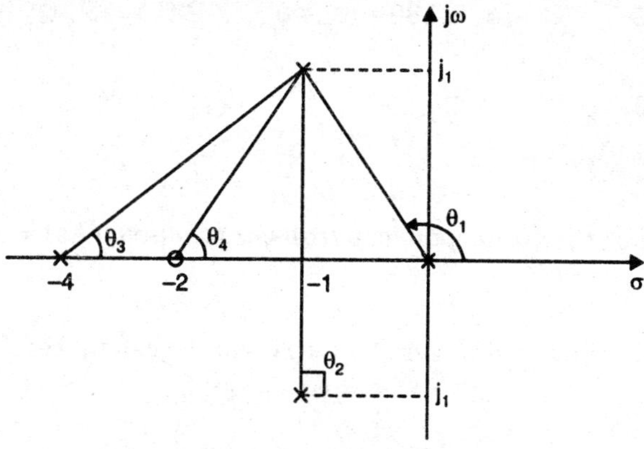

Fig. 5.9

$$\phi_D = 180 - (\theta_1 + \theta_2 + \theta_3 - \theta_4)$$

where $\theta_1 = 180 - \tan^{-1} \dfrac{1}{1} = 135°$

$\theta_2 = 90°$

$\theta_3 = \tan^{-1} \dfrac{1}{3}$

$\theta_4 = \tan^{-1}(1) = 45°$

Case II: Angle of arrival

There are two poles and two zeros. The angle of arrival at zero z_1 can be calculated by adding the angles from poles and zeros.

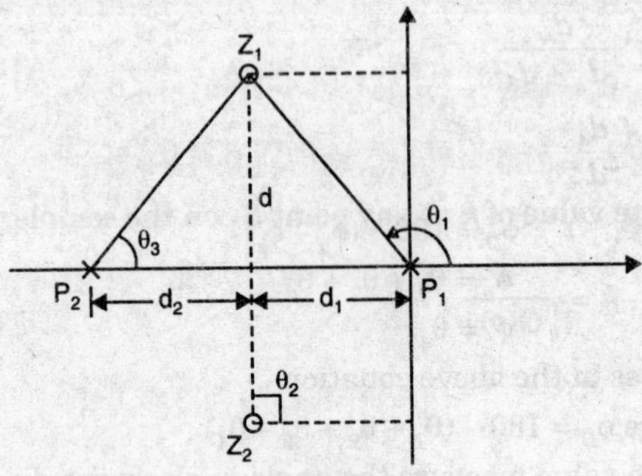

Fig 5.10

Angle at pole P_1 is $\theta_1 = 180 - \tan^{-1} \dfrac{d}{d_1}$

Z_2 is $\theta_2 = 90°$

P_2 is $\theta_3 = \tan^{-1} \dfrac{d}{d_2}$

Angle of arrival $\qquad \phi_A = 180 - (\phi_z - \phi_p)$

where $\quad \phi_z = \theta_2$

$\phi_p = \theta_1 + \theta_3$

$\phi_A = 180 - (\theta_2 - \theta_1 - \theta_3)$

Example 5.6. *For the given open loop transfer function,* $G(s) = \dfrac{k(s^2 + 2s + 2)}{s(s+2)}$. *Calculate the angle of arrival.*

Solution: Angle of arrival at zero $(-1 + j1)$ can be calculated from Fig. 5.11.

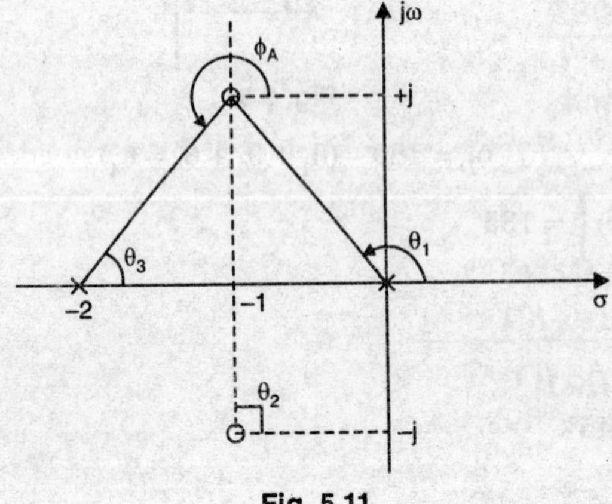

Fig. 5.11

Angle of arrival $\phi_A = 180 - (\theta_2 - \theta_1 - \theta_3)$

$$\theta_1 = 180 - \tan^{-1}\frac{1}{1} = 135°$$

$$\theta_2 = 90°$$

$$\theta_3 = \tan^{-1}\frac{1}{1} = 45°$$

$$\phi_A = 180° - (90° - 45° - 135°) = +270°$$

Step 10: Absolute value of k at any point S_1 on the complex root loci determined from

$$k = \frac{1}{|G(s)|}$$

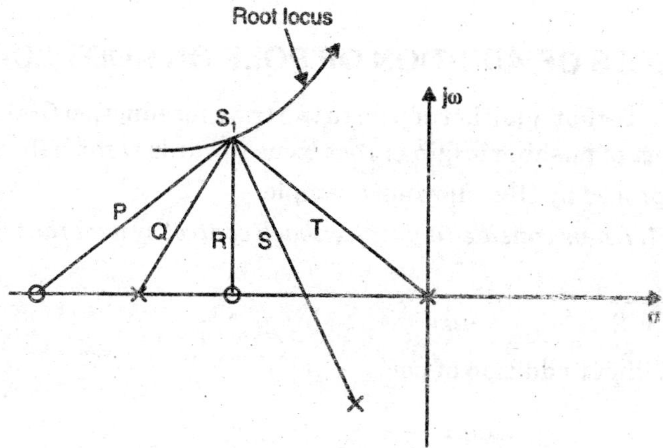

Fig. 5.12(a)

From the Fig. 5.12(a) the value of k is calculated as

$$k = \frac{\text{(product of length of vectors drawn from poles of } G(s)\,H(s) \text{ to } s_1)}{\text{(product of length of vectors drawn from zeros of } G(s)\,H(s) \text{ to } s_1)}$$

From Fig. 5.12(a) the value of k at s_1 is

$$k = \frac{QST}{PR}$$

Alternate method

From Fig. 5.12(b) the value of k is calculated for a given damping ratio δ as

$$kG(s)\Big|_{s = \sigma_1 + j\omega_1} = 1$$

$$\therefore \qquad k = \frac{1}{G(s)\Big|_{s = \sigma_1 + j\omega_1}}$$

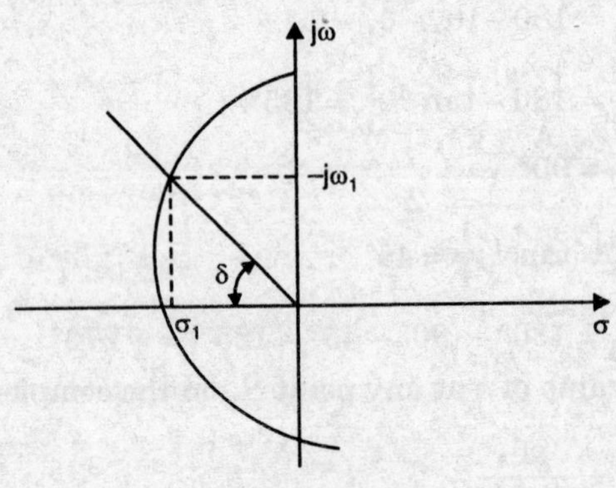

Fig. 5.12(b)

5.5 THE EFFECTS OF ADDITION OF POLE ON ROOT LOCUS

In general we can state that addition of pole to the transfer function $G(s)\,H(s)$ in the left half of s-plane has the effect of pushing origin of root locus towards right half of the s-plane.

This can be proved by the following example.

Example 5.7. *Let us consider a unit feedback control system that has open loop transfer function* $G(s) = \dfrac{k}{s(s+4)}$.

Solution: Without addition of pole

$$G(s) = \frac{k}{s(s+4)}$$

$$G(s)\,H(s) = \frac{k}{s(s+4)} \qquad \text{since } H(s) = 1 \text{ (unit feedback)}$$

(*i*) No. of poles, $p = 2$, are $0, -4$

No. of zeros, $z = 0$

No. of root loci = order of the system = 2

The points between 0 and – 4 on the real axis lie on the root locus, represented by thick line.

(*ii*) Asymptotic angles $\alpha_1, \alpha_3 = \pm \dfrac{n\pi}{p-z}$

$$= \pm \pi/2, \pm 3\pi/2$$

(*iii*) Centroid, $\sigma = \dfrac{\sum p - \sum z}{p-z} = \dfrac{0-4}{2} = -2$

(*iv*) Breakaway point $\dfrac{d}{ds}\,(G(s)\,H(s)) = 0$

$$\frac{d}{ds}\left(\frac{k}{s(s+4)}\right) = 0$$

$$\frac{d}{ds}[s^2 + 4s] = 0$$

$$2s + 4 = 0$$

$$\therefore \qquad s = -2$$

(*v*) Intersection point on $j\omega$ axis

The characteristic equation is

$$s^2 + 4s + k = 0.$$

Routh array is

$$\begin{array}{c|cc} s^2 & 1 & k \\ s & 4 & 0 \\ s^0 & k & \end{array}$$

It does not intersect the imaginary axis.

The root locus plot for the system is shown in the Fig. 5.13.

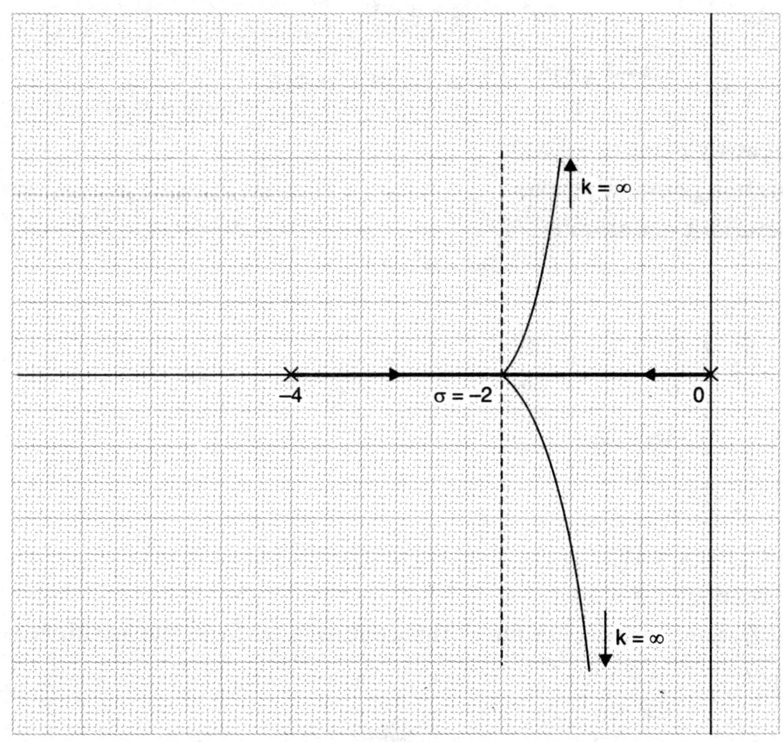

Fig. 5.13 Before adding a pole to the transfer function $G(s) = \dfrac{k}{s\,(s+4)}$

When a pole – 2 is added to G(s) H(s)

$$\therefore \qquad G(s)\,H(s) = \frac{k}{s(s+2)(s+4)}$$

(*i*) No of poles, $p = 3$, are $0, -2, -4$

No of zeros, $z = 0$

No of root loci = Order of the system = 3

The points between 0 to – 2 and – 4 to – ∞ on the real axis lie on the root locus, represented by thick line.

(*ii*) Angle of asymptotes $\alpha_1, \alpha_3, \alpha_5 = \pm \pi/3, \pm \pi, \pm 5\pi/3$.

(*iii*) Centroid $\sigma = \dfrac{\sum p - \sum z}{p - z} = \dfrac{0-2-4}{3} = -2$

(*iv*) Breakaway point $\dfrac{d}{ds}\left[\dfrac{k}{s(s+2)(s+4)}\right] = 0$

$$\frac{d}{ds}[s^3 + 6s^2 + 8s] = 0$$

$$3s^2 + 12s + 8 = 0$$

Roots are $s_1, s_2 = \dfrac{-12 \pm \sqrt{144-96}}{6} = -3.15, -0.845$

Valid breakaway point is – 0.845

(*v*) Intersection point on $j\omega$ axis

The characteristic equation is

$$s^3 + 6s^2 + 8s + k = 0$$

Routh array is

s^3	3	8
s^2	6	k
s	$\dfrac{48-3k}{6}$	
s^0	k	

For stability, $0 < k < 16$;

At critical condition, $k = 16$

To find the frequency of oscillation at this condition we form the auxiliary equation from Routh table

$$6s^2 + k = 0$$

or $$6s^2 + 16 = 0$$

or $$6\omega^2 + 16 = 0$$

$$6\omega^2 = -16$$

$$\omega^2 = -2.66$$

or $$\omega = \pm j\, 1.631$$

The root locus intersect the $j\omega$ axis at $\omega = 1.631$ rad/sec. The root locus plot for the system is given in the Fig. 5.14.

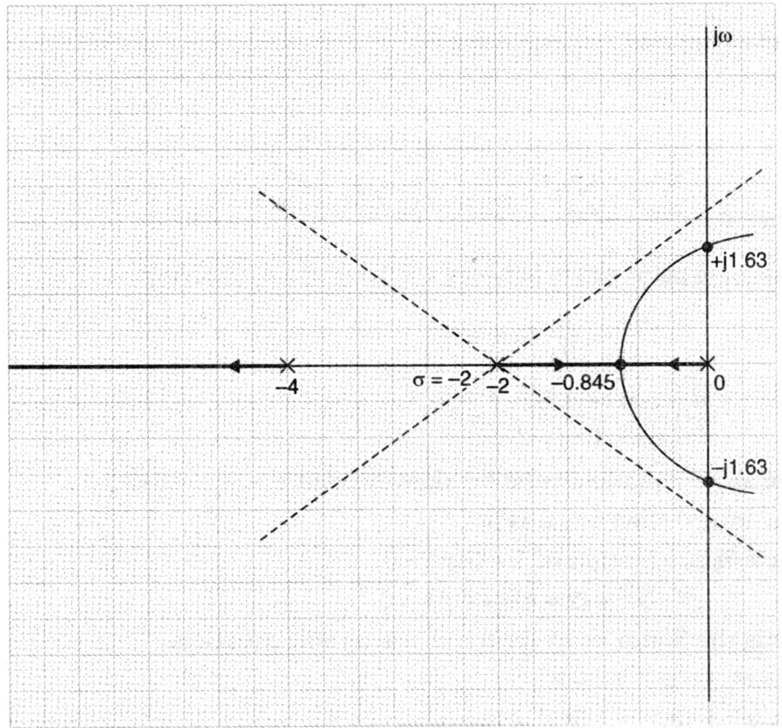

Fig. 5.14 Adding a pole to the transfer function $G(s) = \dfrac{k}{s\,(s+4)\,(s+2)}$

From Fig. 5.13 and Fig. 5.14, it is observed that effect of addition of a pole on left side of s-plane will shifts the breakaway point towards the right compared to first case. So effect of pole shifts the root locus towards imaginary axis, decreases the relative stability, makes the system more oscillatory in nature and decreases the operating values of k.

Problem 5.12. *Sketch the root locus for the characteristic equation*

$$s(s + 1)\,(s + 2) + k\,(s + 1.5) = 0$$

Solution: Write in standard form

$$1 + \frac{k(s + 1.5)}{s(s + 1)\,(s + 2)} = 0$$

(*i*) Poles are at $s = 0, -1$, and -2

Zero is at $s = -1.5$

No. of root loci $= p - z = 3 - 1 = 2$, two terminating at ∞.

Poles and zeros are marked by x and 0 respectively in the Fig. 5.15.

The points between 0 to – 1 and – 1.5 to – 2 on the real axis lie on the root locus, represented by thick line.

(*ii*) angle of asymptotes $\alpha_1, \alpha_3, \alpha_5 = \pm \dfrac{n\pi}{p-z}$

$$= \pm \frac{\pi}{2}, \pm 3\frac{\pi}{2}, \pm 5\frac{\pi}{2}$$

(*iii*) Centroid, $\sigma = \dfrac{\sum p - \sum z}{p-z} = \dfrac{0-1-2(-1.5)}{3-1} = \dfrac{-1.5}{2} = -0.75$

(*iv*) A breakaway point exists between $s = 0$ and $s = -1$. Solving

$$\frac{d}{ds}G(s) = 0$$

$$\frac{d}{ds}\left[\frac{k(s+1.5)}{s(s+1)(s+2)}\right] = 0$$

Solving the above equation, valid breakaway point is $s_1 = -0.542$

(*v*) Point of intersection on $j\omega$ axis

The characteristic equation is $(1 + G(s)) = 0$

$$s^3 + 3s^2 + (2+k)s + 1.5k = 0$$

Constructing the Routh array for the characteristic equation.

s^3	1	$2+k$
s^2	3	$1.5k$
s	$\dfrac{6+1.5k}{3}$	
s^0	$1.5k$	

For stability, $0 < k < 4$;

At critical condition $k = 4$

To find the frequency of oscillation at this condition, we form the auxiliary equation from Routh table

$$3s^2 + 1.5k = 0$$

$$3s^2 + 6 = 0 \qquad \text{or} \quad 3\omega^2 + 6 = 0$$

$$3s^2 = -6$$

$$s^2 = -2 \Rightarrow \pm j\omega$$

The root locus intersect the $j\omega$ axis at $\omega = 1.414$ rad/sec. The root locus plot for the system is given in the Fig. 5.15.

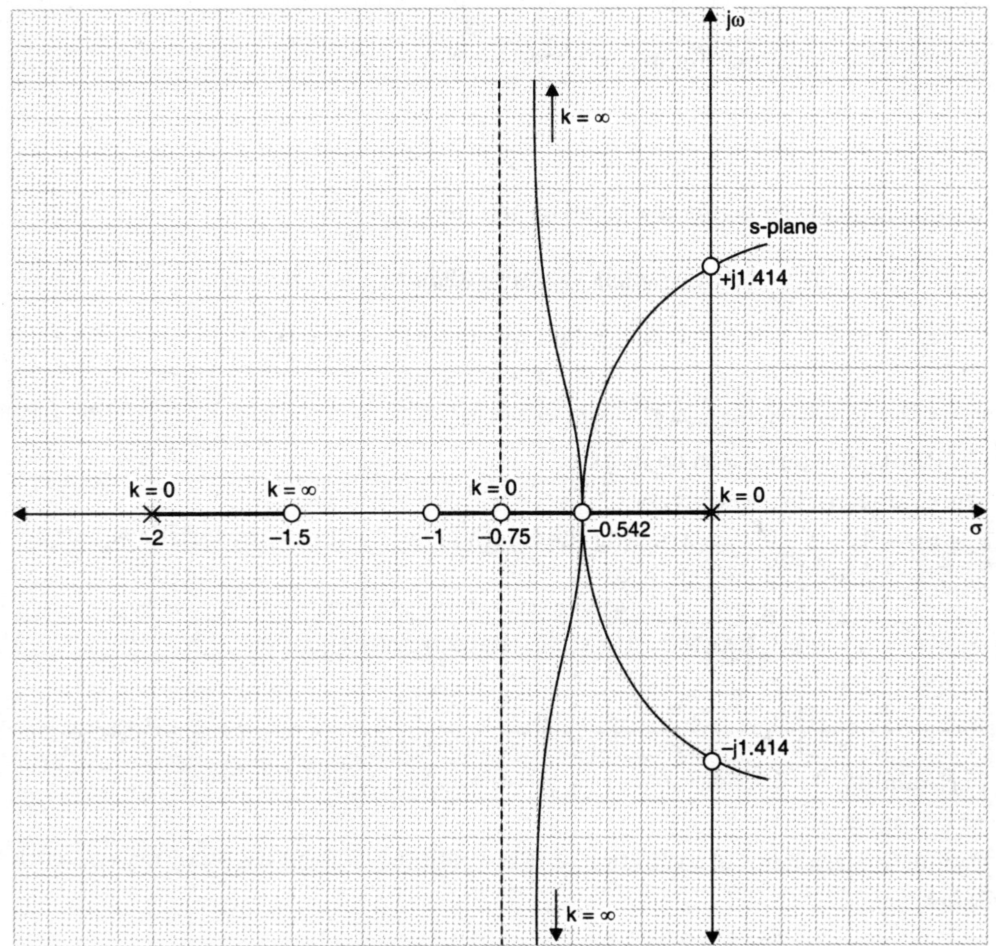

Fig. 5.15 Root locus of characteristic equation $s(s + 1)(s + 2) + k(s + 1.5) = 0$

Problem 5.13. *The open loop transfer function of a unity feedback control system is given by* $G(s) = \dfrac{k}{s(s+3)^2}$. *Sketch the root locus plot of the closed loop system for positive values of k and there from determine the value of k that would make the system.*

Solution: (*i*) No. of poles = 3, which are 0, – 3, – 3 ;

No. of zeros = 0

No. of root loci = 3, all terminating at ∞

The points between 0 and – 3 on the real axis lie on the root locus, represented by thick line.

(*ii*) Angle of asymptotes $\alpha_n = \pm \dfrac{n\pi}{p-z}$

$\alpha_1, \alpha_3, \alpha_5 = \pm \pi/3, \pm \pi, \pm 5\pi/3$

(*iii*) The centroid of asymptotes, $\sigma = \dfrac{\sum p - \sum z}{p - z}$

$$= \frac{-3-3}{3} = -2$$

(*iv*) The breakaway points are given by the roots of

$$\frac{d}{ds}(s^3 + 6s^2 + 9s) = -(3s^2 + 12s + 9)$$

or
$$s^2 + 4s + 3 = 0 \Rightarrow (s+1)(s+3) = 0$$
$$s = -1, s = -3$$

Valid breakaway point is – 1

(*i*) Points of intersection on $j\omega$ axis

The characteristic equation is $1 + G(s) = 0$

$s^3 + 6s^2 + 9s + k = 0$. Constructing the Routh array for the characteristic equation.

s^3	1	9
s^2	6	k
s	$\dfrac{54-k}{6}$	
s^0	k	

For stability, $0 < k < 54$;

At critical condition, $k = 54$

To find the frequency of oscillation at this condition we form the auxiliary equation from Routh table

$$6s^2 + k = 0$$
$$6s^2 + 54 = 0$$

or
$$6s^2 = -54$$
$$s^2 = -9 \Rightarrow \pm j\omega$$

The root locus intersect the $j\omega$ axis at $\omega = 3$ rad/sec. The root locus plot for the system is given in the Fig. 5.16.

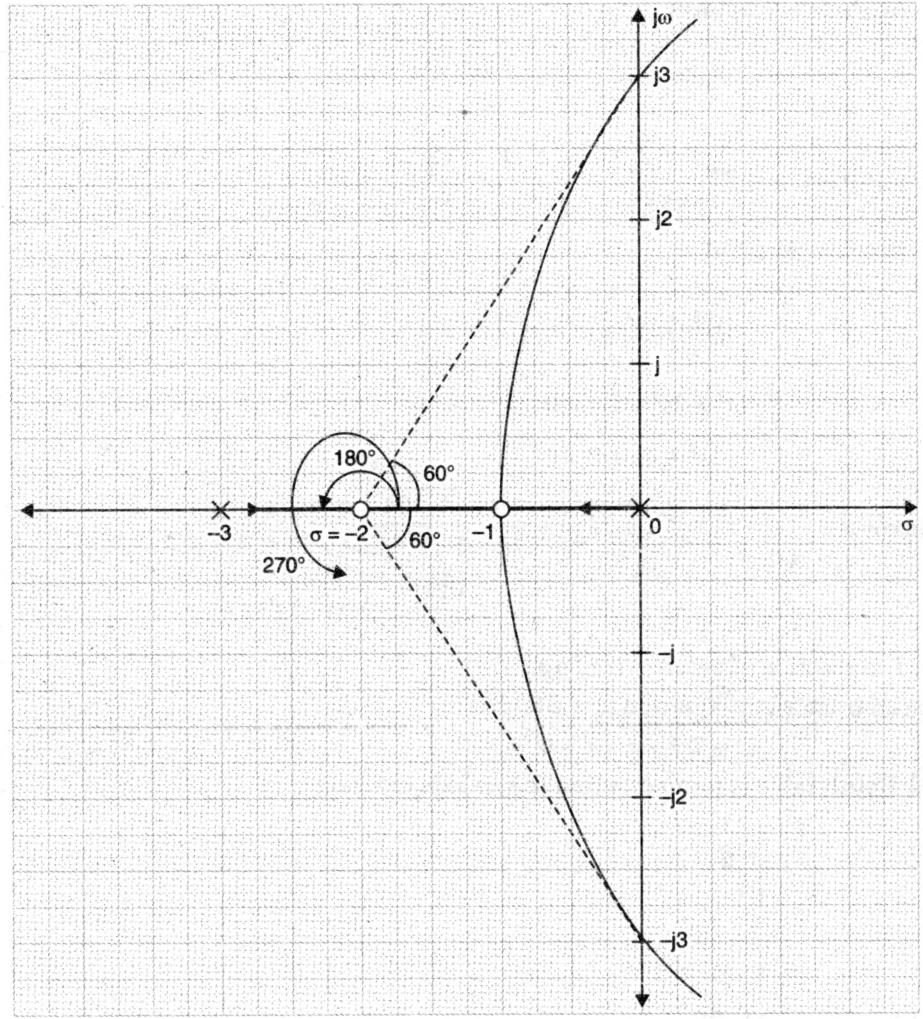

Fig. 5.16 Root locus of open loop transfer function $G(s) = \dfrac{k}{s(s+3)^2}$.

Problem 5.14. *Sketch the root locus for a given unit feedback control system whose open loop transfer function is*

$$G(s) = \frac{k}{s(s^2 + 2s + 2)}$$

Solution: (*i*) No. of poles = 3 which are 0, – 1 ± *j*1

No. of zeros = 0

No. of root loci = 3 all terminating at ∞

The points between 0 and – ∞ on the real axis lie on the root locus, represented by thick line.

(*ii*) Angle of asymptotes

$$\alpha_1, \alpha_3, \alpha_5 = \pm \frac{n\pi}{p-z} = \pm \pi/3, \pm \pi, \pm 5\pi/3.$$

(*iii*) Centroid, $\sigma = \dfrac{\sum p - \sum z}{p-z} = \dfrac{-1-1}{3} = \dfrac{-2}{3} = -0.666$

(*iv*) Breakaway points.

$$\frac{d}{ds}\left[\frac{k}{s(s^2+2s+2)}\right] = 0$$

$$\frac{d}{ds}(s^3 + 2s^2 + 2s) = 0$$

$$3s^2 + 4s + 2 = 0$$

The roots $\quad s_1, s_2 = \dfrac{-4 \pm \sqrt{16-24}}{6}$

$$= \frac{-2}{3} \pm j\frac{\sqrt{2}}{3} = -0.667 \pm j0.471$$

(*v*) Intersection of points with $j\omega$ axis.

The characteristic equation $[1 + G(s)H(s)] = 0$

i.e., $\quad s^3 + 2s^2 + s + k = 0$

The Routh table is formed to find the $j\omega$ axis crossing

s^3	1	1
s^2	2	k
s	$\dfrac{2-k}{2}$	
s^0	k	

For stability $k > 0$ and

$$\frac{2-k}{2} > 0 \quad \text{or} \quad k < 2$$

Then given system is stable when $\quad 0 < k < 2$

At maximum value of $k = 2$, auxiliary characteristic equation is

$$2s^2 + k = 0$$

$$2s^2 + 2 = 0$$

$$s = \pm j1 \quad \textit{i.e.,} \quad \omega = 1 \text{ rad/sec.}$$

(*vi*) Angle of departure

From $s = 0$ pole, $\quad \theta_1 = 180° - \tan^{-1}\left(\dfrac{1}{1}\right) = 135°$

From $s = (-1 - j1)$ pole, $\quad \theta_2 = 90°$

$\therefore \quad \theta_p = 135° + 90°$

Angle of departure $\theta_D = 180° - (\phi_p + \phi_z)$

$\theta_D = 180° - (135° + 90°) = -45°$ $(\because \ \theta_z = 0)$

The root locus of the system is sketched as shown in Fig. 5.17.

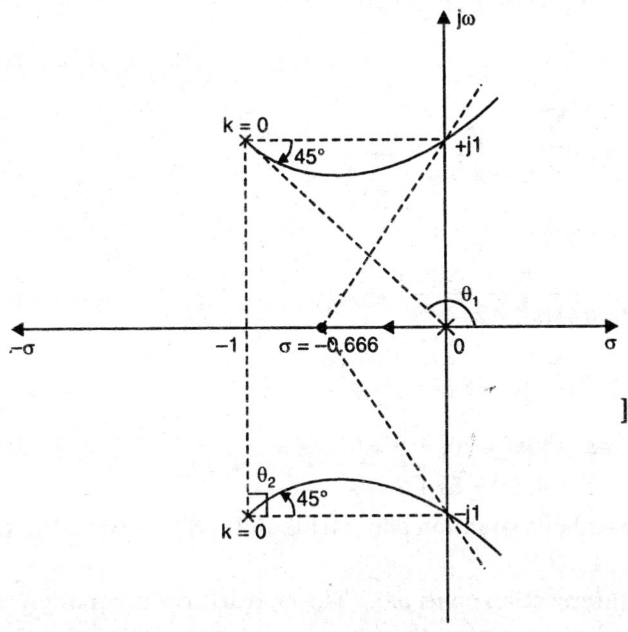

Fig. 5.17 Root locus of open loop transfer function $G(s) = \dfrac{k}{s(s^2 + 2s + 2)}$

Problem 5.15. *The open loop transfer function of a feedback control system is given by*

$$G(s)\,H(s) = \frac{k}{s(s+4)(s^2 + 2s + 2)}$$

(*a*) *Judge whether the system is stable or not. If not determine the range of values of k for which the system behaves stable.*

(*b*) *For the above system, sketch the root locus plot, indicating the point where the asymptotes of the loci meet (Breakaway point need not be calculated).*

(*c*) *How does the performance of the system get affected if a zero at – 4 is added to the forward path transfer function G(s) and how does it modify the root locus plot?*

Solution: From problem 5.14, the obtained root locus for the function, $G(s) = \dfrac{k}{s(s^2 + 2s + 2)}$ is shown in Fig. 5.17.

When $s = -4$ poles is added to the open loop transfer function,

$G(s)$ becomes $\dfrac{k}{s(s+4)(s^2 + 2s + 2)}$

(*i*) There are 4 root loci, starting at open loop poles.

No. of poles = 4, which are $0, -4, -1 \pm j$

No. of zeros = 0

The points between 0 and – 4 on the real axis lie on the root locus, represented by thick line.

(*ii*) Angle of asymptotes $\alpha_1, \alpha_3, \alpha_5, \alpha_7 = \pm \dfrac{n\pi}{p-z}$

$$= \pm \pi/4, \pm 3\pi/4, \pm 5\pi/4, \pm 7\pi/4$$

(*iii*) Centroid, $\sigma = \dfrac{\sum p - \sum z}{p-z} = \dfrac{0-4-1-1}{4-0} \Rightarrow \dfrac{-6}{4} = -1.5$

(*iv*) Breakaway point

$$\frac{d}{ds}\left[\frac{k}{s(s+4)(s^2+2s+2)}\right] = 0$$

$$\frac{d}{ds}[s^4 + 6s^3 + 10s^2 + 8s + k] = 0$$

$$4s^3 + 18s^2 + 20s + 8 = 0$$

$$s^3 + 4.5s^2 + 5s + 2 = 0$$

By solving the above equation using trail and error method, the valid breakaway point, $s = -2.9$.

(*i*) Points of intersection on $j\omega$ axis. The characteristic equation is

$$1 + G(s)H(s) = 0,$$

$$s^4 + 6s^3 + 10s^2 + 8s + k = 0$$

s^4	1	10	k
s^3	6	8	
s^2	52/6	k	
s	$8 - \dfrac{36k}{52}$		
s^0	k		

For stability $0 < k < 11.56$

$k = 1.56$ the closed loop system is stable and root loci intersect the $j\omega$ axis at

$$\frac{52}{6}s^2 + 11.56 = 0$$

$$s^2 = \frac{-11.56 \times 6}{52}$$

$$s = \pm j1.15;\ \omega = 1.15 \text{ rad/sec}$$

(*v*) Angle of departure, $\theta_D = 180 - (\phi_p + \phi_z)$

$$\phi_p = \theta_1 + \theta_2 + \theta_3$$

From $s = 0$, $\theta_1 = 180 - \tan^{-1} 1/1 = 135°$

For $s = (-1 - j1)$, $\theta_2 = 90°$

$s = -4$, $\theta_3 = \tan^{-1}(1/3) = 18.41°$

$\therefore \qquad \phi_D = 180 - (\theta_1 + \theta_2 + \theta_3) = -63.4° \qquad (\because \; \theta_2 = 0)$

The root locus plot for the system is shown in Fig. 5.18.

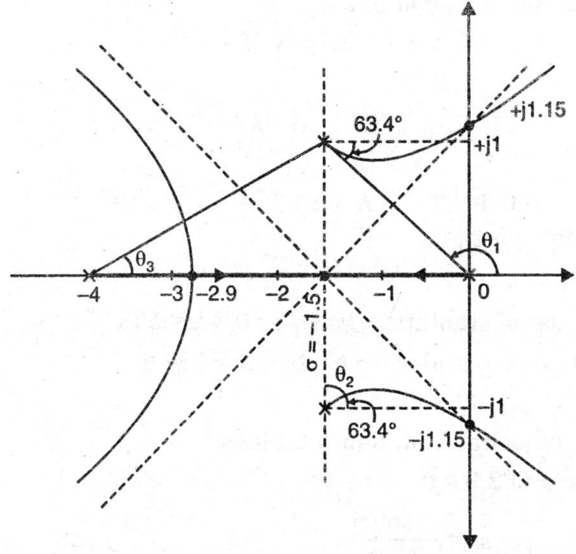

Fig. 5.18 Root locus of open loop transfer function $G(s)\,H(s) = \dfrac{K}{s(s+4)\,(s^2+2s+2)}$.

Problem 5.16. *Open loop transfer function* $G(s) = \dfrac{k}{s(s+4)(s^2+8s+32)}$.

Solution:

(*i*) There are 4 root loci, all terminating at ∞

No. of poles = 4, $\quad 0, -4, -4 \pm j4$

No. of zeros = 0

The points between 0 and – 4 on the real axis lie on the root locus, represented by thick line.

(*ii*) Angle of asymptotes $\alpha_1, \alpha_3, \alpha_5, \alpha_7 = \pm \dfrac{n\pi}{p-z}$

$$= \pm \pi/4, \pm 3\pi/4, \pm 5\pi/4, \pm 7\pi/4$$

(*iii*) Centroid, $\sigma = \dfrac{\sum p - \sum z}{p-z} = \dfrac{0-4-4-4}{4} = -3$

(*iv*) Breakaway point $= \dfrac{d}{ds}\,G(s) = 0$

$$\frac{d}{ds}\,(s^2+4s)\,(s^2+8s+32) = 0$$

$$\frac{d}{ds}\,(s^4+12s^3+64s^2+128s) = 0$$

$$4s^3 + 36s^2 + 128s + 128 = 0$$

Gives the breakaway point as $s = -2$ (nearly)

(*v*) Points of intersection on $j\omega$ axis.

$$1 + G(s) = s^4 + 12s^3 + 64s^2 + 128s + k = 0$$

Routh array is

s^4	1	64	k
s^3	3	32	
s^2	160/3	k	
s	$32 - 9k/160$		
s^0	k		

The range of values of stability is given by $0 < k < 568.9$.

Closed loop system is unstable for $k < 0$, $k > 568.9$

At $k = 568.9$

The auxiliary polynomial from Routh table is

$$53.3s^2 + 568.9 = 0$$

$$s^2 = \frac{-568.9}{53.3} \Rightarrow s = \pm j3\ 26$$

$$\omega = 3.26 \text{ rad/sec}$$

The root loci intersect the $j\omega$ axis at 3.26 rad/sec.

Root locus of the given system is as shown in the Fig. 5.19.

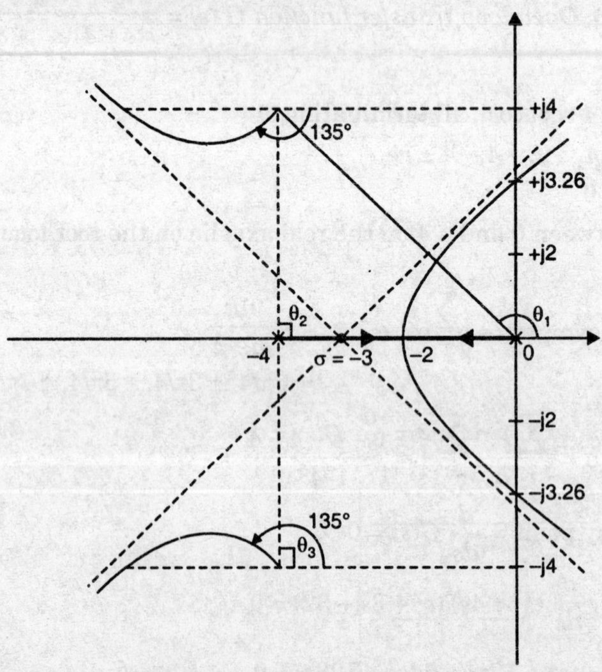

Fig. 5.19 Root locus of open loop transfer function $G(s) = \dfrac{k}{s(s+4)(s^2+8s+2)}$.

(*vi*) angle of departure $\phi_D = 180° - (\phi_p + \phi_z)$

$$\phi_D = \theta_1 + \theta_2 + \theta_3$$

Angle of departure at pole $-4 + j4$

From $s = 0$ pole, $\theta_1 = 180° - \tan^{-1}(4/4) = 135°$

At $s = -4$, $\theta_2 = 90°$

At $s = -4 - j4$, $\theta_3 = 90°$

$$\phi_D = 180° - (\phi_p + \phi_z) = 180° - (\theta_1 + \theta_2 + \theta_3)$$

$$= 180° - (135° + 90° + 90°) = -135° \qquad (\because \ \theta_z = 0)$$

Measure the angle of departure $-135°$ in clockwise direction from the pole reference of pole $(-4 + j4)$.

Problem 5.17. *A unity feedback control system has the open loop transfer function* $G(s) = \dfrac{k}{s^2(s+2)(s+5)}$. *Sketch the root locus diagram indicating clearly the breakaway points, asymptotes and their centroid and values of k and ω at the imaginary axis intersection. For what value of k is the closed loop system stable?*

If the feedback is changed to H(s) = 1 + 2s, for what values of k is the closed loop system now stable?

Solution:

The open loop transfer function is, $G(s) = \dfrac{k}{s^2(s+2)(s+5)}$

(*i*) Finite poles P are at $s = 0, 0, -2, -5$; zeros are not present.

Hence number of asymptotes = order of the system = 4.

The points between -2 and -5 on the real axis lie on the root locus, represented by thick line.

(*ii*) And angle of asymptotes are at $\alpha_1, \alpha_3 = \pm \dfrac{n\pi}{p - z} = \pm 45, \pm 135$

(*iii*) Centroid is at $\sigma = \dfrac{\sum p - \sum z}{p - z} = \dfrac{-2-5}{4} = -1.75$

(*iv*) Breakaway points are given by $\dfrac{d}{ds}G(s) = 0$

i.e.,
$$\frac{d}{ds}\left[\frac{k}{s^2(s+2)(s+5)}\right] = 0$$

$$\frac{d}{ds}\left[\frac{k}{s^4 + 7s^3 + 10s^2}\right] = 0$$

i.e.,
$$-k(4s^3 + 21s^2 + 20s) = 0$$

$$-ks[(s+4)(s+5)] = 0$$

i.e., $s = 0, s = -4, s = -5$

The valid breakaway points are $s = 0, s = -4$.

(*v*) The characteristic equation is given by $1 + G(s)\,H(s) = 0$

$$1 + \frac{k}{s^2(s^2 + 7s + 10)} = 0$$

$$s^4 + 7s^3 + 10s^2 + k = 0$$

The Routh array is

s^4	1	10	k
s^3	7	0	0
s^2	10	k	0
s	$\frac{-7k}{0}$	0	
s^0	k	0	

The Routh array gives the condition for stability

$$\frac{-7k}{10} \geq 0 \quad \text{and} \quad k \geq 0$$

At $k = 0$ the system is marginally stable.

The root locus is sketched for the value of $k = 0$ as shown in the Fig. 5.20.

The auxiliary equation from the Routh array is

$$10s^2 + 0 = 0 \quad \text{or} \quad s = 0$$

It is not intersecting the imaginary axis.

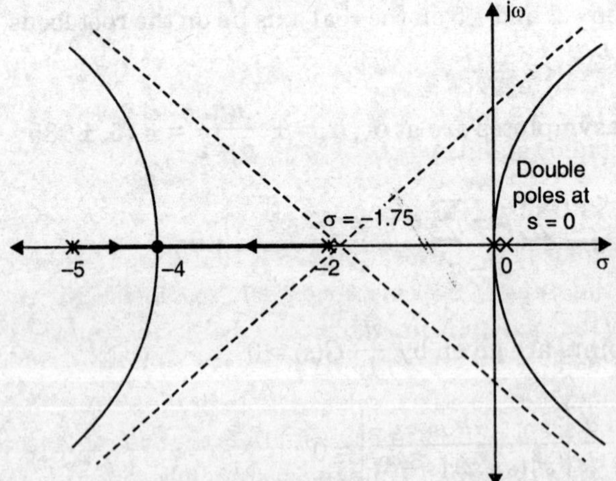

Fig. 5.20 Root locus of open loop transfer function $G(s) = \dfrac{k}{s^2(s+2)(s+5)}$.

Addition of zero (1+2s) of the given open loop transfer function

$$\therefore \qquad G(s)\,H(s) = \frac{k(1+2s)}{s^2\,(s+2)\,(s+5)}$$

The characteristic equation is

$$s^2\,(s^2 + 7s + 10) + k\,(1 = 2s) = 0$$

$$s^4 + 7s^3 + 10s^2 + 2ks + k = 0$$

The Routh array is

$$\begin{array}{c|ccc} s^4 & 1 & 10 & k \\ s^3 & 7 & 2k & 0 \\ s^2 & \dfrac{70-2k}{7} & k & \\ s & \dfrac{k(4k-91)}{2k-70} & 0 & \\ s^0 & k & & \end{array}$$

The condition for stability is given by

$$70 - 2k > 0$$

$$2k < 70$$

$$k < 35$$

$$\frac{91-40}{70-2k} > 0$$

$$91 - 4k > 0$$

$$4k < 91$$

$$k < \frac{91}{4} \quad \text{or} \quad k < 25, \quad \text{and} \quad k > 0$$

Hence the condition for stability is $0 < k < 25, 35$.

5.5.1. Stability Analysis using Root Locus

As the stability is related to the location of the closed loop poles or the zeros of the characteristic polynomial in the s-plane, the existence of the root locus branch in the right half of the s-plane, establishes that the system is unstable. The root locus plot in Fig. 5.20 shows that for $0 < k < \infty$, the root locus does not go into the right half of the s-plane. Thus it is a very stable system. But for the system whose root locus plot is given in Fig. 5.19, the system has roots in the left half of the s-plane for $0 < k < 568.9$ and at $k = 568.9$ the system is critically stable. Only for $k > 568.9$ the system becomes unstable. Thus the system has conditional stability.

Hence the root locus portrait is very convenient method of finding stability of a closed loop system. However the complexity of drawing the root locus plot increases with order of the system. But this is a convenient method for systems upto order six or seven.

Problem 5.18. *A unity feedback system has an open loop function* $G(s) = \dfrac{k}{s(s^2 + 4s + 13)}$.

Make a rough sketch of root locus plot by determining the following (a) Centroid, number and angle of asymptotes (b) angle of departure of root loci from the poles (c) Breakaway points if any (d) points of intersection with jω axis and (e) maximum value of k for stability.

Solution: The open loop transfer function is $G(s) = \dfrac{k}{s(s^2 + 4s + 13)}$

(*i*) Poles are at $s = 0$ and $s^2 + 4s + 13 = 0$

$$s = \frac{-4 \pm \sqrt{16 - 52}}{2} = \frac{-4 \pm j6}{2} = -2 \pm j3$$

Finite zeros are nil.

Number of asymptotes = order of the system = 3

The points between 0 and $-\infty$ on the real axis lie on the root locus, represented by thick line.

(*ii*) Angle of asymptotes is ± 60, 180.

(*iii*) Centroid, $\sigma = \dfrac{\sum p - \sum z}{p - z} = \dfrac{-2-2}{3} = \dfrac{-4}{3} = -1.33$

(*iv*) Angle of departure

Angle of departure from pole $-2 + j3$ is

From $s = 0$, pole $\theta_1 = 180 - \tan^{-1} 3/2$

From $-2 - j3$ pole is $\theta_2 = 90°$

$$\theta_D = 180° - (\theta_1 + \theta_2)$$

$$= 180° - 180° + \tan^{-1} 3/2 - 90° \Rightarrow \tan^{-1} 3/2 - 90° = -33.69°$$

(*v*) Breakaway points are given by

$$\frac{d}{ds}[G(s)\,H(s)] = 0$$

$$\frac{d}{ds}\left[\frac{k}{s^3 + 4s^2 + 13s}\right] = 0$$

$$-k\,[3s^2 + 8s + 13] = 0$$

$$s = \frac{-8 \pm \sqrt{64 - 146}}{2} = \frac{-8 \pm j\sqrt{82}}{2} = \frac{-8 \pm j9.1}{2}$$

which do not give possible breakaway points.

(*vii*) The characteristic equation is

$$1 + G(s)\,H(s) = 0$$

$$1 + \frac{k}{s(s^2 + 4s + 13)} = 0$$

$$s^3 + 4s^2 + 13s + k = 0$$

The Routh array is

s^3	1	13	0
s^2	4	k	
s	$\dfrac{52-k}{4}$	0	
s^0	k		

The condition for stability is $k > 0$

$$\frac{52-k}{4} > 0 \quad \text{or} \quad k < 52 \quad \text{or} \quad \text{in short} \quad 0 < k < 52$$

For $k = 52$, the auxiliary equation is

$$4s^2 + 52 = 0$$

$$s^2 = -13$$

$$s = \pm j\sqrt{13} = \pm j3.61$$

Root locus intersect the imaginary axis at $\omega = 3.16$ rad/sec.

Root locus for a given transfer function is as shown in the Fig. 5.21.

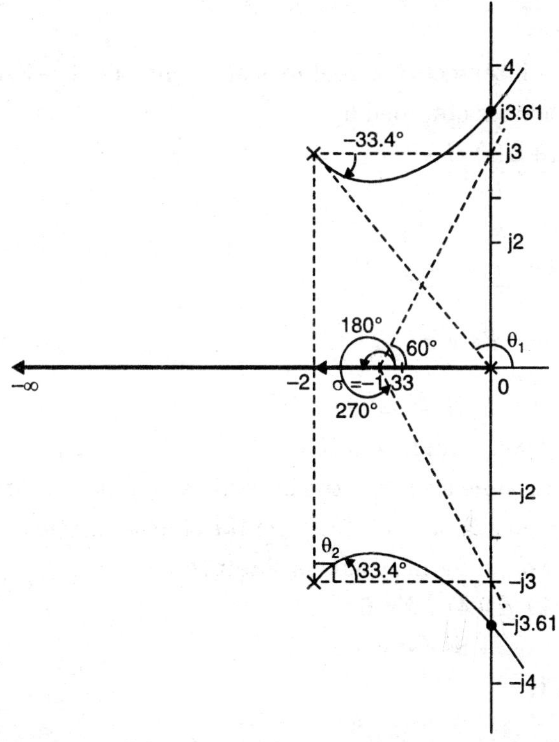

Fig. 5.21 Root locus of open loop transfer function $G(s) = \dfrac{k}{s(s^2 + 4s + 13)}$.

Problem 5.19. *Sketch the root locus plot of unity feedback system with an open loop transfer function* $G(s) = \dfrac{k}{s(s+2)(s+4)}$. *Find the range of k for the system to have damped oscillatory response. Determine the value of k so that the dominant pair of complex poles of the system has a damping ratio of 0.5. Corresponding to this value of k, determine the closed loop transfer function in the factored form.*

Solution: For the given transfer function, we have

The number of zeros, $z = 0$

The number of poles, $p = 3$

The number of asymptotes = 3

The points between 0 to – 2 and – 4 to – ∞ on the real axis lie on the root locus, represented by thick line.

Angle of asymptotes , $\alpha = \dfrac{\pm n\pi}{p - z}$

$$\alpha_1, \alpha_3, \alpha_5 = \pm 60, \pm 180, \pm 300$$

$$\text{Centroid} = \frac{\sum p - \sum z}{p - z} = \frac{(0 - 2 - 4) - 0}{3 - 0} = -2$$

Asymptotic angles intersects the real axis at a centroid of (– 2, 0)

The breakaway point is obtained by

$$\frac{d[1 + G(s)\,H(s)]}{ds} = 0$$

$$\frac{d\left(1 + \dfrac{k}{s(s+2)(s+4)}\right)}{ds} = 0$$

$$\frac{d(s^3 + 6s^2 + 8s + k)}{ds} = 0$$

$$3s^2 + 12s + 8 = 0$$

The roots are = – 0.845 and – 3.154.

– 3.154 is not admissible and thus the breakaway point is located at (– 0.845,0).

Intersection with $j\omega$ axis is obtained from the characteristic polynomial.

Intersection point on imaginary axis is obtained

$$s(s + 2)(s + 4) + k = 0$$

or

$$s^3 + 6s^2 + 8s + k = 0$$

The Routh array is

$$\begin{array}{c|cc} s^3 & 1 & 8 \\ s^2 & 6 & k \\ s & 48 - k & \\ s^0 & k & \end{array}$$

The limiting value of k is 48 for this value auxiliary equation

$$6s^2 + 48 = 0$$

Gives
$$s^2 = -8$$
$$s = \pm j2.82$$

The root locus plot is shown in Fig. 5.22, given that $\delta = 0.5$, $\cos\theta = 0.5$. Thus $\theta = 60°$. The intersection of $\delta = 0.5$ line with the root locus is at given by $-0.8 + j0.95$.

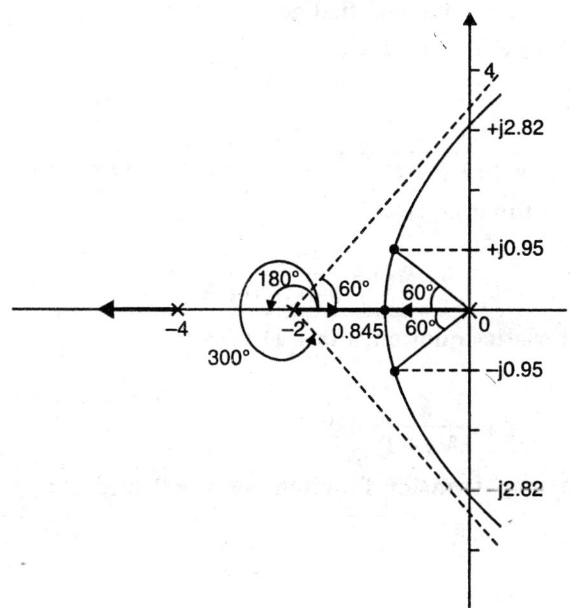

Fig. 5.22 Root locus of unity feedback system wih open loop transfer function $G(s) = \dfrac{k}{s(s+2)(s+4)}$.

The value of k at $-0.8 + j0.95$ is

$$\left.\frac{k}{s(s+2)(s+4)}\right|_{s=(0.8+j0.95)} = 1$$

$$k = (-0.8 + j0.95)(-0.8 + j0.95 + 2)(-0.8 + j0.95 + 4) = 6.33$$

5.6 ROOT CONTOURS

The root locus for a closed loop system was plotted with the variable parameter k is the open loop gain. If the parameter is other than k or more than one parameter is involved, by using the same technique, root contours are obtained. The following example illustrates the construction of the root contours.

Problem 5.20. *The open loop transfer function of a feedback system is*

$$G(s)H(s) = \frac{k}{s(s+1)(s+\alpha)}$$

The pole at s = – α may be treated as another variable for sketching the root contour.

Solution: The characteristic equation is

$$s(s + 1)(s + \alpha) + k = 0$$

The above expression can be modified as

$$s^2 (s + 1) + \alpha s (s + 1) + k = 0$$

This is rewritten as

$$1 + \frac{\alpha s(s+1)}{s^2(s+1)+k} = 0$$

Open loop transfer function is

$$T(s) = \frac{\alpha s(s+1)}{s^2(s+1)+k}$$

$\underset{\alpha \to 0}{\text{Lt}}$, the characteristic equation, $s^2(s + 1) + k = 0$

i.e.,
$$1 + \frac{k}{s^2(s+1)} = 0$$

$T(s)$ is the open loop transfer function for $\alpha = 0$ and the root locus plot is shown in Fig. 5.23.

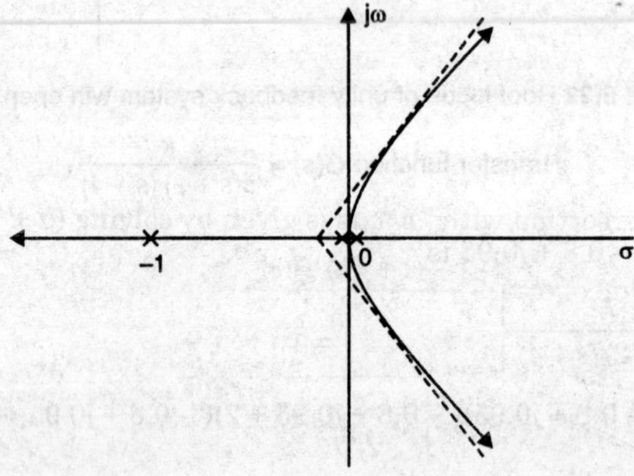

Fig. 5.23

$\underset{\alpha \to \infty}{\text{Lt}}$, the roots are $s = 0$ and $s = -1$.

For $\alpha = 0$, the roots lie on the root locus plot of Fig. 5.23 for different values of k. Thus the root contours start from the roots (*i.e.*, points) on the root locus for say $k = k_1$ and as $\alpha \to \infty$, they move towards the real axis by crossing the $j\omega$-axis and meet at a breakaway point on the real axis and then move towards the zeros at $s = 0$ and $s = -1$ respectively. The segment on the real axis starting from $k = k_1$ will move to a zero at infinity. Thus the root contours are sketched in Fig. 5.23 for $k = k_1$. Similarly for $k = k_2 > k_1$ also the root contour is shown in the Fig. 5.24.

For obtaining the $j\omega$-axis intersection, consider the characteristic equation,

$$s^2(s + 1) + \alpha s\,(s + 1) + k = 0$$

$$s^3 + s^2(\alpha + 1) + \alpha s + k = 0$$

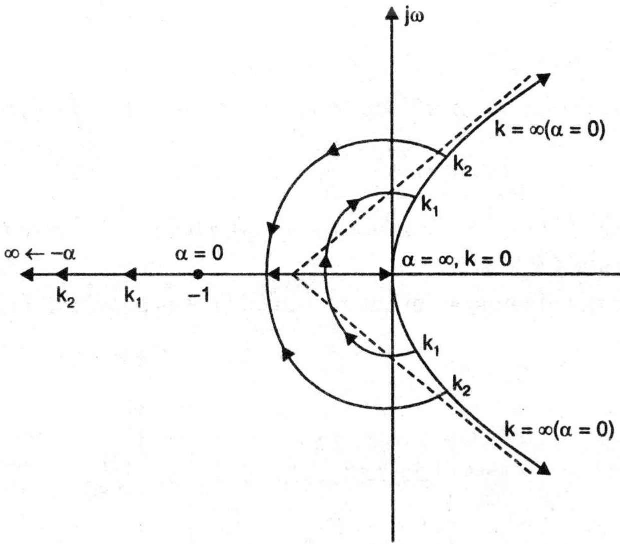

Fig. 5.24

The Routh array is

s^3	1	α
s^2	$1+\alpha$	k
s	$\dfrac{\alpha(\alpha+1)-k}{\alpha+1}$	
s^0	k	

The point of intersection with $j\omega$-axis is given by solving $(\alpha + 1)\,s^2 + k = 0$

The limiting value of $\alpha = \dfrac{-1+(1+4k)^{1/2}}{2}$

The point of intersection $= \pm j \dfrac{k_1}{\left[\dfrac{1+(1+4k)^{1/2}}{2}\right]^{1/2}}$

The root contours are sketched on the root locus plot for the multiple feedback loop systems.

5.7 ROOT SENSITIVITY

The sensitivity of a transfer function $T(s)$ to a parameter (k) variation can be defined as

$$S_K^T = \frac{(\ln T)}{(\ln k)} \frac{\dfrac{\partial T}{T}}{\dfrac{\partial k}{k}} \qquad (5.21)$$

In the same manner the sensitivity of the root $s = -r_k$ of the characteristic polynomial with respect to the variation of the open loop gain (k) is defined as

$$S_k^{-r_k} = \frac{(-r_k)}{(\ln k)} = \frac{\partial(-r_k)}{\frac{\partial k}{k}} \tag{5.22}$$

Problem 5.21. *Given the open loop transfer function of a unity feedback system*

$$G(s) = \frac{k(s+3)}{s(s+2)}$$

Obtain the sensitivity of the dominant conjugate pole of the closed loop transfer function for a variation of k from 4.0 to 4.84.

Solution: For the given system the root locus plot is shown in Fig. 5.25.

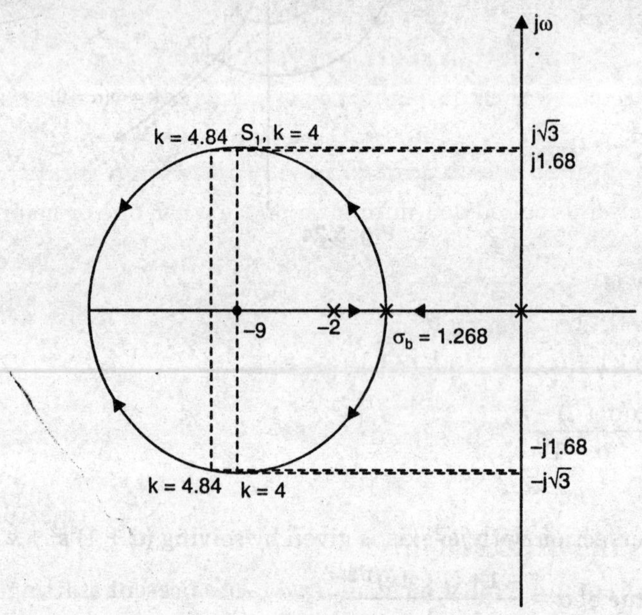

Fig. 5.25

While finding the breakaway point between $s = 0$ and $s = -2$ by solving $\frac{dk}{ds} = 0$, we obtain two values for σ_b as -1.268 and -4.732. The second value can be taken as breakaway points between the zeros at $s = -3$ and $s = -\infty$.

For $k = 4$ the roots are obtained from

$$s^2 + 2s + 4s + 12 = 0$$

and the roots are $s_1 = -3 + j(3)^{1/2}$ and $s_2 = -3 - j(3)^{1/2}$

Similarly for $k = 4.84$

We have $s_1 = -3.42 + j(1.68)$ and $s_1 = -3.42 - j(1.68)$

Thus the sensitivity is

$$S_k^{s_1} = \frac{\partial s_1}{\frac{\partial k}{k}} = \frac{\partial s_1 \times k}{\partial k}$$

$$= \frac{(-0.42 - j0.052) \times 4}{0.84}$$

$$= -2 - j0.247 = 2.015 \angle -172.95°.$$

SHORT QUESTIONS AND ANSWERS

1. What is BIBO stability criterion?

The system is said to be bounded input-bounded output if the output is bounded for a given bounded input.

2. When is the transfer function said to be minimum phase?

The transfer function is called minimum phase when all the poles and zeros are in left half of s-plane.

3. When is the transfer function said to be non-minimum phase?

The transfer function is called non-minimum phase when one or more zeros are in the right half of s-plane.

4. What is the necessary condition for stability?

The necessary condition for stability is that all the coefficients of the characteristic polynomial are positive.

5. What is asymptotic stability?

When a system parameter is perturbed to cause instability in the system, the system must return to stability as time tends to infinity.

6. What is the necessary condition that the characteristic equation of a feedback system satisfies the BIBO stability?

The characteristic equation must have all powers of s present starting from the highest power equal to the order of the polynomial. All terms must have positive and real coefficients. No power of s must be missing.

7. When does the procedure for making the Routh array get terminated?

(*i*) Due to the first element in a row being zero and at least one element being a non zero in that row, the procedure cannot continue.

(*ii*) Due to the presence of $j\omega$-axis poles, there will be row(s) with all zeros and the procedure cannot continue.

This is the cause of unstable system.

8. What is meant by conditional stability?

The range of open loop gain k for which the closed loop poles lie to the left half of the s-plane, is called conditional stability.

9. What is meant by critical or marginal stability?

The closed loop poles present on the $j\omega$-axis cause of sustained oscillations in the system. This is called critical or marginal stability.

10. What are the poles of the system?

The poles of the system is defined as the roots of the denominator polynomial of the transfer function.

11. What are the zeros of the system?

The zeros of the system is defined as the roots of the numerator polynomial of the transfer function.

12. What is the nature of response if the poles are complex?

The response is oscillatory.

13. What is the necessary and sufficient condition for stability?

The necessary and sufficient condition for stability is that all the elements in the first column of Rouths array should be positive.

14. How can you find the number of roots lying on the right side of the s-plane?

It can be found from the number of sign changes in the first column of Rouths array.

15. What is breakaway point?

It is the point on the root locus and is the location of repeated roots of characteristic equation.

16. What is the cause of instability?

The chief cause for instability is the inability of the system elements to respond quickly to input.

17. What is meant by absolute stability?

Absolute stability merely indicates whether a system is stable or not.

18. What are the disadvantages of Routh criterion?

- This method is valid only if the characteristic equation is algebraic and that all the coefficients are real.
- This gives an idea of only the absolute stability, that is stable or not.
- It does not give an indication of the degree of instability and the means of avoiding it.
- It indicates the presence and number of unstable roots but not their values.

19. What are the advantages of root locus?

- Root locus method is a simple graphical procedure of evaluating the roots of the characteristic equation of a system for its poles and zeros.
- This method provides a very satisfactory way to carry out many analysis and design problems in time domain.

20. State the rule for point to lie on root loci.

A point on the real axis lies on root loci if number of poles and number of zeros on the real axis to the right of that point is odd.

21. What is the effect of addition of pole on root locus?

In general we can state that addition of pole to the transfer function $G(s)\,H(s)$ in the left half of s-plane has the effect of pushing the origin of root locus towards right half of the s-plane.

22. What is a root contour?

They are the root loci plotted for the systems with more than one variable parameter and for multiple loop feedback control systems.

OBJECTIVE TYPE QUESTIONS

1. Necessary conditions for stability is
   (a) there is no missing term
   (b) all coefficients of the S variable must be positive
   (c) both (a) and (b)
   (d) for bounded input, output is unbounded
2. Check the stability for the given characteristic equation $s^4 + 3s^3 + 4s^2 + 2s + 1 = 0$ is
   (a) Stable (b) Unstable
   (c) Conditionally stable (d) Marginally stable
3. What is the sufficient condition for stability?
   (a) There is no missing term
   (b) All coefficients of s variable must be positive
   (c) Roots of characteristic equation must lie on left side of s-plane
   (d) all of the above
4. If the roots of characteristic equation lie on left side of S-plane the system is
   (a) Stable (b) Unstable
   (c) Conditionally stable (d) Marginally stable
5. Given the open loop transfer function of unity feedback system $G(s) = \dfrac{s}{s^2 + 2s + 1}$ is
   (a) Stable (b) Unstable
   (c) Conditionally stable (d) Marginally stable
6. If the roots of characteristic equation lie on imaginary axis the system is
   (a) Stable (b) Unstable
   (c) Conditionally stable (d) Marginally stable
7. For the system stability the range of values of k for a given Ufb system with open loop transfer function $G(s) = \dfrac{k}{(s+2)(s+4)(s^2+6s+25)}$
   (a) -200 (b) 666
   (c) $-200 < k < 666$ (d) $-200 > k > 666$

8. If first entry in any row of Routh array is negative the system is

(*a*) Stable (*b*) Unstable

(*c*) Conditionally stable (*d*) Marginally stable

9. The number of changes in first column of Routh array represents.

(*a*) Stability (*b*) Unstability

(*c*) Number of roots lie on right side of s-plane

(*d*) both *b* and *c*

10. When entry in first column of Routh array is zero and rest of the elements are not zero the remedy is

(*a*) put $s = z - 1$ in the characteristic equation

(*b*) put $s = \frac{1}{z}$ in the characteristic equation

(*c*) replace zero element with positive element

(*d*) both (*b*) and (*c*)

(*e*) Take the auxiliary equation

11. Routh array for a system is given below

s^4	1	3	5
s^3	1	2	0
s^2	1	5	
s	– 3		
s^0	5		

The system is

(*a*) Stable (*b*) Unstable

(*c*) Conditionally stable (*d*) Marginally stable

12. Root locus technique is applicable to

(*a*) Time domain (b) Frequency domain

(c) Both (*a*) and (*b*) (*d*) None

13. In root locus technique centroid *x* is equal to

(*a*) $\frac{\sum z - \sum p}{p - z}$ (*b*) $\frac{\sum p - \sum z}{p - z}$

(*c*) $\frac{\sum p - \sum z}{z - p}$ (*d*) $\frac{\sum z - \sum p}{z - p}$

14. In root locus technique angle of departure ϕD is equal to

(*a*) $180 - (\phi_p - \phi_z)$ (*b*) $180 - (\phi_z - \phi_p)$

(*c*) $180 + (\phi_p - \phi_z)$ (*d*) $180 + (\phi_z - \phi_p)$

15. In root locus technique angle of arrival ϕ is equals to

(*a*) $180 - (\phi_p - \phi_z)$ (*b*) $180 - (\phi_z - \phi_p)$

(*c*) $180 + (\phi_p - \phi_z)$ (*d*) $180 + (\phi_z - \phi_p)$

16. No. of root loci is equal to

(*a*) No. of zeros (*b*) Order of the system

(*c*) No.of poles (*d*) Both (*b*) and (*c*)

17. The transfer function of a system is $\frac{k}{a_3s^3 + a_2s^2 + a_1s + a_0}$. For the system to be absolutely stable,

(*a*) $a_3, a_2, a_1, a_0 > 0$ and $a_2 a_1 - a_3 a_0 > 0$ (*b*) $a_3, a_2, a_1, a_0 > 0$ and $a_2 a_1 - a_3 a_0 < C$

(*c*) $a_3, a_2, a_1, a_0 > 0$ and $a_2 a_1 - a_3 a_0 = 0$ (*d*) $A_2, a_0 > 0$ and $a_3 a_1 < 0$

18. Routh's array for a system is given below

s^4	1	3	5
s^3	1	2	0
s^2	1	5	
s^1	– 3		
s^0	5		

The system is

(*a*) Stable (*b*) Unstable

(*c*) Marginally stable (*d*) Conditionally stable

19. The number of sign changes in the entries in the first column of Routh's array denotes

(*a*) the number of zeros of the system in the RHP

(*b*) the number of roots of characteristic polynomial in RHP

(*c*) the number of open-loop poles in RHP

(*d*) the number of open-loop zeros in RHP

20. Consider a characteristic equation $s^4 + 3s^3 + 5s^2 + 6s + k + 10 = 0$. The condition for stability is

(*a*) $k > 5$ (*b*) $-10 \leq k$

(*c*) $k > -4$ (*d*) $-10 < k < -4$

21. The characteristic equation of a unity feedback system is given by $s^3 + s^2 + 4s + 4 = 0$

(*a*) The system has one pole in the RH-s plane

(*b*)The system has no poles in the RH-s plane

(*c*) The system is asymptotically stable

(*d*) The system exhibits oscillatory behaviour

22. An electromechanical closed-loop control system has the following characteristic equation $s^3 + 6k\,s^2 + (k + 2)s + 8 = 0$ where K is the forward gain of the system. The condition for closed-loop stability is

(*a*) $k = 0.528$ (*b*) $k = 2$

(*c*) $k = 0$ (*d*) $k = -2.528$

23. The case in the Routh table in which particular row elements are zero, show that

(*a*) a differentiation has to be carried out and conveys no other information

(*b*) it is a special case in Routh array

(*c*) whether the system is stable or not

(*d*) some roots are distributed symmetrically about the origin

24. The system represented by its transfer function has some poles lying on the imaginary axis, it is

(*a*) unconditionally stable (*b*) Conditionally stable

(*c*) unstable (*d*) Marginally stable

25. In the first column of the Routh array, an element was found to be zero. The first column element above this zero and below this zero has the same sign. This condition indicates that the system

(*a*) is stable (*b*) is unstable

(*c*) has all the roots in LHP except one (*d*) has some roots on the $j\omega$-axis

26. The statements that holds good for relative stability analysis is :

(*a*) Routh array method cannot be used

(*b*) Graphical methods can be used and Routh array cannot be used

(*c*) Graphical methods cannot be used

(*d*) Both graphical as well as Routh array can be used

27. The root-locus plot is shown alongside. What is the transfer function?

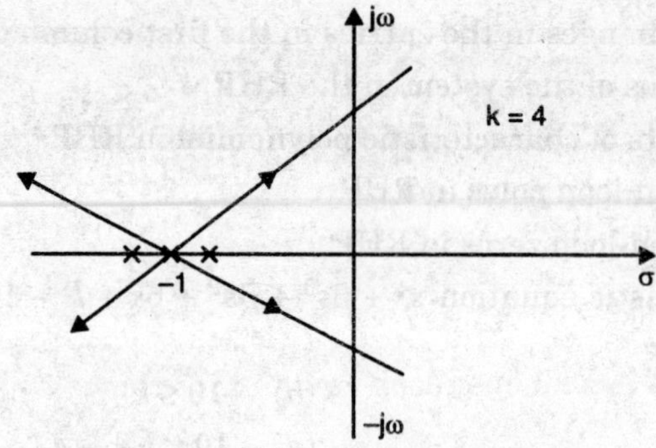

(*a*) $\dfrac{4}{s+1}$ (*b*) $\dfrac{4}{(s+1)^2}$

(*c*) $\dfrac{4}{(s+1)^3}$ (*d*) $\dfrac{4}{(s+1)^4}$

28. The asymptotes and the break point coincide at $s = -2$. The transfer function can be

(*a*) $\dfrac{k}{(s+1)(s+2)}$ (*b*) $\dfrac{k(s+2)}{(s+1)(s+3)}$

(*c*) $\dfrac{k}{(s+1)(s+2)(s+3)}$ (*d*) $\dfrac{k}{(s+2)^3}$

29. The transfer function is $\frac{k}{(s+1)(s+2)(s+3)}$. The break point will lie between

(*a*) 0 and – 1 (*b*) – 1 and – 2

(*c*) – 2 and – 3 (*d*) Beyond – 3

30. A unity feedback system has an open-loop transfer function $G(s) = \frac{k}{s(s^2+4s+13)}$. The centroid of the asymptotes of the root locus plot lies at

(*a*) – 4 (*b*) $-\left(\frac{4}{3}\right)$

(*c*) – 13 (*d*) – 10

31. The open-loop transfer function of an unity feedback system is given by $G(s) = \frac{5}{(s+1)(s+2)(s+3)}$. The number of asymptotes of the root locus plot that tend to infinity is given by

(*a*) 3 (b) 1

(*c*) 2 (*d*) 4

32. The open-loop transfer function of an unity feedback system is given by $G(s) = \frac{k}{s(s+1)(s+2)}$. The breakaway point of the root locus plot is given by

(*a*) – 0.423 (*b*) – 0.523

(*c*) – 0.700 (*d*) – 0.5

33. A unity feedback system has an open-loop transfer function $G(s) = \frac{k}{s(s^2+4s+13)}$. The angle of asymptotes are given by

(*a*) 45°, 135°, 225° (*b*) 60°, 180°, 300°

(*c*) 90°, 180°, 270° (*d*) 45°, 90°, 135°

34. The open-loop transfer function of a feedback system in $G(s)\,H(s) = \frac{k}{s(s+4)(s^2+4s+20)}$ the four branches of root-locus originate at

(*a*) – 2, – 3, – 1, + *j*4, – 1, – *j*4 (*b*) – 1, – 2, – 3 + *j*4, – 3 – *j*4

(*c*) 0, – 4, – 2 - *j*4, – 2 – *j*4 (*d*) 0, – 2, – 1 + *j*4, – 1 – *j*4

35. The open-loop transfer function of a feedback system is $G(s)\,H(s) = \frac{k(s+1)(s+2)}{s(s+3)(s+4)}$. The real axis segments of the root locus lies between

(*a*) 0 and – 1 ; – 2 and – 3 ; – 4 and – ∞ (*b*) – 1 and – 2 ; – 3 and – 4

(*c*) – 1 and – 1.5 ; – 3.5 and – 4 (*d*) 1 and 2 ; 3 and 4

36. The main objective of drawing the root locus is

(*a*) To find the time response of the system

(*b*) To find the frequency response of the system

(*c*) To find the roots of the characteristic equation for different values of system parameters

37. There are three zeros and two poles of $GH(s)$. There will be

(*a*) Three root loci
(*b*) Two root loci
(*c*) Five root loci
(*d*) One root locus

38. The root loci

(*a*) Start from zeros and end at poles
(*b*) Start from and end at infinity
(*c*) Start from poles and end at zeros and infinity
(*d*) Start from zeros and end at poles and infinity

39. For the root locus, the phase angle criterion is

(*a*) Odd multiple of 180°
(*b*) Even multiple of 180°
(*c*) Odd multiple of 90°
(*d*) None of the above

40. For positive value of k and negative feedback, the root loci exist on the real axis only in those parts

(*a*) Where odd number of poles and zeros are present to the right of a point.

(*b*) Where even number of poles and zeros are present to the right of the point.

(*c*) Where odd number of poles and zeros are present to the left of the point.

(*d*) Where even number of poles and zeros are present to the left of the point.

41. For negative feedback system, for complementary root locus (k varied from 0 to $-\infty$), the phase angle criterion is

(*a*) Odd multiple of 180°
(*b*) Even multiple of 180°
(*c*) Odd multiple of 90°
(*d*) 270° only

42. The intersection of the asymptote is given by

(*a*) $x = \dfrac{\sum \text{Poles of } GH(s) - \sum \text{Zeros of } GH(s)}{p - z}$

(*b*) $x = \dfrac{\sum \text{Zeros of } GH(s) - \sum \text{Poles of } GH(s)}{p - z}$

(*c*) $x = \dfrac{\sum \text{Poles of } GH(s) + \sum \text{Zeros of } GH(s)}{p - z}$

(*d*) $x = \dfrac{\sum \text{Poles of } GH(s) - \sum \text{Zeros of } GH(s)}{z - p}$

43. The angles which the asymptotes make with the real axis is given by

(*a*) $\phi = \dfrac{(2n+1)\pi}{(p+z)}$
(*b*) $\phi = \dfrac{(2n-1)\pi}{(p+z)}$
(*c*) $\phi = \dfrac{(2n+1)p}{(p-z)}$
(*d*) $\phi = \dfrac{(2n-1)\pi}{(p-z)}$

44. The intersection of root locus with the imaginary axis is obtained

(*a*) By putting $j\omega = 0$ in $GH(s)$

(*b*) By putting real part of $GH(s) = 0$

(*c*) From Routh array of $1 + GH(s)$ find, critical value of K and find ω from the auxiliary equation of a row.

(*d*) From the Routh array get auxiliary equation and get k

45. The break away points are obtained by

(*a*) Putting $1 + GH(j\omega) = 0$ and solving for ω

(*b*) Putting $GH(j\omega) = 0$ and solving for ω

(*c*) Differentiating $1 + GH(s)$ with respect to s and equating $\left(\frac{dK}{ds}\right) = 0$

(*d*) Differentiating $1 + GH(s)$ with respect to ω and equating $\left(\frac{dK}{d\omega}\right) = 0$

46. The spirule is used

(*a*) To draw the root locus only

(*b*) To draw the root locus and calibrate it in terms of variable parameter

(*c*) To find the closed loop roots only

(*d*) To find the damping ratio only

KEY

1. (*c*)	**2.** (*a*)	**3.** (*c*)	**4.** (*a*)	**5.** (*a*)	**6.** (*d*)
7. (*c*)	**8.** (*b*)	**9.** (*c*)	**10.** (*d*)	**11.** (*b*)	**12.** (*a*)
13. (*b*)	**14.** (*a*)	**15.** (*b*)	**16.** (*d*)	**17.** (*a*)	**18.** (*a*)
19. (*b*)	**20.** (*d*)	**21.** (*a*) (*d*)	**22.** (*a*), (*d*)	**23.** (*d*)	**24.** (*d*)
25. (*d*)	**26.** (*d*)	**27.** (*c*)	**28.** (*d*)	**29.** (*b*)	**30.** (*b*)
31. (*c*)	**32.** (*a*)	**33.** (*b*)	**34.** (*c*)	**35.** (*a*)	**36.** (*d*)
37. (*a*)	**38.** (*c*)	**39.** (*a*)	**40.** (*a*)	**41.** (*b*)	**42.** (*a*)
43. (*c*)	**44.** (*c*)	**45.** (*c*)	**46.** (*b*)		

EXERCISE

1. Explain the concept of stability.

2. What are causes of instability?

3. What are the necessary and sufficient conditions of stability for linear time invariant systems?

4. State and explain Hurwitz stability criterion.

5. State and explain Routh stability criterion.

6. What are the limitations of Routh stability criterion?

7. Explain the special cases in Routh stability criterion.
8. What is meant by absolute stability and relative stability?
9. Explain the construction rules for root locus technique.
10. Define and derive the breakaway point on the root locus.
11. What is root contour and how is it constructed?
12. Explain root sensitivity.
13. Determine the stability of the closed loop system whose open loop transfer is $\frac{10(3s+1)}{s(s+1)(1+6s)(1+0.1s)}$, using Routh-Hurwitz criterion.
14. Apply Routh criterion and determine the stability of the system whose characteristic polynomial is $s^4 + s^3 + 5s^2 + s + 14 = 0$.
15. Investigate the stability of the system whose characteristic polynomial is
$$s^4 + 3s^3 + 2s^2 + s + 14 = 0.$$
16. Investigate the stability of the system using Hurwitz criterion, whose characteristic polynomial is $2s^4 + 5s^3 + 3s^2 + 2s + 4 = 0$.
17. A feedback system has the open loop transfer function of $G(s) = \frac{ke^{-s}}{s(s^2+2s+5)}$. Find the limiting values of k for maintaining stability.
18. Find the number of open loop poles, number of asymptotes, centroid and angle of asymptotes of the given system transfer function $G(s) = \frac{k}{(s+2)^3}$.
19. Sketch the root locus plot for the system with open loop transfer
$$G(s) = \frac{k(s+4)}{(s+1)(s^2+6s+13)}.$$
20. Obtain the angle of departure of the root locus of the closed loop system at a complex poles of $G(s) = \frac{k}{s(s^2+2s+2)}$.

6 Frequency Domain Analysis

6.1 INTRODUCTION

Frequency domain analysis analyses the system response w.r.t. frequency or how the system response varies w.r.t. frequency. The term frequency response refers to the steady state response of a system to a sinusoidal input. Industrial control systems are often designed by use of frequency response methods. Many techniques are available in the frequency response methods for the analysis and design of control systems.

This frequency domain analysis gives the relationship between transfer function and frequency response of a linear system. If we apply a sine wave input to a stable linear system, the steady state output will also be a sine wave of the same frequency. The output will in general, differ from the input in both magnitude and phase. The frequency response can be calculated by replacing 's' in transfer function by '$j\omega$' (complex frequency)

i.e.,

$$G(s)\Big|_{s - j\omega} = G(j\omega) \tag{6.1}$$

$$M\, e^{j\theta(\omega)} = M \angle\theta \tag{6.2}$$

where **Magnitude (M)** is the ratio of the amplitudes of the output and input sinusoids and is called the magnitude or gain.

Phase angle (θ) is the angle by which the output leads the input, both M and θ are fractions of angular frequency ω

$$G(s) = \frac{k(s - z_1)(s - z_2)\ldots\ldots}{(s - p_1)(s - p_2)\ldots\ldots}$$

$$G(j\,\omega) = \frac{k(j\omega - z_1)(j\omega - z_2)\ldots\ldots}{(j\omega - p_1)(j\omega - p_2)\ldots\ldots} = M \angle\theta \tag{6.3}$$

θ = Sum of arguments of vectors from zeros – sum of arguments of vectors from poles.

Both magnitude M and phase angle θ play a very important role in frequency domain analysis.

6.2 FREQUENCY DOMAIN SPECIFICATIONS

In our study so far, we wrote the differential equation of the system, solved it by algebraic methods after Laplace transformation. The nature of roots of the equation indicated the stability of the system. Solution by differential equation is a cumbersome method. If the solved equations do not meet the specifications required, it is not easy to decide the manner in which the system should be changed to obtain the desired results. So we proceed to consider methods such as graphical methods.

The chief aim of designer is to meet performance specifications. Basically there are two types of performance specifications: (1) Time domain specification and (2) Frequency domain specification.

The time domain specification are discussed in chapter 4. The frequency domain specifications are the study of variable parameter as a function of frequency of the sinusoidal input signal. We can study the system performance over a range of frequencies and restrict the operation of the desired frequency range or alternatively choose the best frequency range.

Frequency domain specifications are

(*i*) Gain margin and Phase margin

(*ii*) Resonant peak (M_p) and Resonant frequency (ω_r)

(*iii*) Delay time (t_d)

(*iv*) Bandwidth (BW) and Cut off rate

(*i*) **Gain margin and phase margin.** These factors are the measures of relative stability and are related to the closeness of the closed loop poles to the $j\omega$ axis.

The **gain margin** is defined as the reciprocal of the open loop transfer function evaluated at the frequency ω_{pc} at which the phase angle is – 180°.

$$\text{Gain margin} = 20 \log_{10} \frac{1}{|G(j\omega)H(j\omega)|} \text{ dB} \tag{6.4}$$

Gain margin gives the relative distance between $(-1 + j0)$ point and the $P(j\omega)$ or $G(j\omega) H(j\omega)$ plot.

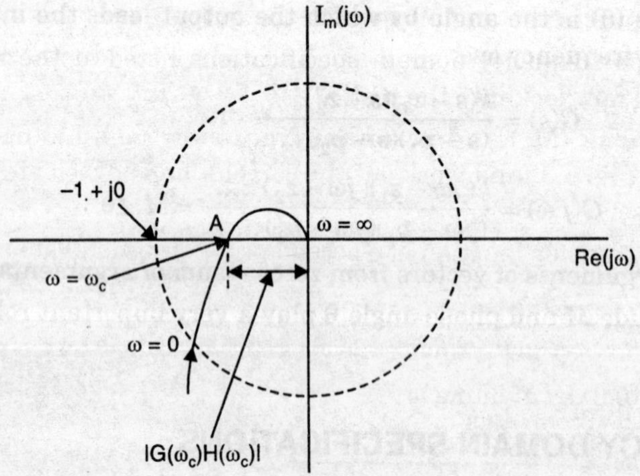

Fig. 6.1

A rigorous definition of **phase margin** is the angle (ϕ_{PM}) at gross crossover frequency ω_{gc} at which magnitude equals to one (or) 0 dB.

In simple language it means ϕ_{PM} is the angle between the negative real axis and the radius vector joining the origin to the gain crossover frequency, ω_{gc}. The radius vector is $|G(j\omega) H(j\omega)|$.

The gain crossover frequency ω_{gc} is that frequency at which $|G(j\omega) H(j\omega)| = 1$, that is, the point of intersection of polar plot and $(-1, j0)$ circle.

Phase margin $\phi_{PM} = 180° + \phi$ (6.5)

where $\phi = \angle G(j\omega) H(j\omega)$ and $|G(j\omega) H(j\omega)| = 1$

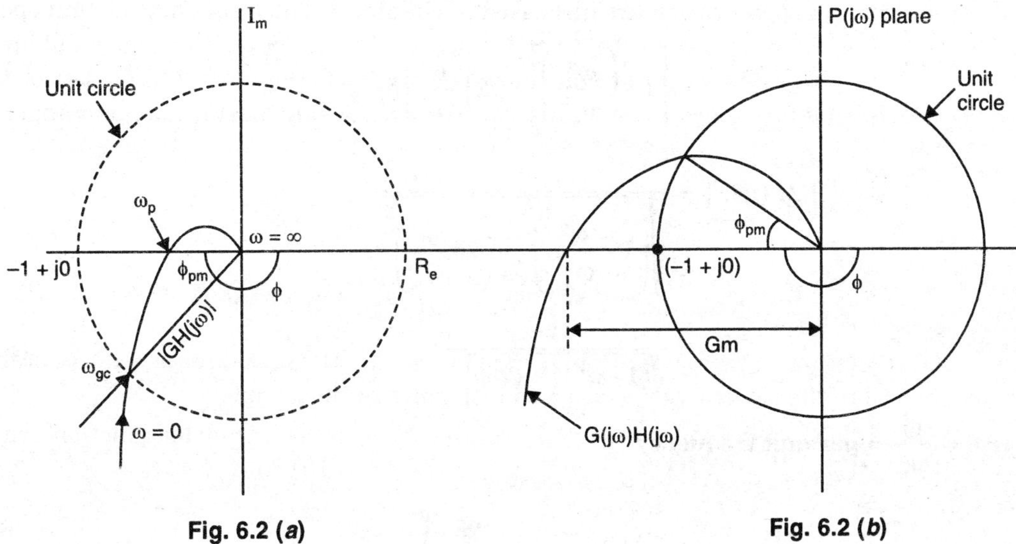

Fig. 6.2 (a) Fig. 6.2 (b)

Please note we are measuring ϕ in the counter clockwise direction, hence it is negative. So for stability ϕ_{PM} should be positive, that is, ϕ is less than 180°. If ϕ_{PM} is negative, $\phi > 180°$, the system is unstable as shown in Fig. 6.2 (*a*). In Fig. 6.2 (*b*) the *GM* is greater than $(-1, +j0)$, hence negative. So negative gain and phase margins indicate unstable systems.

Among the seven frequency domain specifications listed at the beginning, two have been defined. We shall now look into the rest.

(*ii*) **Resonant peak (M_p) and resonant frequency (ω_r).** Let us take a second order system is as shown in Fig. 6.3. And work out the correlation between step transient response and frequency response.

The closed loop transfer function of a second order sinusoidal steady state is

$$M(j\omega) = \frac{C(j\omega)}{R(j\omega)} = \frac{\omega_n^2}{(j\omega)^2 + 2\delta\omega_n (j\omega) + \omega_n^2} \qquad (6.6)$$

Rewriting eqn. (6.6) by dividing ω_n^2

$$M(j\omega) = \frac{1}{1 + j2\left(\dfrac{\omega}{\omega_n}\right)\delta - \left(\dfrac{\omega}{\omega_n}\right)^2}$$

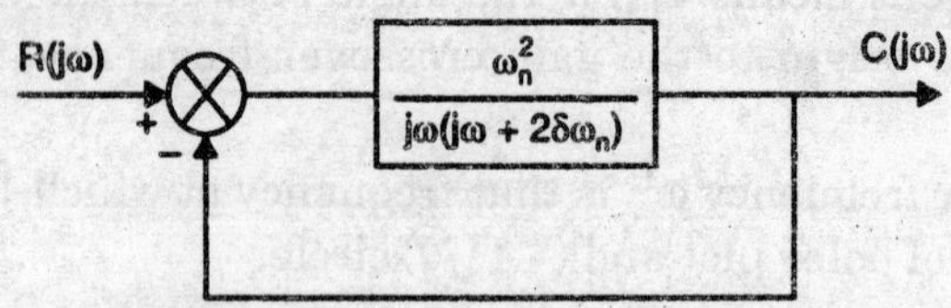

Fig. 6.3

$$= \frac{1}{\left[1-\left(\frac{\omega^2}{\omega_n^{\,2}}\right)\right] + j2\left(\frac{\omega}{\omega_n}\right)\delta}$$

$$|\, M(j\omega) \,| = \frac{1}{\sqrt{\left(1-\frac{\omega^2}{\omega_n^{\,2}}\right)^2 + \left(2\delta\,\frac{\omega}{\omega_n}\right)^2}} \tag{6.7}$$

$$= \frac{1}{\sqrt{(1-u^2)^2 + (2\delta u)^2}} \tag{6.7(a)}$$

where $u = \frac{\omega}{\omega_n}$ = per unit frequency

$$\angle M(j\omega) = \phi_M = -\tan^{-1}\frac{2\delta\,\frac{\omega}{\omega_n}}{\left(1-\frac{\omega^2}{\omega_n^{\,2}}\right)} \tag{6.8}$$

The resonant frequency ω_r is that frequency at which $M(j\omega)$ is maximum. So we will obtain the maximum by differentiating eqn. (6.7 (*a*)) and equating to zero.

$$\frac{d(M(j\omega))}{du} = \frac{-1}{2}\,(1-u^2)^2 + (2\delta\, u)^2]^{-1/2} \times (4u^3 - 4u + 8u\,\delta^2) = 0$$

or $\quad 4u^3 - 4u + 8u\,\delta^2 = 0$

or $\quad 4u^2 - 4 + 8\delta^2 = 0$

or $\quad u^2 - 1 + 2\delta^2 = 0$

The roots are

$$u_1 = 0 \text{ and } u_2 = \sqrt{1-2\delta^2} \quad \text{or} \quad \frac{\omega}{\omega_n} = \sqrt{1-2\delta^2}$$

$$u_1 = 0 \text{ means the slope of the curve at } \frac{\omega}{\omega_n} = 0$$

$$\therefore \quad \text{Resonant frequency } \omega_r = \omega_n\sqrt{1-2\delta^2} \tag{6.9}$$

ω_r is the resonant frequency at which resonant peak occurs.

The maximum value of $M(j\omega)$ is called the peak resonance M_p. Substituting eqn. (6.9) in eqn. (6.7).

$$\therefore \qquad M_p = \frac{1}{2\delta\sqrt{1-\delta^2}} \qquad (6.10)$$

As ω_r is real quantity, eqn. (6.9) should be real and $2\delta^2 < 1$.

$$2\delta^2 < 1$$

$$\delta < \frac{1}{\sqrt{2}} = 0.707 \qquad (6.11)$$

While resonant frequency ω_r is a function of both ω_n and δ, M_p is dependent only on δ. Substituting the value $\delta = \frac{1}{\sqrt{2}}$ in eqn. (6.10), $M_p = 1$. This means that for values of $\delta > 0.707$, the first root $u_1 = 0$ or $\omega_r = 0$ becomes valid.

Relation between phase margin and damping ratio (δ)

Equation (6.8) gives the relationship between phase margin ϕ_{PM} and δ.

$$| M(j\omega) | = \phi_m = -\tan^{-1} \frac{2\delta \frac{\omega}{\omega_n}}{\left(1-\frac{\omega^2}{\omega_n{}^2}\right)}$$

Substituting $\frac{\omega}{\omega_n} = \sqrt{1-2\delta^2}$

$$\phi_m = -\tan^{-1} \frac{2\delta\sqrt{1-2\delta^2}}{1-(1-2\delta^2)} = -\tan^{-1} \frac{2\delta\sqrt{1-2\delta^2}}{2\delta^2}$$

$$= -\tan^{-1} \frac{\sqrt{1-2\delta^2}}{\delta}$$

Phase margin, $\qquad \phi_{PM} = 180° + \phi_m$

$$\phi_{PM} = 180° - \tan^{-1} \frac{\sqrt{1-2\delta^2}}{\delta} \qquad (6.12)$$

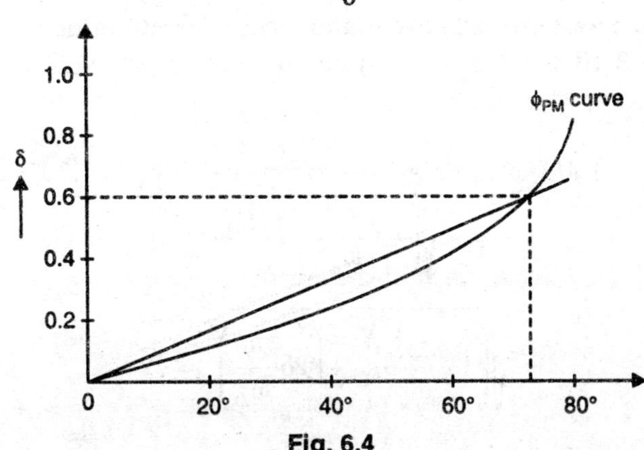

Fig. 6.4

From Fig. 6.4, the damping ratio δ almost varies linearly with ϕ_{PM}. It is about 0.6 at 60°. The peak resonance M_p, from eqn. (6.10), varies inversely as δ. This variation is the same as in transient response for unit step in time domain.

i.e., maximum overshoot $= e^{-\pi\delta/\sqrt{1-\delta^2}}$

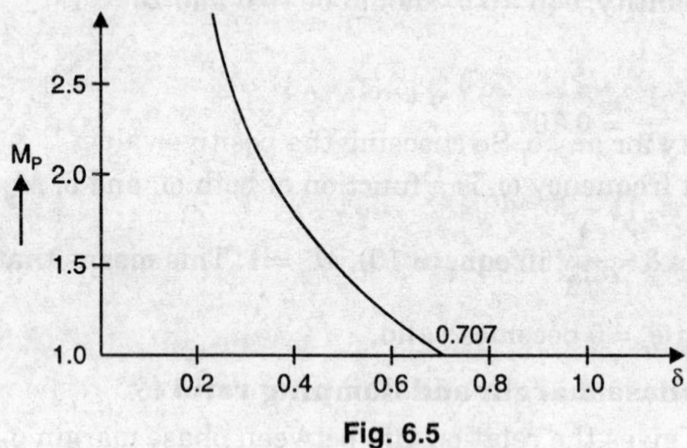

Fig. 6.5

For values of $\delta < 0.4$, M_p shoots up sharply becoming very high (>> 1) and may damage the system. In time domain, overshoot does not exceed 1.

In time domain, $\omega_d = \omega_n \sqrt{1-\delta^2}$

In frequency domain, $\omega_r = \omega_n \sqrt{1-2\delta^2}$

When δ is small, ω_d is almost the same as ω_r. So if the damping ratio is small, the resonant frequency ω_r is an indication of the speed of response of the system. So in frequency response systems the damping ratio used is $0.4 < \delta < 0.7$. This gives a value of M_p, $1 < M_p < 1.4$.

(*iii*) **Delay time $\mathbf{T_d}$**. Delay time is a measure of the speed of response

$$T_d = -\frac{dr}{d\omega} \tag{6.13}$$

where r = argument of $\dfrac{C(j\omega)}{R(j\omega)} = G(j\omega)$

$T_d(\omega)$ is specified for the frequencies of interest.

(*iv*) **Bandwidth and cut off rate**: Bandwidth (BW) is defined as the frequency at which $M(j\omega)$ drops 70.7% or 3 dB down from one frequency gain as shown in the Fig 6.6 (*a*). From eqn. (6.7)

$$\therefore \qquad |M(j\omega)| = \frac{1}{\sqrt{\left(1-\dfrac{\omega^2}{\omega_n^2}\right)^2 + \left(2\delta\dfrac{\omega}{\omega_n}\right)^2}} = 0.707$$

or

$$\sqrt{\left(1-\frac{\omega^2}{\omega_n^2}\right)^2 + \left(2\delta\frac{\omega}{\omega_n}\right)^2} = \sqrt{2}$$

Putting $\frac{\omega}{\omega_n} = u$ and rearranging

$$u^4 + 2u^2 (2\delta^2 - 1) - 1 = 0$$

$$u_{1,2} = \frac{-2(2\delta^2 - 1) \pm \sqrt{4(2\delta^2 - 1)^2 + 4}}{2}$$

$$= (1 - 2\delta^2) \pm \sqrt{4\delta^4 - 4\delta^2 + 2}$$

ω is a real quantity for any δ. So choosing the positive value

$$BW = \omega_n \left[1 - 2\delta^2 + \sqrt{4\delta^4 - 4\delta^2 + 2}\,\right]^{1/2} \quad (6.14)$$

The range $0 \le \omega \le \omega_C$ (cutoff frequency) at which the magnitude of the closed loop drops to 3 dB is the bandwidth.

It may be noted from eqn. (6.14) as δ decreases from units, BW increases as also M_p (vide eqn. (6.10). So bandwidth is an indication of the speed of response of a system; wider bandwidth means more noise. So the choice of bandwidth is a compromise.

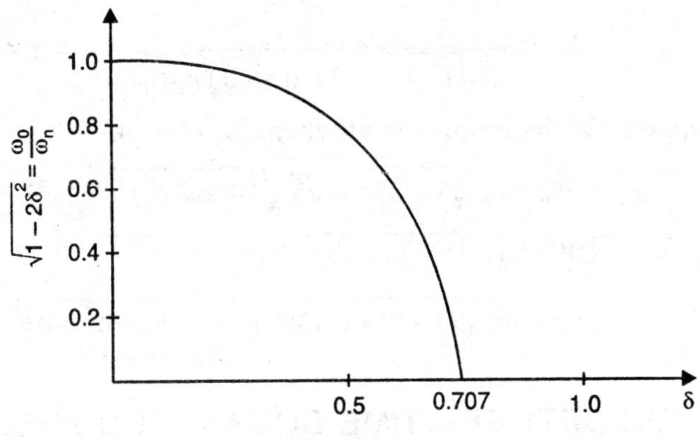

Fig. 6.6 (a)

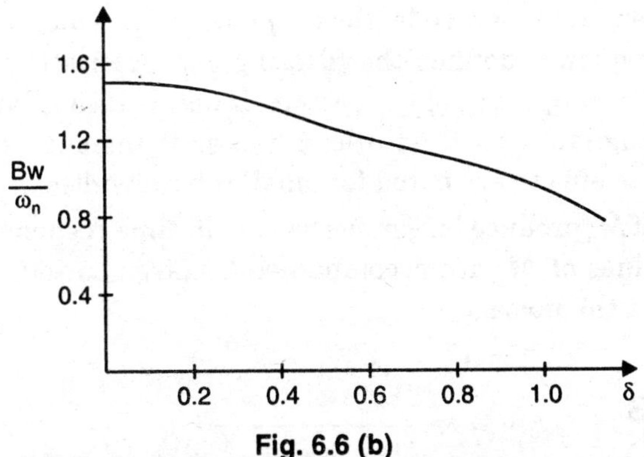

Fig. 6.6 (b)

(*v*) **Cut-off rate.** It is the frequency rate at which the magnitude ratio decreases beyond the cut off frequency ω_C.

Problem 6.1. *Determine the resonant frequency ω_r, resonant peak M_p and bandwidth for the system whose transfer function is*

$$G(j\omega) = \frac{5}{5 + j2\omega + (j\omega)^2}.$$

Solution:
$$\frac{C(j\omega)}{R(j\omega)} = \frac{5}{5(1 + j2/5\omega + 1/5(j\omega)^2)} = \frac{1}{1 + j2/5\omega + 1/5\,(j\omega)^2}$$

The transfer function is of the form

$$\frac{C(j\omega)}{R(j\omega)} = \frac{1}{1 + j2\delta(\omega/\omega_n) + (j\omega/\omega_n)^2}$$

$$\left(\frac{\omega}{\omega_n}\right)^2 = \left(\frac{\omega}{\sqrt{5}}\right)^2 \ ; \omega_n = \sqrt{5}$$

$$2\delta\ \omega/\omega_n = 2/5\ \omega,\ \delta = \sqrt{5}/5$$

$$\delta = 1/\sqrt{5} = 0.45$$

Peak value,
$$M_p = \frac{1}{2\delta\sqrt{1-\delta^2}} = \frac{1}{2 \times 0.45\sqrt{1-(0.45)^2}} = 1.24$$

Resonant frequency is the frequency at which M_p occurs.

$$\therefore \qquad \omega_r = \omega_n\sqrt{1-2\delta^2} = \sqrt{5}\sqrt{1-2(0.45)^2} = 1.725 \text{ rad/sec.}$$

Bandwidth,
$$BW = \omega_n\sqrt{1-2\delta^2 + (2-4\delta^2+4\delta^4)^{1/2}}$$

$$= \sqrt{5}\sqrt{(1-(2 \times 0.45)^2 + [2-4(0.45)^2 + 4(0.45)^4]^{1/2}} = 3.0992.$$

6.3 CORRELATION BETWEEN TIME DOMAIN AND FREQUENCY DOMAIN SPECIFICATIONS

Bandwidth of the system, represented by the frequencies for which | $M(j\omega)$ | falls below 70.7% of its value at zero frequency defines the filtering characteristic of the system. It is also a measure of the transient properties of the system. If bandwidth is large, the systems responds to higher frequency signals and thus allows faster transient response and gives larger overshoots, the opposite effects are noted for smaller bandwidths.

Larger values of M_p produce larger overshoots in time response and to make the system less stable. Larger values of M_p are accompanied by large cut-off rate, which distinguishes between the signal and the noise.

6.4 BODE PLOTS

Bode analysis plays a very important role in finding the stability of the system. The bode analysis is an improvement of Nyquist analysis. The magnitude of $GH(j\omega)$ and phase angle of $GH(j\omega)$ are plotted as a function of frequency ω, but using logarithmic scales because it has the advantage of plotting over wider ranges than linear scales. Besides, logarithmic graphs in most cases turn out linear. Another advantage of logarithmic plot is that product terms are converted to sum terms.

The *dB* magnitude verses log ω plot is called **Bode magnitude plot.** It is also called as log-modulus plot.

The phase angle versus log ω plot is called **Bode phase angle plot.**

$$dB = 20 \log_{10} G$$

G is expressed as a ratio $\frac{A}{B}$. If $A > B$ the dB value is plus. If $A < B$, the dB value is minus.

Example: $dB = 20 \log 1/10 = -20$

$dB = 20 \log 2 = 6$

Transfer functions lend themselves easy on working with logarithmic scales because it is the ratio of output and input. A transfer function can be expressed as a ratio of polynomials.

$$G(j\omega) = \frac{k(1+j\omega/z_1)(1+j\omega/z_2)\ldots\ldots(1+j\omega/z_m)}{(j\omega)^{\gamma}(1+j\omega/p_1)\ldots\ldots(1+j\omega/p_n)} \tag{6.15}$$

γ is positive including zero, $m < n$

In frequency domain, four basic factors appear in a transfer function

(*i*) constant factor K

(*ii*) derivative and integral factor $(j\omega)^{\pm n}$

(*iii*) first order factor $(j\omega T + 1)^{\pm m}$

(*iv*) second order factor $\left[1 + j\frac{2\delta\omega}{\omega_n} + \left(\frac{j\omega}{\omega_n}\right)^2\right]^{\pm p}$

We shall consider each one as follows.

(*i*) **Constant factor K.** From the eqn. (6.15), the constant includes the constant of all factors.

Magnitude $= 20 \log_{10} K$ dB

Phase angle $\phi = 0$

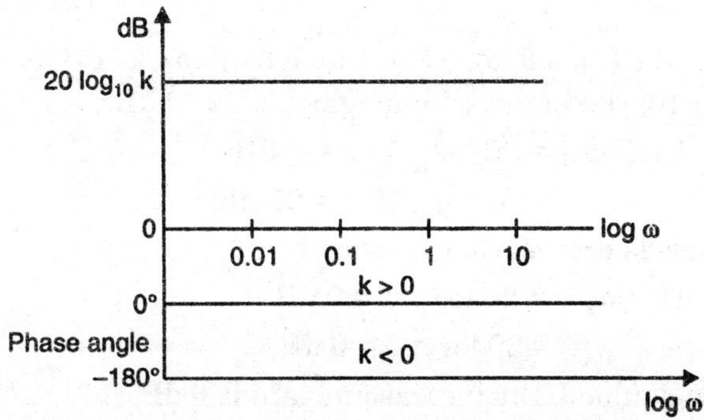

Fig. 6.7

(ii) Derivative and integral factor $(j\omega)^{\pm 1}$:

If we take function $G(j\omega) = j\omega$

$$| G(j\omega) | = 20 \log | j\omega | = 20 \log \omega \quad (6.16)$$

$$| G(j\omega) | = \tan^{-1} \frac{\omega}{0} = 90° = \pi/2 \quad (6.17)$$

If we plot the magnitude on a logarithmic paper $|G(j\omega)|$ will vary linearly with ω.

If the derivative is $(j\omega)^n$

$$| G(j\omega) |^n = n\, 20 \log (j\omega) = 20\, n \log \omega \quad (6.18)$$

Phase angle = n × 90°, over the entire frequency range.

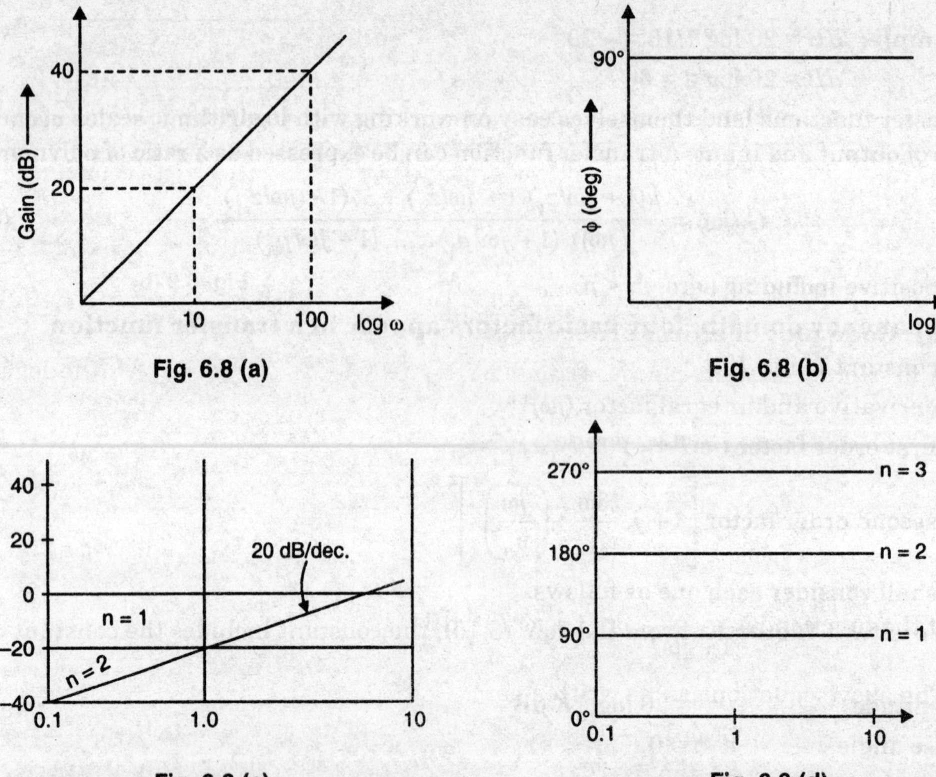

Fig. 6.8 (a) Fig. 6.8 (b)

Fig. 6.8 (c) Fig. 6.8 (d)

The curves in Figs. 6.8 (*a*), 6.8 (*b*), 6.8 (*c*) and 6.8 (*d*) are 20 dB/dec curves. For a ten time increase in frequency 1 : 10, the increase in magnitude is 20 dB.

$$| G(j\omega) | = 20 \log_{10} 1 \quad = 0 \text{ dB}$$

$$= 20 \log_{10} 10 \quad = 20 \text{ dB}$$

Another term in use is octave, that is twice 1 : 2

$$| G(j\omega) | = 20 \log_{10} 1 = 0 \text{ dB}$$

$$= 20 \log_{10} 2 = 6 \text{ dB}$$

If the frequency is doubled, the increase in gain is 6 dB, 6 dB/octave.

As two slopes represents the same function,

$G(j\omega) = j\omega$ they must be equivalent.

$\therefore$ 20 db/dec = 6 dB/octave.

Integrative: $\dfrac{1}{(j\omega)^n}$ or $(j\omega)^{-n}$

$$|G(j\omega)^n| = -n\ 20 \log \omega \tag{6.19}$$

In general $\angle \mathbf{G(j\omega)^{\pm n} = \pm n \angle tan^{-1}\, j\omega = \pm n \times 90°}$

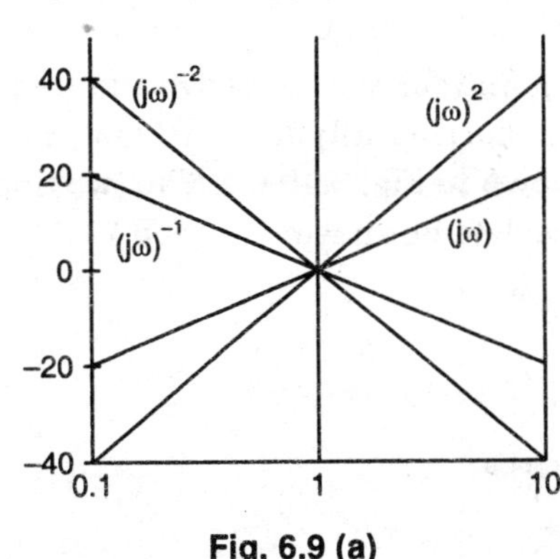

Fig. 6.9 (a)

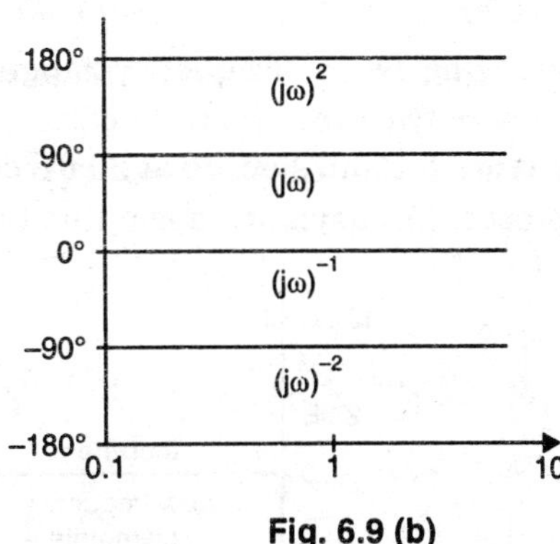

Fig. 6.9 (b)

(*iii*) **Bode plot of a first order function**

A first order system can be written in the general form, $(1 + j\omega T)^{\pm 1}$, the logarithmic magnitude is

$$|G(j\omega)|\ \text{dB} = 20 \log_{10} |1 + j\omega T|\ \text{dB} \tag{6.20}$$

$$= 20 \log_{10} \sqrt{1 + \omega^2 T^2}\ \text{dB}$$

$$|G(j\omega)|\ \text{dB} = 10 \log (1 + \omega^2 T^2) \tag{6.21}$$

If the function is $\dfrac{1}{1 + j\omega T}$ or $(1 + j\omega T)^{-1}$ dB

The above equation can be written as

$$|G(j\omega)| = -10 \log_{10} (1 + \omega^2 T^2)\ \text{dB}$$

The phase angle $\phi = \tan^{-1} (\omega T)$

(*a*) If in eqn. (6.21), $\omega T << 1$ or $\omega << 1/T$ such as at low frequencies, it reduces to magnitude $G(j\omega) = 10 \log 1 = 0$ dB

(*b*) If $\omega T >> 1$ as at high frequencies

magnitude $G(j\omega) = -10 \log \omega^2 T^2 = -20 \log \omega T$

As ωT increases in decades of 10, we get a 20 dB/dec curve.

This frequency $\omega = 1/T$ at which the slope from 0 dB to – 20 dB/dec occurs is called corner frequency, denoted by ω_c

i.e., $\omega_c = 1/T$

ωT	*magnitude in* dB	*phase angle* ϕ
0	0	0°
0.01	0.002	+ 0.6
0.1	0.04	+ 5.7
1.0	+ 3.0	+ 45°
10	+ 20	+ 84.5
100	+ 40	+ 89.4
1000	+ 80	+ 90°

For values below the frequency range $0 < \omega < 1/T$ and for values above the frequency range $1/T < \omega < \infty$, two straight lines can be drawn—one horizontally for 0 dB and the other commencing at $\omega T = 1$ and sloping at 20 dB decade as shown in Fig. 6.10 (*a*). The two lines are called asymptotes. (An asymptote is a line that is approached but never reached.)

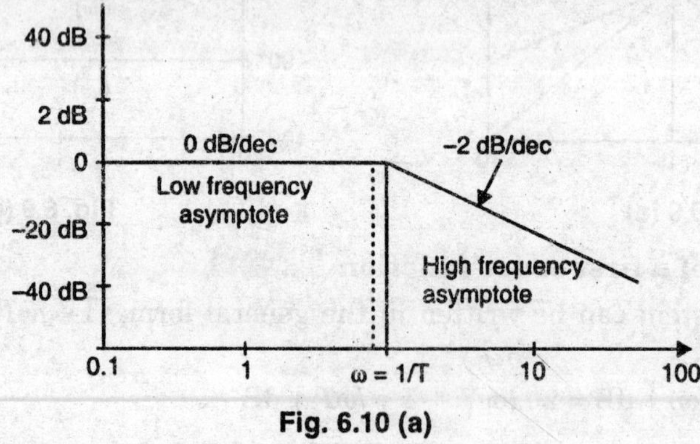

Fig. 6.10 (a)

Phase angle plot

For $G(j\omega) = \dfrac{1}{(1 + j\omega T)}$ $\qquad \angle G(j\omega) = -\tan^{-1} \omega T$

For simple zero, $G(j\omega) = (1 + j\omega T)$ $\qquad \angle G(j\omega) = + \tan^{-1} \omega T$

For various values of ωT, given in above table. At ωT equals to 1 phase angle is 45°, at 0.1, 5.7°, at 10, 84.3° and at 1000, 90°. Phase angle plot for $(1 + j\omega T)$ is as shown in the Fig. 6.10 (*b*). The asymptotic phase angle rises from 0° at 0.1 to +90° at +10, passing through +45° at 1. Below 0.1 and above 10 it is horizontal at 90°.

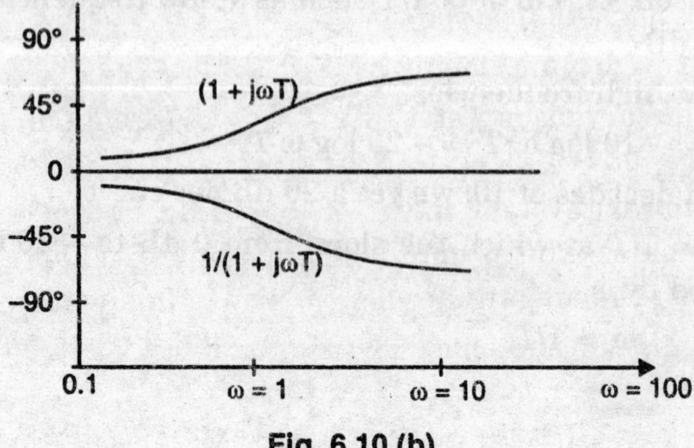

Fig. 6.10 (b)

(*iv*) **Bode plot of second order function**

The transfer function of a second order system is

$$G(j\omega) = \frac{1}{1 + j2\delta\dfrac{\omega}{\omega_n} + \left(\dfrac{\omega}{\omega_n}\right)^2} \qquad (6.22)$$

The response of the system to any frequency ω is determined by δ and ω_n. By equating the denominator of above equation to zero, we obtain the roots of the characteristic equation. They are real and unequal if $\delta > 1$(over damped) and real and equal if $\delta = 1$ (critically damped). So the denominator can be written as a product of two first order terms with real poles.

$$G\,(j\omega) = \frac{1}{(1 + j\omega T_1)(1 + j\omega T_2)} \qquad (6.23)$$

If under damped, *i.e.*, $0 < \delta < 1$, the roots are complex and it cannot be factored to two first order terms.

$$\text{Mag } |\, G\,(j\omega)\,|\ \text{dB} = 20\log_{10}\left|\frac{1}{1 + 2\delta\left(\dfrac{j\omega}{\omega_n}\right) + \left(j\dfrac{\omega}{\omega_n}\right)^2}\right|$$

$$= -\,20\log\sqrt{1 - \left(\frac{\omega}{\omega_n}\right)^2 + 2\delta\left(\frac{\omega}{\omega_n}\right)^2} \qquad [(6.24\ (a)]$$

$$\text{Phase} \qquad \phi\,(j\omega) = -\tan^{-1}\frac{2\delta(\omega/\omega_n)}{1 - \left(\dfrac{\omega}{\omega_n}\right)^2} \qquad [6.24\ (b)]$$

(*i*) At low frequencies, $\omega < \omega_n$ or $\omega/\omega_n << 1$

Magnitude = – 20 log 1 = 0 dB

The low frequency asymptote is a horizontal line at 0 dB.

(*i*) At high frequencies, $\omega/\omega_n >> 1$, eqn. 6.24 (*a*) reduces to

Magnitude $= -\,20\log(\omega/\omega_n)^2$ dB

$= -\,40\log\omega/\omega_n$

The magnitude is a straight line sloping at – 40 dB/decade.

The low and high frequency asymptotes meet at $\omega = \omega_n$, where magnitude is 0 dB. ω_n is the corner frequency.

As in the case of first order system, in the second order system also, the actual values differ from the asymptotic values. The asymptotic values hold good for $\omega < 0.1\ \omega_n$ and $\omega > 10\ \omega_n$. Between 0.1 ω_n and 10 ω_n, the magnitudes depends on damping ratio.

A plot of ω/ω_n and magnitude and phase in drawn in Fig. 6.11 and Fig. 6.12 respectively.

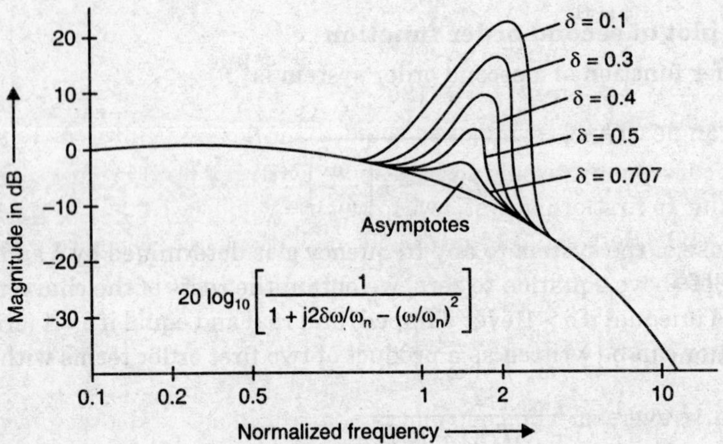

Fig. 6.11

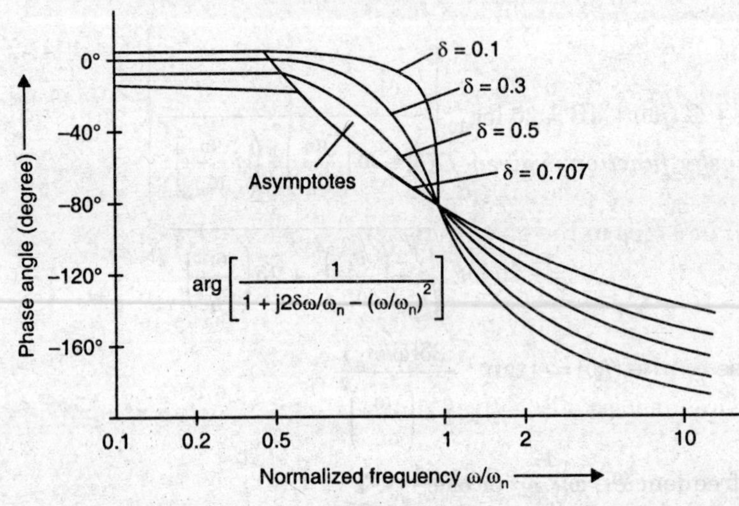

Fig. 6.12

Where $\omega/\omega_n = 1$, eqn. (6.24) is

Magnitude $= -20 \log 2\delta$

$\delta = 0.1$; Magnitude = 14 dB

$\delta = 0.5$; Magnitude = 0 dB

$\delta = 0.707$; Magnitude = – 3 dB

So in the vicinity of corner frequency ω_n, the magnitude dB peaks for small values of $\delta < 0.5$. This is due to the resonant nature of the under damped system.

The curves are drawn on the normalized scale of ω/ω_n so that they are applicable for any value of ω_n.

On the same scale ω/w_n, the phase angles are drawn as per eqn. (6.24 (*b*)). The phase angle depends on both δ and ω_n.

$\omega = 0, \qquad \phi = 0$

$\omega = 1, \qquad \phi = -\tan^{-1}\infty = -90°$

$\omega = \infty, \qquad \phi = -\tan^{-1}\infty = -180°$

The phase angle curve is symmetric about the inflexion point at $\phi = -90°$.

The actual curves may be said to correspond with asymptotic curves for $\omega/\omega_n < 0.1$ and > 10. This is similar to first order system. However, the horizontal axes are at 0° and – 180° instead of 0° and – 90°. Between 0.1 and 10, the deviations between actual curve and asymptotic curves are large.

For the system $1 + j\left(2\delta\frac{\omega}{\omega_n}\right) - \left(\frac{\omega}{\omega_n}\right)^2$.

The signs are reversed. The bode plot is a mirror image of the reciprocal function.

Problem 6.2. *Sketch the Bode plot and determine the following:*

(*a*) *Gain crossover frequency*

(*b*) *Phase crossover frequency*

(*c*) *Gain margin*

(*d*) *Phase margin*

For the transfer function is given, $G(s) = \dfrac{10}{s(1+0.4s)(1+0.1s)}$.

Solution: Put $s = j\omega$ in the given s-domain transfer function, we can get

$$\therefore \qquad G(j\omega) = \frac{10}{j\omega(1+j0.4\omega)(1+j0.1\omega)}$$

(*i*) **Magnitude plot**

The corner frequencies of the given function are

$$\omega_1 = \frac{1}{0.4} = 2.5 \text{ rad/sec}$$

$$\omega_2 = \frac{1}{0.1} = 10 \text{ rad/sec}$$

The various factors of $G(j\omega)$ are given in Table 6.1.

Table 6.1

Sl. No.	*Factor*	*Corner frequency*	*Slope (dB / dec)*	*Change in slope*
1.	$\frac{10}{j\omega}$	–	– 20	–
2.	$\frac{1}{1+j0.4\omega}$	$\omega_1 = 2.5$	– 20	– 20 – 20 = – 40
3.	$\frac{1}{1+j0.1\omega}$	$\omega_2 = 10$	– 20	– 40 – 20 = – 60

Consider a low and high frequencies ω_l *and* ω_h, such that $\omega_l < \omega_1$ and $\omega_h > \omega_2$ respectively.

Assume, $\omega_1 = 0.1$ rad/sec and $\omega_h = 50$ rad/sec

Magnitude (M) = magnitudes of $G(j\omega)$ in dB and determine M at ω_l, ω_1, ω_2, ω_h.

At $\omega = \omega_l$; $\quad M = 20 \log \left|\frac{10}{j\omega}\right| = 20 \log \frac{10}{0.1} = 40$ dB

$\omega = \omega_1$; $\quad M = 20 \log \left|\frac{10}{j\omega}\right| = 20 \log \frac{10}{2.5} = 12$ dB

$\omega = \omega_2$; $\quad M = \left[\text{slope from } \omega_1 \text{ to } \omega_2 \times \log\left(\frac{\omega_2}{\omega_1}\right)\right] + M\ (\text{at } \omega = \omega_1)$

$$= -40 \times \log \frac{10}{2.5} + 12 = -12 \text{ dB}$$

$\omega = \omega_h$; $\quad M = \left[\text{slope from } \omega_2 \text{ to } \omega_h \times \log\left(\frac{\omega_h}{\omega_2}\right)\right] + M\ (\text{at } \omega = \omega_2)$

$$= -60 \log \frac{50}{10} + (-12) = -54 \text{ dB}$$

Let the points *m, n, o* and *p* be the points corresponding to frequencies ω_l, ω_1, ω_2 and ω_h respectively on the semi log sheet for magnitude plot. Select the scale for magnitude plot as

x-axis = 1 unit is 0.1 rad/sec

y-axis = 1 unit is 20 dB

Draw the straight lines to join the above points and indicate the slope as the respective region as shown in Fig. 6.13.

(*ii*) **Phase plot.** The phase angle of the given function is given by

$$\phi = \angle G(j\omega) = -90° - \tan^{-1} 0.4\,\omega - \tan^{-1} 0.1\,\omega$$

The phase angles of $G(j\omega)$ with different values of 'ω' are tabulated in table 6.2

Table 6.2

ω *(rad / sec)*	ϕ *(deg)*
0.1	– 93°
1	– 118
2.5	– 150
4	– 170
10	– 210
20	– 236

On the same graph, select the scale for phase plot as:

x-axis = 1 unit is 0.1 rad/sec, same as magnitude plot

y-axis = 1 unit is 20°

Indicate the calculated phase angle on the graph sheet. Draw the smooth curve to joins the points as shown in Fig. 6.13.

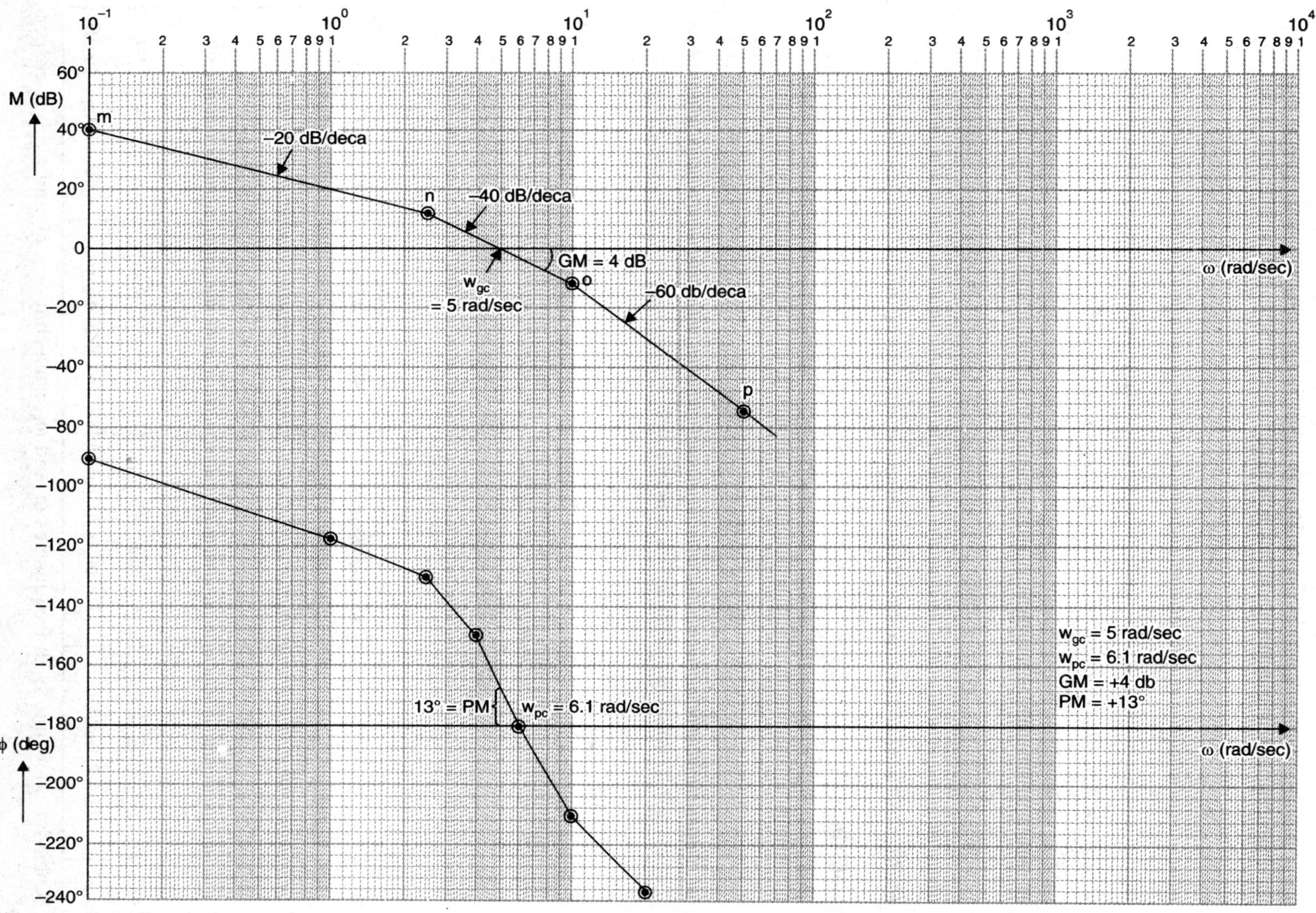
M (dB)
φ (deg)
ω (rad/sec)
ω (rad/sec)
m
n
o
p
–20 dB/deca
–40 dB/deca
–60 db/deca
GM = 4 dB
w_{gc} = 5 rad/sec
w_{pc} = 6.1 rad/sec
13° = PM
w_{gc} = 5 rad/sec
w_{pc} = 6.1 rad/sec
GM = +4 db
PM = +13°

Fig. 6.13

From the Bode plot shown in Fig. 6.13, the following results are obtained

(*a*) Gain crossover frequency (ω_{gc}) = 5 rad/sec

(*b*) Phase crossover frequency (ω_{pc}) = 6.1 rad/sec

(*c*) Gain margin = 4 dB

(*d*) Phase margin = 13°

The gain margin and phase margin are positive; therefore the given system is stable.

Problem 6.3. *Sketch the Bode plot and determine the following:*

(*a*) *Gain crossover frequency*

(*b*) *Phase crossover frequency*

(*c*) *Gain margin*

(*d*) *Phase margin,*

for the transfer function given by $G(s) = \dfrac{100(1+0.1s)}{s(1+0.25)(1+0.5s)}$.

Solution:

Put $s = j\omega$ in the given s-domain transfer function, we can get

$$\therefore \quad G(j\omega) = \frac{100(1+j0.1\omega)}{j\omega(1+j0.2\omega)(1+j0.5\omega)}$$

(*i*) **Magnitude plot**

The corner frequencies of the given function are

$$\omega_1 = \frac{1}{0.1} = 10 \text{ rad/sec}$$

$$\omega_2 = \frac{1}{0.2} = 5 \text{ rad/sec}$$

$$\omega_3 = \frac{1}{0.5} = 2 \text{ rad/sec}$$

The various factors of $G(j\omega)$ are given in table 6.3.

Table 6.3

Sl. No.	*Factor*	*Corner frequency*	*Slope (dB/dec)*	*Change in slope*
1.	$\frac{100}{j\omega}$	–	– 20 dB	– 20
2.	$\frac{1}{1+j0.5\omega}$	$\omega_1 = 2$	– 20 dB	– 20 – 20 = – 40
3.	$\frac{1}{1+j0.2\omega}$	$\omega_2 = 5$	– 20 dB	– 40 – 20 = – 60
4.	$(1 + 0.1j\omega)$	$\omega_3 = 10$	+ 20 dB	– 60 + 20 = – 40 dB

Consider a low and high frequencies ω_l and ω_h, such that $\omega_l < \omega_1$ and $\omega_h > \omega_2$ respectively.

Assume, $\omega_l = 1$ rad/sec.

$\omega_h = 100$ rad/sec.

$\therefore$ Magnitude $(M) = |\ G(j\omega)\ |$ in dB and determine M at ω_l, ω_1, ω_2 and ω_h.

At $\omega = \omega_l$, $\quad M = 20 \log \left(\dfrac{100}{\omega_l}\right) = 40$ dB

$\omega = \omega_1$, $\quad M = \left[\text{slope from } \omega_1 \text{ to } \omega_1 \times \log \dfrac{\omega_1}{\omega_1}\right] + M \text{ at } (\omega = \omega_l)$

$$= -20 \log \left(\frac{2}{l}\right) + 40 = 34 \text{ dB}$$

$\omega = \omega_2$, $\quad M = \left[\text{slope from } \omega_1 \text{ to } \omega_2 \times \log \dfrac{\omega_2}{\omega_1}\right] + M \text{ (at } \omega = \omega_1)$

$$= -40 \times \log 5/2 + 34 = 18 \text{ dB}$$

$\omega = \omega_3$, $\quad M = \left[\text{slope from } \omega_2 \text{ to } \omega_3 \times \log \dfrac{\omega_3}{\omega_2}\right] + M \text{ at } (\omega = \omega_2)$

$$= \left[-60 \times \log\left(\frac{10}{5}\right)\right] + 18 = 0 \text{ dB}$$

$\omega = \omega_h$, $\quad M = \left[\text{slope from } \omega_3 \text{ to } \omega_h \times \log \dfrac{\omega_h}{\omega_3}\right] + M \text{ (at } \omega = \omega_3)$

$$= \left(-40 \log \frac{100}{10}\right) + 0 = -40 \text{ dB}$$

Let the points *m*, *n*, *o*, *p* and *q* be the points corresponding to frequencies ω_l, ω_1, ω_2, ω_3 and ω_h respectively on the graph sheet for magnitude plot. Select the scale on semi log for magnitude plot as :

x-axis = 1 unit is 0.1 rad/sec

y-axis = 1 unit is 20 dB

Draw the straight lines to join the above points and indicate the slope as the respective region as shown in Fig. 6.14.

(*ii*) **Phase plot.** The phase angle of the given function is given by

$$\phi = \angle G(j\omega) = \tan^{-1}(0.1\,\omega) - 90° - \tan^{-1}(0.2\,\omega) - \tan^{-1}(0.5)\,\omega$$

The phase angles of $G\,(j\omega)$ with different values of 'ω' are tabulated in table 6.4.

Table 6.4

ω *(rad / sec)*	ϕ *(deg)*
1	122.16
2	145.49°
5	176.63°
10	187.12°

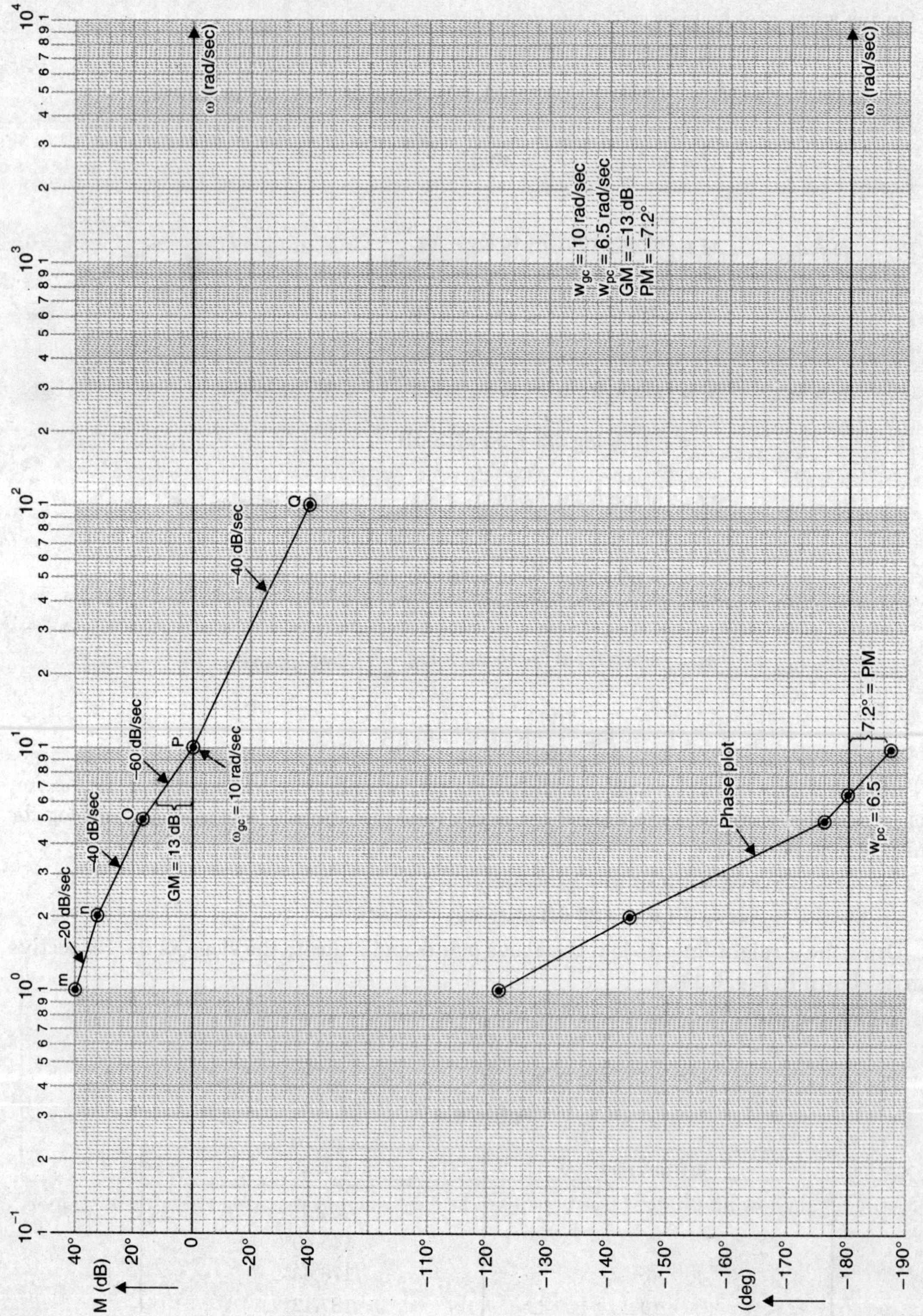

ω (rad/sec)
w_{gc} = 10 rad/sec
w_{pc} = 6.5 rad/sec
GM = –13 dB
PM = –7.2°
–20 dB/sec
–40 dB/sec
–60 dB/sec
–40 dB/sec
m
n
O
P
Q
GM = 13 dB
ω_{gc} = 10 rad/sec
Phase plot
7.2° = PM
w_{pc} = 6.5
M (dB)
φ (deg)

Fig. 6.14

On the same graph sheet select the scale for phase plot as:

x-axis = 1 unit is 0.1 rad/sec, same as magnitude plot

y-axis = 1 unit is 10°

Indicate the calculated phase angle on the graph sheet. Join the points by a smooth curve as shown in Fig. 6.14.

From the Bode plot shown in Fig. 6.14, the following results are obtained:

(*a*) gain crossover frequency (ω_{gc}) = 10 rad/sec

(*b*) phase crossover frequency (ω_{gc}) = 6.5 rad/sec

(*c*) gain margin = – 13 dB

(*d*) phase margin = – 7.2°

Problem 6.4. *Sketch the Bode plot and determine the following:*

(*a*) *gain crossover frequency*

(*b*) *phase crossover frequency*

(*c*) *gain margin*

(*d*) *phase margin*

for the transfer function given as $G(s) = \dfrac{10(1+s)e^{-0.1s}}{s(1+0.2s)}$.

Solution: Put $s = j\omega$ in the given s-domain transfer function, we can get

$$G(j\omega) = \frac{10(1+j\omega)e^{-j0.1\omega}}{j\omega(1+j0.2\omega)}$$

(*i*) **Magnitude plot.** The corner frequencies of the given function are

$$\omega_1 = \frac{1}{1} = 1 \text{ rad/sec}$$

$$\omega_2 = \frac{1}{0.2} = 5 \text{ rad/sec}$$

The various factors of $G(j\omega)$ are given in table 6.5.

Table 6.5

Sl. No.	*Factor*	*Corner frequency*	*Slope (dB / dec)*	*Change in slope*
1.	$\dfrac{10}{j\omega}$	–	– 20	–
2.	$1 + j\omega$	$\omega_1 = 1$	20	– 20 + 20 = 0
3.	$\dfrac{1}{1+0.2j\omega}$	$\omega_2 = 5$	– 20	– 20

Consider the low and higher frequencies ω_l and ω_h, such that $\omega_l < \omega_1$ and $\omega_h > \omega_2$ respectively.

Assume, $\omega_l = 0.5$ rad/sec

$\omega_h = 50$ rad/sec

Magnitude (M) = $|G(j\omega)|$ in dB and determine M at ω_l, ω_1, ω_2, ω_h.

At $\quad \omega = \omega_l, \qquad M = 2\log\left(\dfrac{10}{j\omega}\right) = 20\log\left(\dfrac{10}{0.5}\right) = 26 \text{ dB}$

$$\omega = \omega_1, \qquad M = \left[\text{slope from } \omega_l \text{ to } \omega_1 \times \log\frac{\omega_1}{\omega_l}\right] + M\,(\text{at } \omega = \omega_l)$$

$$= -20 \times \log\frac{1}{0.5} + 26 = 20 \text{ dB}$$

$$\omega = \omega_2, \qquad M = \left[\text{slope from } \omega_1 \text{ to } \omega_2 \times \log\frac{\omega_2}{\omega_1}\right] + M\,(\text{at } \omega = \omega_1)$$

$$= 0 \times \log(5/1) + 20 = 20 \text{ dB}$$

$$\omega = \omega_h, \qquad M = -20 \times \log\left(\frac{50}{5}\right) + 20 = 0 \text{ dB}$$

Let the points m, n, o and p be the points corresponding to frequencies ω_l, ω_1, ω_2 and ω_h respectively on the semi log sheet for magnitude plot. Select the scale on graph for magnitude plot as :

x-axis = 1 unit is 0.1 rad/sec

y-axis = 1 unit is 10 dB

Draw the straight lines to join the above points and indicate the slope on the respective region as shown in Fig. 6.15.

(*ii*) **Phase plot.** The phased angle of the given function is given by

$$\phi = \angle G(j\omega) = -0.1\omega \times \frac{180}{\pi} - 90° + \tan^{-1}\omega - \tan^{-1}(0.2\,\omega)$$

The phase angles of $G(j\omega)$ with different values of 'ω' are tabulated in table 6.6.

Table 6.6

ω (*rad/sec*)	ϕ (*deg*)
0.5	72.01
1	62
5	84.95
10	126.43
20	193.4
50	371.89

On the same graph sheet select the scale for phase plot as

On $\quad x$-axis = 1 unit is 0.1 rad/sec, same as magnitude plot

y-axis = 1 unit is 40°

Indicate the calculated phase angles on the graph sheet; join the points by a smooth curve as shown in Fig. 6.15.

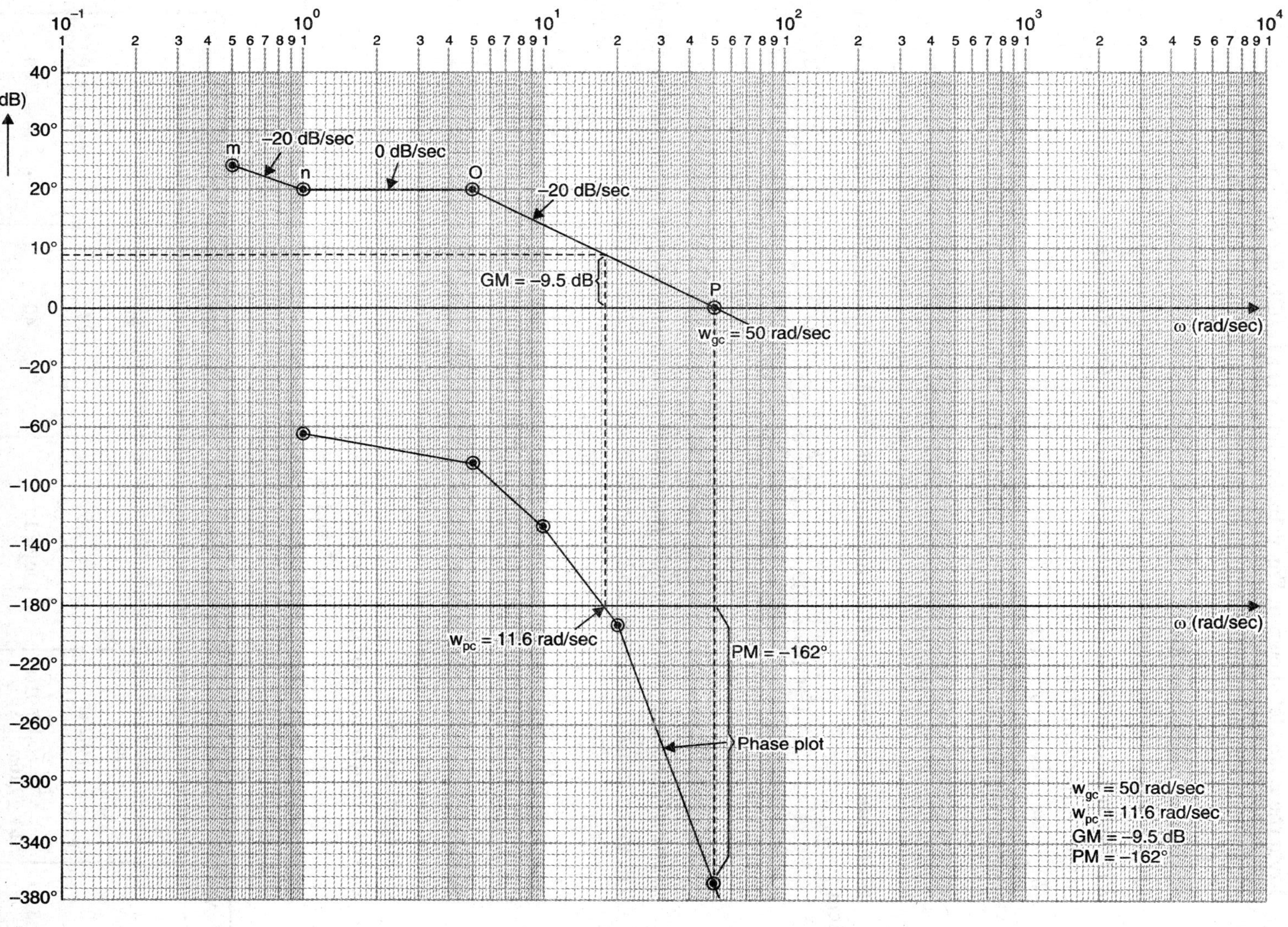

10^{-1}
10^{0}
10^{1}
10^{2}
10^{3}
10^{4}
(dB)
40°
30°
20°
10°
0
–20°
–60°
–100°
–140°
–180°
–220°
–260°
–300°
–340°
–380°
m
–20 dB/sec
n
0 dB/sec
O
–20 dB/sec
GM = –9.5 dB
P
w_{gc} = 50 rad/sec
ω (rad/sec)
w_{pc} = 11.6 rad/sec
ω (rad/sec)
PM = –162°
Phase plot
w_{gc} = 50 rad/sec
w_{pc} = 11.6 rad/sec
GM = –9.5 dB
PM = –162°

Fig. 6.15

From the Bode plot shown in Fig. 6.15, the following results are obtained:

(*a*) gain crossover frequency (ω_{gc}) = 50 rad/sec

(*b*) phase crossover frequency (ω_{pc}) = 11.6 rad/sec

(*c*) gain margin = – 9.5 dB

(*d*) phase margin = – 162°

Problem 6.5. *Sketch the Bode plot and determine the following:*

(*a*) *gain crossover frequency*

(*b*) *phase crossover frequency*

(*c*) *gain margin*

(*d*) *phase margin*

for the transfer function given $G(s) = \dfrac{75(1+0.2s)}{s(s^2+16s+100)}$.

Solution: Denominator of $G(s)$ is compared with denominator of standard second order system

i.e.,
$$s^2 + 16s + 100 = s^2 + 2\,\delta\omega_n\, s + \omega_n^2$$
$$\omega_n^2 = 100, \qquad 2\,\delta\omega_n = 16$$
$$\omega_n = 10,$$
$$\delta = \frac{16}{2\omega_n} = \frac{16}{2.10} = 0.8$$

The given transfer function can be modified as

$$G(s) = \frac{75(1+0.25)}{s\times 100\left(\dfrac{s^2}{100}+\dfrac{16}{100}s+1\right)} = \frac{0.75\,(1+0.2s)}{s(0.01s^2+0.16s+1)}$$

Put $s = j\omega$
$$G(j\omega) = \frac{0.75(1+j0.2\omega)}{j\omega(0.01(j\omega)^2+j0.16\omega+1)}$$

(*i*) **Magnitude plot**

Corner frequencies of given transfer function are

$$\omega_1 = 5 \text{ and } \omega_2 = 10$$

The various factors of $G(j\omega)$ are given in table 6.7.

Table 6.7

Sl. No.	*Factor*	*Corner frequency*	*Slope (db / dec)*	*Change in slope*
1.	$\dfrac{0.75}{j\omega}$	–	– 20	–
2.	$1 + j\,0.2\,\omega$	$\omega_1 = 5$	20	– 20 + 20 = 0
3.	$\dfrac{1}{1+j0.16\omega} - 0.01\omega^2$	$\omega_2 = 10$	– 40	0 – 40 = – 40

Consider a low and higher frequencies ω_l and ω_h, such that $\omega_l < \omega_1$ and $\omega_h > \omega_2$ respectively.

Assume, $\omega_l = 0.5$ rad/sec

$\omega_h = 100$ rad/sec

$\therefore$ Magnitude $(M) = |G(j\omega)|$ in dB and determine M at $\omega_l, \omega_1, \omega_2, \omega_h$

At $\omega = \omega_l$, $$M = 20 \log \left(\frac{0.75}{j\omega_1}\right) = 20 \log \left(\frac{0.75}{0.5}\right) = 3.5 \text{ dB}$$

$\omega = \omega_1$, $$M = 20 \log \left(\frac{0.75}{j\omega}\right) = 20 \log \left(\frac{0.75}{5}\right) = -16.5 \text{ dB}$$

$\omega = \omega_2$, $$M = \left[\text{slope change from } \omega_1 \text{ to } \omega_2 \times \log \frac{\omega_2}{\omega_1}\right] + M \text{ (at } \omega = \omega_1)$$

$$= 0 + (1 - 6.5) = -16.5 \text{ dB}$$

$\omega = \omega_h$, $$M = \left[\text{slope from } \omega_2 \text{ to } \omega_h \times \log \frac{\omega_h}{\omega_2}\right] + M \text{ (at } \omega = \omega_2)$$

$$= -40 \,.\, \log \left(\frac{100}{10}\right) + (-16.5) = -56.5 \text{ dB}$$

Let the points *m, n, o* and *p* be the points corresponding to frequencies ω_1, ω_1, ω_2 and ω_h respectively on the graph sheet for magnitude plot. Select the scale on semi log sheet for magnitude plot as:

x-axis = 1 unit is 0.1 rad/sec

y-axis = 1 unit is 10 dB

Draw the straight lines to join the above points and indicate the slope on the respective region as shown in Fig. 6.16.

(*ii*) **Phase plot.** The phase angle of given transfer function is given by

$$\phi_1 = \angle G(j\omega) = -90 - \tan^{-1}(0.2\,\omega) - \tan^{-1}\left(\frac{0.16\omega}{1 - 0.01\omega^2}\right), \text{ for } \omega \le \omega_h \text{ i.e, } \omega_n = 10$$

$$\phi_2 = \tan^{-1} 0.2\,\omega - 90° - \tan^{-1}\left(\frac{0.16\omega}{1 - 0.01\omega^2} + 180\right), \text{ for } \omega > \omega_h \text{ i.e., } \omega > 10$$

The phase angles of $G(j\omega)$ with different values of 'ω' are tabulated in table 6.8.

Table 6.8

ω (rad/sec)	$\tan^{-1} 0.2\,\omega$ (deg)	$\tan^{-1} \dfrac{0.16\omega}{1-0.01\omega^2}$ (deg)	ϕ (deg)
0.5	5.7	4.6	– 88°
5	45°	46.8	– 92°
10	63.4°	90°	– 116°
50	54.3	161.6	– 168°
100	87.1	– 9.2 + 180 = 170.8	– 174°

On the same graph sheet select the scale for phase plot as follows:

x-axis = is 0.1 rad/sec. same as magnitude plot

y-axis = is 20°

Indicate the calculated phase angle on the graph sheet. Join the points by a smooth curve as shown in Fig. 6.16.

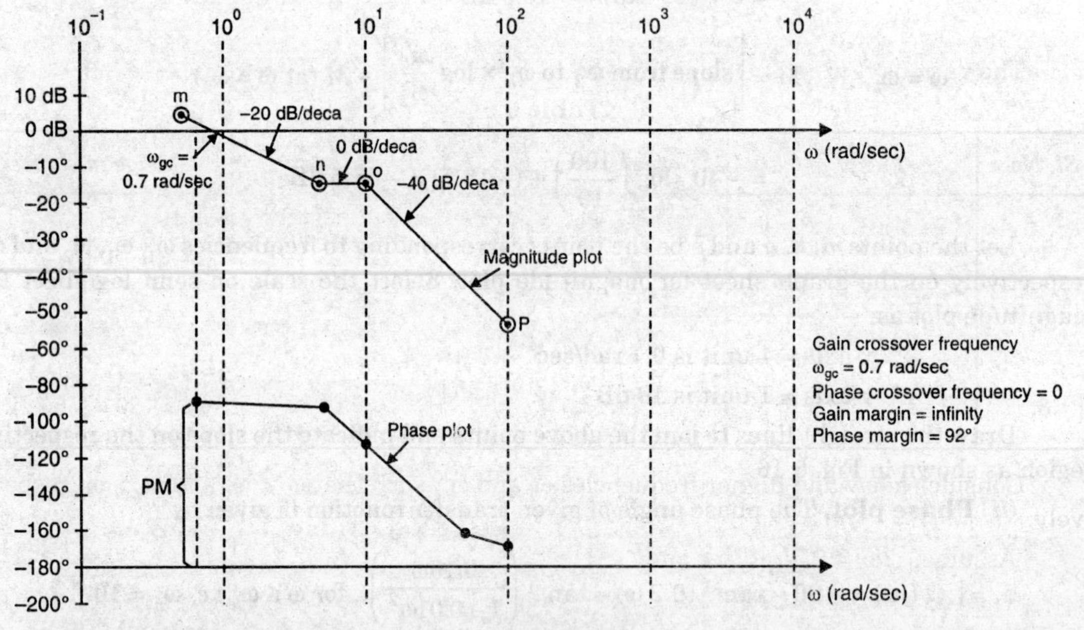

Fig. 6.16

From the Bode plot shown in Fig. 6.16, the following results are obtained as:

(*a*) Gain crossover frequency (ω_{gc}) = 0.7 rad/sec

(*b*) Phase crossover frequency (ω_{pc}) = 0

(*c*) Gain margin = ∞ dB

(*d*) Phase margin = 92°

Problem 6.6. *Sketch the Bode plot and determine the following:*

(*a*) *gain crossover frequency*

(*b*) *phase crossover frequency*

(c) *gain margin*

(d) *phase margin*

for the transfer function given $G(s) = \dfrac{K}{s(s+1)(s+2)}$ *and determine the K for stability.*

Solution: Let $K = 1$.

$$\therefore \qquad G(s) = \frac{1}{s(s+1)(s+2)}$$

Put $s = j\omega$ in the above s-domain transfer function, we can get

$$G(j\omega) = \frac{1}{2j\omega(1+j\omega)(1+0.5j\omega)}$$

(*i*) **Magnitude plot.** The corner frequencies of the given transfer function are

$$\omega_1 = \frac{1}{1} = 1 \text{ rad/sec}$$

$$\omega_2 = \frac{1}{0.5} = 2 \text{ rad/sec}$$

The various factors of $G(j\omega)$ are given in table 6.9.

Table 6.9

Sl. No.	*Factor*	*Corner frequency*	*Slope (dB / dec)*	*Change in slope*
1.	$\dfrac{0.5}{j\omega}$	–	– 20	–
2.	$\dfrac{1}{(1+j\omega)}$	$\omega_1 = 1$	– 20	– 20 – 20 = – 40 dB
3.	$\dfrac{1}{(1+0.5j\omega)}$	$\omega_2 = 2$	– 20	– 40 – 20 = – 60 dB

Consider a low and higher frequencies ω_l and ω_h, such that $\omega_l < \omega_1$ and $\omega_h > \omega_2$ respectively.

Assume, $\omega_l = 0.5$ rad/sec and

$\omega_h = 10$ rad/sec

$\therefore$ Magnitude $(M) = (G(j\omega))$ in dB and determine M at $\omega_l, \omega_1, \omega_2, \omega_h$.

At $\omega = \omega_l$;,

$$M = \left[\text{slope from } \omega_l \text{ to } \omega_1 \times \log \frac{\omega_1}{\omega_1}\right] + M\ (\text{at } \omega = \omega_l)$$

$$= -20 \times \log \frac{1}{0.5} = -6 \text{ dB}$$

$\omega = \omega_2$,

$$M = \left[\text{slope from } \omega_1 \text{ to } \omega_2 \times \log \frac{\omega_2}{\omega_1}\right] + M\ (\omega = \omega_1)$$

$$= -40 \times \log\left(\frac{2}{1}\right) - 6 = -18 \text{ dB}$$

$$\omega = \omega_h, \qquad M = \left[\text{slope from } \omega_2 \text{ to } \omega_h \times \log \frac{\omega_h}{\omega_2}\right] + M(\omega = \omega_2)$$

$$= -60 \times \log\left(\frac{10}{2}\right) - 18 = -60 \text{ dB}$$

Let the points *m, n, o,* and *p* be the points corresponding to frequencies ω_l, ω_1, ω_2 and ω_h respectively on the graph sheet for magnitude plot. Select the scale on semi log for magnitude plot as fallows

x-axis = 1 unit is 0.1 rad/sec

y-axis = 1 unit is 20 dB

Draw the straight lines to join the above points and indicate the slope on the respective region as shown in Fig. 6.17.

(*ii*) **Phase plot.** The phase angle of given transfer function is given by

$\phi = \angle G(j\omega) = -90 - \tan^{-1}\omega - \tan^{-1}(0.5\,\omega)$

The phase angles of $G(j\omega)$ with different values of 'ω' are tabulated in table 6.10.

Table 6.10

ω *(rad / sec)*	ϕ *(deg)*
0.5	130.6
1	161.5
2	198.43
10	252.97

On the same graph sheet select the scale for phase plot as follows.

On *x*-axis = 1 unit is 0.1 rad/sec, same as magnitude plot

y-axis = 1 unit is 20°

Indicate the calculated phase angle on the graph sheet joining the points by a smooth curve as shown in Fig. 6.17.

For the Bode plot shown in Fig. 6.17, the following results are obtained

(*a*) gain crossover frequency, (ω_{gc}) = 0.5 rad/sec

(*b*) phase crossover frequency, (ω_{pc}) = 1.5 rad/sec

(*c*) gain margin = – 14 dB

(*d*) phase margin = 52°

The Bode plot drawn at $K = 1$, gain margin obtained as 14 dB. That is the gain to be raised by 14 dB to put the system at the stability limit. This can be achieved by shifting the magnitude curve 14 dB.

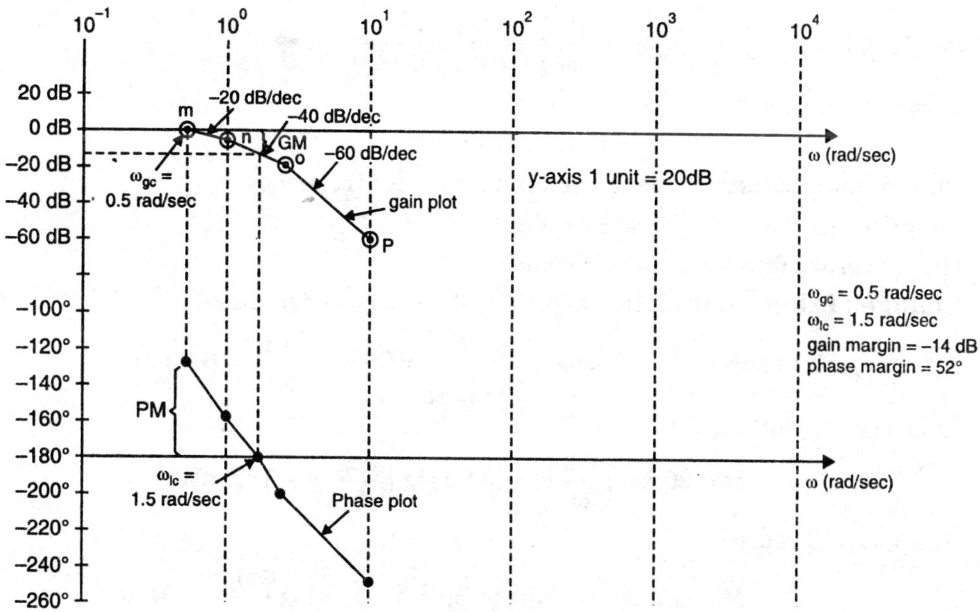

Fig. 6.17

We know that constant '*K*' can add 20 log *K* to every point of the magnitude plot. Now equate the vertical distance by which the magnitude plot is to be shifted to 20 log *K* and solve for *K*.

$$\therefore \quad 20 \log K = 14$$

$$K = 10^{14/20} = 5$$

∴ The range of *K* is $0 < K < 5$ is the stability limit.

Problem 6.7. *The open loop transfer function of a certain unity feedback system is*

$$G(s) = \frac{K}{s(s+1)(s+5)}$$ *Construct Bode plot and determine*

(*a*) *Limiting value of K for the system to be stable.*

(*b*) *Value of K for gain margin to be 10 dB*

(*c*) *Value of K for phase margin to be 45°.*

Solution:
$$G(s) = \frac{K}{s(s+1)(s+5)}$$

The given transfer function is converted into standard function as

$$G(s) = \frac{0.2\,K}{s(1+s)(1+0.2s)}$$

Let $K = 1$

$$\therefore \quad G(s) = \frac{0.2}{s(1+s)(1+0.25)}$$

Put $s = j\omega$, $G(j\omega) = \dfrac{0.2}{j\omega(1+j\omega)(1+0.2j\omega)}$

Corner frequencies, $\omega = 1$ and 5

i.e., $\omega_1 = 1, \omega_2 = 5$ rad/sec

Choose lower corner frequency and higher corner frequency

Lower corner frequency $\omega_l = 0.1$ rad/sec

Higher corner frequency $\omega_h = 10$ rad/sec

Change of slope. Calculate magnitude at each corner frequency

At $\omega = \omega_l = 0.1$ rad/sec, $M = 20 \log \left(\dfrac{0.2}{\omega}\right)_{\omega = 0.1} = 20 \log \left(\dfrac{0.2}{0.1}\right) = 6.02$ dB

At $\omega = \omega_1 = 1$ rad/sec,

$$M = 20 \log \left(\frac{0.2}{\omega}\right)_{\omega = 1} = 20 \log \left(\frac{0.2}{1}\right) = -13.9 \text{ dB}$$

At $\omega = \omega_2 = 5$ rad/sec,

$$M = \text{change of slope from } \omega_1 \text{ to } \omega_2 \times \log \left(\frac{\omega_2}{\omega_1}\right) + M \text{ at } \omega = \omega_1$$

$$= -40 \log \left(\frac{5}{1}\right) + (-13.9) = -41.85 \text{ dB}$$

At $\omega_h = 10$ rad/sec, $M = \text{change of slope from } \omega_2 \text{ to } \omega_h \times \log \left(\dfrac{\omega_h}{\omega_2}\right) + \text{M at } \omega = \omega_2$

$$= -60 \log \left(\frac{10}{5}\right) + (-41.85) = -59.91 \text{ dB}$$

(*i*) Draw the magnitude Vs frequency plot which gives the magnitude plot as shown in Fig. 6.18.

(*ii*) Calculate the phase angle for the phase plot.

Phase angle $\phi = -90 - \tan^{-1}(\omega) - \tan^{-1}(0.2\,\omega)$

The phase angle of $G(j\omega)$ with different values of 'ω' are tabulated in table 6.11.

Table 6.11

Frequency (rad / sec)	*Phase angle* (ϕ)
0	– 90
0.1	– 96.85
1	– 146.30
5	– 213.6
10	– 237.72

(*iii*) Draw the phase angle *Vs* frequency on same graph which gives the phase plot as shown in Fig. 6.18.

(*iv*) From the graph shown in Fig. 6.18

$$\omega_{gc} = 0.24 \text{ rad/sec}$$

$$\omega_{pc} = 2.3 \text{ rad/sec}$$

Gain margin at $\omega_{pc} = 28$ dB

Phase margin at $\omega_{gc} = 68°$

(*v*) (*a*) but for the system to be stable, 20 log $K = 28$ dB

$$\log K = \frac{28}{20} = 1.4$$

$$K = 10^{1.4}$$

$$K = 25.11$$

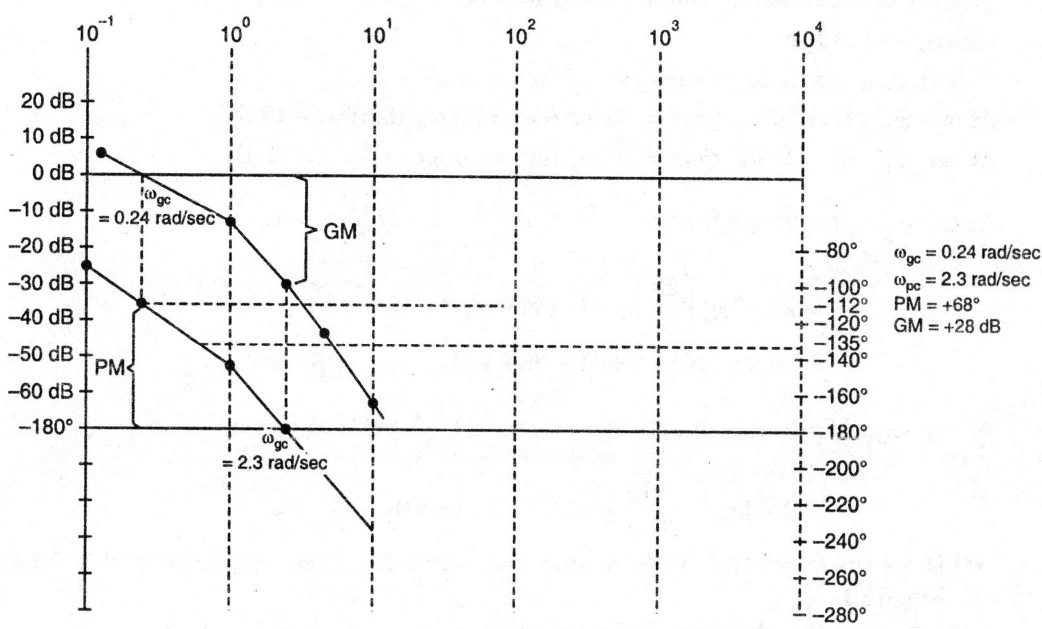

Fig. 6.18

(*b*) For the gain margin 10 dB to be lifted to the gain 28 dB, the graph has to be lifted up by 28 –10 = 18 dB

$\therefore$ 20 log $K = 18$

$K = 7.94$

(*c*) Phase margin, $\phi_m = 180 + \phi_{gc}$, where ϕ_{gc} is the phase of $G(j\omega)$ at $\omega = \omega_{gc}$

When $\phi_m = 45°$, $\phi_{gc} = \phi_m - 180 = 45 - 180 = -135°$

With $K = 1$, the dB gain at $\phi = -135°$ is – 9 db. The gain should be made zero to have a P.M of 45°. Hence to every point of magnitude plot at dB gain of 9 dB should be added. The corrected magnitude plot is obtained by shifting plot with K = 1 by 9 dB upwards.

$\therefore$ 20 log $K = 9$

$K = 10^{9/20} = 2.81.$

Problem 6.8. *Sketch the Bode plot for the transfer function* $G(s) = \dfrac{Ks^2}{(1+0.5s)(1+0.05s)}$.

Determine the system gain K for the gain crossover frequency to be 5 rad/sec.

Solution: Let $K = 1$, then

$$G(s) = \frac{s^2}{(1+0.5s)(1+0.05s)}$$

Put $s = j\omega$ in the given s-domain transfer function, we can get

$$G(s) = \frac{(j\omega)^2}{(1+j0.5\omega)(1+j0.05\omega)}$$

Corner frequencies are 2 and 20 rad/sec.

Assume $\omega_l = 0.5$ rad/sec and $\omega_h = 50$ rad/sec.

Change of slope

Calculate magnitude at each corner frequency

At $\omega = \omega_l$, $M = 20 \log (j\omega)^2 = 20 \log (\omega)^2 = 20 \log (0.5)^2 = -12$ dB

At $\omega = \omega_1$, $M = 20 \log (j\omega)^2 = 20 \log (\omega^2) = 20 \log (2)^2 = 12.04$ dB

At $\omega = \omega_2$, $M = [\text{slope from } \omega_1 \text{ to } \omega_2 \times \log\left(\frac{\omega_2}{\omega_1}\right) + M \text{ at } \omega = \omega_1$

$$= 20 \times \log\left(\frac{20}{2}\right) + M \text{ at } \omega = \omega_1$$

$$= 20 \log (10) + 12.04 = 32.04 \text{ dB}$$

At $\omega = \omega_h$, $M = (\text{slope change } \omega_2 \text{ to } \omega_n \times \log\left(\frac{\omega_n}{\omega_2}\right) + M \text{ at } \omega = \omega_2$

$$= 0 \times \log\left(\frac{-50}{20}\right) + 32.04 = 32.04 \text{ dB}$$

(*i*) Draw the magnitude *Vs* frequency plot which gives the magnitude plot as shown in Fig. 6.19.

(*ii*) Calculate the phase angle for the phase plot.

Phase angle, $\phi = \angle G(j\omega) = (80 - \tan^{-1}(0.5\omega) - \tan^{-1}(0.05\omega)$

The phase angle of $G(j\omega)$ with different values of 'ω' are tabulated in table 6.12.

Table 6.12

ω *(rad / sec)*	ϕ *(degree)*
0.5	164.53
1	150.57
2	129.28
10	74.74
20	50.71
50	24.09

(*iii*) Drawing the phase angle *Vs* frequency on same graph gives the phase plot as shown in Fig. 6.19.

Calculation of K. From the Bode plot as shown in Fig. 6.19, for given gain crossover frequency 5 rad/sec, the gain is

i.e., At $\omega = 5$ rad/sec, the gain = 20 dB

We can shift every point to – 20 dB to get the value of K at gain margin $\omega = 5$ rad/sec.

$$\therefore \quad 20 \log K = -20 \text{ dB}$$

$$K = 0.1.$$

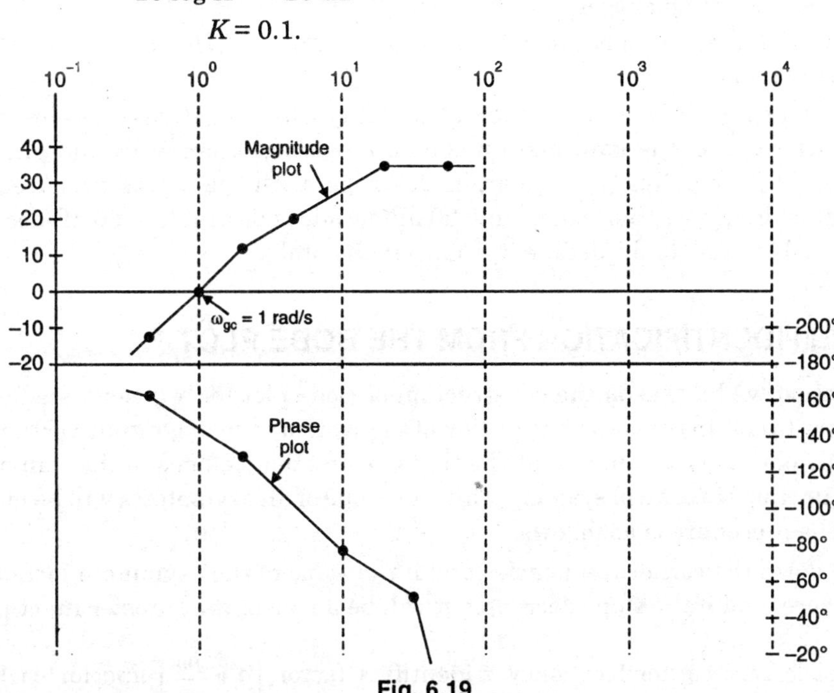

Fig. 6.19

6.5 RELATIVE STABILITY FROM BODE PLOTS

The relative stability is measured with the parameters-gain margin and phase margin. In the polar plot as plot passes through the point $(-1 + j0)$, the system becomes critical stable. Also as the polar plot approaches $(-1 + j0)$ the relative stability decreases. It can also be easily determined from Bode plots as shown in Fig. 6.20.

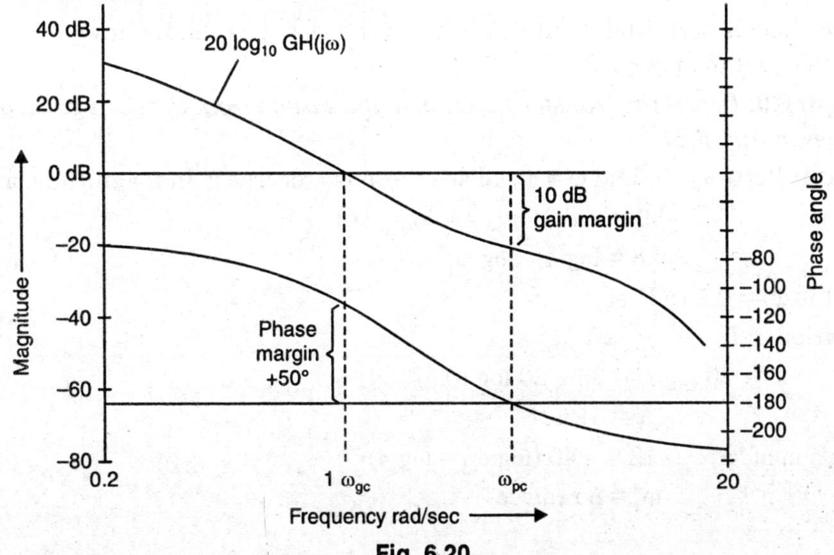

Fig. 6.20

To recapitulate, the definition of gain margin is the magnitude of the reciprocal of the open loop transfer function at phase crossover frequency of ω_{pc} when the phase is – 180°. As 0 dB at phase cross over frequency ω_{pc}.

Phase margin is the number of degrees $\angle GH(j\omega)$ (or) arg $GH(j\omega)$ is before – 180° at the gain crossover frequency ω_{gc}.

Positive gain and phase margins indicate stability while negative values instability. For satisfactory performance, the gain margin should be greater than 6 dB and phase margin between 30° and 60°. For this phase margin the slope at the gain crossover frequency ω_{gc} should be greater than – 40 dB/decade, but – 20 dB/decade is desirable. – 60 dB/decade makes the system unstable. At – 40 dB/decade it is marginally stable.

6.6 SYSTEM IDENTIFICATION FROM THE BODE PLOT

In the last section we have seen the construction of Bode plot for a given transfer function. Now we shall try to obtain the transfer function of the system from magnitude versus frequency plot of the Bode plot. It is assumed that the transfer function belongs to the clan of the open loop transfer functions of control systems. Thus the slope of all asymptotes will be multiplies of ± 20 dB/dec. The procedure is as follows:

- Note down the corner frequencies and if the slope of the asymptote increases it will be a zero, and if the slope decreases it will be a pole of the transfer function.
- For each zero corner frequency z identify a factor $\left(1+\frac{j\omega}{z}\right)$ and for each pole frequency p, identify the factor $\left(1+\frac{j\omega}{p}\right)$ in the frequency transfer function $G(j\omega)$.
- Note down the slope of the lowest frequency asymptote. If it is not zero, then it gives the order of the pole at the origin, and is used to obtain the value of the constant K in $G(j\omega)$.
- If the slope is zero, and it runs parallel to the $j\omega$ axis, then its intercept on the $M(\omega)$ axis is equal to 20 log (K).

Problem 6.9. *Derive the transfer function of the system from the data given on the Bode diagram shown in Fig. 6.21.*

Solution: Between ω_1 and $\omega = 4$ rad/sec there is a decrease in magnitude of 36 dB.

$$\therefore \quad -36 = -40\,(\log_{10} 4 - \log_{10} \omega_1)$$

$$0.8 = \log 4 - \log \omega_1$$

From this $\omega_1 = 0.5$ rad/sec

Calculation of K

$$20 \log K = 36 + 20 \log (0.5)$$

$$K = 31.62$$

(or) Calculation of 'ω_2', $-12 = -40\,(\log \omega_2 - \log 4)$

$$\omega_2 = 8 \text{ rad/sec}$$

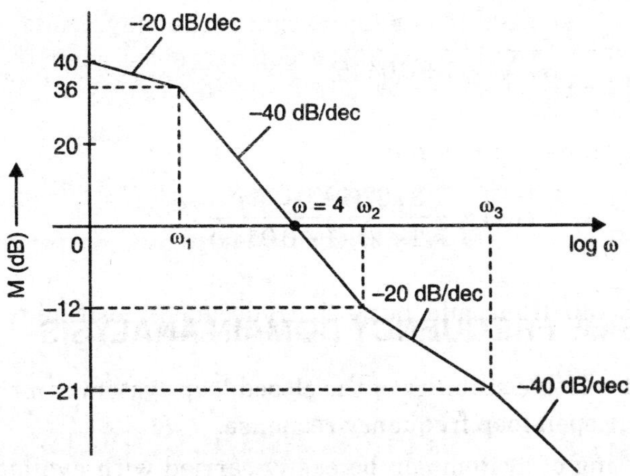

Fig. 6.21

In between ω_2 and ω_3 the rate of decay is 20 rad/sec and the magnitude between ω_2 to ω_3 rad/sec is

$$-21-(-12) = -20\,[\log(\omega_3) - \log(\omega_2)]$$

$$-21+12 = -20\,[\log(\omega_3) - \log(8)]$$

$$79 = +20\,[\log(\omega_3) - \log(8)]$$

$$\frac{9}{20} = \log(\omega_3) - \log(8)$$

$$0.45 = \log(\omega_3) - \log(8)$$

$$\log \omega_3 = 0.45 + \log(8)$$

and $$\omega_3 = 22.5 \text{ rad/sec}$$

The first line has a slope of – 20 dB/dec indicating term $\frac{1}{s}$ and it is not passing through ω =1 rad/sec, *i.e.*, the term is $\frac{K}{s} = \frac{31.62}{s}$.

The second term has a slope of – 14.0 dB/dec. That means it indicates the pole at the value of ω_1 = 0.5 rad/sec.

The term equal to $\dfrac{1}{\left(1+\dfrac{s}{0.5}\right)} = \dfrac{1}{(1+2s)}$ and the third term has a slope of – 20 dB/dec, which means the slope + 20 dB/dec is added to the previous slope.

∴ It is zero and the value of ω_2 is 8 rad/sec, the term is $\left(1+\dfrac{s}{8}\right) = (1+0.125s)$

and the fourth term has a slope of – 40 rad/sec which means – 20 db/dec is added to the previous one which means this term is pole and the value of ω_3 = 22.5 rad/sec.

$\therefore$ The term is $\dfrac{1}{\left(1+\dfrac{s}{22.5}\right)} = \dfrac{1}{(1+0.0445)}$

$\therefore$ The complete transfer function

$$G(s) = \frac{31.62(1+0125s)}{s(1+2s)(1+0.044s)}$$

6.7 ADVANTAGES OF FREQUENCY DOMAIN ANALYSIS

- The absolute and relative stability of the closed loop system can be estimated from the knowledge of their open loop frequency response.
- The practical testing of system can be easily carried with available sinusoidal signal generators and precise measurement equipments.
- The transfer function of complicated systems can be determined experimentally by frequency response tests.
- The design and parameter adjustment of the open loop transfer function of a system for specified closed loop performance is carried out more easily in frequency domain.
- When the system is designated by the use of frequency response analysis, the effects of noise disturbance and parameters variations are relatively easy to visualize and incorporate corrective measures.
- The frequency response analysis and designs can be extended to certain nonlinear systems.

SHORT QUESTIONS AND ANSWERS

1. **What are the advantages of frequency domain analysis?**
   - The absolute and relative stability of the closed loop system can be estimated from the knowledge of their open loop frequency response.
   - The transfer function of complicated systems can be determined experimentally by frequency response tests.
   - The design and parameter adjustment of the open loop transfer function of a system for specified closed loop performance is carried out more easily in frequency domain.
   - The frequency response analysis and designs can be extended to certain nonlinear systems.
2. **What are the frequency domain specifications?**

   Frequency domain specifications are
   - Gain margin and Phase margin
   - Resonant peak and Resonant frequency
   - Delay time
   - Bandwidth and Cut off rate.

3. **Define gain margin.**

   It is defined as the reciprocal of the open loop transfer function evaluated at the frequency ω_{pc} at which the phase angle is – 180°.

4. **Define phase margin.**

   It is the angle (ϕ_{PM}) at gross crossover fr. quency ω_{gc} at which magnitude equals to one (or) 0 dB.

5. **Define the resonance peak.**

   The maximum value of $M(j\omega)$ is called the peak resonance.

6. **Define the resonance frequency.**

   The resonant frequency ω_r is the frequency at which $M(j\omega)$ is maximum.

7. **Define bandwidth.**

   Bandwidth (BW) is defined as the frequency at which $M(j\omega)$ drops 70.7% or 3 dB down from one frequency gain.

8. **Define the cutoff rate.**

   It is the frequency rate at which the magnitude ratio decreases beyond the cut off frequency ω_c.

9. **Define the Bode magnitude plot.**

   The dB magnitude versus log ω plot is called Bode magnitude plot. It is also called as log-modulus plot.

10. **Define the Bode phase angle plot.**

    The phase angle versus log ω plot is called Bode phase angle plot.

11. **What is bode plot?**

    The bode plot is the logarithmic plot which consists of two plots, one is magnitude plot and other is phase angle plot.

12. **What is meant by asymptotes?**

    In Bode plot the low frequency and higher frequency approximations can be represented by straight lines are called asymptotes.

13. **Define corner frequency.**

    It is the frequency at which the slope of asymptotes will change.

OBJECTIVE TYPE QUESTIONS

1. If the phase ϕ which is the angle between radial line connecting a pole and origin is equal to 45°, then the peak overshoot is

   (*a*) 0.5% (*b*) 1.2%

   (*c*) 3% (*d*) 4.32%

2. The demoralized bandwidth for a particular value of ω_n and damping factor ζ is.

   (*a*) $\omega_n \sqrt{1-\delta^2}$ (*b*) $(\omega_n + \delta^2)^{3/2}$

   (*c*) $\omega_n\ [1 - 2\delta^2 \sqrt{2 - 4\delta^2 + 4\delta^4}\ ]$ (*d*) $\{\omega_n\ [1 - 2\delta^2 + \sqrt{2 - 4\delta^2 + 4\delta^4}\ ]\}^{1/2}$

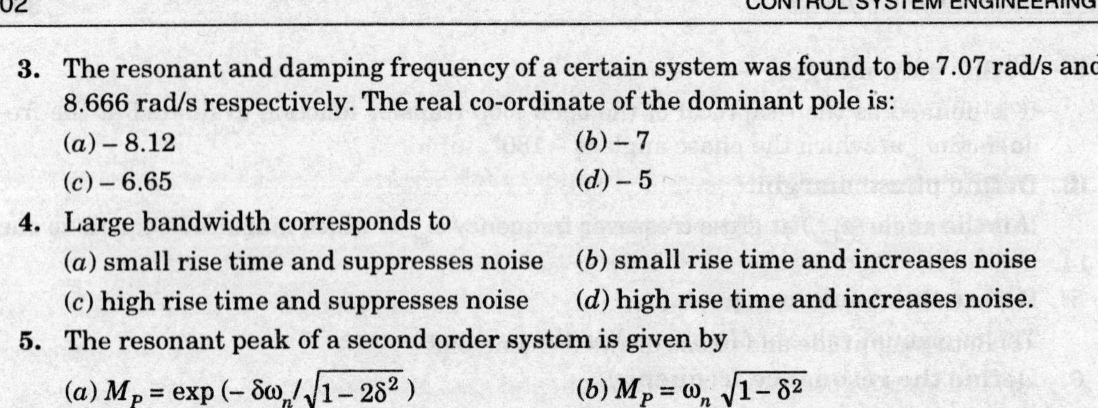

3. The resonant and damping frequency of a certain system was found to be 7.07 rad/s and 8.666 rad/s respectively. The real co-ordinate of the dominant pole is:

(*a*) – 8.12 (*b*) – 7

(*c*) – 6.65 (*d*) – 5

4. Large bandwidth corresponds to

(*a*) small rise time and suppresses noise (*b*) small rise time and increases noise

(*c*) high rise time and suppresses noise (*d*) high rise time and increases noise.

5. The resonant peak of a second order system is given by

(*a*) $M_P = \exp(-\delta\omega_n/\sqrt{1-2\delta^2})$ (*b*) $M_P = \omega_n\sqrt{1-\delta^2}$

(*c*) $M_P = \dfrac{1}{\delta\sqrt{1-2\delta^2}}$ (*d*) $M_P = \dfrac{1}{2\delta\sqrt{1-2\delta^2}}$

6. In frequency domain analysis, response varies w.r.t.

(*a*) time (*b*) frequency

(*c*) time and frequency (*d*) constant.

7. Magnitude value of the given open loop transfer function $G(s) = \dfrac{1}{s(s+1)}$ is

(*a*) $\sqrt{2}$ (*b*) $\dfrac{1}{\sqrt{2}}$

(*c*) $\sqrt{3}$ (*d*) $\dfrac{1}{\sqrt{3}}$

8. Phase of the given open loop transfer function $G(s) = \dfrac{1}{s(s+1)}$ is

(*a*) 180° (*b*) – 180°

(*c*) – 135° (*d*) 90°

9. Which one is not a frequency domain specification?

(*a*) resonant frequency (*b*) resonant peak

(*c*) bandwidth (*d*) rise time.

10. At corner frequency the slope of Bode plot

(*a*) changes (*b*) zero

(*c*) constant (*d*) infinity

11. Magnitude in decibels of given transfer function $G(s) = \dfrac{1}{s+2}$ is

(*a*) $20\log\left|\dfrac{j\omega}{2}+1\right|$ (*b*) $10\log\left|\dfrac{j\omega}{2}+2\right|$

(*c*) $-20\log\left|\dfrac{j\omega}{2}+1\right|$ (*d*) $-10\log\left(\dfrac{j\omega}{2}+1\right)$

12. Phase value of prob. 6 is

(*a*) $\tan^{-1} \omega/2$ (*b*) $\tan^{-1} \omega$

(*c*) $-\tan^{-1} \omega/2$ (*d*) Infinity.

13. Why should the magnitude values be taken in decibels

Ans. In order to extend the frequency range.

14. One of the disadvantages of Bode plot is

(*a*) Magnitude is taken in decibels (*b*) Phase is degrees

(*c*) Both magnitude and phase draws on same plot

(*d*) It does not specify the phase margin and gain margin.

15. Gain crossover frequency at which magnitude is

(*a*) zero (*b*) Unity

(*c*) Infinity (*d*) Constant.

16. Phase crossover frequency at which phase is

(*a*) 90° (*b*) 180°

(*c*) – 180° (*d*) 0°

17. Gain margin is the reciprocal of

(*a*) magnitude (*b*) Phase

(*c*) Both *a* and *b* (*d*) None

18. A system is stable when

(*a*) $\omega_{gc} = \omega_{pc}$ (*b*) $\omega_{gc} < \omega_{pc}$

(*c*) $\omega_{gc} > \omega_{pc}$ (*d*) $\omega_{gc} = \omega_{pc} = 0$

19. For the system to be stable

(*a*) gain margin should be positive (*b*) Phase margin is positive

(*c*) both (*a*) and (*b*) (*d*) Gain margin and phase margin are negative.

20. System is conditionally stable when

(*a*) $\omega_{gc} = \omega_{pc}$ (*b*) $\omega_{gc} < \omega_{pc}$

(*c*) $\omega_{gc} > \omega_{pc}$ (*d*) $\omega_{gc} = \omega_{pc} = 0$

KEY

1. (*d*)	**2.** (*d*)	**3.** (*d*)	**4.** (*b*)	**5.** (*d*)	**6.** (*b*)
7. (*b*)	**8.** (*c*)	**9.** (*d*)	**10.** (*b*)	**11.** (*c*)	**12.** (*c*)
13. (*a*)	**14.** (*c*)	**15.** (*b*)	**16.** (*c*)	**17.** (*a*)	**18.** (*b*)
19. (*c*)	**20.** (*a*)				

EXERCISE

1. Define frequency domain specifications.
2. Derive the expressions for frequency domain specifications.
3. Explain the relation between phase margin and damping ratio.
4. Discuss the correlation between time domain and frequency domain specifications.
5. Explain the system identification from the Bode plot.
6. What are the advantages of frequency domain specifications.
7. Determine the value of the gain constant K for the system with open loop transfer function $G(s) = \dfrac{K}{s(1+0.1s)(1+0.01s)}$, so that it has a phase margin of about 45°. For this value of K, find the new gain margin.
8. Given the open loop transfer function of a unity feedback system $G(s) = \dfrac{1}{s(1+s)(1+2s)}$. Draw the Bode plot and measure from the plot the frequency at which the magnitude is 0 dB?
9. Given the open loop transfer function with unity feedback as $G(s) = \dfrac{75(1+02s)}{s(s^2+16s+100)}$. Draw the Bode plot.
10. Given the open loop transfer function with unity feedback as $G(s) = \dfrac{Ke^{-10s}}{s(1+s)(1+10s)}$. Draw the bode plot and determine the gain K for the gain crossover frequency to be 5 rad/sec?
11. Determine the transfer function of the system whose asymptotic logarithmic magnitude ploy is shown in Fig. 6.22.

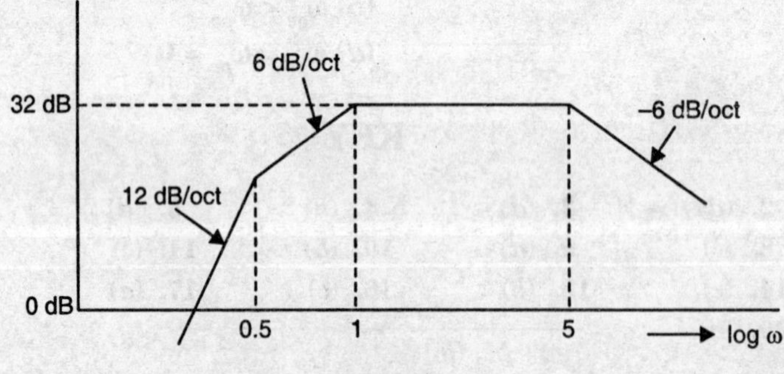

Fig. 6.22

7 Polar and Nyquist Plots

7.1 POLAR PLOTS

Introduction

A polar plot is another type of frequency domain plot of transfer functions, whereas Bode plot is one such frequency domain plot. All the information available in Bode plot is also available in the polar plot. The polar plot plane is a two dimensional plane, the radius vector represents the magnitude of the transfer function. The polar plot is the locus of the system transfer function in the frequency domain for $0 < \omega < \infty$ and every point on the polar plot has value of frequency. An arrow is used to point the direction of increase of frequency. Polar plots constructed for the range $\infty < \omega < 0$ will be the mirror image of the polar plot for the positive values of ω.

For the inverse polar plot of a transfer function, the inverse of the given transfer function is considered and its polar plot is drawn. For certain studies on the stability of control systems the inverse polar plots are conveniently used. Nyquist stability criterion uses the polar plot of a transfer function and because of this, the polar plots are also known as Nyquist plots.

Polar plots

The open loop function of a feedback control system is obtained substituting $s = j\omega$ in the transfer function $G(s)$ in the s-domain. It is expressed in the standard form

$$GH(j\omega) = \frac{K(j\omega + z_1)\ldots\ldots(j\omega + z_n)}{(j\omega)^n (j\omega + p_1)\ldots\ldots(j\omega + p_m)} \tag{7.1}$$

The magnitude of $GH(j\omega)$ is given by

$$|GH(j\omega)| = M(\omega) = \frac{K(\omega^2 + z_1)\ldots\ldots(\omega^2 + z_2^2)\ldots\ldots(\omega^2 + z_n{}^2)}{(\omega)^n(\omega^2 + p_1^2)\ldots\ldots(\omega^2 + P_m{}^2)}$$

$$= \frac{K.m_1(\omega).m_2(\omega)\ldots\ldots m_n(\omega)}{(\omega)^n M_1(\omega) M_2(\omega)\ldots\ldots M_m(\omega)} \tag{7.2}$$

And the phase angle $\phi(\omega)$ is given by

$$\phi(\omega) = \tan^{-1}\left(\frac{\omega}{z_1}\right) + \tan\left(\frac{\omega}{z_2}\right) \ldots\ldots\ldots - n \times 90 - \tan^{-1}\left(\frac{\omega}{p_1}\right) - \tan^{-1}\left(\frac{\omega}{p_2}\right) - \ldots\ldots\ldots - \tan^{-1}\left(\frac{\omega}{p_m}\right) \quad (7.3)$$

$M(\omega)$ and $\phi(\omega)$ are first computed from eqns. (7.2) and (7.3) for the asymptotic values of $\omega = 0$ and $\omega = \infty$. For intermediate points on the polar curve the values of ω at zeros and poles of the transfer function $GH(j\omega)$ is considered.

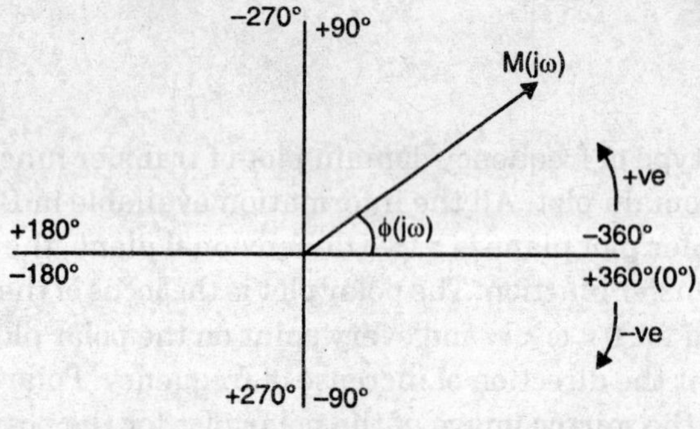

Fig. 7.1

Figure 7.1 shows the representation of $M(\omega)$ and $\phi(\omega)$ of the transfer function $G(j\omega)$ on the $G(j\omega)$ plane in polar co-ordinates.

$\phi(\omega)$ is always measured from 0° line which is the reference line. All the angles measured in the anticlockwise direction are assigned to be positive. That is

$$GH(j\omega) = M(\omega) \angle \phi(\omega) \quad (7.4)$$

is in polar co-ordinates in the ω-domain. Alternately

$$GH(j\omega) = \pm \text{Re}\,\{GH(j\omega)\} \pm j\,\text{Im}\,\{GH(j\omega)\} \quad (7.5)$$

This is represented in Fig. 7.2.

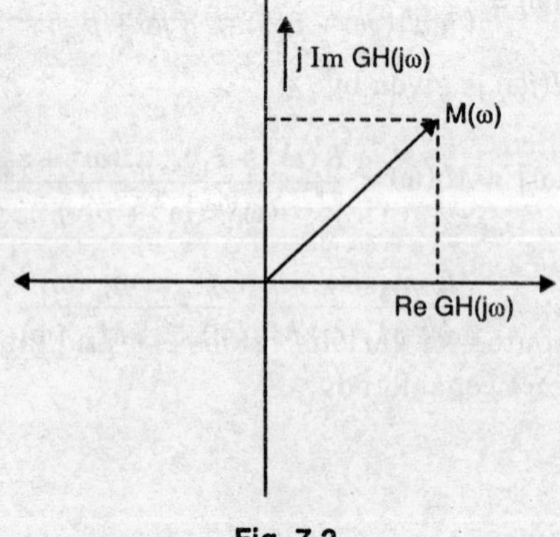

Fig. 7.2

7.1.1 Polar Plots for Typical Transfer Functions

(*a*) Let $G(s) = 1/s$, then

$$G(j\omega) = 1/j\omega$$
$$= 1/\omega \angle -90° \quad (7.6)$$

$M(\omega)$ versus $\phi(\omega)$ is plotted for various values of ω in the range $0 < \omega < \infty$ as per the table 7.1. To start with, asymptotic values are considered for ω.

Table 7.1

ω *(rad/sec)*	$M(\omega)$	$\phi(\omega)$ *(deg)*
0	∞	– 90
∞	0	– 90
0.1	10	– 90
1.0	1.0	– 90
10	0.1	– 90

The polar plot is shown in Fig. 7.3.

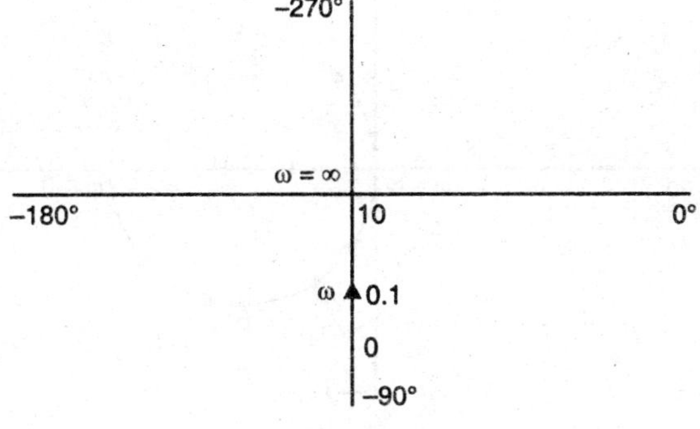

Fig. 7.3

(*a*) It is seen that as ω increases from zero to infinity, the locus of $GH(j\omega)$ moves along the line of – 90°. For the range $-\infty < \omega < 0$, the plot is the mirror image of the polar plot for the range $0 < \omega < \infty$. The negative values for ω have no physical meaning. The direction of arrow on the plot shows the increases of ω.

(*b*) Let $G(s) = \dfrac{1}{s+p}$, then

$$G(j\omega) = \frac{1}{j\omega + p} = \frac{1}{(\omega^2 + p^2)^{1/2}} \angle -\tan^{-1}\left(\frac{\omega}{p}\right) \quad (7.7)$$

$M(\omega)$ and $\phi(\omega)$ are computed for various values of ω and tabulated in table 7.2. To start with, asymptotic value for ω are considered.

Table 7.2

ω *(rad/sec)*	$M(\omega)$	$\phi(\omega)$ *(deg)*
0	$\frac{1}{p}$	0
∞	∞	– 90
$\frac{p}{2}$	$\frac{0.89}{p}$	– 26.56
p	$\frac{0.707}{p}$	– 45
$2p$	$\frac{0.44}{p}$	– 63.43

The polar plot is shown in Fig. 7.4 for $0 < \omega < \infty$. For the range $-\infty < \omega < 0$, the polar plot is shown in dotted lines.

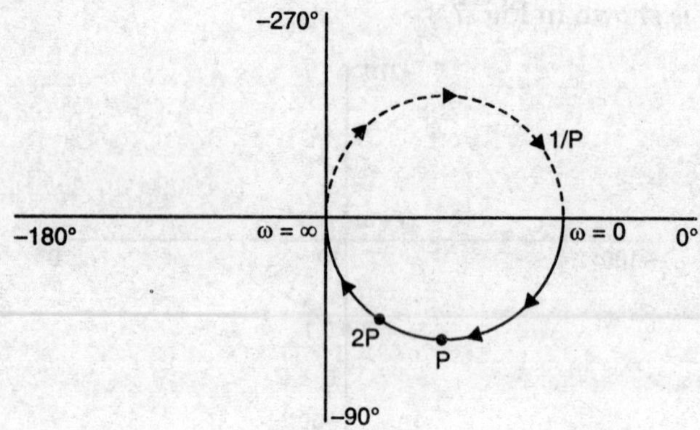

Fig. 7.4

(*c*) Let $G(s) = (s + z)$, then

$$G(j\omega) = (j\omega + z) = [\omega^2 + z^2]^{1/2} \angle \tan^{-1}(\omega/z) \tag{7.8}$$

$M(\omega)$ and $\phi(\omega)$ are calculated for $0 < \omega < \infty$ and are tabulated in table 7.3. To start with, asymptotic value for ω are considered.

Table 7.3

ω *(rad/sec)*	$M(\omega)$	$\phi(\omega)$ *(deg)*
0	z	0
∞	∞	90
$\frac{z}{2}$	$1.11z$	26.56
z	$1.414z$	45
$2z$	$2.236z$	63.43
$10z$	$10.04z$	84.28

The polar plot is shown in Fig. 7.5. The polar plot for $-\infty < \omega < 0$ is plotted in dotted lines.

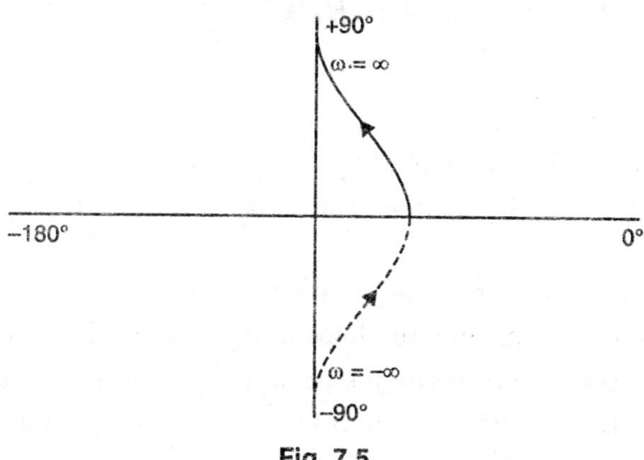

Fig. 7.5

7.1.2 Computation of Frequency (ω) at a Given Point on the Polar Plot

The specified point is marked on the polar plot. A straight line is drawn joining the point to the origin of the polar plane of $GH\,(j\omega)$. The length of the line gives the magnitude $GH\,(j\omega)$ and the angle measured from the 0° line gives $\phi(\omega)$ at the marked point. By analytically solving the expressions for the magnitude or the phase angle, and by equating with the respective measured quantities, ω is obtained.

7.1.3 Procedure to Sketch the Polar Plot

Step 1: Put $s = j\omega$ in the given transfer function $G(s)$ and convert it into $G(j\omega)$

Step 2: Evaluate the magnitude $|G(j\omega)|$ and phase $\angle G(j\omega)$ at $\omega \to 0$ and $\omega \to \infty$.

Step 3: At various values of ω, $0 < \omega < \infty$, evaluate magnitude and phase angle.

Step 4: The frequency at which the polar plot intersects the real and imaginary axis are found by evaluating the imaginary and real part of magnitude of $G(j\omega)$ equals to zero.

7.1.4 Calculation of Gain Crossover Frequency and Phase Crossover Frequency

In Bode plot we have defined that the frequency at which the magnitude crosses 0 dB is called gain cross order frequency. In polar plots we are not taking the magnitude in dB. So gain cross over frequency is the frequency at which the magnitude crosses unity (or) one corresponding to 0 dB.

Find the phase angle ϕ for a given transfer function. Measure the phase angle in clockwise direction from reference line 0°, which will intersect the polar plot at a point at which frequency $\omega = \omega_{gc}$.

Draw a circle with radius 1 and centre as origin. The circle intersects the – 180° axis at a point $\omega = \omega_{pc}$.

7.1.5 Calculation of Gain Margin and Phase Margin from Polar Plot

Magnitude $\qquad M = |G(j\omega)\,H(j\omega)|_{\omega = \omega_{pc}}$

Gain margin, $GM = 20 \log_{10} 1/M$

Phase margin, $\phi_m = 180 + \angle G(j\omega)\, H(j\omega)\big|_{\omega = \omega_{pc}}$

Direction of plot

Starting point at $\omega \to 0$ angle is $-\phi_1$

Terminating point at $\omega \to \infty$ angle is ϕ_2

Rotation of plot $= \phi_1 + \phi_2$ if negative, clockwise direction, for positive value anti clockwise direction.

Stability determination from polar plot

$\omega_{gc} < \omega_{pc}$, Gain margin and phase margin are positive, system is stable

$\omega_{gc} > \omega_{pc}$, Gain margin and phase margin are negative, system is unstable

$\omega_{gc} = \omega_{pc}$, Gain margin and phase margin are zero, system is conditionally stable

Problem 7.1. *Express (– 5.4 + j8) is polar form.*

Solution:

Magnitude $M = \sqrt{(-5.4)^2 + 8^2} = 9.65$

Phase $\phi = \tan^{-1} 8/-5.4 = -55.98°$

The sign in real part indicates 2nd quadrant

$\therefore$ $\phi = 180° - 55.98° = 124.02°$

Polar form $= 9.65 \angle 124.02°$ (or) $- 9.65 \angle 55.98°$

Problem 7.2. *Sketch the polar plot for a given open loop transfer function*

$$G(s) = \frac{10}{s(s+1)(s+3)}.$$

Solution:

Put $s = j\omega$ in $G(s)$

$$G(j\omega) = \frac{10}{j\omega(j\omega+1)(j\omega+3)}$$

Find the magnitude and phase at $\omega \to 0$ and $\omega \to \infty$

$$|\,G(j\omega)\,| = \frac{10}{\omega\,(\sqrt{\omega^2+1}\,)\sqrt{(\omega^2+9)}}$$

$$\phi = -\tan^{-1}\omega - \tan^{-1}\omega/3 - 90°$$

At $\omega \to 0$, $M = \infty$ $\quad \phi = -90°$

and At $\omega \to \infty$, $M = 0$ $\quad \phi = -270°$

At various values of ω, evaluate M and ϕ and draw the polar plot on normal graph sheet as shown in Fig. 7.6.

Gain crossover frequency (ω_{gc})

$$|\,GH(j\omega)\,| = 1$$

$$\frac{10}{\omega\sqrt{\omega^2+1}\sqrt{(\omega^2+9)}} = 1$$

$\Rightarrow$

$$10 = \omega^2 (\omega^2 + 1) (\omega^2 + 9).$$

Let $\omega^2 = x$

$$x(x + 1) (x + 9) = 10$$

$$x^3 + 10x^2 + 9x - 10 = 0$$

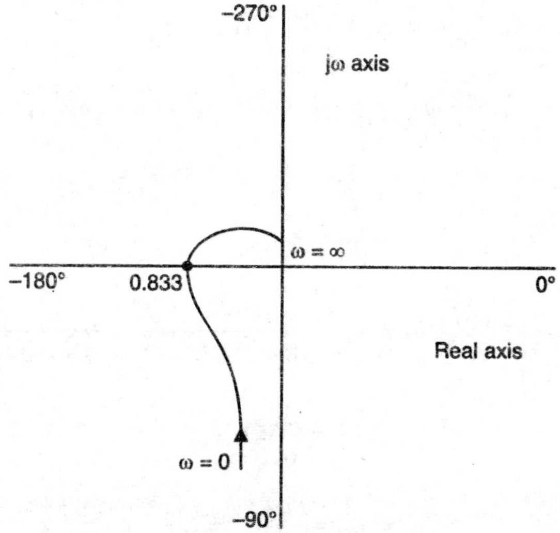

Fig. 7.6 Polar plot for a given transfer function $G(s) = \dfrac{10}{s(s+1)(s+3)}$

Constructing the Routh table for x variable of characteristic equation.

x^3	1	9
x^2	10	– 10
	1	– 1
x	10	0
x^0	– 1	

Taking auxiliary equation

$$10x^2 - 10 = 0$$

$$x^2 = 1, x = 1$$

$\therefore$

$$\omega_{gc} = 1$$

$$\phi|_{\omega_{gc}} = -90^\circ - \tan^{-1}(1) - \tan^{-1}\frac{1}{3}$$

$$= -90^\circ - 45^\circ - 18.36^\circ$$

$$= -153.26^\circ \text{ (or) } 26^\circ.7$$

$$P_m = 180^\circ + \phi$$

$$= 180^\circ - 153.26^\circ = 26.7^\circ$$

Phase crossover frequency ω_{pc}

$$\phi = -90^\circ - \tan^{-1}\omega - \tan^{-1}\frac{\omega}{3} = -180^\circ$$

$$-\tan^{-1}\frac{(\omega + \omega/3)}{1-\omega^2/3} = -90^\circ$$

or

$$\frac{\omega + \omega/3}{1-\omega^2/3} = \tan 90^\circ = \frac{1}{0}$$

$$1 - \frac{\omega^2}{3} = 0$$

$$\omega^2 = \sqrt{3}$$

$$\omega_{pc} = \sqrt{3} = 1.732$$

$$M = \frac{10}{\sqrt{3}\,(\sqrt{3+1})\,(\sqrt{3+9})} = \frac{10}{\sqrt{3}\times\sqrt{4}\times\sqrt{12}}$$

$$= \frac{10}{12} = \frac{5}{6} = 0.833$$

Gain margin $= \dfrac{1}{M} = \dfrac{1}{0.833} = 1.2$

$\omega_{gc} < \omega_{pc}$, phase margin and gain margin both are positive so the given system is stable.

Problem 7.3. *Given the transfer function of a system*

$$G(s) = \frac{1}{(1-s)(1+2s)}$$

Determine whether the polar plot of the system crosses the imaginary axis of the polar plane. If so, determine the frequency at which it crosses and the corresponding magnitude of $G(j\omega)$.

Solution: Putting $s = j\omega$, the frequency domain transfer function is

$$G(j\omega) = \frac{1}{(1-j\omega)(1+j2\omega)}$$

The magnitude $M(\omega)$ and phase angle $\phi(\omega)$ are given by

$$M(\omega) = \left[\frac{1}{(1+\omega^2)(1+4\omega^2)}\right]$$

$$\phi(\omega) = -\tan^{-1}\omega - \tan^{-1}(2\omega)$$

$M(\omega)$ and $\phi(\omega)$ are calculated and tabulated in table 7.4 for $0 < \omega < \infty$.

Table 7.4

ω (rad/sec)	$M(\omega)$	$\phi(\omega)$ (deg)
0	1	0
∞	0	– 180
0.5	0.632	– 71.56
1	0.316	– 108.43
10	0.00496	– 171.41

From the values of $\phi(\omega)$ it is seen that the plot crosses the imaginary axis in the range $0 < \omega < \infty$. The polar plot is shown in Fig. 7.7.

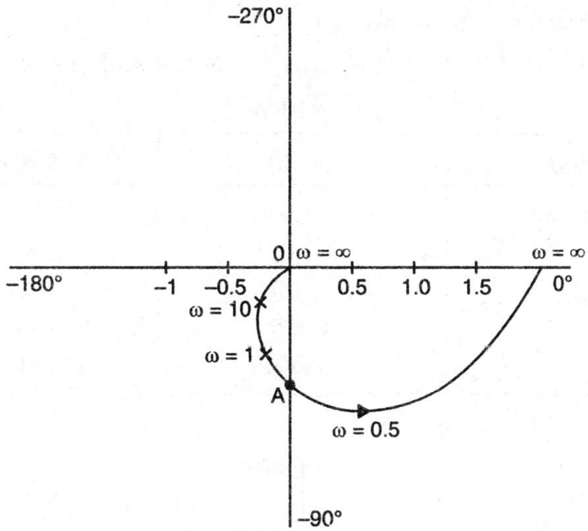

Fig. 7.7 Polar plot for open loop transfer function $G(s) = \dfrac{1}{(1+s)(1+2s)}$

From the plot is seen that it crosses the imaginary axis at point A. Let the frequency at A be ω_A. For obtaining value for ω_A we shall use the phase angle expression and equate $\phi(\omega_A)$ to $-90°$.

That is, $\quad \phi(\omega_A) = -90° = -\tan^{-1}(\omega_A) - \tan^{-1}(2\omega_A)$

Taking the tangents of the angle on both the sides, we have

$$\tan(-90°) = -\infty = -\frac{[\omega_A + 2\omega_A]}{[1 - 2\omega_A^2]}$$

Therefore $[1 - 2\,\omega_A^2]$ must be zero so that $\omega_A = 0.707$ rad/sec.

Substituting for ω_A in $M(\omega_A)$, we get the magnitude at the point of crossing the imaginary axis as 0.471. Thus can be seen in the Fig. 7.7, by measuring the length OA on the polar plot.

Problem 7.4. *Given the transfer function of system*

$$G(s) = \frac{1}{s(1+s)^2}$$

sketch the polar plot and find the frequency when $|G(j\omega)| = 1$ and the corresponding phase angle.

Solution: Putting $s = j\omega$, the frequency domain transfer function is

$$G(j\omega) = \frac{1}{(j\omega)(1+j\omega)^2}$$

We have $M(\omega) = \dfrac{1}{\omega(1+\omega^2)}$

$$\phi(\omega) = -90^\circ - \tan^{-1}(\omega)$$

$M(\omega)$ and $\phi(\omega)$ are calculated for the range $0 < \omega < \infty$ and are tabulated in table 7.5.

Table 7.5

ω *(rad/sec)*	$M(\omega)$	$\phi(\omega)$ *(deg)*
0	∞	– 90
∞	0	– 270
0.5	1.6	– 143.13
1	0.5	– 180
10	0.0009	– 258.57

The polar plot is drawn in Fig. 7.8.

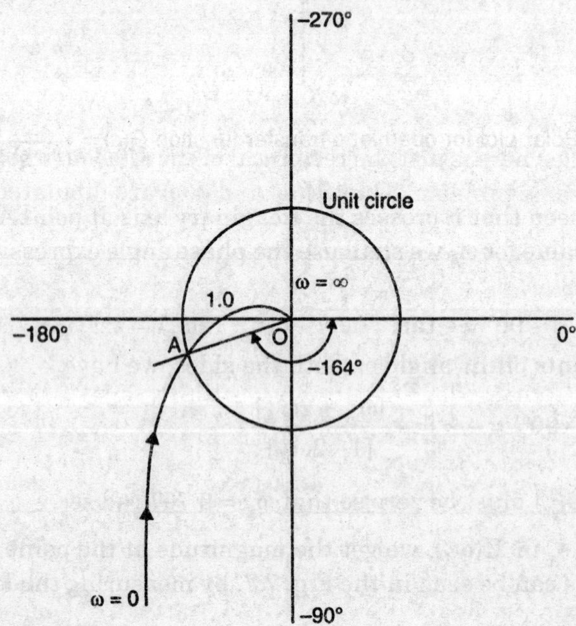

Fig. 7.8 Polar plot for transfer function $G(s) = \dfrac{1}{s(s+1)^2}$.

To obtain the frequency at which $M(\omega)$ is equal to one a unit circle with radius equal to one and centre at origin is drawn on the polar plot. Let the point of intersection of the polar plot with the unit circle be A. Let the frequency at A be ω_A.

The angle which the radius vector OA makes with 0° line (ϕ_A) is measured. It can be analytically determined as follows.

$$M(\omega_A) = 1 = \frac{1}{\omega_A(1+\omega_A^2)}$$

Cross multiplying, we get

$$\omega_A^3 + \omega_A - 1 = 0$$

Solving for ω_A, gives $\omega_A = 0.755$.rad/sec.

And on substitution in $\phi(\omega)$, we get $\phi_A = -164.1°$

These values are seen to be fairly in agreement with the measured value on the polar plot.

7.2 INVERSE POLAR PLOT

These are polar plots made for $G^{-1}(j\omega)$ and the procedure is just the same for $G(j\omega)$. These plots are found to be very useful during the stability studies of the system using Nyquist stability criterion, which will be discussed in the later sections. The following example illustrates the construction of inverse polar plot.

Problem 7.5. *Obtain the inverse polar plot for the transfer function in problem 7.5.*

Solution: For the given transfer function, we have

$$G^{-1}(j\omega) = (j\omega)\,(1 + j\omega)^2$$

for which the magnitude and phase angle expressions are

$$M(\omega) = \omega(1 + \omega)^2$$

$$\phi(\omega) = 90° + 2\tan^{-1}(\omega)$$

So, it seen that $M(\omega)$ here is just the reciprocal of the $M(\omega)$ in problem 7.5 and $\phi(\omega)$ is just negative of $\phi(\omega)$ in the same problem. Thus $M(\omega)$ and $\phi(\omega)$ are tabulated in table 7.6.

Table 7.6

ω *(rad/sec)*	$M(\omega)$	$\phi(\omega)$ *(deg)*
0	∞	90
∞	0	270
0.5	0.625	143.13
1	2	180
10	1010.00	258.57

The polar plot is given in Fig. 7.9.

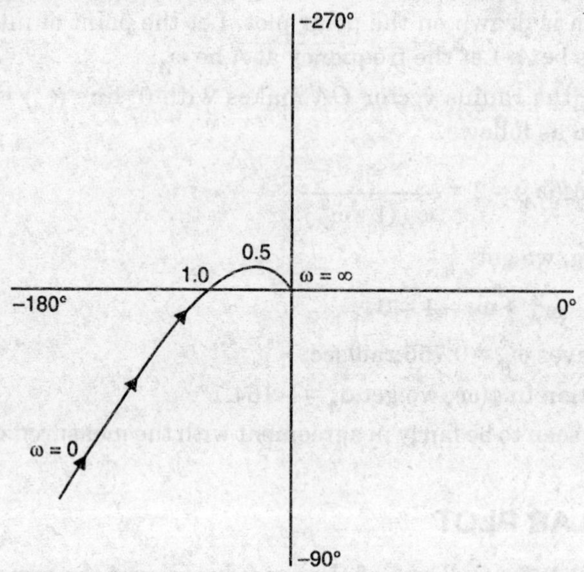

Fig. 7.9 Inverse polar plot for transfer function $G(s) = s(1 + s)^2$

7.3 EFFECT OF ADDING A POLE (OR) ZERO TO A TRANSFER FUNCTION ON ITS POLAR PLOT

Case I: Let us consider the open loop transfer function of system given by

$$G(s) = \frac{1}{(1 + s\tau)} \tag{7.9}$$

Then $$G(j\omega) = \frac{1}{(1 + j\omega\tau)}$$

The polar plot of the transfer function is shown in Fig. 7.10 (a).

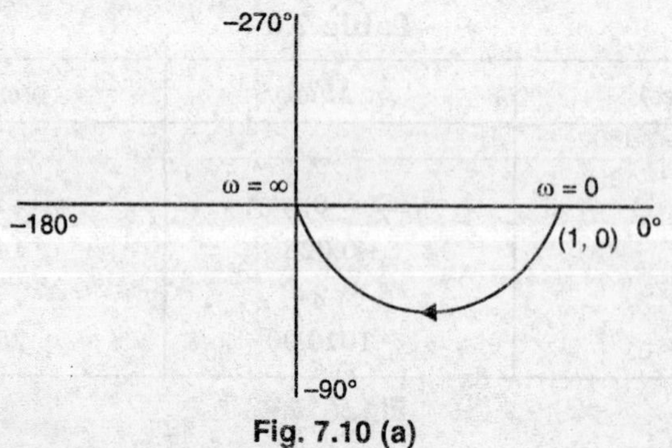

Fig. 7.10 (a)

Case II: A pole is added at the origin and the new transfer function is

$$G(s) = \frac{1}{s(1 + s\tau)} \tag{7.10}$$

The corresponding frequency domain transfer function is

$$G(j\omega) = \frac{1}{(j\omega)(1 + j\omega\tau)}$$

And the polar plot is given in Fig. 7.10 (*b*).

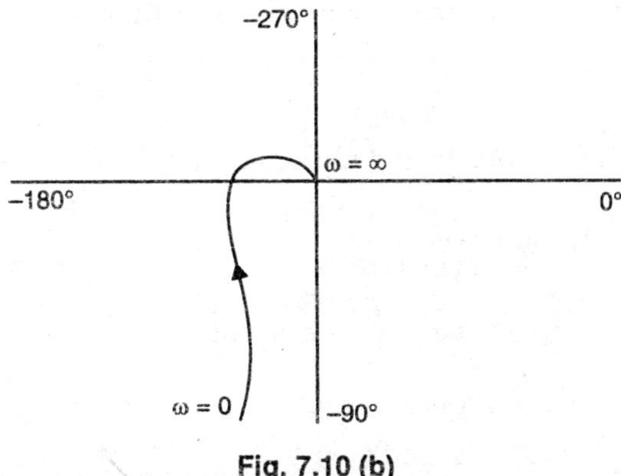

Fig. 7.10 (b)

It is seen from the plot in Fig. 7.10 (*b*) that the plot at ω = 0 and ω = ∞ are shifted by 90° in the clockwise direction.

Case III: Let one more pole at the origin be added to the transfer function in eqn (7.10) so that

$$G(s) = \frac{1}{s^2(1 + s\tau)} \tag{7.11}$$

And the corresponding frequency domain transfer function is

$$G(j\omega) = \frac{1}{(j\omega)^2(1 + j\omega\tau)}$$

For this function the polar plot is shown in Fig. 7.10 (*c*).

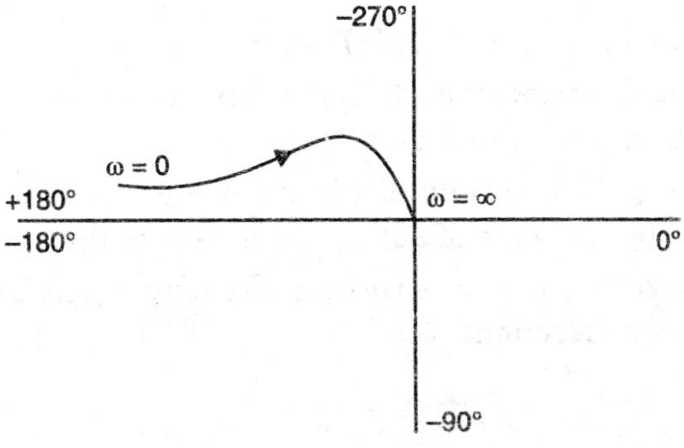

Fig. 7.10 (c)

It is seen the polar plot at ω = 0 and ω = ∞ are shifted by 90° in the clockwise direction. That the angle of shifting is 180° with respect to the transfer function.

Case IV: Now let us consider a zero at $s = -z$ added to the transfer function in eqn. (7.9). The new transfer function is

$$G(s) = \frac{(s + z)}{(1 + s\tau)}$$

And the corresponding frequency domain transfer function is

$$G(j\omega) = \frac{(j\omega + z)}{(1 + j\omega\tau)}$$

For this function the magnitude and phase angle are

$$M(j\omega) = \left| \frac{(z^2 + \omega^2)}{(1 + \omega^2\tau^2)} \right|$$

$$\phi(\omega) = \tan^{-1}\frac{\omega}{z} - \tan^{-1}\omega\tau$$

Then we have $M(0) = z$ and $M(\infty) = 1$ and

$$\phi(0) = 0° \text{ and } \phi(\infty) = 0°$$

The polar plot of the system is shown in Fig. 7.10 (*d*).

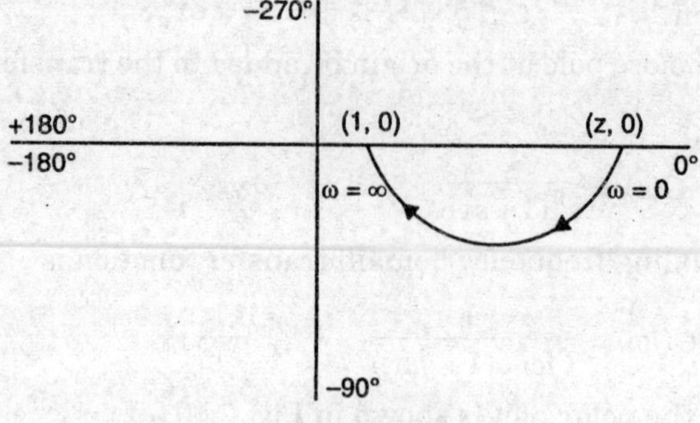

Fig. 7.10 (d)

These results are summarized as follows:

- An addition of a non zero pole to a transfer function rotates the polar plot by an angle of 90° in the clockwise direction as $\omega \to \infty$.
- An addition of a pole at origin to a transfer function rotates the polar plot both at $\omega \to 0$ and $\omega \to \infty$ by an angle of 90° in the clockwise direction.
- The addition of a zero is to rotate the high frequency portion of the polar plot by 90° in the anticlockwise direction.

7.4 NYQUIST PLOTS

For stability, the basic condition was laid down that the roots of characteristic equation must have negative real parts or negative real parts of a complex root $\sigma + j\omega$. The absolute stability of the system can be obtained from its characteristic polynomial by Routh-Hurwitz criterion. The Nyquist criterion employs a different approach for finding the stability of the system. The criterion is useful for finding the stability of the closed loop system from an open loop frequency response without knowing the roots of closed loop system.

Nyquist criterion

Consider the closed loop system shown in Fig. 7.11.

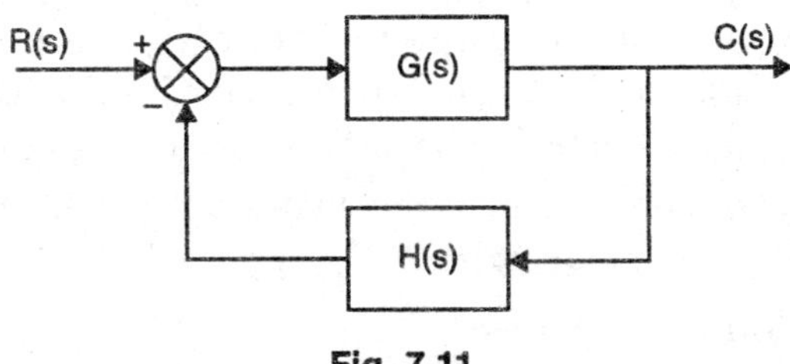

Fig. 7.11

The closed loop transfer function is

$$\frac{C(s)}{R(s)} = \frac{G(s)}{1+G(s)\,H(s)}$$

The denominator $(1 + G(s)\,H(s))$ was made equal to zero and the nature of roots by this characteristic equation determined the absolute stability. The roots were plotted in s-plan and if they were in the left half of the s-plane the system was termed stable.

The basic algebra theorem states that a polynomial of degree n has m zeros and can be expressed as a product of n linear factors. As $G(s)\,H(s)$ is a product, and hence a polynomial, it can be split into its factors.

$$F(s) = \frac{N(s)}{D(s)} = \frac{K\,(s+z_1)\,(s+z_2)\,......\,(s+z_m)}{(s+p_1)(s+p_2)\,......\,(s+p_n)}$$

$N(s)$ = Numerator of the polynomial

$D(s)$ = Denominator of the polynomial

$n \le m$, $\quad -z_1, -z_2, \;......\; -z_m$ are zeros and factors of $N(s)$

$\quad -p_1, p_2, \;.......\; -p_n$ are poles and factors of $D(s)$

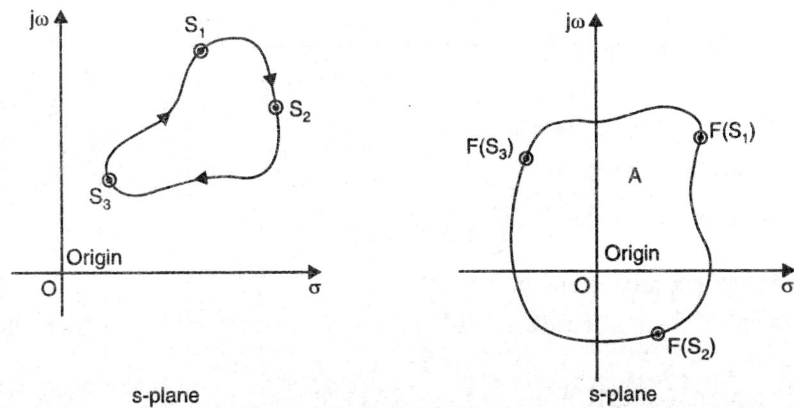

Fig. 7.12 (a)

The pole-zero map was a Cartesian system with σ as real axis and $j\omega$ as imaginary axis. The $x - y$ plane of pole-zero was called the s-plane and the left of s-plane was the absolute stable region.

For each value of s, there will result a value of $F(s)$, which is a complex number. So we have to transform the s-plane to the s-plane. If we join the series values of $s(s_o, s_1$, etc.) we obtain a contour. We transform each of complex values of s_o, s_1 etc. into the s-plane and map them. If we join $F(s_1)$, $F(s_2)$, etc. we obtain another contour. If the s contour is a closed one, $F(s)$ contour is also closed. But $F(s)$ contour will be of a different shape and location.

The contour in the $F(s)$ plane is called the Nyquist plot. The Nyquist curve is obtained by plotting the complex values in polar form.

Nyquist stability criterion is stated as:

If the counter $F(s)$ of the open loop transfer function corresponding to the Nyquist counter in the s-plane encircle the point $(-1 + j0)$ (is called the critical point) in the counter clock wise direction as many number of times as the number of right half s-plane poles of $G(s)H(s)$, then the closed loop system is stable.

7.4.1 Procedure for Construction of Nyquist Plot

Step 1: If $G(s) = A(\omega) + jB(\omega)$ [7.12(*a*)]

In polar form, $F(j\omega) = | G(j\omega) \angle\phi(\omega)$ [(7.12(*b*)]

$$| G(j\omega) | = \sqrt{A^2 + B^2}, \ \phi(j\omega) = -\tan^{-1}\frac{B(\omega)}{A(\omega)} \tag{7.13}$$

Evaluate both magnitude and phase at $\omega \to 0$ and $\omega \to \infty$ and draw the magnitude and phase on same plot.

Step 2: Draw a semi circle as shown in Fig. 7.12 (*b*) which represents 3 paths. Inner semi circle is multiplied with variable '*s*' which represents the path 4.

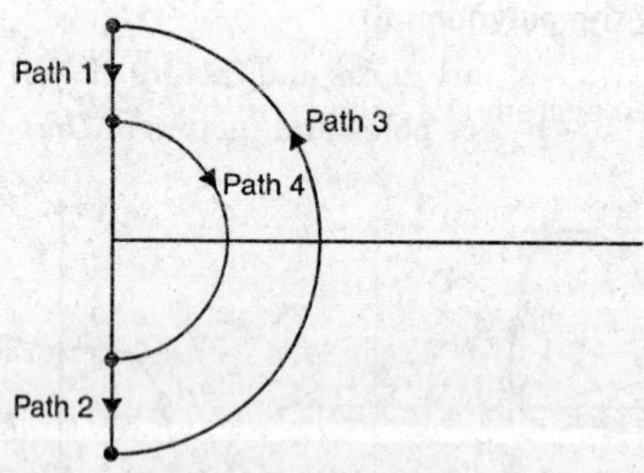

Fig. 7.12 (b)

Step 3: Path 2 is a mirror image of path 1

Step 4: For path 3 substitute $s \to Re^{j\theta}$, $R \to \infty$, $\theta \to -\pi/2$ to $\pi/2$, $1 + Ts = Ts$ in the given transfer function.

Step 5: For path 4 substitute $s \to Re^{j\theta}$, $R \to 0, \theta = \pi/2$ to $-\pi/2$, $1 + sT = 1$ in the given transfer function and draw the plot.

A closed curve begins and ends at the same point. After getting the closed curve (or) plot, the Nyquist plot intersects the real axis at $\omega = \omega_{gc}$ at which the phase is $-180°$. Find the phase crossover frequency ω_{pc} at which the magnitude $|GH(j\omega)|$ is equal to 1.

Step 6: In order to find the stability of closed loop and open loop system, find number of poles that lie on right side of s-plane which is p.

If the plot encircles the point $(-1 + j0)$ the number of encircles equals the number of closed loop poles. If the plot does not encircle the point $(-1 + j0)$, number of encircles equals to zero.

$$N = p - z \tag{7.14}$$

No. of encircles $= N$

If $p = 0$, the open loop system is stable

$z = 0$, the closed loop system is stable

Problem 7.6. *The open loop transfer function of a feedback system is* $G(s)\,H(s) = \dfrac{K(1+s)}{(1-s)}$. *Comment on its stability using Nyquist plot.*

Solution:
$$G(s)\,H(s) = \frac{K(1+s)}{(1-s)}$$

Put $s \to j\omega$ Then
$$G(j\omega)\,H(j\omega) = \frac{K(1+j\omega)}{(1-j\omega)}$$

$$M = \frac{K\sqrt{1+\omega^2}}{\sqrt{1+\omega^2}} = K$$

$$\phi = \tan^{-1}\omega - \tan^{-1}(-\omega) \Rightarrow 2\tan^{-1}\omega$$

As $\omega \to 0$, $\phi = 0°$

$\omega \to \infty$, $\phi = 180°$

$180° - 0° = +180°$; for positive value the direction of plot is in anticlockwise direction.

Nyquist plot for a given system is as shown in Fig. 7.13.

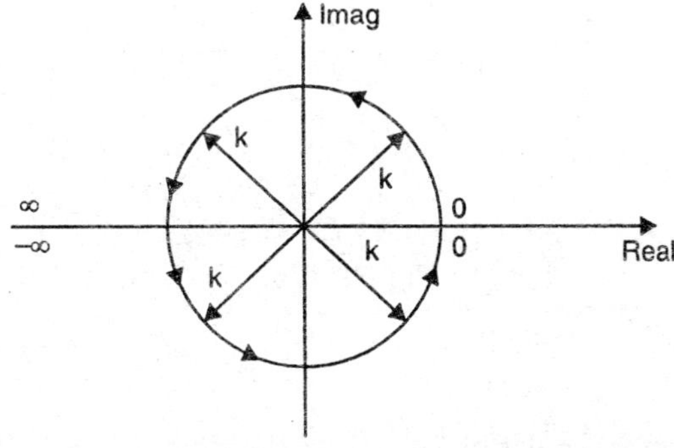

Fig. 7.13

Magnitude M is constant for all values of ω. So the plot rotates in anti clockwise direction with constant magnitude K.

Case (i): For $K > 1$, the plot encircles the point $-1 + j0$.

$\therefore$ No. of encircles N = number of closed loops.

$N = +1$ (+ for anti clockwise direction)

There is a one pole lying on right of s-plane.

$p = 1$ implies that the open loop system is unstable.

$$\therefore \quad p = N - z$$

$$1 = 1 - z$$

$z = 0$ implies that the closed loop system is stable.

Case (ii): For $K < 1$, the plot will not encircle the point $-1 + j0$

So number of encircles $N = 0$

Right side pole $p = 1$

$$p = N - z$$

$$1 = 0 - z \Rightarrow z \neq 0$$

This implies that the closed loop system is unstable.

Problem 7.7. *Draw the Nyquist plot for the open loop system* $G(s) = \dfrac{K(s+3)}{s(s-1)}$ *and find its stability.*

Solution: Put $s \to j\omega$ in the given transfer function $G(s)$

$$G(j\omega) = \frac{K(j\omega + 3)}{j\omega(j\omega - 1)}$$

Path 1: $M = \dfrac{K\sqrt{\omega^2 + 9}}{\omega\sqrt{1 + \omega^2}}$

angle $\phi = \tan^{-1} \omega/3 - 90° - \tan^{-1}(-\omega)$

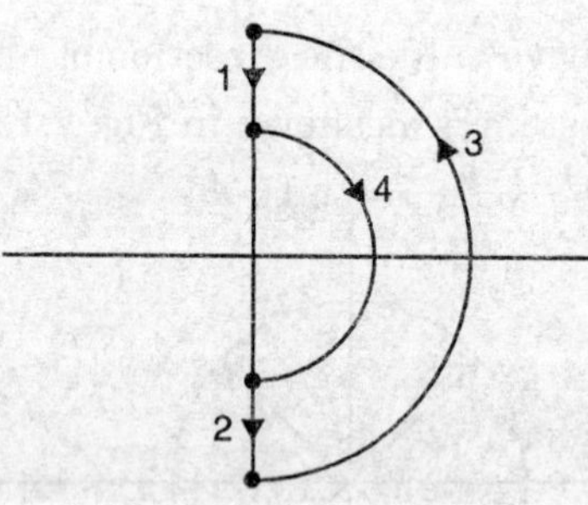

Fig. 7.14 (a)

At $\omega \to 0$, $M = \infty$ $\phi = -90°$

At $\omega \to \infty$, $M = 0$ $\phi = 90°$

Path 2: Is the mirror image of path 1

Path 3: $s \to Re^{j\theta}$, $R \to \infty$, $\theta = -\pi/2$ to $\pi/2$, $1 + sT = T$

$$G(s) = \frac{3K\,(s/3 + 1)}{s(s-1)} = \frac{3K\,s/3}{s(-s)} \Rightarrow \frac{-K}{s}$$

$$= \frac{-K}{\text{Re}^{j\theta}} = \frac{-Ke^{-j\theta}}{\infty} = 0$$

Path 4: $s \to Re^{j\theta}, R \to 0, \theta = \pi/2 \text{ to } -\pi/2, 1 + sT = 1$

$$G(s) = \frac{3K(s/3+1)}{s(s-1)} = \frac{3K \times 1}{s(-1)} = \frac{-3K}{s} = -\infty e^{-j\theta}$$

$$= -\infty\, e^{-j\pi/2} \text{ to } -\infty e^{j\pi/2}$$

Nyquist plot for a given system is shown in Fig. 7.14 (*b*).

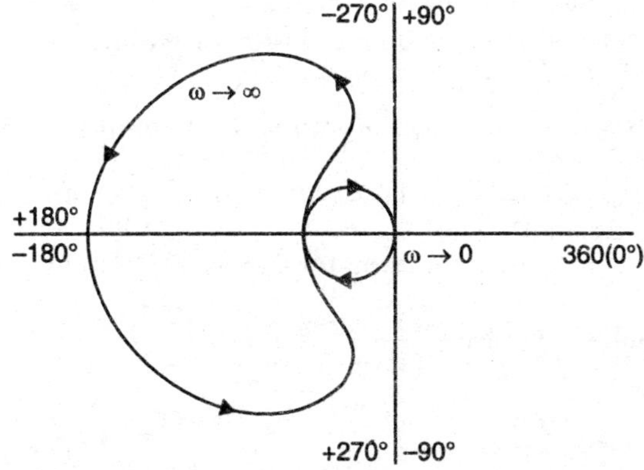

Fig. 7.14 (b)

Phase crossover frequency ω_{pc} at which $\phi = -180°$

$$\tan^{-1}\omega/3 - 90 + \tan^{-1}\omega = -180°$$

$$\tan^{-1}\frac{\left(\frac{\omega}{3}+\omega\right)}{\left(1-\frac{\omega^2}{3}\right)} = -90°$$

$$\frac{\frac{4\omega}{3}}{1-\omega^2/3} = \tan(-90°) = \frac{1}{0}$$

$$1 - \frac{\omega^2}{3} = 0$$

$$\omega^2 = 3 \quad \Rightarrow \quad \omega = \sqrt{3}$$

$\therefore$ $$\omega_{pc} = 1.732$$

Magnitude at $\omega_{pc} = \sqrt{3}$ is

$$M = \frac{K\sqrt{(\sqrt{3})^2+9}}{\sqrt{3}\times\sqrt{4}} \Rightarrow \frac{K\sqrt{12}}{\sqrt{3}\times\sqrt{4}} = K$$

Gain margin $= \dfrac{1}{K}$

For $K > 1$, Nyquist plot encircles the point $-1 + j0$. So number of encircles equals to number of closed loops. Loops are in anti clockwise direction. So

$$N = 2$$

There is one pole which lies on right side of s-plane. So the open loop system is unstable, $P = 1$.

Equation for stability, $p = N - z$

$$1 = 2 - z$$

$$z = 1$$

∴ There is a one zero which lies on right side of s-plane, so the closed loop system is unstable.

Problem 7.8. *A unity feedback control system has an open loop transfer function given by* $G(s)\,H(s) = \dfrac{100}{s(s+5)(s+2)}$. *Draw Nyquist diagram and determine stability.*

Solution: Put $s \to j\omega$ in the given transfer function $G(s)\,H(s)$

Path 1: $G(j\omega)\,H(j\omega) = \dfrac{100}{j\omega(j\omega+5)(j\omega+2)}$

$$\text{Magnitude, } M = \frac{100}{\omega\sqrt{\omega^2+25}\,\sqrt{\omega^2+4}}$$

$$\text{Phase, } \qquad \phi = -90° - \tan^{-1}\omega/5 - \tan^{-1}\omega/2$$

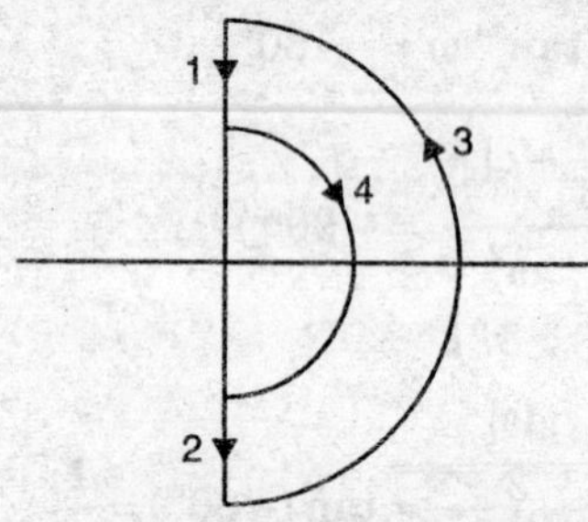

Fig 7.15 (a)

At $\omega \to 0$, $\quad M = \infty \quad \phi = -90°$

$\omega \to \infty$, $\quad M = 0 \quad \phi = -270°$

Angle $-270 + 90 = -180$ is negative so it rotates in clockwise direction.

Path 2: is mirror image of path 1.

Path 3: Put $s \to Re^{j\theta}$, $R \to \infty$, $1 + sT = sT$, $\theta = +\pi/2$ to $-\pi/2$

$$G(s) = \frac{100}{s(s+5)(s+2)}$$

$$G(s) = \frac{100}{s(s/5+1)(s/2+1)} = \frac{100}{s^3/10} = \frac{1000}{s^3} = \frac{1000}{(Re^{j\theta})^3} = 0$$

Path 4: Put $s \rightarrow Re^{i\theta}$, $R \rightarrow 0$, $1 + sT = 1$, $\theta = -\pi/2$ to $+\pi/2$

$$G(s) = \frac{100}{s(s+5)(s+2)} = \frac{100}{10s \times 1 \times 1} = \frac{10}{s} = \frac{10}{Re^{j\theta}}$$

$$= +\infty\, e^{+j\pi/2} \text{ to } +\infty\, e^{-j\pi/2}$$

Nyquist plot for a given system is as shown in Fig. 7.15 (*b*).

Phase crossover frequency at which $\phi = -180°$

$$-90 - \tan^{-1} \omega/5 - \tan^{-1} \omega/2 = -180°$$

$$-\tan^{-1} \frac{(\omega/5 + \omega/2)}{1 - \omega^2/10} = -90°$$

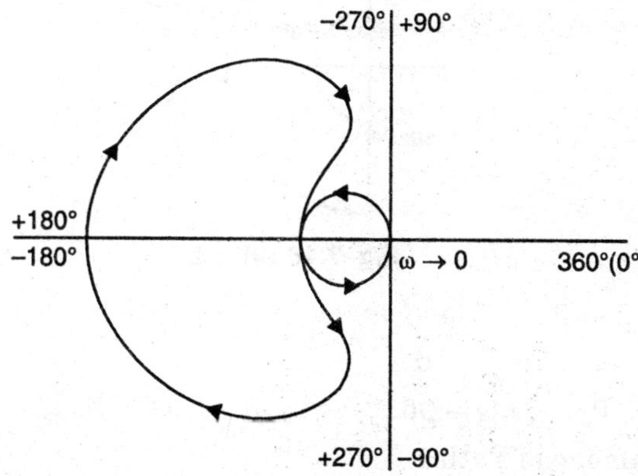

Fig. 7.15 (b)

$$\frac{\omega/5 + \omega/2}{1 - \omega^2/10} = \frac{1}{0} \Rightarrow 1 - \frac{\omega^2}{10} = 0$$

$$\omega_{pc} = \sqrt{10} = 3.16 \text{ rad/sec}$$

$$\text{Magnitude at } \omega_{pc} = \frac{100}{3.16\sqrt{(3.16)^2 + 25}} = \sqrt{(3.16)^2 + 4} = 1.43$$

$$\text{Gain margin} = \frac{1}{1.43} = 0.69$$

0.69 will not encircle the point $(-1 + j0)$

∴ Number of encircles $N = 0$

There is no pole lying on right side of *s*-plane, $p = 0$

∴ Open loop system is stable

Equation for stability $p = N - z$

$$0 = 0 - z \Rightarrow z = 0$$

∴ Closed loop system is stable.

Problem 7.9. *Draw the Nyquist plot for a given transfer function*

$$G(s) = \frac{(s+2)}{(s+1)(s-1)}$$

Path 1. Substitute $s = j\omega$ in a given transfer function $G(s)$

$$G(j\omega) = \frac{j\omega + 2}{(j\omega + 1)(j\omega - 1)}$$

Magnitude $$M = \frac{\sqrt{\omega^2 + 4}}{\sqrt{\omega^2 + 1}\sqrt{\omega^2 - 1}}$$

$$\phi = \tan^{-1} \omega/2 - \tan^{-1} \omega + \tan^{-1} \omega$$
$$= \tan^{-1} \omega/2$$

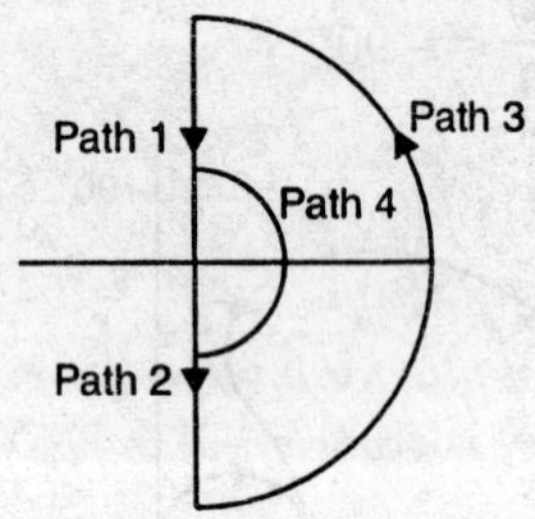

Fig. 7.16 (a)

Evaluate	$\omega \to 0$	$\omega \to \infty$
M	-2	0
ϕ	0	$+90$

Path 2. is mirror image of Path 1

Path 3. Put $s \to Re^{j\theta}$, $R \to \infty$, $\theta \to \pi/2$ to $-\pi/2$ $G(j\omega) = (1 + sT) = sT$

$$= \frac{2\left(\frac{s}{2} + 1\right)}{s(-s)} = \frac{-1}{s} = \frac{-1}{Re^{j\theta}} = 0$$

Path 4. Put $s \to Re^{j\theta}$, $R \to 0$, $\theta = -\pi/2$ to $\pi/2$, $1 + sT = 1$

$$G(s) = \frac{2\left(\frac{s}{2} + 1\right)}{(s + 1)(s - 1)} = -2$$

Magnitude 2, $\theta = \pi/2$ to $-\pi/2$ as shown in Fig. 7.16 (b).

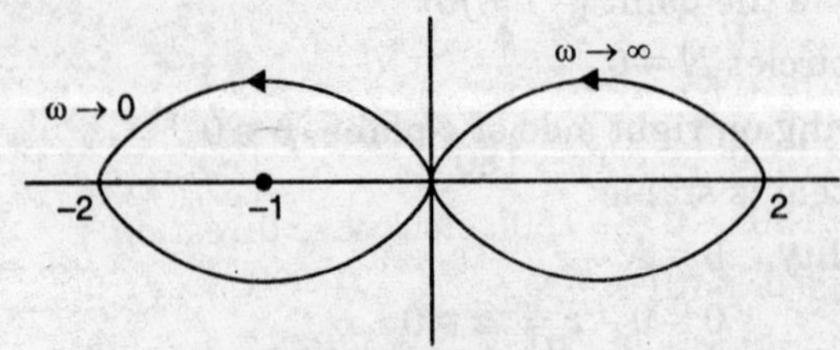

Fig. 7.16 (b)

$$\phi = \tan^{-1} \omega/2 = -180°$$

$$\omega_{pc} = 0$$

$$|G(j\omega)|_{\omega_{pc}} = 2$$

Gain margin = 1/2 = 0.5 does not encircle the point $-1 + j0$ so the nunber of encircle is equal to zero.

$$N = 0$$

p = Number of poles that lie on right side of s-plane is 1

$\therefore \quad N = p - z$

$0 = 1 - z \quad \Rightarrow \quad z = 1 \qquad z \neq 0$

One zero lies on right side of s-plane so the closed loop system is unstable.

$p \neq 0$, so the open loop system is unstable.

Problem 7.10. *Draw the Nyquist plot for a given open loop transfer function and test the stability. Find Gain margin and phase crossover frequency.*

$$G(s) = \frac{1}{(s+1)(s+2)}$$

Path 1. Substitute $s \to j\omega$ in the given open loop transfer function

$$G(j\omega) = \frac{1}{(j\omega+1)(j\omega+2)} \qquad M = \frac{1}{\sqrt{\omega^2+1}\sqrt{\omega^2+4}}$$

Phase, $\phi = -\tan^{-1}\omega - \tan^{-1}\omega/2$

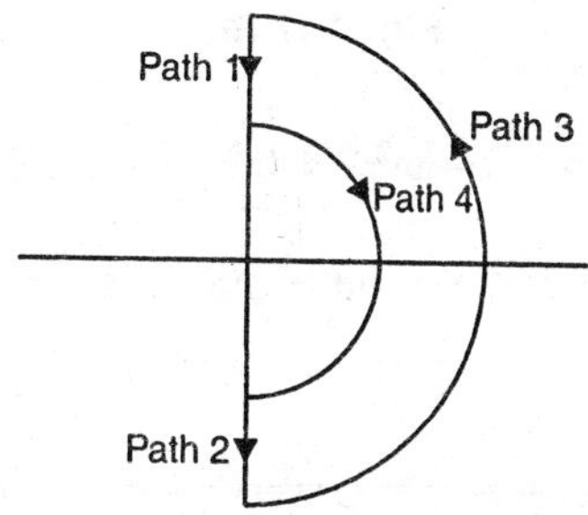

Fig. 7.17 (a)

Evaluate	M	ϕ
$\omega \to 0$,	-0.5	$0°$
$\omega \to \infty$,	0	$-180°$

Direction of rotation $-180° - 0 = -180°$, clockwise direction.

Path 2. is mirror image of path 1

Path 3. $s \to Re^{j\theta}$, $\theta \to -\pi/2$ to $\pi/2$, $\quad R \to \infty$, $1 + sT = sT$

$$G(s) = \frac{1}{2(s+1)(s/2+1)} \quad \Rightarrow \quad \frac{1}{2 \times s \times s/2} = \frac{1}{s^2}$$

$$G(s) = \left.\frac{1}{R^2 e^{j2\theta}}\right|_{R=\infty} = 0$$

Path 4. $s \to Re^{i\theta}, R \to 0, \theta = -\dfrac{\pi}{2}$ to $\dfrac{\pi}{2}$, $1 + ST = 1$

$$G(s) = \frac{1}{2} = 0.5$$

magnitude 0.5, $\theta = -\dfrac{\pi}{2}$ to $\dfrac{\pi}{2}$ as shown in Fig. 7.17 (*b*).

Nyquist plot for a given system is as shown in Fig. 7.17 (*b*).

Phase crossover frequency at which $\phi = -180°$

$$-\tan^{-1}\omega - \tan^{-1}\omega/2 = -180°$$

$$\tan^{-1}\frac{(\omega + \omega/2)}{1-\omega^2/2} = 180°$$

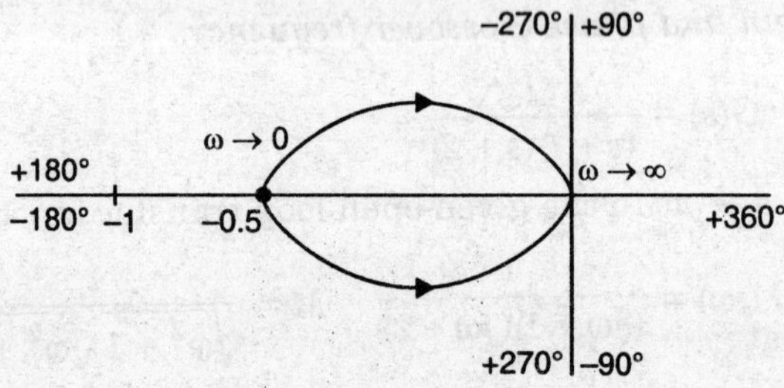

Fig. 7.17 (b)

$$\frac{3\omega/2}{1-\omega^2/2} = 0 \Rightarrow 1 - \omega^2/2 = 0$$

$$\omega^2 = \sqrt{2}$$

$$\omega_{pc} = 1.414$$

$$M = \frac{1}{\sqrt{(\sqrt{2})^2 + 1}\,\sqrt{(\sqrt{2})^2 + 4}} \Rightarrow \frac{1}{\sqrt{3} \times \sqrt{6}} = 0.235$$

Gain margin = 1/0.235 = 4.25

Point 4.25 encircles the point $-1 + j0$ so the number of encircles is equal to number of closed loops. Clockwise loops are taken as negative, anti clockwise loops are taken as positive.

$$N = -1$$

Number of poles lie on right side of s-plane $p = 0$ for stability

$$N = p - z$$

$$-1 = 0 - z \Rightarrow z = 1 \Rightarrow z \neq 0$$

$p = 0$ so the open loop system is stable

$z \neq 0$ so the closed loop system is unstable.

Problem 7.11. $G(s) = \dfrac{1}{s(s+1)(s+2)}$. *Draw the Nyquist plot and test the stability. Find* ω_{pc}, ω_{gc}, *PM and GM.*

Solution: Put $s \to j\omega$ in $G(s)$

$$G(j\omega) = \frac{1}{j\omega(j\omega+1)(j\omega+2)}$$

$$\text{Magnitude } M = \frac{1}{\omega\sqrt{\omega^2+1}\,\sqrt{\omega^2+4}}$$

$$\text{Phase } \phi = -\tan^{-1}\omega - \tan^{-1}\omega/2 - 90°$$

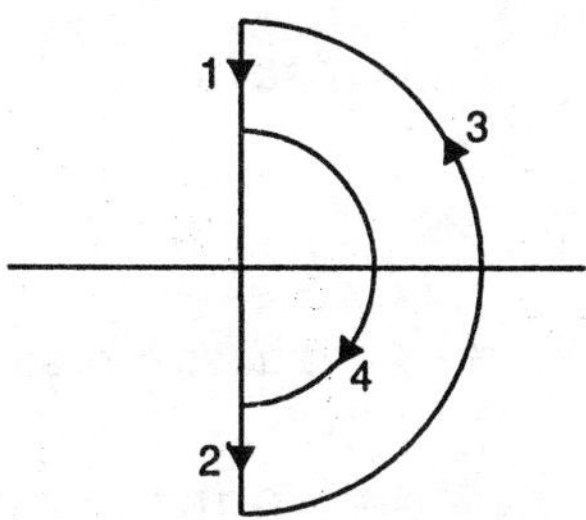

Fig. 7.18 (a)

Evaluate M and ϕ at

	M	ϕ
$\omega \to 0,$	∞	$-90°$
$\omega \to \infty,$	0	$-270°$

Direction of Rotation = $-270° - (-90°) = -180°$ negative, so clockwise direction.

Path 2. is mirror image of Path 1

Path 3. $s \to Re^{j\theta}$, $R \to \infty$, $\theta \to \pi/2$ to $-\pi/2$, $1 + sT = sT$

$$G(s) = \frac{1}{2s \times s \times s/2} = 1/s^3 = \left.\frac{1}{R^3 e^{j3\theta}}\right|_{R\to\infty} = 0$$

Path 4. $s \to Re^{j\theta}$, $R \to 0$, $\theta = -\pi/2$ to $\pi/2$, $1 + sT = 1$

$$G(s)\ \frac{1}{2s \times 1 \times 1} = 1/2s = \frac{1}{2Re^{j\theta}} = +\infty\ \angle\pi/2 \text{ to } -\pi/2$$

Path 4 is due to the presence of multiplication of s in denominator. The plot is closed towards right because $G(Re^{j\theta})$ in path 4 is $+\infty$ and the direction is from $\pi/2$ to $-\pi/2$.

Nyquist plot for a given system is as shown in Fig. 7.18 (*b*).

Phase crossover frequency at which phase angle $\phi = -180°$

$$-\tan^{-1}\omega - \tan^{-1}\omega/2 - 90° = -180°$$

$$-\tan^{-1}\frac{(\omega+\omega/2)}{1-\omega^2/2} = -90°$$

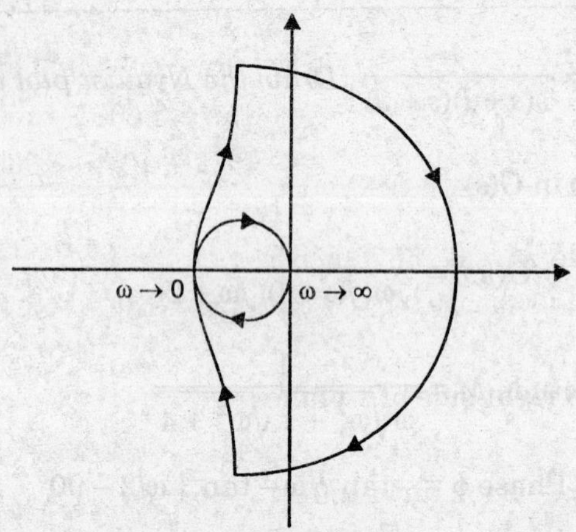

Fig. 7.18 (b)

$$\frac{3\omega/2}{1-\omega^2/2}=\frac{1}{0} \quad \Rightarrow \quad 1-\omega^2/2=0$$

$$\omega = \omega_{pc} = \sqrt{2} = \text{phase crossover frequency.}$$

Magnitude $$M=\frac{1}{\sqrt{2}\,\sqrt{3}\,\sqrt{6}}=0.166$$

Gain margin = 1/0.166 = 6

Gain crossover frequency at which magnitude

$$M=\frac{1}{\omega\sqrt{\omega^2+1}\,\sqrt{\omega^2+4}}=1$$

Squaring on both sides and cross multiplying

$$\omega^2(\omega^2+1)(\omega^2+4)=1$$

$$\omega^2(\omega^4+5\omega^2+4)-1=0$$

$$\omega^6+5\omega^4+4\omega^2-1=0$$

$$x^3+5x^2+4x-1=0 \quad (\text{Since } \omega^2=x)$$

x^3	1	4
x^2	5	– 1
x	$\frac{21}{5}$	0
x^0	– 1	

$$5x^2-1=0$$

$$x^2=1/5$$

$$x=1/\sqrt{5}=0.447$$

$$\omega=\sqrt{x}=\omega_{gc}=\sqrt{0.447}=0.668 \text{ rad/sec}$$

gain crossover frequency, ω_{gc} = 0.668 rad/sec.

$$\phi = -\tan^{-1}\omega - \tan^{-1}\omega/2 - 90^\circ$$
$$= -\tan^{-1} 0.668 - \tan^{-1} 0.668/2 - 90^\circ$$
$$= -33.7 - 18.46 - 90^\circ = -142.16^\circ$$

Phase margin $= 180^\circ + \phi = 180^\circ - 142.16^\circ = 37.84^\circ$

The point magnitude $M = 0.1667$ encircles the point $(-1 + j0)$ as number of encircles equals to number of closed loops. So here we have two closed loops in clockwise direction.

No. of encircles $= -2$

No. of poles lying on right side of s-plane $p = 0$

$\therefore \quad N = p - z$

$-2 = 0 - z \quad \Rightarrow \quad z = 2$ (or) $z \neq 0$

There are 2 zeros which lie on right side of s-plane, so the closed loop system is unstable.

$p = 0$, there is no pole which lies on right side of s-plane, so the open loop system is stable.

Problem 7.12. *Draw the Nyquist plot for a given open loop transfer function*

$$G(s) = \frac{(s+2)}{s(s+1)(s+4)}$$

Solution: We will get four paths for s variable in denominator put $s \to j\omega$

Path 1. Evaluate the magnitude and phase at $\omega \to 0$ and $\omega \to \infty$.

$$G(j\omega) = \frac{(j\omega + 2)}{j\omega(j\omega + 1)(j\omega + 4)}$$

$$M = \frac{\sqrt{\omega^2 + 4}}{\omega\sqrt{\omega^2 + 1}\sqrt{\omega^2 + 16}}$$

$$\phi = \tan^{-1}\omega/2 - 90^\circ - \tan^{-1}\omega - \tan^{-1}\omega/4$$

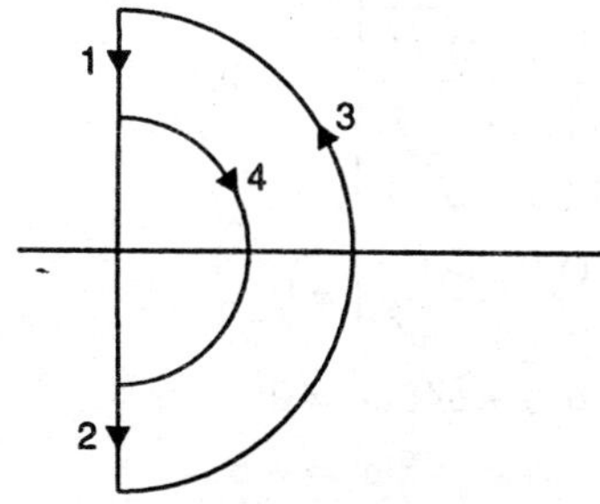

Fig. 7.19 (a)

	M	ϕ
$\omega \to 0$,	∞	-90°
$\omega \to \infty$,	0	-180°

Direction of rotation $-180^\circ - (-90^\circ) = -90^\circ$ which is clockwise.

Path 2. is mirror image of Path 1

Path 3. $s \to Re^{j\theta}, R \to \infty, 1 + sT = Ts\ \theta \to -\pi/2$ to $\pi/2$

$$G(s) = \frac{2s/2}{s \times s \times s/4} = \frac{1}{s^2} = \left.\frac{1}{R^2 e^{j2\theta}}\right|_{R \to \infty} = 0$$

Path 4. $s \to Re^{j\theta}, R \to 0, 1 + sT = 1, \theta = -\pi/2$ to $+\pi/2$

$$G(s) = \frac{2 \times 1}{s \times 1 \times 4} = \left.\frac{1}{2Re^{j\theta}}\right|_{R \to \infty}$$

$$= \infty\, e^{-j\pi/2} \quad \text{to } e^{+j\pi/2}$$

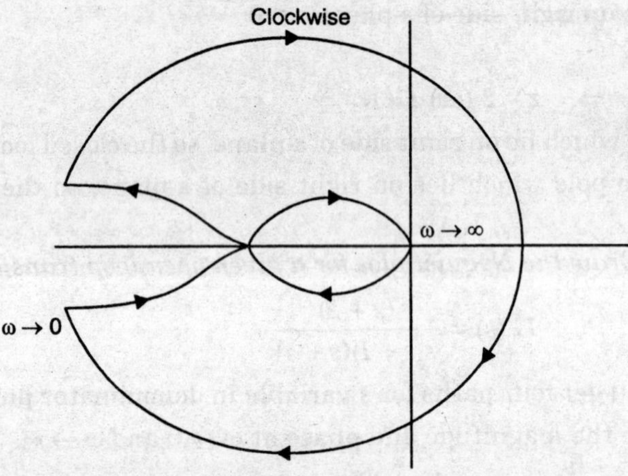

Fig. 7.19 (b)

Phase crossover frequency ω_{pc} at which phase angle $\phi = -180°$

$$\tan^{-1} \omega/2 - (\tan^{-1} \omega + \tan^{-1} \omega/4) = -180°$$

$$\tan^{-1} \omega/2 - \left(\tan^{-1} \frac{\omega + \omega/4}{1 - \omega^2/4}\right) = -90°$$

$$\tan^{-1} \frac{\omega/2 - \left(\dfrac{\omega + \omega/4}{1 - \omega^2/4}\right)}{1 + \dfrac{\omega}{2}\left(\dfrac{\omega + \omega/4}{1 - \omega^2/4}\right)} = -90°$$

$$1 + \frac{\omega}{2}\left(\frac{5\omega}{4 - \omega^2}\right) = 0$$

$$2(4 - \omega^2) + \omega(5\omega) = 0$$

$$3\omega^2 = -8$$

$$\omega_{pc} = \sqrt{\frac{-8}{3}} = \pm j\ 1.63 \text{ rad/sec}$$

$$M = \frac{\sqrt{(1.63)^2 + 4}}{1.63\sqrt{(1.63)^2 + 1}\ \sqrt{(1.63)^2 + 16}} = 0.1926$$

$$\text{Gain margin} = \frac{1}{M} = 5.22$$

7.5 M CIRCLE AND N CIRCLE

So far we have analyzed system stability from open loop frequency response plot, $1 + G(j\omega) H(j\omega)$. From the same plots we can determine the closed loop frequency response also. The transfer function of a closed loop system with unity feedback

$$G(j\omega) = \frac{G(j\omega)}{1+G(j\omega)} = \frac{A+jB}{1+A+jB}$$

Magnitude $$M(j\omega) = \left|\frac{G(j\omega)}{1+G(j\omega)}\right| = \frac{\sqrt{A^2+B^2}}{\sqrt{(1+A)^2+B^2}}$$

Phase $$\phi(j\omega) = \tan^{-1}\frac{\operatorname{Im} G(j\omega)}{\operatorname{Re} G(j\omega)}$$

$$= \angle \frac{A+jB}{1+A+jB}$$

Omitting $j\omega$,

$$M^2 [(1+A)^2 + B^2] = A^2 + B^2$$

Rearranging,

$$A^2 (1 - M^2) + B^2 (1 - M^2) - 2M^2 A = M^2 \quad (7.15)$$

If $M = 1$, $A = -\frac{1}{2}$, $B = 0$

This is the equation of a straight line parallel to y-axis passing through $\left(-\frac{1}{2}, j0\right)$

If $M \neq 1$, we will obtain a series of circles for different values of M.

Dividing by $1 - M^2$ of eqn (7.15) and adding $[M^2/M^2 - 1)^2]^2$ on both sides

$$A^2 + B^2 - \frac{2M^2}{1-M^2} A + \left[\frac{M^2}{(1-M^2)^2}\right]^2 = \frac{M^2}{1-M^2} + \left(\frac{M^2}{1-M^2}\right)^2$$

Simplifying,

$$\left(A - \frac{M^2}{1-M^2}\right)^2 + B^2 = \left(\frac{M}{1-M^2}\right)^2. \quad (7.16)$$

Equation (7.16) is the equation of a circle, $x^2 + y^2 = a^2$.

Draw a circle where the coordinates of the centre are

$$\frac{M^2}{1-M^2}, 0 \quad (7.17)$$

The radius of the circle is

$$r = \left|\frac{M}{1-M^2}\right| \quad (7.18)$$

So we will obtain a family of circles for different values of M in the $G(j\omega)$ plane. They are called constant M circles shown in Fig. 7.20.

(1) $M < 1$, as M becomes smaller and smaller, the radius becomes smaller and smaller and converges to the origin.

$M = 0.7 \qquad r = 1.37$

$M = 0.3 \qquad r = 0.33$

These circles lie to the right of $M = 0$ locus.

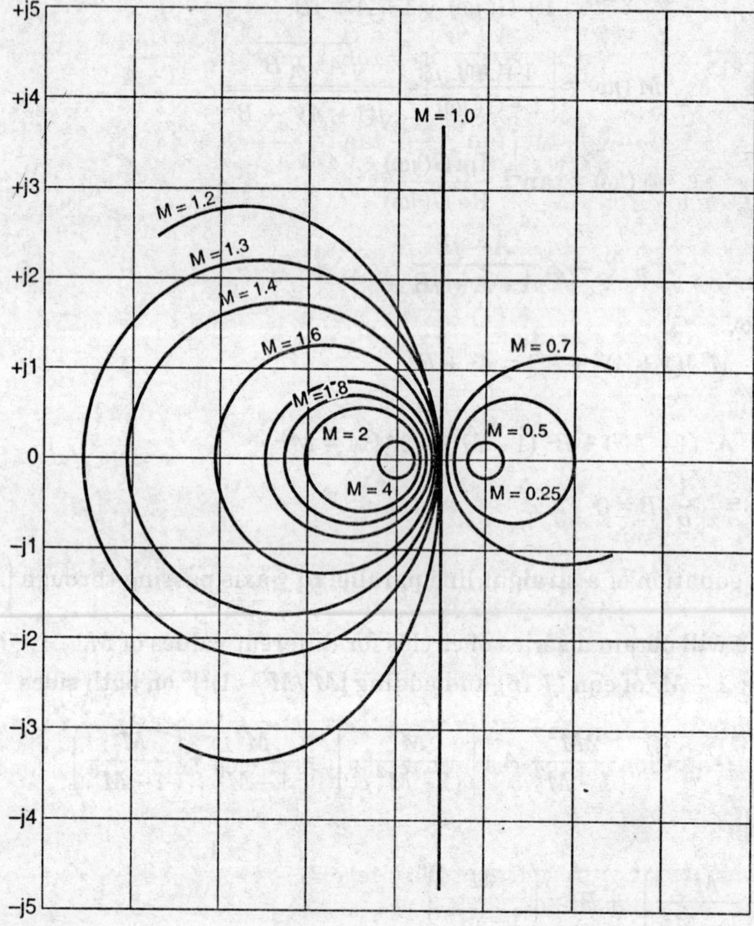

Fig. 7.20 M circles.

(2) $M = 1$, the radius becomes infinite. It is the line parallel to the y-axis at $\left(\frac{1}{2}, j0\right)$

(3) $M > 1$, the radius gets smaller and smaller from infinity and converges at $(-1, j0)$ points as M becomes higher and higher.

$M = 2, \qquad r = 0.67$

$M = 4, \qquad r = 0.27$

$M = 6, \qquad r = 0.17$

These circles lie to the left of the line $\left(-\frac{1}{2}, 0\right)$.

The centre of all the circles lie on the real $G(j\omega)$ axis.

The resonance peak M_p is given by the largest value of M of the M circle tangent to the polar plot or tangent to the smallest circle. The resonant frequency is at the point of tangency.

Constant phase angle loci (N circles)

Phase angle $$\phi = \angle \frac{A + jB}{1 + A + jB} = \tan^{-1}\frac{B}{A} - \tan^{-1}\frac{B}{1+A}$$

$$\tan\phi = \tan\left[\tan^{-1}\frac{B}{A} - \tan^{-1}\frac{B}{1+A}\right]$$

Applying the identity

$$\tan(A - B) = \frac{\tan A - \tan B}{1 + \tan A \tan B}$$

$$N = \tan\phi = \frac{\frac{B}{A} - \frac{B}{A+B}}{1 + \frac{B}{A}\left(\frac{B}{A+B}\right)} = \frac{B}{A^2 + A + B^2}$$

or $$A^2 + A + B^2 - \frac{Y}{N} = 0$$

Adding the term $\left(\frac{1}{4} + \frac{1}{4}N^2\right)$ to both sides the above equation becomes

$$\left(A + \frac{1}{2}\right)^2 + \left(B - \frac{1}{2N}\right)^2 = \frac{1}{4} + \frac{1}{4N^2} \quad (7.19)$$

This is the equation of the circle and a family of circles can be drawn for different values of N.

Drawing a N circle with centre of the circle $= \left(-\frac{1}{2}, \frac{1}{2N}\right)$ (7.20)

Radius of the circle $= \left(\frac{N^2 + 1}{4N^2}\right)^{1/2}$ (7.21)

This is as shown in Fig. 7.21.

If $\phi = 45°$, then $\tan\phi = 1$, centre and radius of the circle are $(-0.5, -0.5)$, 0.707.

Equation (7.19) is satisfied when $A = 0, B = 0$ and $A = -1, B = 0$. Regardless of the value of N each circle passes through the origin and $-1 + j0$ point.

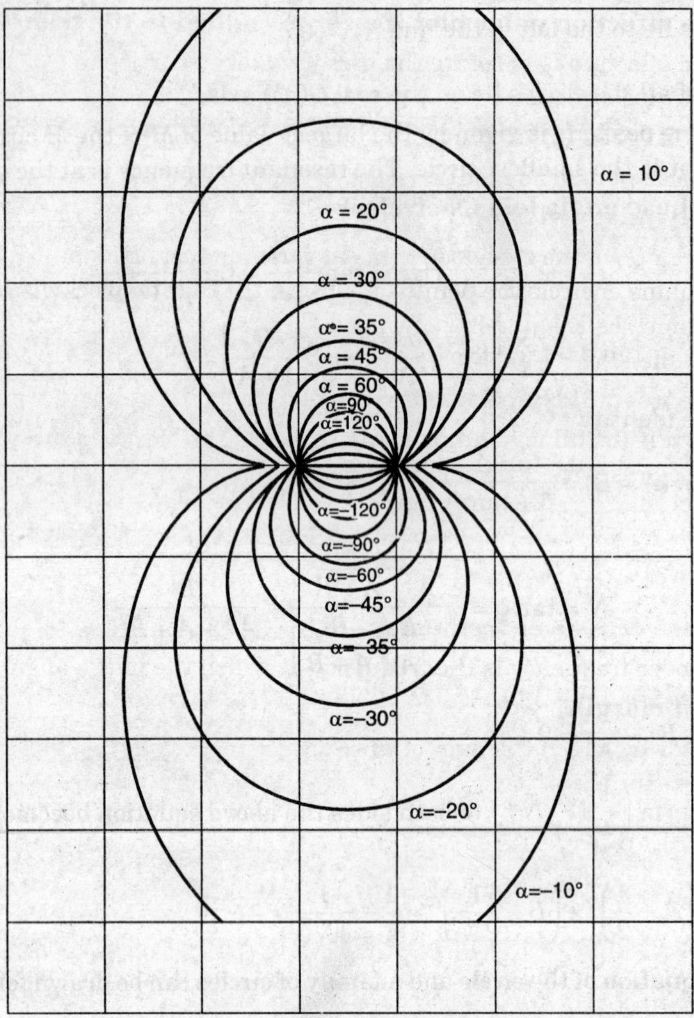

Fig. 7.21 N circles.

SHORT QUESTIONS AND ANSWERS

1. **What is polar plot?**

   The polar plot is the locus of the system transfer function in the frequency domain for $0 < \omega < \infty$.

2. **What is the effect on polar plot if a non-zero pole is added to the transfer function?**

   Addition of a non zero pole to the transfer function results in further rotation of the end points of polar plot through an angle of 90º.

3. **What is the effect on polar plot if a pole is added to the transfer function?**

   When a pole at the origin is added to the transfer function, it rotates the entire polar plot by a further angle of 90º.

4. **What is the effect on polar plot if a zero is added to the transfer function?**

   The addition of a zero is to rotate the high frequency portion of the polar plot by 90° in the anticlockwise direction.

5. **Define the inverse polar plot.**

   These are polar plots made for $G^{-1}(j\omega)$ and the procedure is just the same for $G(j\omega)$.

6. **State the Nyquist criterion.**

   If the counter $F(s)$ of the open loop transfer function corresponding to the Nyquist counter in the s-plane encircle the point$(-1, +j0)$ in the counter clockwise direction as many number of times as the number of right half s-plane poles of $G(s)H(s)$, the closed loop system is stable.

7. **What is the need of Nyquist criterion?**

   This criterion is useful for finding the stability of the closed loop system from an open loop frequency response without knowing the roots of closed loop system.

8. **Define gain crossover frequency.**

   Gain crossover frequency is the frequency at which the magnitude crosses unity (or) one corresponding to 0 dB.

9. **Define phase crossover frequency.**

   Phase crossover frequency is the frequency at which the phase angle is 180°.

10. **Define gain margin.**

    It is the factor by which the gain of the system can be increased to drive the system to the verge of instability.

11. **Define phase margin.**

    It is the additional phase lag at the gain cross over frequency required to bring the system to the verge of instability.

12. **What is the value of gain and phase margins for a good relative stability?**

    For a good relative stability the gain margin is about 6 dB and a phase margin is about 30° to 60°.

13. **What is Nichols chart?**

    The transformation of the constant M and N circles to log magnitude and phase angle coordinates is known as Nichols chart.

14. **What is constant M circle for M = 1?**

    A straight line parallel to the y-axis at $x = -\frac{1}{2}$.

OBJECTIVE TYPE QUESTIONS

1. One of the advantage of polar plots is

   (*a*) It will extend the frequency range

   (*b*) Will give the separate magnitude and phase plots

(*c*) Will give the magnitude and phase on single plot

(*d*) Will not specify the gain margin and phase margin.

2. Nyquist plots will give the stability of

(*a*) absolute stability (*b*) conditional stability

(*c*) marginal stability (*d*) relative stability.

3. In Nyquist plot stability equation is

(*a*) $N = p - z$ (*b*) $N = z$

(*c*) $N = p$ (*d*) $N = z$.

4. In $N = p - z$, p represents

(*a*) No. of pole

(*b*) No.of poles on left side of s-plane

(*c*) No.of poles right side of s-plane

(*d*) No.of poles equal to zero.

5. If the intersection point on real axis encircles the point $-1 + j0$ number point of encircles equals to

(*a*) Zero (*b*) One

(*c*) Infinity (*d*) Number of closed loops.

6. If the intersection point on real axis will not encircles the point $-1 + j0$ number of encircles equals to

(*a*) Zero (*b*) One

(*c*) Infinity (*d*) No. of closed loops.

7. In Nyquist criteria, open loop system is stable when

(*a*) $N = p - z$ (*b*) $p = 0$

(*c*) $z = 0$ (*d*) $p = z$.

8. In Nyquist criterion, closed loop system is stable when

(*a*) $N = p - z$ (*b*) $p = 0$

(*c*) $z = 0$ (*d*) $p = z$.

9. In a polar plot, the curve was found to cross the negative real axis at – 1.2. Then

(*a*) The gain margin is 1.2 and the system is stable

(*b*) The gain margin is 0.833 and the system is stable

(*c*) The gain margin is 1.2 and the system is unstable

(*d*) The gain margin is 0.83 and the system is unstable.

10. The polar plot of a closed-loop system with a transfer function $\dfrac{G}{1+GH}$ is drawn for

(*a*) G (*b*) $1 + GH$

(*c*) GH (*d*) $\dfrac{G}{1+GH}$

11. The polar plot of a transfer function is shown alongside. It can be stated that

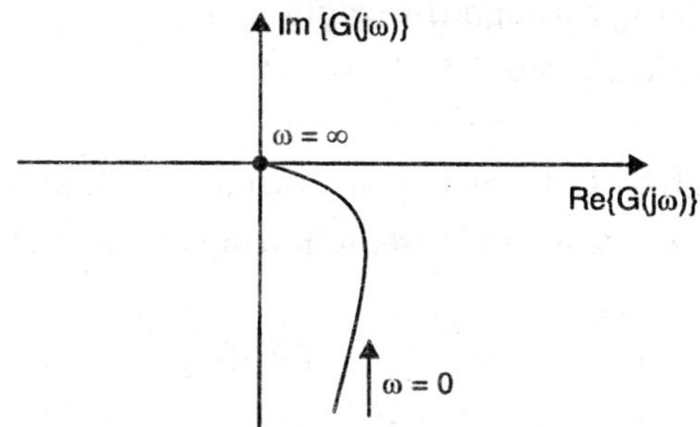

(*a*) The finite zero is closer to the origin than the finite pole

(*b*) The finite pole is closer to the origin than the zero is

(*c*) The system is unstable

(*d*) The system is non-minimum phase.

12. The polar plot is drawn

(*a*) Decibel versus ω (*b*) Decibel versus log ω

(*c*) Decibel versus phase angle

(*d*) Magnitude and phase incorporated in the *x*-*y* plane.

13. For a transfer function with pure transportation lag, the polar plot is

(*a*) a semi circle with centre at (– 1, 0) and radius 1 in the clockwise direction

(*b*) a semi circle with centre at (– 1, 0) and radius 1 in the anti-clockwise direction

(*c*) a circle with origin as the centre and radius 1

(*d*) Polar plot does not exist.

14. By substituting $s = j\omega$, the frequency response plot gives

(*a*) Transient response of the system (*b*) Steady state response of the system

(*c*) Initially transient and then steady state response

(*d*) None of the above.

15. The polar plot of a system with transfer function $G(s) = \dfrac{K}{s(s+T)}$ for positive T and negative K will be

(*a*) in the first quadrant (*b*) in the second quadrant

(*c*) in the third quadrant (*d*) in the fourth quadrant.

16. Which of the following transfer function is a non-minimum phase transfer function?

(*a*) $\dfrac{(s+1)}{(s+2)(s+3)(s+4)}$ (*b*) $\dfrac{(s+1)(s+2)}{s(s+3)(s+4)}$

(*c*) $\dfrac{10(s+1)}{s^2(s+3)}$ (*d*) $\dfrac{(s-1)}{(s+2)(s+3)}$

17. In the $G^{-1}(j\omega)$ plane, the constant phase angle loci are

(*a*) straight lines passing through the origin

(*b*) straight lines passing through (– 1, 0) point

(*c*) straight lines passing through (1, 0) point

(*d*) some pass through (– 1, 0), some pass through (1, 0) and some pass through origin.

18. The inverse polar plot is the plot of the following sinusoidal transfer function

(*a*) $G(1/j\omega)$ (*b*) $\dfrac{1}{G(j\omega)}$

(*c*) $\dfrac{1}{G(1/j\omega)}$ (*d*) None of these.

19. A stable feedback control system has open-loop transfer function with a pole at RHP and zero at RHP. The corresponding Nyquist plot will

(*a*) encircle (– 1, *j*0) point in counter clockwise direction once

(*b*) have anti-clockwise encirclement of the (– 1, *j*0) point once

(*c*) Encircle (–1, *j*0) point at many times in the counter clockwise direction as the number of LHP poles of the closed-loop transfer function

(*d*) Not encircle (– 1, *j*0) point at all.

20. Nyquist plot can be used

(*a*) Only to find the closed-loop poles in the right half plane

(*b*) Ascertain the stability only

(*c*) To find the open-loop poles in the right half plane

(*d*) To find the number of closed-loop poles in the left half plane

21. Nyquist plot of a system is shown alongside. What is the type of the system?

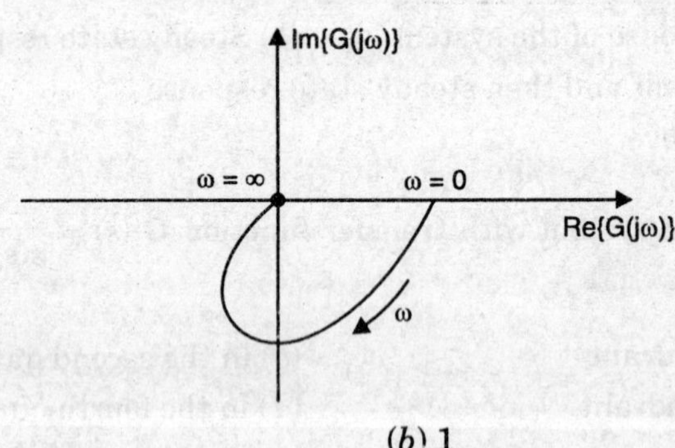

(*a*) 0 (*b*) 1

(*c*) 2 (*d*) 3.

22. Nyquist plot of a system is shown below in the diagram. The system is

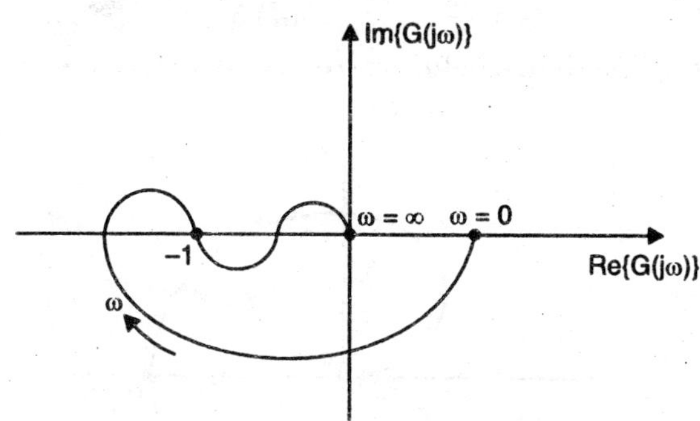

(*a*) Marginally stable (*b*) Conditionally stable
(*c*) Stable (*d*) Unstable.

23. The unit circle of the Nyquist plot transforms into 0 dB. Line of the amplitude plot of the Bode diagram is

(*a*) At zero frequency (*b*) At low frequency
(*c*) At high frequency (*d*) At any frequency .

24. A unit feedback system has an openloop transfer function

$G(j\omega) = \dfrac{K}{j\omega(j0.2\omega + 1)(j0.05\omega + 1)}$. The phase crossover frequency of the Nyquist plot is given by

(*a*) 1 rad/sec (*b*) 100 rad/sec
(*c*) 10 rad/sec (*d*) 0.1 rad/sec.

25. If the Nyquist plot of a certain feedback system crosses the negative real axis of – 0.1 point, the gain margin of the system is given by

(*a*) 0.1 (*b*) 10
(*c*) 100 (*d*) 200.

26. The Nyquist plot (for positive frequencies) of the open-loop transfer function is shown in figure below. The gain margin is

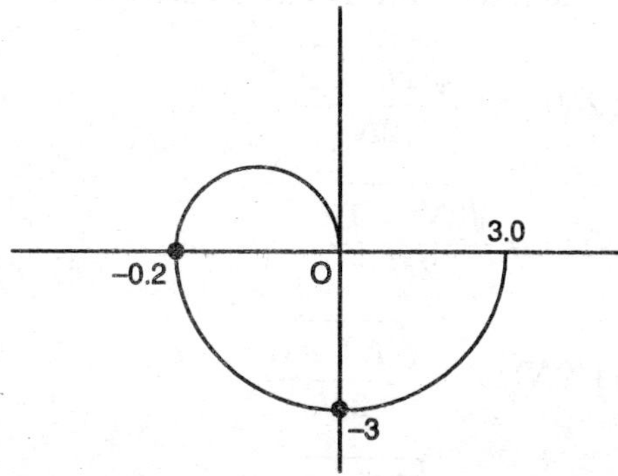

(*a*) 2 (*b*) 3

(*c*) 4 (*d*) 5.

27. The Nyquist plot in the figure below corresponds to the system whose transfer function is

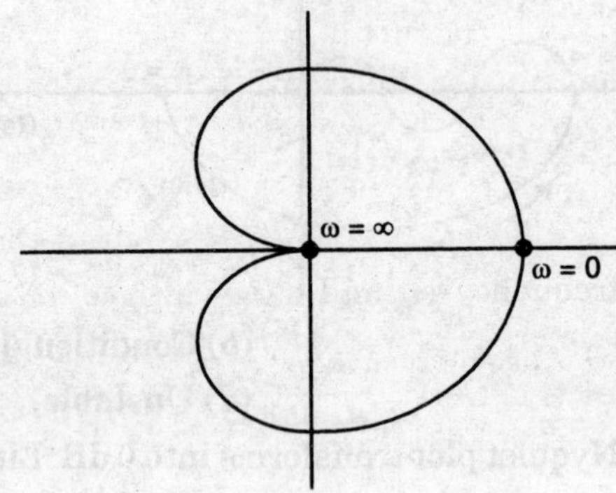

(*a*) $\dfrac{1}{(s+1)^3}$ (*b*) $\dfrac{1}{(s+1)^2}$

(*c*) $\dfrac{1}{s^2+2s+2}$ (*d*) $\dfrac{1}{(s+1)}$

28. If the gain margin of a certain feedback system is given as 20 dB, the Nyquist plot will cross the negative real axis at the point

(*a*) $s = -0.05$ (*b*) $s = -0.2$

(*c*) $s = -0.1$ (*d*) $s = -0.01$

29. The centre and radius of a constant *M* circles are given by

(*a*) $x = \dfrac{M}{(1-M^2)}, y = 0 \,; r = \dfrac{M}{(1-M^2)}$ (*b*) $x = \dfrac{M}{(M^2-1)}, y = 0 \,; r = \dfrac{M}{(M^2-1)}$

(*c*) $x = \dfrac{M^2}{(M^2-1)}, y = 0 \,; r = \dfrac{M}{(M^2-1)}$ (*d*) $x = -\dfrac{M^2}{(M^2-1)}, y = 0 \,; r = \dfrac{M}{(M^2-1)}$

30. The centre and radius of a constant *N* circles are given by

(*a*) $x = (-1/2), y = (1/2N) \,; r = \dfrac{\sqrt{(N^2+1)}}{2N}$

(*b*) $x = (1/2, y = (1/2N) \,; r = \dfrac{\sqrt{(N^2+1)}}{2N}$

(*c*) $x = -(1/2), y = -(1/2N) \,; r = \dfrac{\sqrt{(N^2+1)}}{2N}$

(*d*) $x = -(1/2), y = (1/2N) \,; r = \dfrac{\sqrt{(N^2+1)}}{N}$

31. In the $G^{-1}(j\omega)$ plane, the constant M circle has the following centre and radius

(a) $(x, y) = (-1, 0)$; $r = M$
(b) $(x, y) = (+1, 0)$; $r = M$
(c) $(x, y) = (-1, 0)$; $r = 1/M$
(d) $(x, y) = (+1, 0)$; $r = 1/M$

32. Undamped natural frequency ω_n and resonance frequency ω_r of a unity feedback system with open-loop transfer function $G(s) = \dfrac{\omega_n^2}{s(s+2\zeta\omega_n)}$; $\zeta < 1/\sqrt{2}$, are related as

(a) $\omega_n = \omega_r$
(b) $\omega_n > \omega_r$
(c) $\omega_n < \omega_r$
(d) None of these.

33. Undamped natural frequency ω_n and bandwidth ω_b of a unity feedback system with open-loop transfer function $G(s) = \dfrac{\omega_n^2}{s(s+2\zeta\omega_n)}$ are related as

(a) $\omega_n = \omega_b$
(b) $\omega_n > \omega_b$
(c) $\omega_n < \omega_r$
(d) None of these.

34. From the Nichol's chart, one can determine the following quantities pertaining to a closed-loop system

(a) Magnitude and phase
(b) Bandwidth
(c) Magnitude only
(d) Phase only.

35. Nichol's chart is derived from

(a) Constant M circles
(b) Constant N circles
(c) Constant M and N circles
(d) Polar plot.

36. Nichol's chart is drawn

(a) *Decibel versus phase angle*
(b) Decibel versus ω
(c) Decibel versus log ω
(d) Magnitude versus phase angle

KEY

1. (c)	**2.** (d)	**3.** (a)	**4.** (c)	**5.** (d)	**6.** (a)
7. (b)	**8.** (c)	**9.** (d)	**10.** (c)	**11.** (a)	**12.** (d)
13. (c)	**14.** (d)	**15.** (a)	**16.** (d)	**17.** (a)	**18.** (b)
19. (d)	**20.** (b)	**21.** (a)	**22.** (b)	**23.** (d)	**24.** (c)
25. (b)	**26.** (d)	**27.** (b)	**28.** (c)	**29.** (d)	**30.** (a)
31. (c)	**32.** (b)	**33.** (c)	**34.** (a) and (b)	**35.** (c)	**36.** (a)

EXERCISE

1. Explain the procedure for constructing the polar plots.
2. Discuss the calculation of Gain crossover frequency and phase crossover frequency with respect to the polar plots.
3. Illustrate the effect of addng a pole or zero to a transfer function on its polar plot.

4. State and explain the Nyquist stability criterion.
5. Explain the procedure to construct the Nyquist plot.
6. Write the relative merits and demerits of both Nyquist and Bode plots.
7. What is meant by constant M circle and N circle?
8. Draw the polar plot for the transfer function $G(s) = \dfrac{1}{s^2(1+s)(1+2s)}$. Determine the frequency at which the plot crosses the real axis and the corresponding $|G(j\omega)|$.
9. Sketch the direct and inverse polar plots for the system with transfer function $G(s) = \dfrac{1}{(1+s)(1+2s)}$. Also find frequency at which $|G(j\omega)| = 1$ and the corresponding phase angle.
10. Given the open loop transfer function $G(s) = \dfrac{1}{(1+s)(2+s)}$, sketch the Nyquist diagram and investigate the stability of the closed loop system.
11. Draw the Nyquist plot for a given transfer function $G(s) = \dfrac{3}{s(1+s)(1+2s)}$.
12. Given the open loop transfer function $G(s) = \dfrac{30}{(1+s+s^2)(1+2s)}$, sketch the Nyquist plot and investigate the open loop and closed loop systems stability.

8 Compensation

8.1 INTRODUCTION

Control systems are designed to do specific tasks or to achieve specific objectives of the system. The performance of a control system is usually measured in terms of performance specifications. The general specifications of a control system are stability or relative stability, accuracy and speed of the response.

Peak time, peak overshoot, damping ratio, resonant frequency, resonant peak, bandwidth, natural frequency, error coefficient, gain margin and phase margin are called design specifications. Generally those specifications are presented in terms of precise numerical values, for which the system will be designed.

8.2 WHAT IS A COMPENSATOR?

The compensator is a controller plus something more as it improves the stability margin. The compensator modifies the open loop transfer function and enables the eigenvalues to get located at the desired locations on the s-plane. Since basic plant or the process need not be electrical, the compensators can also be electrical, mechanical, pneumatic or hydraulic etc. The first stage of the system compensation is to obtain the transfer function of the compensation device. The choice of the components to be used depends upon the system structure. Realization by electrical components is quite common in many control systems.

The basic advantage of compensators are for example, if system specifications are changed replacement of entire system is very expensive. If specifications are achieved by a subsystem called compensator it will be less expensive.

Compensator is required for a system in order to stabilize an absolutely unstable system and to achieve a specified performance even for a stable system.

8.3 MODES OF CONNECTIONS OF COMPENSATORS

Compensators are classified as three types depending upon the where the compensator is introduced in a plant. They are

1. Series compensator (cascade)
2. Parallel (shunt) compensator
3. Series–parallel compensator.

8.3.1 Series Compensator

If the subsystem called compensator is connected in series with the main system in forward path then the scheme is called series compensation. The block diagram representation is shown in Fig. 8.1.

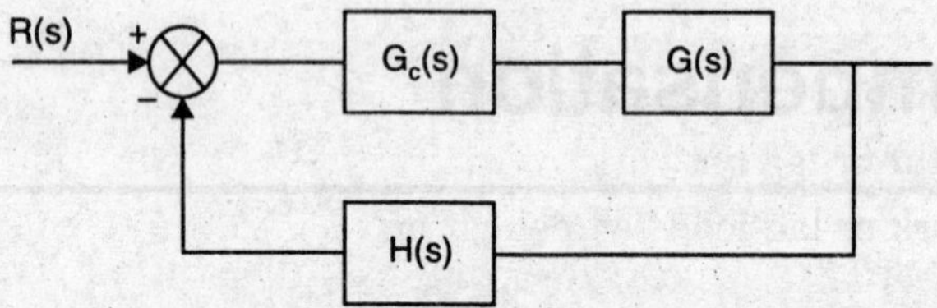

Fig. 8.1 Series compensator

where $G_c(s)$ = compensator

$G(s)$ = main system or plant

This scheme is also called cascade compensation.

8.3.2 Parallel Compensation

If the compensator is introduced in parallel with the main plant or one element of the plant as shown in Fig.8.2, such compensation is called feedback compensation or parallel compensation. The block diagram of parallel compensation is shown in Fig. 8.2.

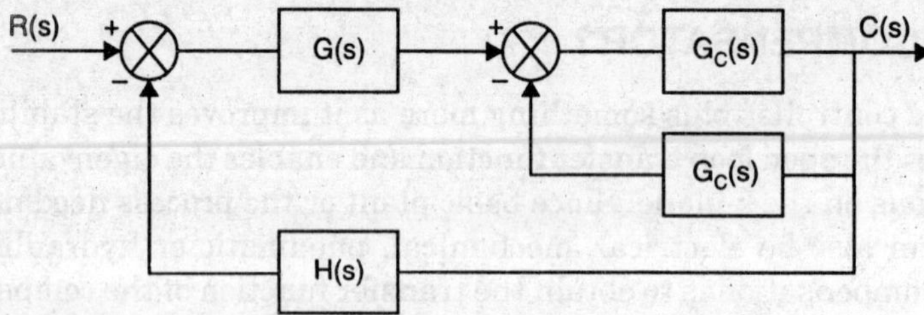

Fig. 8.2 Parallel compensator

8.3.3 Series-Parallel Compensation

Sometimes it is required to provide both types of compensation *i.e.*, feedback as well as cascade which is called series–parallel compensation. The block diagram of series-parallel compensation is shown in Fig. 8.3.

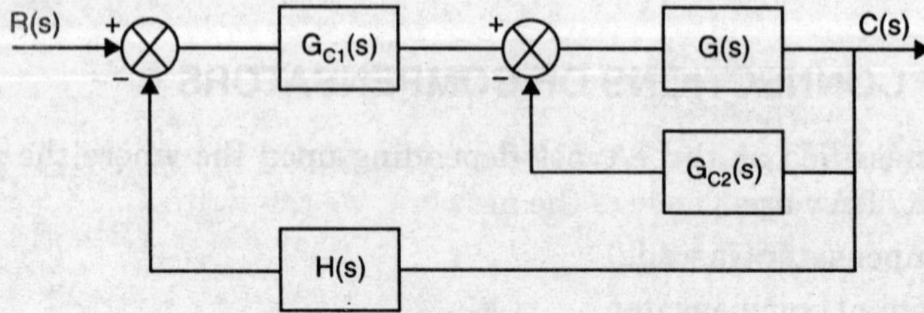

Fig. 8.3 Series and Parallel compensator

8.4 TYPES OF COMPENSATOR

There are different compensators like electrical, mechanical, pneumatic, hydraulic or combinations of various types.

Depending upon the function, compensating networks (electrical) are classified as three types. They are

1. Lead network or lead compensator
2. Lag compensator or lag network
3. Lag-lead network or lag-lead compensation.

8.5 REALIZATION OF BASIC ELECTRICAL COMPENSATORS

8.5.1 Lead Compensator

If we use lead compensator it improves the system response in following aspects:

- Speeds up transient response
- Increases the system error constant and
- Increases the margin of stability.

Consider an electrical network as shown in Fig. 8.4, which is a lead compensation network. When a sinusoidal input is applied to the lead network, output of the network leads input by some angle.

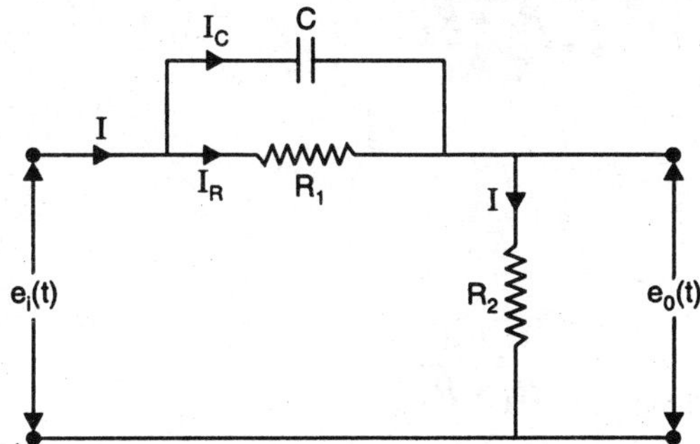

Fig. 8.4 Lead network

Lead network transfer function in standard form is

$$G_c(s) = \frac{s+Z_c}{s+P_c} = \frac{s+\dfrac{1}{T}}{s+\dfrac{1}{\alpha T}}\ , \ \alpha < 1 \tag{8.1}$$

By applying *KCL* at output node of the network, we can write

$$I = I_c + I_R$$

$$\frac{e_0(t)}{R_2} = \frac{e_i(t) - e_0(t)}{R_1} + C\frac{d(e_i(t) - e_0(t))}{dt}$$

By applying Laplace transform to above equation

$$\frac{1}{R_2} E_0(s) = \frac{E_i(s)}{R_1} - \frac{E_0(s)}{R_1} + sC\,[E_i(s) - E_0(s)]$$

$$E_i(s)\left[\frac{1}{R_1} + sC\right] = E_0(s)\left[\frac{1}{R_2} + \frac{1}{R_1} + sC\right]$$

$$\frac{E_0(s)}{E_i(s)} = \frac{\frac{1}{R_1} + sC}{\frac{1}{R_2} + \frac{1}{R_1} + sC}$$

$$= \frac{\left(\frac{1}{R_1 C} + s\right)}{\left(s + \frac{1}{R_2 C} + \frac{1}{R_1 C}\right)}$$

$$= \frac{\left(s + \frac{1}{R_1 C}\right)}{s + \frac{R_1 + R_2}{R_1 R_2 C}}$$

$$= \frac{s + \frac{1}{R_1 C}}{s + \frac{1}{\frac{R_1 R_2 C}{R_1 + R_2}}} \tag{8.2}$$

By comparing eqns. (8.1) and (8.2)

$$T = R_1 C \quad \text{and}$$

$$\alpha = \frac{R_2}{R_1 + R_2}$$

where T and α are lead compensator parameters.

(*i*) Pole and zero plot of lead compensator

From the transfer function of lead compensator we have zero at $s = \frac{-1}{T}$ and a pole at $s = \frac{-1}{\alpha T}$. In lead components zero is closer to the origin than the pole as shown in Fig. 8.5.

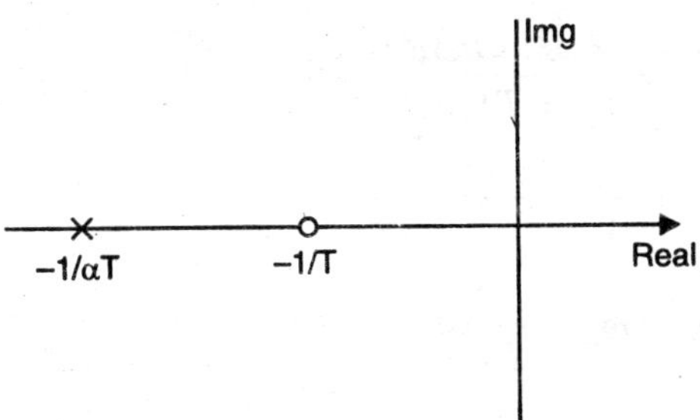

Fig. 8.5 Pole-zero plot of lead compensator

(ii) **Maximum leading angle θ_m**

Transfer function of lead compensator is

$$\frac{E_0(s)}{E_i(s)} = \frac{s+\frac{1}{T}}{s+\frac{1}{\alpha T}} = \frac{\alpha(1+Ts)}{(1+\alpha Ts)}$$

If s is replaced with $j\omega$

$$\therefore \qquad \frac{E_0(j\omega)}{E_i(j\omega)} = \frac{\alpha(1+j\omega T)}{(1+j\omega\alpha T)}$$

Magnitude of the above transfer function is

$$M = \frac{\alpha\sqrt{1+(\omega T)^2}}{\sqrt{1+\omega^2\alpha^2T^2}}$$

Phase angle of the transfer function is

$$\theta = -\tan^{-1}\omega\,\alpha T + \tan^{-1}\omega T \qquad (8.3)$$

To get condition for maximum lead angle θ_m we have to equate first derivative of θ to be zero

i.e.,

$$\frac{d\theta}{d\omega} = 0$$

$$0 = -\frac{\omega\alpha T}{1+\omega^2\alpha^2T^2} + \frac{\omega T}{1+\omega^2T^2}$$

$$\omega T\,(1+\omega^2\alpha^2T^2) - \omega\alpha T\,(1+\omega^2T^2) = 0$$

$$(1+\omega^2\alpha^2T^2) - \alpha\,(1+\omega^2T^2) = 0$$

$$1+\omega^2\alpha^2T^2 - \alpha + \alpha\omega^2T^2 = 0$$

$$\omega^2\alpha^2T^2 = 1$$

Where $\omega = \omega_m$, where maximum phase lead occurs. ω_m is geometric mean of two corner frequencies of the transfer function.

By taking tan on both sides of eqn. (8.3)

$$\tan\theta = \tan\,[\tan^{-1}\omega T - \tan\omega\alpha T]$$

$$= \frac{\omega T - \alpha\omega T}{1 + (\omega T)(\alpha\omega T)}$$

$$= \frac{\omega T(1-\alpha)}{1+\omega^2\alpha^2T^2}$$

At maximum angle, $\omega = \omega_m = \dfrac{1}{T\sqrt{\alpha}}$

$$= \frac{\dfrac{1}{T\sqrt{\alpha}}T(1-\alpha)}{1+\dfrac{1}{T^2\alpha}\alpha T^2}$$

$$\tan\theta_m = \frac{1-\alpha}{2\sqrt{\alpha}}$$

and $$\sin\theta_m = \frac{1-\alpha}{1+\alpha}$$

From above equation we can express α interms of θ_m as

$$\alpha = \frac{1-\sin\theta_m}{1+\sin\theta_m}$$

Magnitude of transfer function at $\omega = \omega_m$ is

$$M = \frac{1}{\sqrt{\alpha}}$$

(iii) **Magnitude and phase plot of lead compensator**

Transfer function of lead compensator is

$$= \frac{s+\dfrac{1}{T}}{s+\dfrac{1}{\alpha T}}$$

$$= \frac{\alpha(1+sT)}{(1+s\alpha T)}, \ \alpha < 1.$$

Corner frequency of first order factor in numerator is $\dfrac{1}{T}$ and slope is 20 dB/dec. And there is first order factor in denominator that offers corner frequency of $\dfrac{1}{\alpha T}$ and slope of – 20 dB/dec. Numerator has less corner frequency. So we have to Start with the numerator factor which gives + 20 dB/dec slope to its magnitude plot. From its corner frequency, it will be extended up to next factor corner frequency which gives the slope to its magnitude curve which is – 20 dB/dec.

Angle of lead compensator increases initially, achieves its maximum θ_m and then again decreases. The Bode plot is shown in Fig. 8.6.

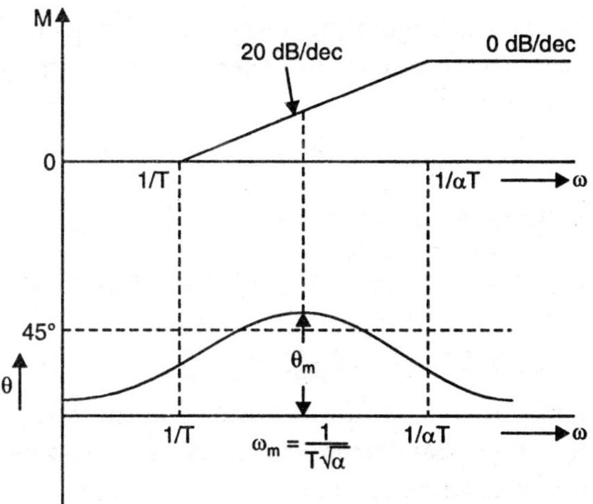

Fig. 8.6

(*iv*) **Functions of lead compensator**

Following are the functions or uses of lead compensator.

- It increases the damping of the closed loop system so that overshoot, rise time and settling time decreases. Thus transient response improves.
- It improves gain and phase margins and also improves the relative stability.
- It also increases the system error constant.

(*v*) **Step by step design procedure for lead compensation in frequency domain:**

The following step by step procedure leads to the design of lead compensator:

Step 1: For the uncompensated system, the value of the open loop gain (K) to satisfy the given study state error constant is computed (*i.e.*, from K_v, K_a etc.).

Step 2: For the above value of K, the phase margin of the system is obtained from its Bode plot.

Step 3: If the phase margin satisfies the prescribed phase margin, then the system does not need the compensation. Otherwise

Step 4: Compute the phase lead(ϕ_L) to be added by the lead compensator to make up for the desired phase margin(ϕ_d).

i.e.,
$$\phi_L = \phi_d - \phi_0 + \epsilon$$

where ϵ = (5°approx) a margin of safety required because the crossover frequency will increase due to compensation.

Step 5: The parameter of the lead compensator is given by

$$\alpha = \frac{1 - \sin \phi_L}{1 + \sin \phi_L}$$

If ϕ_L is more than 60°, it is desirable to use two identical lead compensation in cascade such that ϕ_L for each is $\frac{\phi_L}{2}$.

Step 6: In order to determine the frequency ω_m, where the lead compensator must add $\phi_m = \phi_L$ phase angle, we decide the frequency at which the gain corresponds to the gain of the compensator at $\omega = \omega_m$ *i.e.*, 10 log(1/α).

Thus a frequency ω_m is located on the Bode plot of the uncompensated system at which the gain is $-10 \log \frac{1}{\alpha}$ for a obtained in step 5.

Step 7: The two corner frequencies of the lead compensator are given by

$$\omega_1 = \omega_m \sqrt{\alpha} \quad \text{and} \quad \omega_2 = \frac{\omega_m}{\sqrt{\alpha}}$$

Step 8: Draw the magnitude and the phase plots of the compensated system $G(s) = G_c(s) \times G_{uc}(s)$, where $G_c(s)$ is the transfer function of the compensator and $G_{uc}(s)$ is the transfer function of the uncompensated system.

If the phase margin is still low increase the value of ∈ and repeat from step 4.

Step 9: Check additional specification on the compensated system performance like bandwidth, redesign for another choice of ω_m till this specification is met. It may be noted that additional specification can only be met if it is consistent.

Problem 8.1. *Design a suitable compensator for the system shown in Fig.8.7 to satisfy the following specifications :*

(*i*) *The phase must be greater than 45°*

(*ii*) *Velocity error constant to remain the same*

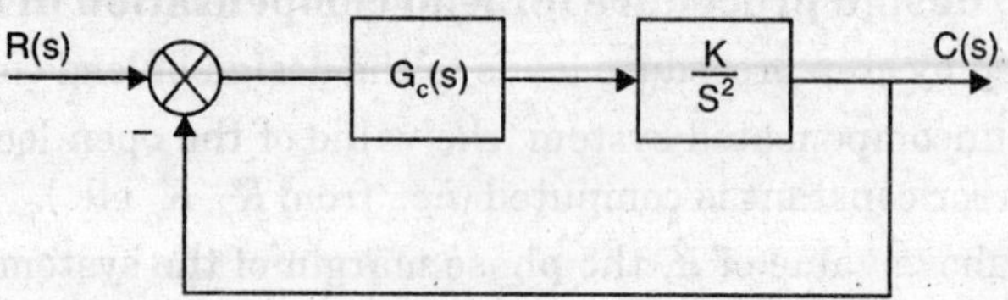

Fig. 8.7

Solution: For the uncompensated system the $K_v = \infty$. It will remain the same as long as the cascading of the compensator to the system does not alter the type of the system and in this case being type 2.

Fig 8.8 shows the Bode plot for the uncompensated system. It is seen from the graph that the gain crossover frequency is 1 rad/sec and the phase of the angle of the system at all values of ω is 180°. Thus phase margin is 0°.

Hence, it requires a phase lead network to improve the phase margin.

The phase angle contribution of lead compensator must be 45° (at the least). Since the effect of adding the compensator shifts the gain crossover frequency, a tolerance of 5° is added to phase lead required.

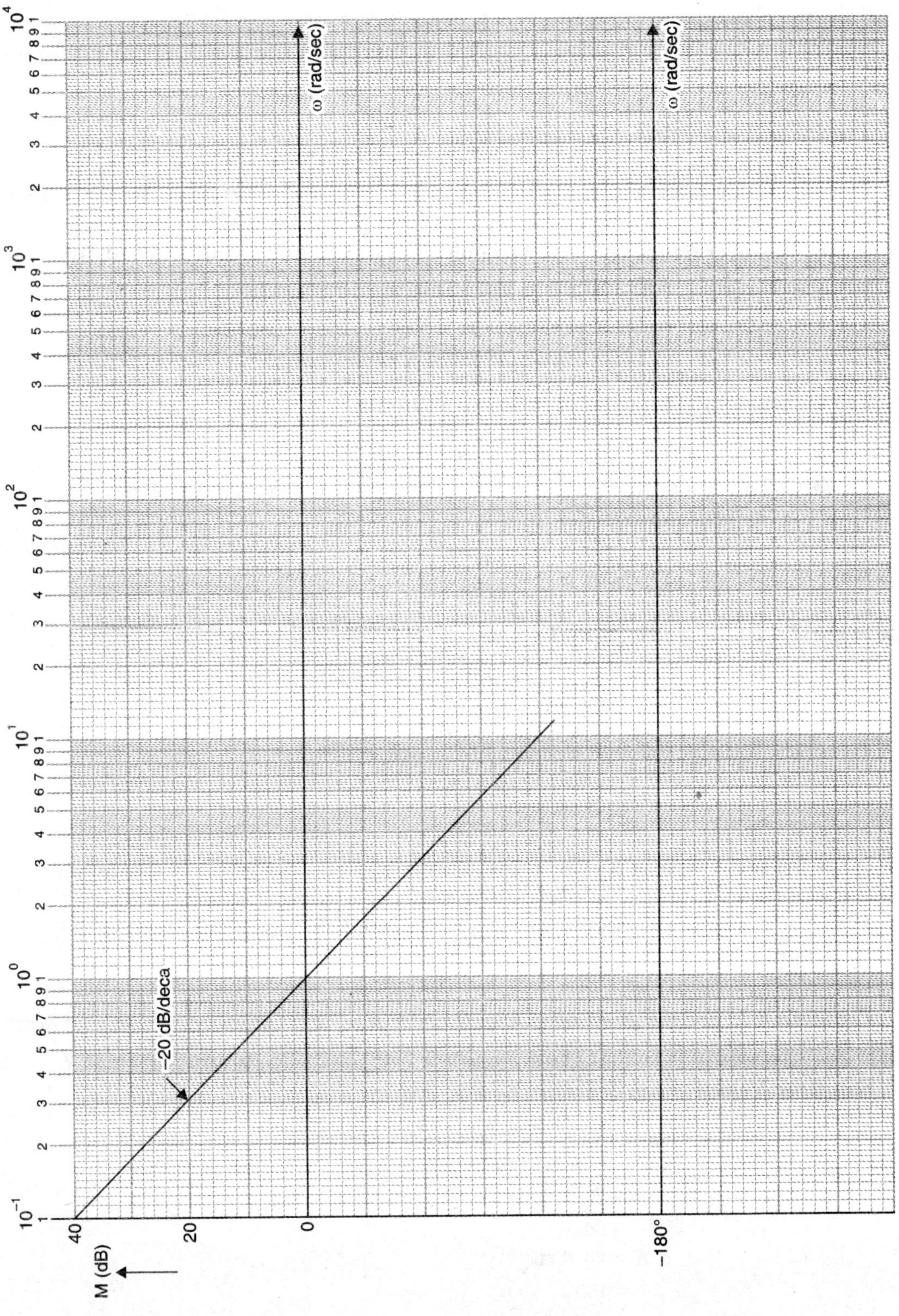

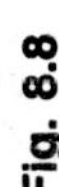
M (dB)
40
20
0
−180°
−20 dB/deca
ω (rad/sec)
ω (rad/sec)
10^{-1}
10^{0}
10^{1}
10^{2}
10^{3}
10^{4}

Fig. 8.8

Thus the phase angle contribution is $\phi_m = 45° + 5° = 50°$.

and $$\alpha = \frac{1 - \sin 50°}{1 + \sin 50°} = 0.1325$$

The value of α is considered satisfactory.

The magnitude contribution at the frequency at which ϕ_m is 50° is $10 \log \left(\frac{1}{\alpha}\right)$. Hence, the new gain crossover frequency corresponding to a gain of $- 10 \log \frac{1}{\alpha}$ (*i.e.*, 8.778 dB) is seen from the Bode plot of Fig. 8.8 to be 1.6 rad/sec. Thus $\omega_m = 1.6$ rad/sec.

The lower corner frequency of the compensator

$$\omega_1 = 1.6 \times \sqrt{0.1325} = 0.5824 \text{ rad/sec}$$

The higher corner frequency of the compensator is

$$\omega_2 = \frac{1.6}{\sqrt{0.1325}} = 4.39 \text{ rad/sec}$$

Thus the compensator transfer function is

$$G_c(j\omega) = \alpha \frac{1 + j\omega T}{1 + j\omega\alpha T}$$

where $\omega_1 = 1/T = 0.5824$ and $\omega_2 = 1/\alpha T = 4.39$

and $$G_c(s) = \frac{(s + 1/\tau)}{(s + 1/\alpha\tau)} = \frac{(s + 0.5824)}{(s + 4.39)}$$

Therefore, transfer function of the compensated system is

$$G(s) = \frac{(s + 0.5824)}{s^2 (s + 4.39)}$$

It has the same value of K_v. The phase margin corresponding to $\omega_{gc} = 1.6$ rad/sec.

$$P.M.\Big|_{\omega = \omega_{gc}} = 180° + \left(- 180° + \tan^{1} \frac{1.6}{0.5824} - \tan^{-1} \frac{1.6}{4.39}\right) = 50° \text{ is acceptable.}$$

Problem 8.2. *For the given open loop transfer function,* $G(s) = \frac{K}{s(s + 4)(s + 6)}$ *design suitable lead compensation so that phase margin is* $\geq 40°$ *and velocity error constant,* $K_v \geq 20$

Solution: Velocity error constant, $K_v = 20$ (given)

$$K_v = \operatorname*{Lt}_{s \to 0} \frac{s.K}{s(s + 4)(s + 6)}$$

$$= \frac{K}{24}$$

$$K = 24 \times 20$$

$$= 480$$

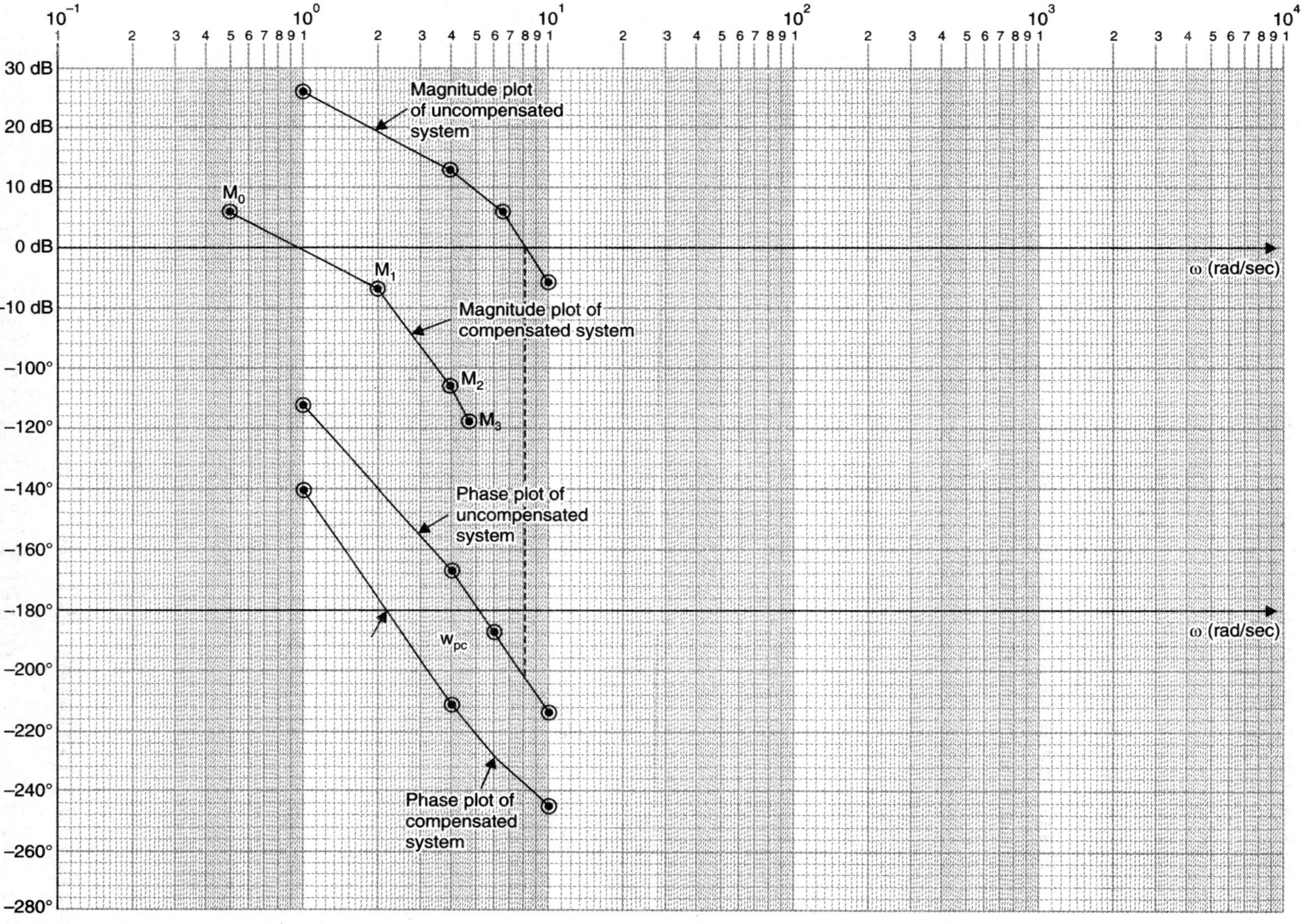
Magnitude plot of uncompensated system
Magnitude plot of compensated system
Phase plot of uncompensated system
Phase plot of compensated system
M_0
M_1
M_2
M_3
w_{pc}
ω (rad/sec)
ω (rad/sec)
30 dB
20 dB
10 dB
0 dB
–10 dB
–100°
–120°
–140°
–160°
–180°
–200°
–220°
–240°
–260°
–280°
10^{-1}
10^0
10^1
10^2
10^3
10^4

Fig. 8.9

Transfer function of uncompensated system is

$$G_{uc}(s) = \frac{480}{s(s+4)(s+6)}$$

In standard form, $G_{uc}(s) = \dfrac{20}{s(0.25s+1)(0.17s+1)}$

Figure 8.9 shows the Bode plot for the uncompensated system. It is seen from the graph that the gain crossover frequency is 8 rad/sec, the phase margin, $\phi_{pm} = -32$.

It has not satisfied the desired phase margin.

Hence, it requires a phase lead network to improve the phase margin.

The phase angle contribution of lead compensator must be 40° (at the least). Since the effect of adding the compensator shifts the gain crossover frequency, a tolerance of 5° is added to the phase lead required.

Thus the phase angle contribution is $(\phi_m) = \phi_d - \phi + \epsilon = 40 - (-32)° + 5 = 77°$

The required phase margin is more than 60°. Two cascade sections of lead compensator is required.

$$\phi_m = \frac{77}{2} = 38.5°$$

$$\alpha = \frac{1 - \sin\phi_m}{1 + \sin\phi_m}$$

$$= \frac{1 - \sin 38.5°}{1 + \sin 38.5°} = 0.23$$

The value of α is considered satisfactory.

The magnitude contribution at the frequency at which ϕ_m is 38.5° is $10 \log\left(\frac{1}{\alpha}\right)$. Hence, the new gain crossover frequency corresponding to a gain of $-10 \log \frac{1}{\alpha}$ (*i.e.*, – 6.38 dB) is seen from the Bode plot to be 10 rad/sec, thus $\omega_{gc} = 10$ rad/sec.

The lower corner frequency of the compensator

$$\omega_1 = 10 \times \sqrt{0.23} = 4.79 \text{ rad/sec}$$

The higher corner frequency of the compensator is

$$\omega_2 = \frac{10}{\sqrt{0.23}} = 20.85 \text{ rad/sec}$$

Thus the compensator transfer function is

$$G_c(j\omega) = \left(\alpha \times \frac{1 + j\omega T}{1 + j\omega\alpha T}\right)^2 \quad \text{(since two cascade lead sections)}$$

where $\omega_1 = 1/T = 4.79$ and $\omega_2 = 1/\alpha T = 20.85$

and

$$G_c(s) = \left(\frac{(s + 1/\tau)}{(s + 1/\alpha\tau)}\right)^2 = \left(\frac{(s + 4.79)}{(s + 20.85)}\right)^2$$

Therefore, transfer function of compensated system is

$$G(s) = \frac{20}{s(0.25s+1)(0.17s+1)}\left(\frac{(s+4.79)}{(s+20.85)}\right)^2$$

Phase angle of compensated system at gain crossover frequency is ω_{gc} = 10 rad/sec is

$$\phi\Big|_{\omega=\omega_{gc}} = -90° - \tan^{-1}(0.25\,\omega) - \tan^{-1}(0.17\,\omega) + 2\left(\tan^{-1}\frac{\omega}{4.79} - \tan^{-1}\frac{\omega}{20.85}\right)$$

$$= -90° - 68° - 59° + 2(64.4° - 25.62°) = -139.4°$$

The phase margin of compensated system corresponding to ω_{gc} = 10 rad/sec is

180° – 139.4° = 40.5° is acceptable.

8.5.2 Lag Compensator

If lag compensator is inserted in a plant it improves the steady state response of the plant.

Consider an electrical network as shown in Fig. 8.10, which is a lag compensation network. When a sinusoidal input is applied to the lead network, output of the network lags input by same angle.

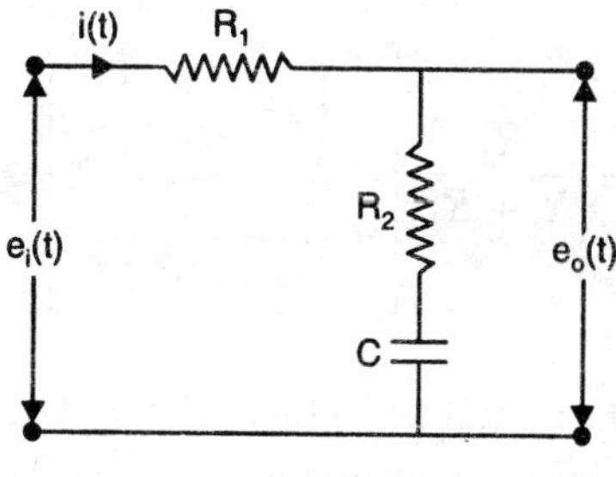

Fig. 8.10

Standard form by lag compensator transfer function is

$$G_c(s) = \frac{s+z_c}{s+p_c} = \frac{s+\dfrac{1}{T}}{s+\dfrac{1}{\beta T}};\ \beta > 1,\ T > 0 \qquad (8.4)$$

Let us obtain transfer function of network shown in Fig. 8.10. Loop equation of above lag network is

$$e_i(t) = R_1 i(t) + R_2 i(t) + \frac{1}{C}\int i(t)\,dt$$

Taking Laplace transform

$$E_i(s) = I(s)\left[R_1 + R_2 + \frac{1}{sC}\right]$$

and

$$E_o(t) = i(t)\,R_2 + \frac{1}{C}\int i(t)\,dt$$

Taking Laplace transform

$$E_o(s) = I(s)\,R_2 + \frac{1}{sC}\,I(s)$$

$$= I(s)\left[R_2 + \frac{1}{sC}\right]$$

$$\frac{E_o(s)}{E_i(s)} = \frac{I(s)\left[R_2 + \frac{1}{sC}\right]}{I(s)\left[R_1 + R_2 + \frac{1}{sC}\right]}$$

$$= \frac{1 + R_2 Cs}{1 + (R_1 + R_2)\,sC}$$

$$= \frac{R_2 C\left[s + \frac{1}{R_2 C}\right]}{(R_1 + R_2)\,C\left[s + \frac{1}{(R_1 + R_2)\,C}\right]}$$

$$= \frac{R_2}{R_1 + R_2}\left[\frac{\left(s + \frac{1}{R_2 C}\right)}{s + \frac{1}{(R_1 + R_2)\,C}}\right] = \frac{1}{\frac{R_1 + R_2}{R_2}}\,\frac{s + \frac{1}{R_2 C}}{s + \frac{1}{\frac{(R_1 + R_2)}{R_2}}} \qquad (8.5)$$

By comparing eqns. (8.4) and (8.5), we have $T = R_2 C$ and $\beta = (R_1 + R_2)/R_2 > 1$

(*i*) **Pole zero plot of lag compensator**

From the transfer function of lag compensator we have zero at $s = -\frac{1}{T}$ and a pole at $s = \frac{-1}{\beta T}$. The pole is closer to origin than zero as shown in Fig. 8.11.

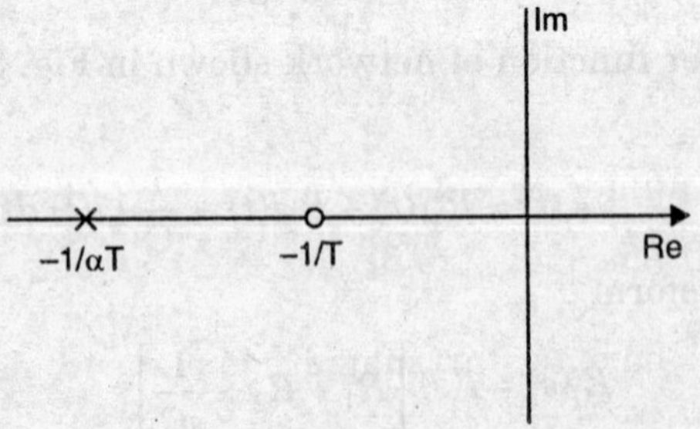

Fig. 8.11

(*ii*) **Maximum lagging angle**

Transfer function of lag compensator is

$$\frac{E_o(s)}{E_i(s)} = \frac{s+\frac{1}{T}}{s+\frac{1}{\beta T}} = \frac{\beta(1+Ts)}{(1+\beta Ts)}$$

In frequency domain we get

$$\frac{E_o\,(j\omega)}{E_i\,(j\omega)} = \frac{(1+Tj\omega)}{(1+\beta Tj\omega)}$$

Magnitude of above transfer function, $M = \dfrac{\sqrt{1+\omega^2T^2}}{\sqrt{1+\omega^2\beta^2T^2}}$

Phase angle of the transfer function is

$$\theta = \tan^{-1}\omega T - \tan^{-1}\omega\beta T$$

$$\tan\theta = \frac{\omega T - \omega\beta T}{1-(\omega T)\,(\omega\beta T)}$$

To find θ_m, we have to equate $\dfrac{d\theta}{d\omega} = 0$

That gives $\qquad \omega_m = \dfrac{1}{T\sqrt{\beta}} = \sqrt{\dfrac{1}{T}\cdot\dfrac{1}{\beta T}}$

The above is very similar to lead compensator

$$\therefore \qquad \tan\theta_m = \frac{1-\beta}{2\sqrt{\beta}}$$

(*iii*) **Magnitude and phase plot of lag compensator**

Transfer function of lag compensator is

$$TF = \frac{s+\frac{1}{T}}{s+\frac{1}{\beta T}} = \frac{\beta(1+sT)}{(1+s\beta T)};\ \beta > 1$$

Corner frequencies of numerator and denominator are $\frac{1}{T}$ and $\frac{1}{\beta T}$; slopes of numerator and denominator are + 20 dB/dec and – 20 dB/dec. Here corner frequency of denominator is less $\left(\frac{1}{T} > \frac{1}{\beta T}\right)$. Therefore, we have to start magnitude plot with denominator.

Angle of lag compensator decreases and achieves its minimum value and again increases. The Bode diagram is shown in Fig. 8.12.

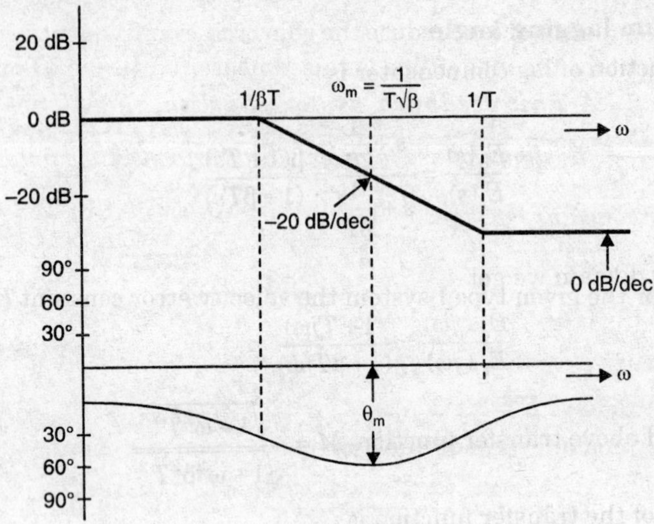

Fig. 8.12

(*iv*) **Functions and uses of lag compensator**

It improves the steady state performance. There is disadvantage of lag compensator if the transient response lasts for longer time.

(*v*) **Step by step procedure for lag compensation in frequency domain**

Step 1: From the steady state requirement, the error constant (K_a) is computed from type of the uncompensated system and Bode plot is drawn keeping the value of open loop gain to give the error constant as obtained above.

Step 2: From the plot, the phase margin of the uncompensated system is measured. If this phase margin is not as per the frequency domain specification desired, the lag compensator is to be desired.

Step 3: Locate the frequency ω_{c2}, where the phase angle curve makes a angle contribution of ϕ_2. When ϕ_2 is measured above the – 180° line, allow 5° – 10° for the phase angle lag contributed at ω_{c2} by the compensation network.

Step 4: Choose the upper corner frequency $\omega_2 = \dfrac{1}{T}$ of the compensator in the range $\dfrac{1}{2}$ to $\dfrac{1}{10}$ of ω_{c2}.

Step 5: Measure the gain of the lag compensated system at ω_{c2} and attenuation required to make ω_{c2} as the new gain crossover frequency = 20 log β.

Step 6: From steps 4 and 5 the parameter τ and β of the compensator are obtained.

Step 7: Thus the compensator transfer function $G(s)$ is obtained.

Step 8: Redraw the Bode plot with the $G_c(j\omega)$ included in cascade with the given system as the open loop transfer function is given by the product of $G_{uc}(j\omega)$ and $G_c(j\omega)$, and check whether the specifications are satisfied.

Step 9: If there are any additional specifications, check up whether they are also satisfied. If not choose another value of τ and redesign the compensator.

The effect of compensator is to reduce the gain crossover frequency on the high frequency region in the Bode plot for the magnitude is raised nearer to the 0 dB lines.

Problem 8.3. *A unit feedback system has an open loop transfer function* $G(s) = \dfrac{K}{s(s+1)(0.2s+1)}$. *Design a phase lag compensator to meet the following specifications.*

Velocity error constant = 8

Phase margin ≥ 40°.

Solution: For the given type I system the velocity error constant K_v is given by

$$\underset{s \to 0}{\text{Lt}}\ sG(s) = K$$

Since $K_v = 8$, we have $K = 8$.

The frequency domain transfer function of the given system is

$$G_{uc}(j\omega) = \frac{8}{(j\omega)(j\omega+1)(0.2j\omega+1)}$$

The Bode plot for the system is drawn in Fig. 8.13.

For drawing the Bode plots, the following are compensated for the uncompensated system drawn in Fig. 8.13, we see that lower frequency asymptote of $|G_{uc}(j\omega)|$ at – 20 dB/dec slope upto ω = 1 rad/sec (*i.e.*, gain crossover frequency).

From ω = 1 rad/sec to ω = 5 rad/sec, the slope is – 40 dB/dec

For ω = 5 rad/sec onwards it is – 60 dB/dec

The phase angle computation is shown in the table 8.1.

Table 8.1

ω *(rad/sec)*	$\phi(\omega)$ *for uncompensated system* $= -90° - \tan^{-1}\omega - \tan^{-1} 0.2\,\omega$
0.1	– 96.85
1	– 146.3
5	– 213.96
10	– 237.72

At the gain crossover frequency of ω = 2.9 rad/sec the phase margin of the uncompensated system is

$$\phi_{pm} = -16°$$

This does not satisfy the specification. Hence lag compensator is to be designed.

The phase angle lag to be provided by the lag compensator is

$$\phi_2 = \phi_d + \epsilon = 40° + 5° = 45°$$

Here ϵ is chosen to be 5°.

From the phase angle ($\phi_{uc}(j\omega)$ *Vs* ω) characteristic of the uncompensated system it is observed that at ω = 0.6 rad/sec the phase angle contribution of the system is 45°.

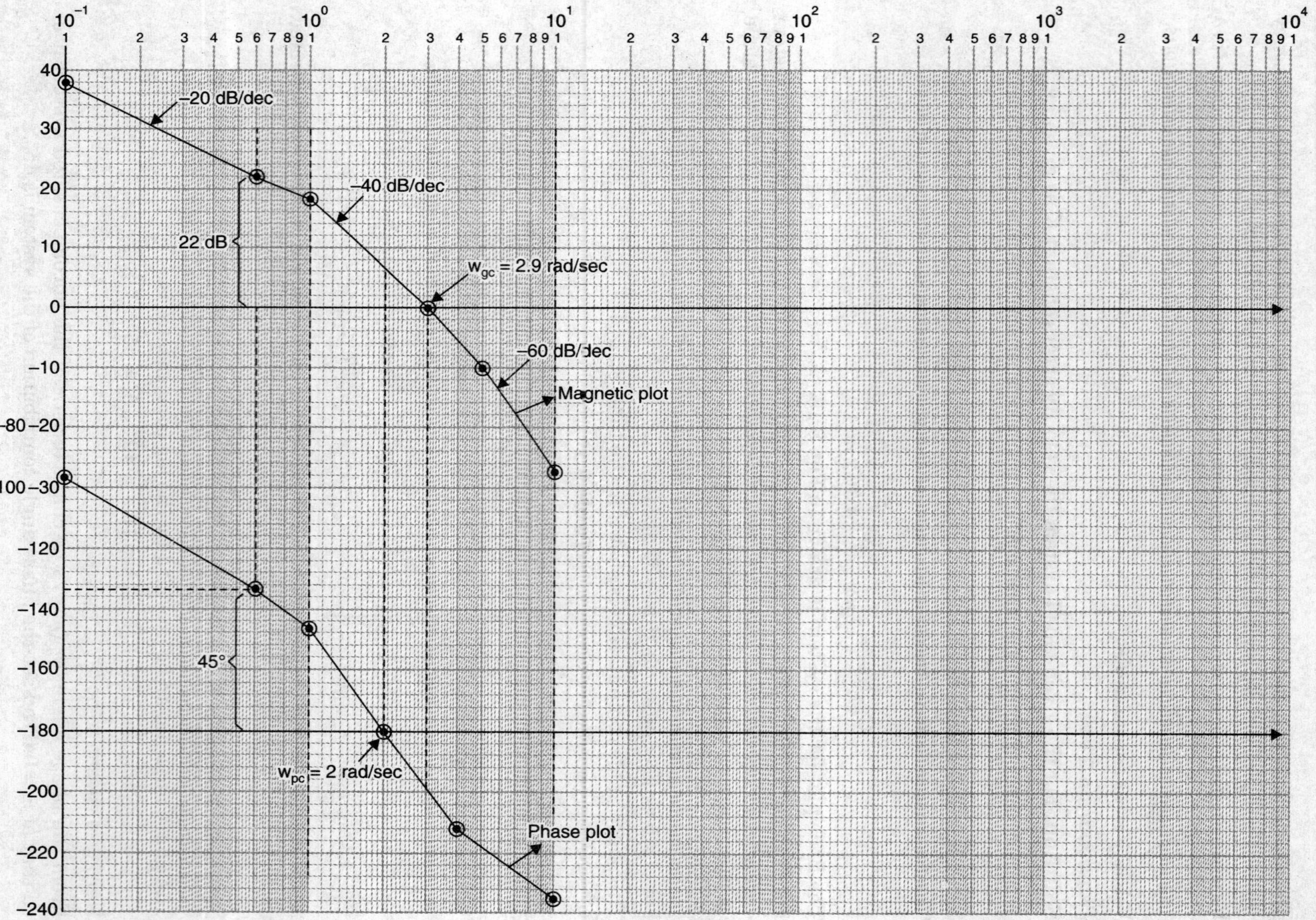

Fig. 8.13. Lag compensator for G(s) =

Hence,

$$\omega_{c2} = \omega_{gcn} = 0.6 \text{ rad/sec}$$

From the Bode plot of Fig. 8.13, the gain of the frequency at ω_{c2}, $20 \log \beta = 22$ dB

Thus, $$\beta = 10^{\frac{22}{20}} = 12.58$$

the upper corner frequency of the lag compensator ω_2 at of one-tenth of ω_{c2}. Thus

$$\tau = \frac{10}{0.6} = 16.6$$

Hence the compensated transfer function is

$$G_{comp}(s) = \frac{\beta(1 + s\tau)}{(1 + s\beta\tau)} = \frac{12.58(1 + s \times 16.6)}{1 + s(208.828)}$$

The open loop transfer function of the compensated system is

$$G_c(s) = G_{comp}(s)\, G_{uc}(s)$$

$$= \frac{12.58(1 + s \times 16.6)}{1 + s(208.828)} \times \frac{8}{(s)\,(s+1)\,(0.2s+1)}$$

Form the bode plot 8.13, the phase margin of the compensated system is found to be 46° is acceptable. The new gain crossover frequency is $\omega_{gc} = 0.6$ rad/sec.

Problem 8.4. *A unity feedback system has an open loop transfer function* $G(s) = \dfrac{K}{s(s+3)(s+10)}$. *Design a suitable lag compensation so that phase margin is ≥ 45° and velocity error constant,* $K_v \geq 15$.

Solution: I. Calculation of gain K

Velocity error constant, $K_v = 15$ (given)

$$= \underset{s \to 0}{\text{Lt}}\; G_{(s)}\, H_{(s)},$$

$$K_v = \underset{s \to 0}{\text{Lt}}\; s \frac{K}{s(s+3)\,(s+10)} = \frac{K}{30} = 15$$

$\therefore$ $$K_v = 450$$

The frequency domain transfer function of the given system is

$$G_{uc}(j\omega) = \frac{450}{j\omega\,(3 + j\omega)\,(10 + j\omega)}$$

In standard from, $$G(j\omega) = \frac{15}{s(1 + j\,0.3\omega)\,(1 + j0.1\omega)}$$

II. Magnitude plot

Corner frequencies, $\omega_1 = 3$ and $\omega_2 = 10$

Sl.No.	Factor	Corner frequency	Slope	Charge in slope
1	$\frac{15}{j\omega}$	–	– 20	–
2	$\frac{1}{1+j0.3\omega}$	$\omega_1 = 3$	– 20	– 40
3	$\frac{1}{1+j0.1\omega}$	$\omega_2 = 10$	– 20	– 60

Consider a low and higher frequencies ω_l and ω_h, such that $\omega_l < \omega_1$ and $\omega_h > \omega_2$ respectively.

Assume, $\omega_l = 0.5$ rad/sec

$\omega_h = 50$ rad/sec

Magnitude $(M) = |G(j\omega)|$ in dB and determine M at $\omega_l, \omega_1, \omega_2, \omega_h$.

At $\omega = \omega_l$, $M = 20 \log_{10} \left|\frac{15}{j\omega}\right| = 20 \log \frac{15}{0.5} = 295$ dB

$\omega = \omega_1$, $M = 20 \log \left(\frac{15}{3}\right) = 20 \log 5 = 14$ dB

$\omega = \omega_2$, $M = \text{slope from } \omega_1 \text{ to } \omega_h \times \log_{10}\left(\frac{\omega_h}{\omega_2}\right) + M \text{ at } \omega = \omega_1$

$$= 40 \log_{10}\left(\frac{10}{3}\right) + 14 = -18 \text{ dB}$$

$\omega = \omega_h$ $M = -60 \log_{10}\left|\frac{50}{10}\right| + M \text{ at } \omega = \omega_2$

$$= -60 \log_{10} 5 + (-18) = -32 \text{ dB}$$

Let the points m, n, o and p be the points corresponding to frequencies ω_1, ω_l, ω_2 and ω_h respectively on the semi log sheet for magnitude plot. Select the scale on graph for magnitude plot as :

x-axis = 1 unit is 0.1 rad/sec

y-axis = 1 unit is 10 db

Draw the straight lines to join the above points and indicate the slope on the respective region as shown in Fig. 8.14.

Phase plot

The phase angle of given function is

$$\phi = \angle G(j\omega) = -90 - \tan^{-1} 0.3\omega - \tan^{-1} 0.1\omega$$

ω	0.1	0.5	1	1.5	2	4	8	10
ϕ	– 92.3	– 101	– 112	– 122	– 132	– 162	– 196	– 206

On the same graph sheet select the scale for phase plot as.

On x-axis = 1 unit is 0.1 rad/sec, same as magnitude plot

y-axis = 1 unit is 20°

Indicate the calculated phase angles on the graph sheet; join the points by a smooth curve as shown in Fig. 8.14.

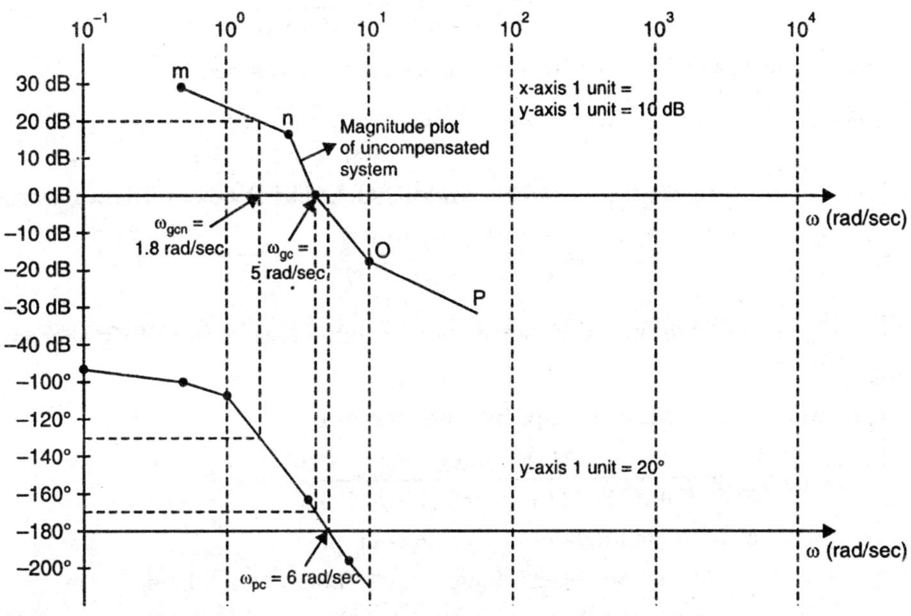

Fig. 8.14

III. Phase margin of uncompensated system

At the gain crossover frequency ω_{gc} = 5 rad/sec, the phase margin (P.M.) is

$$\phi_{pm} = 180° - 172° = 8°$$

This P.M. is not satisfying the specifications. Hence lag compensation is to be designed.

IV. Determine the phase angle lag to be provided by the lag compensation

$$\phi_2 = \phi_d + \in = 45° + 5° = 50°$$

where ϕ_d = desired P.M.

From the bode plot of the uncompensated system, it is observed that at ω = 1.8 rad/sec, the phase angle contribution is 50°

i.e., $$\omega_2 = 1.8 \text{ rad/sec}$$

V. Calculate the transfer function of lag compensation

The gain at the ω_2, $20 \log_{10} \beta = 18$ dB

$$\text{Log}_{10}\, \beta = \frac{18}{20} = 0.9$$

$\therefore$ $$\beta = 10^{0.9} = 1.995 \simeq 2.45$$

The zero of the compensator is located at a frequency of one-tenth of ω_2

$\therefore$ $$Z_c = \frac{1}{T} = \frac{\omega_{gcn}}{10} = \frac{2.45}{10}$$

or $$T = \frac{10}{2.45} = 4$$

Pole of compensated transfer function is

$$G_{com(s)} = \beta\left(\frac{1+sT}{1+s\beta T}\right) = \frac{2.45(1+4s)}{(1+9.8s)}$$

VI. Open loop transfer function of compensated system is

$$G_{c(s)} = G_{com(s)} \times G_{uc(s)}$$

$$= \frac{1}{2.45} \times \frac{2.45(1+4s)}{(1+9.8s)} \times \frac{15}{(1+0.3s)(1+0.1s)}$$

$$= \frac{15(1+4s)}{(1+9.8s)(1+0.3s)(1+0.1s)}$$

where factor $\frac{1}{2.45}$ is the attenuator in series with compensation to nullify the gain of the compensator.

VII. Calculate the P.M. of compensated system

$$G_{c(j\omega)} = \frac{15(1+j4\omega)}{(1+j9.8\omega)(1+j0.3\omega)(1+j0.1\omega)}$$

$$\phi_c = \text{Phase angle of } G_{c(j\omega)} \text{ at } \omega = \omega_{gcn}$$

$$= [\tan^{-1} 4\omega - \tan^{-1} 9.8\omega - \tan^{-1} 0.3\omega - \tan^{-1} 0.1\omega]/\omega = \omega_{gcn}$$

$$= \tan^{-1}(4\times 1.8) - \tan^{-1}(9.8\times 1.8) - \tan^{-1}(0.3\times 1.8) - \tan^{-1}(0.1\times 1.8)$$

$$= 82° - 86.75° - 28.37° - 10.2°$$

$$= -43.32°$$

$\therefore$ P.M. of compensated system, $\phi_m = 180 + \phi_c = 180 - 43.32 = 136.68°$

The P.M. of compensated system is found to be 136.68° which is acceptable.

8.5.3 Lag-lead Compensator

When plant requires both the transient and steady-state response improvement a lag-lead compensator is used.

The lag-lead compensator transfer function in standard form is

$$G_c(s) = \left(\frac{s+\frac{1}{T_1}}{s+\frac{1}{\beta T_1}}\right)\left(\frac{s+\frac{1}{T_2}}{s+\frac{1}{\alpha T_2}}\right);\ \beta > 1,\ \alpha < 1$$

$$= \frac{\left(s+\frac{1}{T_1}\right)\left(s+\frac{1}{T_2}\right)}{\left(s^2 + s\left(\frac{1}{\alpha T_2}+\frac{1}{\beta T_1}\right)+\frac{1}{\alpha\beta T_1 T_2}\right)} \quad (8.6)$$

The electrical lag-lead network will be as shown Fig. 8.15 (*a*). Let us obtain the transfer function if circuit expressed in frequency or *s*-domain is shown in Fig. 8.15 (*b*).

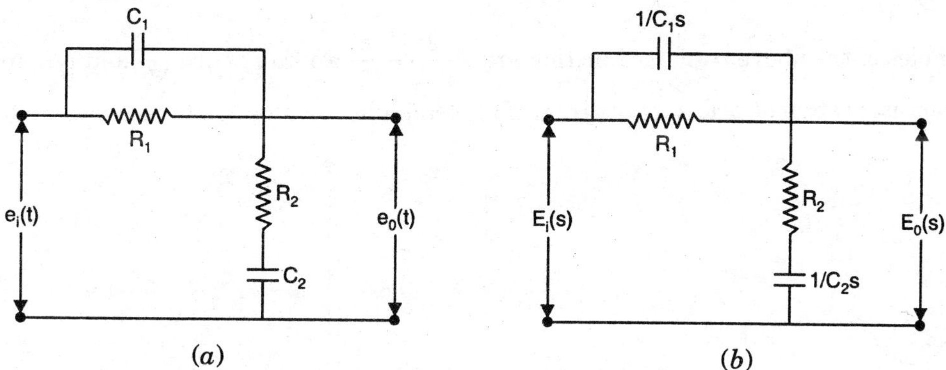

Fig. 8.15

$$E_o(s) = \frac{E_i(s)}{R_2 + \frac{1}{C_2 s} + \left(\frac{\frac{1}{C_1 s} \times R_1}{\frac{1}{C_1 s} \times R_1}\right)} \times \left(R_2 + \frac{1}{C_2 s}\right)$$

$$\frac{E_o(s)}{E_i(s)} = \frac{\left(R_2 + \frac{1}{C_2 s}\right)}{\left(R_2 + \frac{1}{C_2 s} + \frac{\frac{R_1}{C_1 s}}{\frac{1}{C_1 s} + R_1}\right)}$$

By simplifying above equation

$$\frac{E_o(s)}{E_i(s)} = \frac{\left(s + \frac{1}{R_1 C_1}\right)\left(s + \frac{1}{R_2 C_2}\right)}{\left(s_2 + \frac{1}{R_1 C_1} + \frac{1}{R_2 C_1} + \frac{1}{R_2 C_2}\right) s + \frac{1}{R_1 R_2 C_1 C_2}} \qquad (8.7)$$

By comparing eqns. (8.6) and (8.7), we have

$$R_1 C_1 = T_1$$

$$R_2 C_2 = T_2$$

$$\frac{1}{R_1 C_1} + \frac{1}{R_2 C_2} + \frac{1}{R_2 C_2} = \frac{1}{\beta T_1} + \frac{1}{\alpha T_2}$$

and $$R_1 R_2 C_1 C_2 = \alpha\beta\, T_1 T_2$$

From the above equation we can write that $\alpha\beta = 1$

∴ Transfer function can be written as

$$G_c(s) = \frac{(1 + ST_1)(1 + ST_2)}{(1 + S\beta T_1)(1 + ST_2/\beta)}$$

Poles of the above transfer function are $-\frac{1}{\beta T_1} - \frac{\beta}{T_2}$ and zeros are $\frac{-1}{T_1}$ and $\frac{-1}{T_2}$ and the s-plane representation or pole-zero diagram of lag-lead compensator will be as shown in Fig. 8.16.

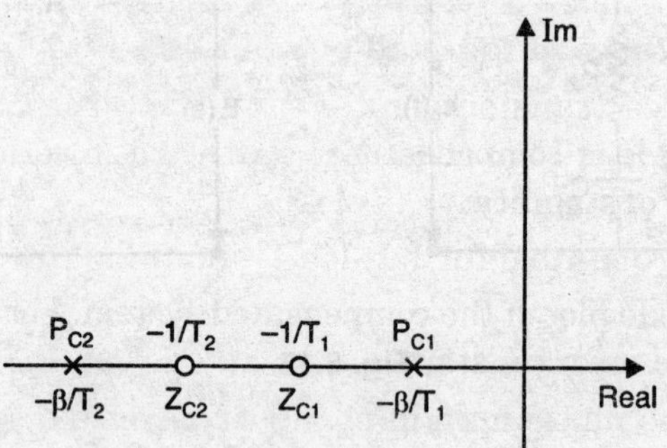

Fig. 8.16

(*i*) Magnitude and phase plot of lag-lead compensator

We have to draw the magnitude plots of different factors in the given transfer function in the increasing order of corner frequencies as shown in Fig. 8.17.

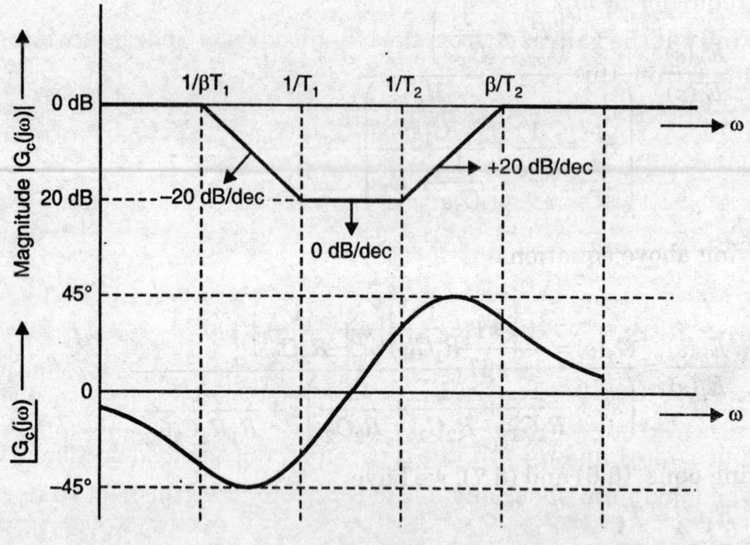

Fig. 8.17

The various corner frequencies in the increasing order are

$$\omega_1 = \frac{\beta}{T_2}$$

$$\omega_2 = \frac{1}{T_2}$$

$$\omega_3 = \frac{1}{\beta T_1}$$

$$\omega_4 = \frac{1}{T_1}$$

(*ii*) **Functions and uses of lag-lead compensator**

When plant requires transient and steady-state responses we can use lag-lead compensator. Basically lag lead-compensator is connected in series. It also increases system error constant and margin of stability.

(*iii*) **Step by step procedure for lag-lead compensator in frequency domain**

Step 1: Draw the Bode plot of the compensated system. For the value of K, which satisfies the specification on the error constant.

Step 2: Measure the phase margin of the compensated system, if it is negative the system is unstable and compensator is required.

Step 3: For the design of lag section first, locate the frequency where the phase angle characteristic contributes a phase angle of ϕ_2, where $\phi_2 = \phi_d + \in$, where ϕ_d = desired phase margin.

Allow $\in$ of 5° to 10° for the phase lag contributed by the lag section of the compensator.

Let this frequency be ω_2.

Step 4: Find out the gain at ω_2 from the magnitude plot and equate it to the attenuation 20 log β. Calculate β from this.

Stem 5: Locate the upper corner frequency of the lag section of the compensator one to ten of ω_{c2}; *i.e.*, $\omega_2 = \dfrac{\omega_2}{10} = \dfrac{1}{T_1}$

Step 6: With β and T_1 determine the transfer function of the lag section as

$$G_1(s) = \frac{1+sT_1}{1+s\beta T_1} \quad \text{or} \quad \frac{1}{\beta}\left(\frac{s+\dfrac{1}{T_1}}{s+\dfrac{1}{\beta T_1}}\right)$$

Step 7: Since $\alpha\beta = 1$ and $\alpha = 1/\beta$, the α parameter of the lead section = $1/\beta$

Step 8: The maximum phase angle lead contributed by the lead section is

$$\phi_m = \sin^{-1}\left(\frac{1-\alpha}{1+\alpha}\right)$$

Step 9: The frequency ω_m where ϕ_m occurs, is at the point, where the gain of the lead compensator is 10 log $\left(\dfrac{1}{\alpha}\right)$ dB in order that it becomes the new gain crossover frequency.

This is measured from Bode plot.

Step 10: $\omega_m = \dfrac{1}{T_2\sqrt{\alpha}}$

Step 11: Compute $\dfrac{1}{T_2} = \omega_m \sqrt{\alpha}$ and $\dfrac{1}{\alpha T_2} = \dfrac{\omega_m}{\sqrt{\alpha}}$

Step 12: The transfer function of the lead section is

$$G_2(s) = \frac{1 + sT_2}{1 + s\alpha T_2}$$

Step 13: The transfer function of lag-lead compensator is

$$G_1(s)G_2(s) = \frac{1 + sT_2}{1 + s\beta T_2} \times \frac{1 + sT_2}{1 + s\alpha T_2}$$

Step 14: The Bode plot for the compensated system is $G_c(s) = G_{uc}(s)\, G_1(s)\, G_2(s)$ is drawn and it is checked against the prescribed specifications.

Problem 8.5. *Consider the system shown in Fig. 8.18*

Compensate the system so that

(*a*) *Velocity error constant K_v is 5 sec^{-1}*

(*b*) *Phase margin not greater than 40°*

(*c*) *Gain margin not greater than 10 dB.*

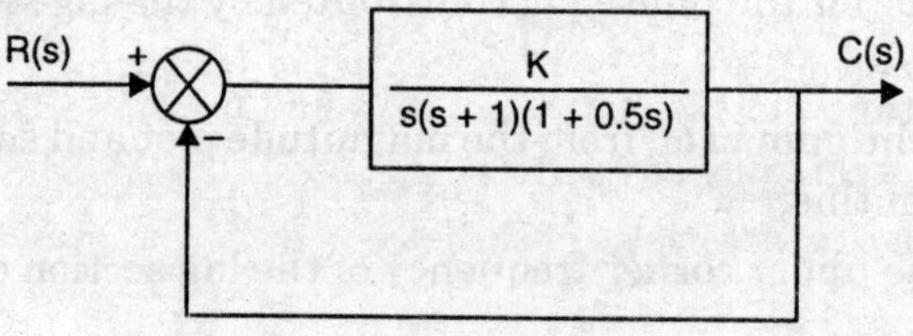

Fig. 8.18

Solution:

$$K_v = \underset{s\to 0}{\text{Lt}}\ sG(s)$$

$$= \underset{s\to 0}{\text{Lt}}\ \frac{K}{s(s+1)(0.5s+1)} = 5\ \text{sec}^{-1}$$

$\therefore$ $K = 5$ is the amplifier gain.

For uncompensated system with $K = 5$, the open loop transfer function is

$$G_{uc}(s) = \frac{5}{s(s+1)(0.5s+1)}$$

Bode-plot for the system is given in Fig. 8.19 (*a*) with the following data obtained for $G_{uc}(j\omega)$ that is

$$G_{uc}(j\omega) = \frac{5}{j\omega(j\omega+1)\,(0.5\,j\omega+1)}$$

The corner frequencies are $\omega = 1$, $\omega = 2$.

The low frequency asymptote $1/j\omega$ passes through $\omega = 1$ with a slope of – 20 dB/sec and the low frequency asymptote of $(1 + j\omega)$ and $(1 + 0.5j\omega)$ and along 0dB line with high frequency asymptote for each with a slope of – 20 dB/sec.

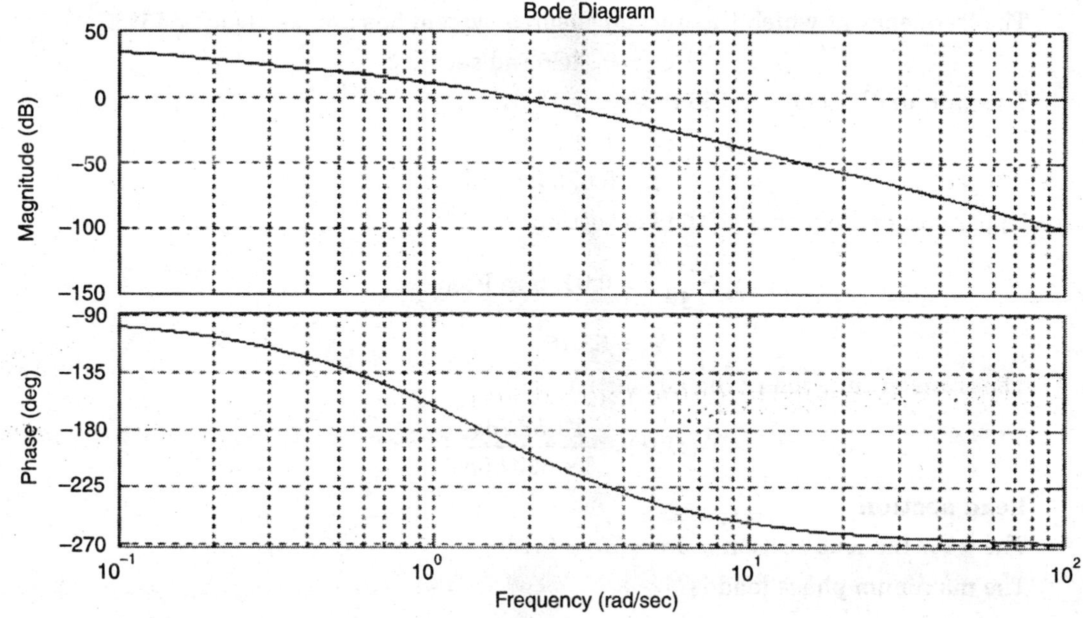

Fig. 8.19 (a)

Thus for the resultant plot $M_{uc}(\omega)$ *Vs* ω, the low frequency asymptote has a slope of – 20 dB/dec and the high frequency asymptote has a slope of – 60 dB/sec.

For the phase angle characteristic of the uncompensated system $\phi(\omega)$ *Vs* ω is as follows.

$$\phi(\omega) = -90 - \tan^{-1}\omega - \tan^{-1} 0.5\,\omega$$

ω *(rad/sec)*	$\phi_{uc}(\omega)$*(deg)*
0	– 90
∞	– 270
0.1	– 98.57
1	– 161.56
2	– 198.43
10	– 252.96

This characteristic is also drawn in Fig. 18.19.

For the uncompensated system gain crossover frequency is ω_{gc} = 2.21 rad/sec and the phase crossover frequency ω_{pc} = 1.6 rad/sec. The gain margin is 6dB and the phase margin is – 27°. Since the phase margin is negative the uncompensated system is unstable. Hence, lag compensator is first designed.

Lag section

The desired phase margin = 40°

The phase lead required for the design of the lag-section is 40° + $\in$

$$\phi_2 = 40° + 5° = 45°$$

The frequency at which the uncompensated system has a phase lead of 45° is

$$\omega_2 = 0.1986 \text{ rad/sec}$$

The gain in dB at $\omega_2 = 0.186$ rad/sec is 15 dB

i.e., $$20 \log \beta = 15 \text{ dB}$$

$$\beta = 5.623 \text{ (say 6)}$$

Upper corner frequency of the lag section is

$$\frac{0.186}{10} = 0.0186 \text{ rad/sec } \frac{1}{T_1}$$

i.e., $$T_1 = 53.76$$

The transfer function of the lag-section is

$$G_1(s) = \frac{1 + 53.76s}{1 + 322.56s}$$

Lead section

The α for the lead section is $1/\beta = 1/6 = 0.166$

The maximum phase lead is

$$\phi_m = \sin^{-1} \frac{(1-\alpha)}{1+\alpha} = 45.61$$

Frequency at which the dB gain

$$= 10 \log (1/\alpha) = 7.781 \text{ dB}$$

is $$\omega_m = 1.5 \text{ rad/sec} = \frac{1}{T_2\sqrt{\alpha}}$$

$\therefore$ $$T_2 = 2.134 \text{ rad/sec}$$

$$\alpha T_2 = 0.354$$

The transfer function of the lead-section is

$$G_2(s) = \frac{(1 + 2.134s)}{(1 + 0.354s)}$$

Thus the transfer function of the lag-lead compensated system is

$$G_c(s) = \frac{5}{s(1+s)(1+0.5s)} \frac{(1+53.76s)}{(1+322.56s)} \frac{(1+2.134s)}{(1+0.354s)}$$

$$= \frac{5 \times 53.76 \times 2.134}{0.5 \times 322.56 \times 0.354} \frac{(s+0.0186)(s+0.468)}{(s+1)(s+2)(s+.0031)}$$

$$= \frac{0.794(s+0.00186)(s+2.272)}{s(s+0.0031)(s+1)(s+2)(s+2.82)}$$

For the compensated system, the Bode-plot is drawn in Fig. 8.19 (*b*). The phase angle characteristic is computed and plotted as given below.

$\omega(rad/sec)$	$\phi(deg)$
0	– 90
∞	– 270
0.5	– 167.12
0.8	– 179.4
1	201.77

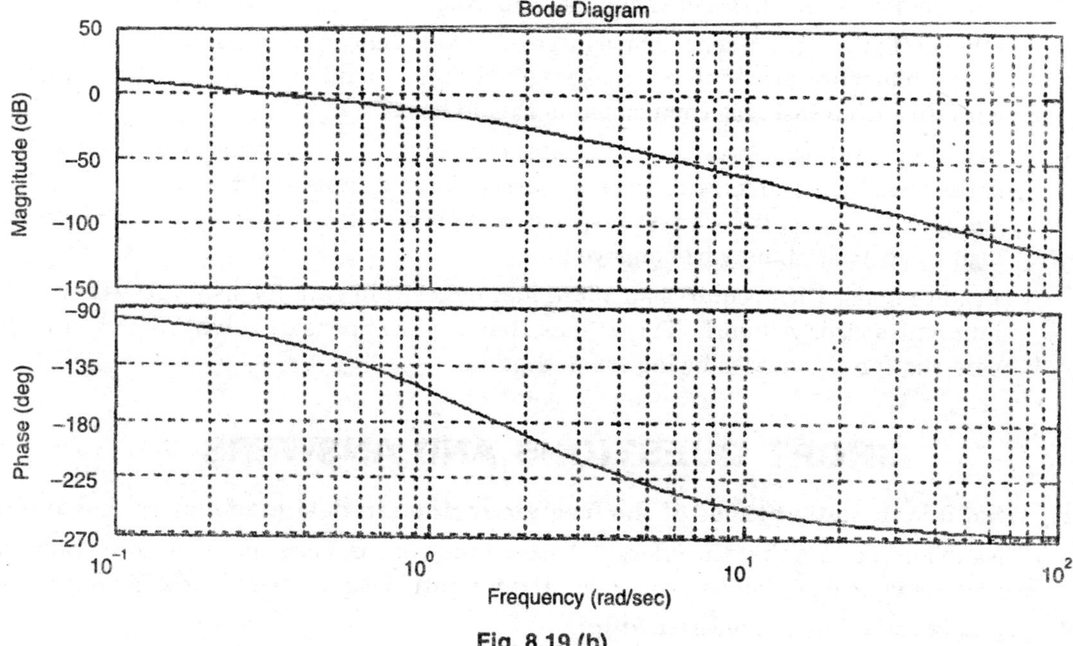

Fig. 8.19 (b).

From the Bode-plot the phase crossover frequency is 0.86 rad/sec and gain crossover frequency is 0.048 rad/sec. Accordingly the gain margin is 29 dB and phase margin is 73°.

Thus the compensated system satisfies the given specification.

Feedback compensation

Although the cascade compensation is economical and quite satisfactory in most cases, feedback compensation may be adopted in the following cases.

(*a*) In a feedback compensator, the feedback signal is obtained from the high level of energy to a lower level, then there is no need of amplification in the feedback path.

(*b*) In non electrical systems suitable compensator devices may not be available.

(*c*) Feedback compensator often provides greater stiffness against load disturbances.

However, the feedback device may not necessarily be an electrical network. It can be a non electrical device like tachometer. Generally, feedback compensator is adopted in minor controls loops.

8.6 COMPARISON OF LEAD AND LAG COMPENSATION

- Lead compensator achieves the desired result through the merits of its phase lead contribution, whereas lag compensation accomplishes the results through the merits of its attenuation property at high frequencies.
- In the frequency domain, lead compensation increases the phase margin and band width, and also yields a higher gain crossover frequency than possible with lag compensation. The bandwidth of the system with lead compensation is always greater than with lag compensation. Therefore, if a large bandwidth or faster response is desired, lead compensation should be employed.

  If noise signals are present, then a large bandwidth may not be desired since it makes the system more susceptible to noise signal because of increase in the high frequency gain. In such a case, lag compensation should be used.
- Lag compensation improves steady state accuracy; however, it reduces the bandwidth. If the reduction of bandwidth is too excessive, the compensated system will exhibit sluggish response. If both fast response and good static accuracy are desired, a lag lead compensation may be employed.
- Lead compensation requires an additional increase in gain to offset the attenuation inherent in this network. This means that lead compensation will require a large gain than that required by lag compensation.

SHORT QUESTIONS AND ANSWERS

1. What are the advantages of the frequency domain design of control systems?

The advantages are that the effect of disturbance, sensor noise and plant uncertainties can be visualized as well as experimental information can be used for design purposes.

2. What is the need of compensator?

Compensator is required for a system in order to

- Stabilize an absolutely unstable system.
- To achieve a specified performance even for a stable system.

3. What are the modes of connections of compensator?

- Series compensator (cascade)
- Parallel (shunt) compensator
- Series-parallel compensator

4. Why Bode plots are commonly used in the frequency domain design?

Bode plot retains the original shape even after compensation. Graphical addition of Bode plot of compensator and un-compensator systems give the Bode plot of compensated system. Gain adjustment just shifts the Bode plot up and down only.

5. What is meant by series compensator?

If the subsystem called compensator is connected in series with the main system in forward path then the scheme is called series compensation.

6. **What is meant by parallel compensator?**

   If the compensator is introduced in parallel with the main plant or one element of the plant it is called feedback compensation or parallel compensation.

7. **What are the different types of compensators?**

   There are different types of compensators like electrical, mechanical, pneumatic, hydraulic or combinations of various types.

8. **What are the different types of electrical compensators?**

   - Lead network or lead compensator
   - Lag compensator or lag network
   - Lag-lead network or lag-lead compensator.

9. **What is the need of lead compensator?**

   If we use lead compensator it improves the system response in following aspects:

   - Speeds up transient response
   - Increases the system error constant and
   - Increases the margin of stability.

10. **What is the need of lag compensator?**

    If lag compensator is inserted in a plant it improves the steady state response of the plant.

11. **What is the need of lag-lead compensators?**

    When a plant requires both the transient and steady-state response improvement a lag-lead compensator is used. It also increases system error constant and margin of stability.

12. **What is a compensator?**

    The compensator is a controller plus something more as it improves the stability margin.

13. **When will a single compensator may fail?**

    For an unstable uncompensated system, lead compensator provides fast response but does not provide enough phase margin, whereas lag compensator stabilizes the system but does not provide enough bandwidth.

14. **What type of compensator is suitable for high frequency?**

    Lag compensator is more suitable.

15. **What are the specifications in frequency domain design are specified?**

    Peak time, peak overshoot, damping ratio resonant frequency, resonant peak, bandwidth, natural frequency, error coefficient, gain margin and phase margin.

OBJECTIVE TYPE QUESTIONS

1. The lead compensator introduces

   (*a*) Phase lead in the system

   (*b*) Attenuation in the system

   (*c*) Amplification in the system

   (*d*) Initially phase lead and then phase lag in the system

2. The lead compensator mainly
   (a) Improves the steady state error
   (b) Improves the transient response
   (c) Improves both steady state and transients equally
   (d) None of the above
3. The lag compensator
   (a) Improves both steady state and transient response
   (b) Improves steady state only
   (c) Improves transients only
   (d) Improves steady state and reduces speed of transient response
4. The lag-lead compensator
   (a) Improves steady state but reduces speed of response
   (b) Improves transient response but no effect on steady state
   (c) Improves both steady state and transient response
   (d) Improves only the transient response
5. The lag network achieves the desired result through its
   (a) Attenuation property at high frequencies
   (b) Attenuation property at low frequencies
   (c) Amplification property
   (d) None of the above
6. A lag network for compensation normally consists of
   (a) R only (b) R and C elements
   (c) R and L elements (d) R, L and C elements
7. The transfer function is $\frac{1+0.5s}{1+s}$. It represents a
   (a) Lead network (b) Lag network
   (c) Lag-lead network (d) Proportional controller
8. A network has a pole at $s = -1$ and a zero at $s = -2$. If this network is excited by sinusoidal input, the output
   (a) Leads the output (b) Lags the input
   (c) Is in phase with input (d) Decays exponentially to zero
9. In lead compensation, zero is
   (a) Nearer to origin (b) Away from the origin
   (c) At the origin (d) Right side of s-plane
10. In lag compensation, pole is
   (a) Nearer to origin (b) Away from the origin
   (c) At the origin (d) Right side of s-plane

KEY

1. (*a*)	**2.** (*b*)	**3.** (*d*)	**4.** (*c*)	**5.** (*a*)	**6.** (*b*)
7. (*b*)	**8.** (*b*)	**9.** (*a*)	**10.** (*a*)		

EXERCISE

1. What is the need of compensation?
2. Explain the modes of connections of compensators.
3. What are types of compensators?
4. Describe the different types of compensation methods
5. What is the need of lead compensator? Derive its transfer function and draw the Bode plot.
6. What is the need of lag compensator? Derive its transfer function and draw the Bode plot.
7. What is the need of lag-lead compensator? Derive its transfer function and draw the Bode plot.
8. Explain the design procedure for lead compensation in frequency domain.
9. Explain the design procedure for lag compensation in frequency domain.
10. Explain the design procedure for lag-lead compensation in frequency domain.
11. Consider a unity feedback system with open loop transfer function $G(s) = \dfrac{K}{s(1+s)(2+s)}$.

    Design a suitable compensator so that the compensated system has

    (*i*) $K_v = 10 \text{ sec}^{-1}$

    (*ii*) Phase margin = 50°

    (*iii*) Gain margin = 10 dB
12. Design a phase lag network for a plant with the open loop transfer function

    $G(s) = \dfrac{20}{s(1+0.2s)^2}$ to have a phase margin of 45°. Verify the performance of the compensated system with the specification.

9 State Variable Analysis

9.1 INTRODUCTION

In the classical control theory, the system studies are based on the transfer function model. The transfer function model was defined for zero initial condition and for linear time invariant (LTIV) system. It is useful to handle effectively SISO systems. It does not provide a complete look into the system *e.g.* for the terminal relationship at the input and output ports.

The transfer function model is based on a mathematical modeling of the physical systems, which is obtained in terms of certain system variables. The system variables are again physical quantities such as position, velocity, force, torque etc. The output variable, which is fed back may not be always an indication of the system variable. That is, the control strategy adopted may satisfy the output requirement but if any of the system variables were to exceed their bounds the transfer function model fails to provide such information. The transfer function can be obtained only for LTIV systems where LT is applicable. The variety of models that can be obtained for a given system is always very limited and may be unique also.

The state variable model overcome most of the above mentioned disadvantages.

9.2 ADVANTAGES OF STATE VARIABLE TECHNIQUES

This technique has the following advantages:

- This approach can be applied to linear or nonlinear, time variant or time invariant systems.
- It is easier to apply where the Laplace transform cannot be applied.
- n^{th} order differential equations can be expressed as 'n' equations of first order whose solutions are easier.
- It is a time domain approach
- This method is suitable for digital computer computation because this is a time domain approach.

  The system can be designed for optimal conditions with respect to given performance indices.

9.3 BASIC DEFINITIONS CONCERNING STATE VARIABLE APPROACH

(*a*) ***State.*** It is an important concept in the system analysis. The state of a dynamic system is the minimal set of variables (known as state variables) such that the knowledge of these variables at the instant $t = t_0$ together with the knowledge of the inputs for $t \geq t_0$, completely determines the behavior of the system for any time $t > t_0$. The state can be regarded as a compact representation of the past history of the system, which can be utilized for predicting its future behaviour in response to an external stimulus.

(*b*) ***State variables.*** State variables are the minimum set of variables, which determine the state of dynamic system. If at least '*n*' variables $[x_1(t), x_2(t), x_3(t), \ldots\ldots, x_n(t)]$ are needed to completely describe the future behaviour of the system, together the initial state and input excitation, then these *n* variables $[x_1(t), x_2(t), x_3(t), \ldots\ldots, x_n(t)]$ are set of state variables. However, once the inputs for time $t \geq t_0$ are specified and knowledge of initial state is known the set of variables $x_1, x_2, x_3, \ldots, x_n$ have characteristics that the future behavior of the system can be determined. Further, note that the state variables need not be physically measurable or observable quantities.

(*c*) ***State vector.*** If '*n*' state variables are necessary to completely determine the behaviour of a given system, then, the '*n*' state variables can be considered as the '*n*' components of the vector called State vector $x(t)$ described in *n* dimensional vector space. The state vector is

$$x(t) = [x_1(t), x_2(t), x_3(t), \ldots\ldots x_n(t)]^T$$

(*d*) ***State space.*** The '*n*' dimensional vector space whose coordinate axis consist of x_1 axis, x_2 axis, ..., x_n axis is called a state space. Any state can be represented by a point in the state space.

(*e*) ***State equations.*** The system equations written in the form of first order differential equations in time domain are known as state equations. The number of state equations and state variables are equal to the order of differential equation required to model it.

(*f*) ***State diagram.*** The pictorial representation of the state model of the system is called as State Diagram. The state diagram of the system can be either in block diagram form or in signal flow graph form.

9.4 STATE VARIABLE REPRESENTATION

Consider a system consisting of '*m*' inputs, '*p*' outputs and '*n*' state variables. The state space representation of a system is visualized as shown in Fig. 9.1. The number of state variables for a system will be equal to the order of the differential equation required to model it. In state space variables are usually represented by $x_1(t)$, $x_2(t)$, $x_3(t)$, ..., $x_n(t)$ and inputs are represented by $u_1(t)$, $u_2(t)$, ... $u_m(t)$ and outputs by $y_1(t), y_2(t)$, ... $y_p(t)$.

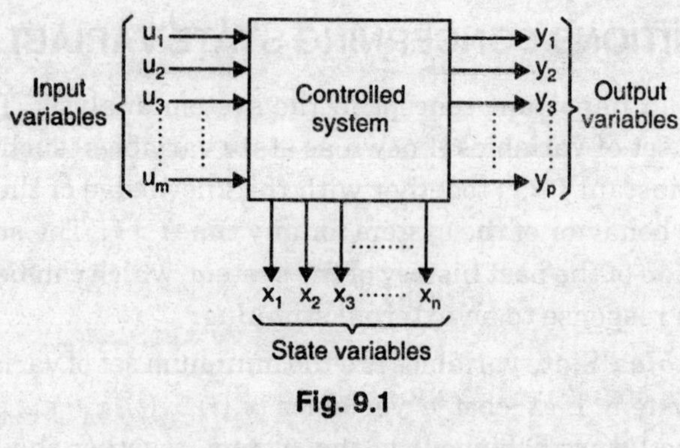

Fig. 9.1

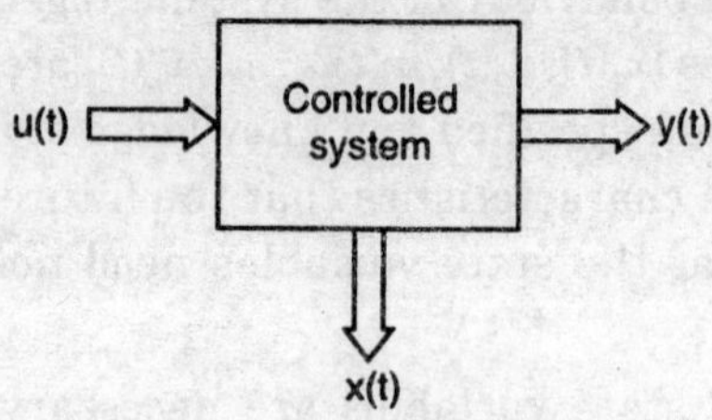

Fig. 9.2

The different variables may be represented by the input vector $u(t)$, output vector $y(t)$ and state vector $x(t)$ as shown in Fig. 9.2.

Where

$$\text{Input vector, } u(t) = \begin{bmatrix} u_1(t) \\ u_2(t) \\ \cdot \\ \cdot \\ u_m(t) \end{bmatrix}$$

$$\text{Output vector, } y(t) = \begin{bmatrix} y_1(t) \\ y_2(t) \\ \cdot \\ \cdot \\ y_p(t) \end{bmatrix}$$

$$\text{State variable vector, } x(t) = \begin{bmatrix} x_1(t) \\ x_2(t) \\ \cdot \\ \cdot \\ x_n(t) \end{bmatrix}$$

9.5 STATE EQUATIONS

The state variable representation can be arranged in the form of n number of first order differential equations as follows:

$$\frac{dx_1}{dt} = \dot{x}_1(t) = f_1\,(x_1, x_2, ...x_n\,;\, u_1, u_2, ...u_m)$$

$$\frac{dx_2}{dt} = \dot{x}_2(t) = f_2\,(x_1, x_2, ...x_n\,;\, u_1, u_2, ...u_m)$$

$$\cdots \quad \cdots \quad \cdots \quad \cdots$$

$$\cdots \quad \cdots \quad \cdots \quad \cdots$$

$$\frac{dx_n}{dt} = \dot{x}_n(t) = f_n\,(x_1, x_2, ...x_n\,;\, u_1, u_2, ...u_m)$$

The 'n' numbers of differential equations may be written in vector notation as

$$\dot{x}(t) = f\,(x(t), u(t))$$

The outputs $y_1(t), y_2(t), ...y_p(t)$ of the system may be given by

$$y_1(t) = g_1\,(x_1, x_2, ...x_n\,;\, u_1, u_2, ...u_m)$$

$$y_2(t) = g_2\,(x_1, x_2, ...x_n\,;\, u_1, u_2, ...u_m)$$

$$\cdots \quad y_p(t) = g_p\,(x_1, x_2, ...x_n\,;\, u_1, u_2, ...u_m)$$

The output equation $y(t)$ may be written in vector notation as

$$y(t) = g\,(x(t), u(t))$$

9.6 STATE MODEL OF LINEAR SYSTEM

The state model of a system consists of state equation and output equation.The state equation of a system is a function of state variables and inputs as defined by equations. For linear time invariant systems are first derivatives of state variables and can be expressed as a linear combination of state variables and inputs.

$$\dot{x}_1 = a_{11}\,x_1 + a_{12}x_2 + ... + a_{1n}x_n + b_{11}u_1 + b_{12}u_2 + ... + b_{1m}u_m$$

$$\dot{x}_2 = a_{21}\,x_1 + a_{22}x_2 + ... + a_{2n}x_n + b_{21}u_1 + b_{22}u_2 + ... + b_{2m}u_m$$

$$\cdots \quad \cdots \quad \cdots \quad \cdots \quad \cdots \quad \cdots \quad \cdots \quad \cdots \quad \cdots \quad \cdots$$

$$\cdots \quad \cdots \quad \cdots \quad \cdots \quad \cdots \quad \cdots \quad \cdots \quad \cdots \quad \cdots \quad \cdots$$

$$\dot{x}_n = a_{n1}\,x_1 + a_{n2}x_2 + ... + a_{nn}x_n + b_{n1}u_1 + b_{n2}u_2 + ... + b_{nm}u_m$$

Where, the coefficients a_{ij} and b_{ij} are arbitrary constants.

The above equations can be expressed in matrix form as

$$\begin{bmatrix} \dot{x}_1 \\ \dot{x}_2 \\ \cdots \\ \dot{x}_n \end{bmatrix} = \begin{bmatrix} a_{11} & a_{12} & \cdots & a_{1n} \\ a_{21} & a_{22} & \cdots & c_{n2} \\ \cdots & \cdots & \cdots & \cdots \\ a_{n1} & a_{n2} & \cdots & a_{nn} \end{bmatrix} \begin{bmatrix} x_1 \\ x_2 \\ \cdots \\ x_n \end{bmatrix} + \begin{bmatrix} b_{11} & b_{12} & \cdots & b_{1m} \\ b_{21} & b_{22} & \cdots & d_{nm} \\ \cdots & \cdots & \cdots & \cdots \\ a_{n1} & a_{n2} & \cdots & a_{nm} \end{bmatrix} \begin{bmatrix} u_1 \\ u_2 \\ \cdot \\ \cdot \\ u_m \end{bmatrix} \qquad (9.1)$$

If vector functions involve time 't', then the system is called Time Varying System. The matrix obtained in eqn. (9.1) can also be written in vector form as

$$\dot{x}_t = A\,x(t) + B\,u(t)$$

where, $x(t)$ = State vector of order $(n \times 1)$

$u(t)$ = Input vector of order $(m \times 1)$

A = System matrix of order $(n \times n)$

B = Input matrix of order $(n \times m)$

The equation, $\dot{x}(t) = Ax(t) + Bu(t)$ is called **State equation.**

The output variables can also be expressed as a linear combination of state variables and inputs as follows:

$$y_1 = c_{11}x_1 + c_{12}x_2 + ... + c_{1n}x_n + d_{11}u_1 + d_{12}u_2 + ... + d_{1m}u_m$$
$$y_2 = c_{21}x_1 + c_{22}x_2 + ... + c_{2n}x_n + d_{21}u_1 + d_{22}u_2 + ... + d_{2m}u_m$$
$$\cdots \quad \cdots \quad \cdots \quad \cdots \quad \cdots \quad \cdots \quad \cdots \quad \cdots \quad \cdots$$
$$\cdots \quad \cdots \quad \cdots \quad \cdots \quad \cdots \quad \cdots \quad \cdots \quad \cdots \quad \cdots$$
$$y_p = c_{p1}x_1 + c_{p2}x_2 + ... + c_{pn}x_n + d_{p1}u_1 + dp_2u_2 + ... + d_{pm}u_m$$

where, the coefficients c_{ij} and d_{ij} are arbitrary constants.

The above equations can be expressed in the matrix form as

$$\begin{bmatrix} y_1 \\ y_2 \\ \cdots \\ y_p \end{bmatrix} = \begin{bmatrix} c_{11} & c_{12} & \cdots & c_{1n} \\ c_{21} & c_{22} & \cdots & c_{2n} \\ \cdots & \cdots & \cdots & \cdots \\ c_{p1} & c_{p2} & \cdots & c_{pn} \end{bmatrix} \begin{bmatrix} x_1 \\ x_2 \\ \cdots \\ x_n \end{bmatrix} + \begin{bmatrix} d_{11} & d_{12} & \cdots & d_{1m} \\ d_{21} & d_{22} & \cdots & d_{2m} \\ \cdots & \cdots & \cdots & \cdots \\ d_{p1} & d_{p2} & \cdots & d_{pm} \end{bmatrix} \begin{bmatrix} u_1 \\ u_2 \\ \cdots \\ u_m \end{bmatrix} \qquad (9.2)$$

The matrix equation (9.2) can also be written in vector form as

$$y(t) = C\,x(t) + D\,u(t)$$

$x(t)$ = State vector of order $(n \times 1)$

$u(t)$ = Input vector of order $(m \times 1)$

$y(t)$ = Output vector of order $(p \times 1)$

C = Output matrix of order $(p \times n)$

D = Transmission matrix of order $(p \times m)$

The equation $y(t) = C\,x(t) + D\,u(t)$ is called **Output Equation.**

The state equation and output equation together is called **state model of the system.** For a linear time invariant system (LTIV) the general form of state model is,

$$\dot{x}(t) = A\,x(t) + B\,u(t) \qquad (9.3)$$

$$y(t) = C\,x(t) + D\,u(t) \qquad (9.4)$$

The block diagram (state diagram) representation of the state model of eqns.(9.3) and (9.4) MIMO is shown in Fig. 9.3.

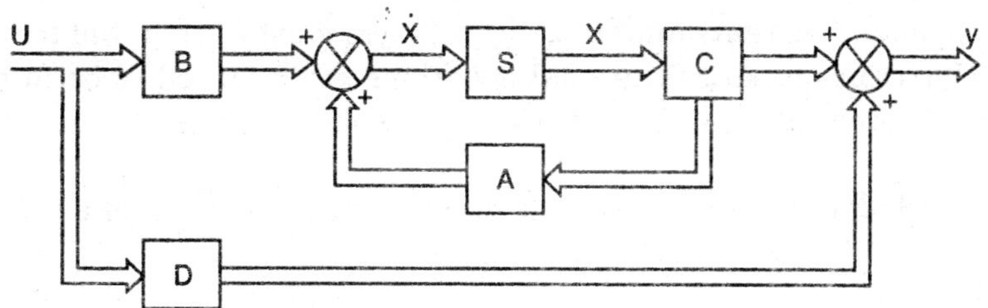

Fig. 9.3

9.7 ELECTRICAL SYSTEM

The electrical system is expressed in the form of a lumped parameter network with voltage and current sources. The state variables are chosen to be currents through the inductors and voltage across the capacitors. The number of state variables is equal to the number of energy storage elements. The energy storage elements are inductance and capacitance.

The current-voltage relation of the basic elements *R*, *L* and *C* are given in table 9.1.

Table 9.1

S.No	*Element*	*Voltage across each element*	*Current through the element*
1	Resistance (*R*)	$v(t) = Ri(t)$	$i(t) = \dfrac{v(t)}{R}$
2	Inductance (*L*)	$v(t) = \dfrac{Ldi(t)}{dt}$	$i(t) = \dfrac{1}{L}\int v(t)dt$
3	Capacitance (*C*)	$v(t) = \dfrac{1}{C}\int (i)t\,dt$	$i(t) = \dfrac{Cdv(t)}{dt}$

9.7.1 Formulation of State Equations

It can be seen that after writing KVL, KCL and *V-I* relationship equations and some algebraic manipulations, it is possible to write state equations for a network. We must choose the particular combination and the particular order in which these equations are invoked. The decision is made on the basis of the following considerations.

(*i*) The inductor currents and capacitor voltages are not eliminated but kept as state variables as we do not want the final equations to contain integrals. The integrals are the ones that arise from the substitution of $i_L = \dfrac{1}{L}\int_0^t V_L\,dt + i_L(t_0)$ for an inductor current in KCL and the substitution of $V_c = \dfrac{1}{C}\int_0^t i\,dt + V_c(t_0)$ for a capacitor voltage in KVL.

(*ii*) The capacitor currents and inductor voltages must be removed from the final set of state variables because the final equations are required to be in the form of first order differential equations. Here the derivative arises from the substitution of

$V_L = L\dfrac{di_L}{dt}$ for an inductor voltage in KVL and $i_c = C\dfrac{dV_c}{dt}$ for a capacitor current in KCL.

Problem 9.1. *Determine the state model of armature-controlled DC motor which is shown in Fig. 9.4.*

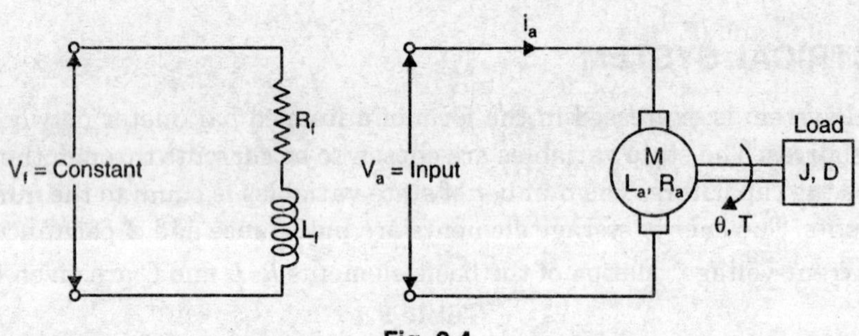

Fig. 9.4

Solution: The equivalent circuit of armature-controlled DC motor is shown in Fig. 9.5.

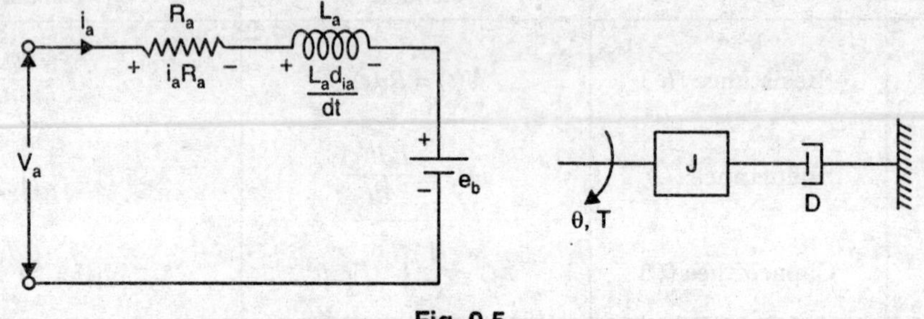

Fig. 9.5

By Kirchhoff's voltage law, we can write

$$i_a R_a + L\frac{di_a}{dt} + e_b = V_a \tag{9.5}$$

The torque T delivered by motor is proportional to the product of armature current i_a, and the air gap flux, which in turn is proportional to the field current.

$\therefore \qquad \phi = K_f i_f$

and $\qquad T = K_1 \phi I_a$

$\therefore \qquad T = K_1 K_f i_f i_a$

In armature-controlled DC motor field, current is held constant.

$$\text{Torque,} \quad T = K_t i_a \tag{9.6}$$

The differential equation governing the mechanical system of motor is given by

$$J\frac{d^2\theta}{dt^2} + D\frac{d\theta}{dt} = T \tag{9.7}$$

The back emf of DC machine is proportional to speed (angular velocity) of shaft (or) derivative of angular displacement.

$$\therefore \qquad e_b \,\alpha\, \frac{d\theta}{dt}$$

Back emf,
$$e_b = K_b \frac{d\theta}{dt} \qquad (9.8)$$

From eqns. (9.5) and (9.8)

$$i_a R_a + L \frac{di_a}{dt} + K_b \frac{d\theta}{dt} = V_a \qquad (9.9)$$

and from eqns. (9.6) and (9.7), we get

$$J \frac{d^2\theta}{dt^2} + D \frac{d\theta}{dt} = K_t\, i_a$$

or
$$J \frac{d\omega}{dt} + D\omega = K_t i_a \left(\because \quad \omega = \frac{d\theta}{dt}\right) \qquad (9.10)$$

Equations (9.9) and (9.10) are differential equations governing armature controlled DC motors.

Let us consider i_a, ω and θ as state variables to model the armature controlled DC motor. The physical variables i_a, ω and θ are related to the general rotation of state variables x_1, x_2 and x_3 and are given as

$$x_1 = i_a \,;\, x_2 = \omega = \frac{d\theta}{dt} \quad \text{and} \quad x_3 = \theta$$

and $V_a = u =$ input variable

Substituting the above state variables in eqns. (9.9) and (9.10)

$$x_1 R_a + L_a \dot{x}_1 + K_b\, x_2 = u$$

or
$$L_a\, \dot{x}_1 = -\, x_1 R_a - K_b\, x_2 + u$$

or
$$\dot{x}_1 = -\frac{R_a}{L_a}\, x_1 - \frac{K_b}{L_a}\, x_2 + \frac{1}{L_a}\, u \qquad (9.11)$$

and
$$J \frac{dx_2}{dt} + Dx_2 = K_t\, x_1$$

or
$$\dot{x}_2 = \frac{K_1}{J}\, x_1 - \frac{D}{J}\, x_2 \qquad (9.12)$$

State variable $x_3 = \theta$

On differentiating x_3 w.r.t. 't', we get

$$\frac{dx_3}{dt} = \frac{d\theta}{dt},$$

$$\therefore \qquad \dot{x}_3 = x_2 \qquad (9.13)$$

Equations (9.11), (9.12) and (9.13) are the state equations of the system.

Arranging the state equations in the matrix form,

$$\begin{bmatrix} \dot{x}_1 \\ \dot{x}_2 \\ \dot{x}_3 \end{bmatrix} = \begin{bmatrix} \dfrac{-R_a}{L_a} & \dfrac{-K_b}{L_a} & 0 \\ \dfrac{K_t}{J} & \dfrac{-D}{J} & 0 \\ 0 & 1 & 0 \end{bmatrix} \begin{bmatrix} x_1 \\ x_2 \\ x_3 \end{bmatrix} + \begin{bmatrix} \dfrac{1}{L_a} \\ 0 \\ 0 \end{bmatrix} u \tag{9.14}$$

Let output, $y = \theta = x_3$

The output equation in the matrix form is

$$y = \begin{bmatrix} 0 & 0 & 1 \end{bmatrix} \begin{bmatrix} x_1 \\ x_2 \\ x_3 \end{bmatrix} \tag{9.15}$$

The state equation (9.14) and output equation (9.15) together constitute the state model of the armature controlled DC motor.

Problem 9.2. *Determine the state model of field controlled DC motor shown in Fig. 9.6.*

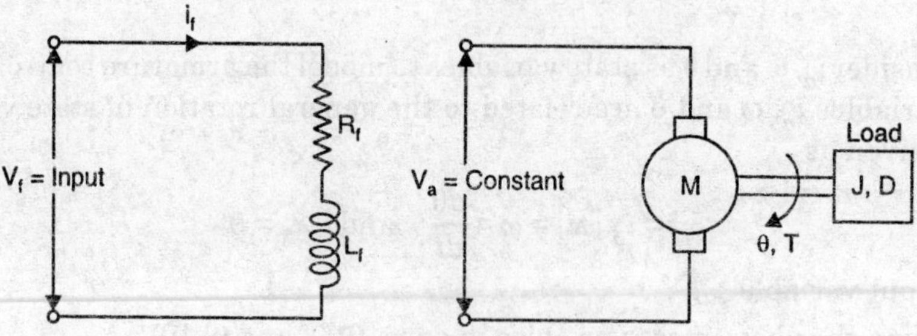

Fig. 9.6

Solution: The equivalent circuit of field controlled DC motor is shown in Fig. 9.7.

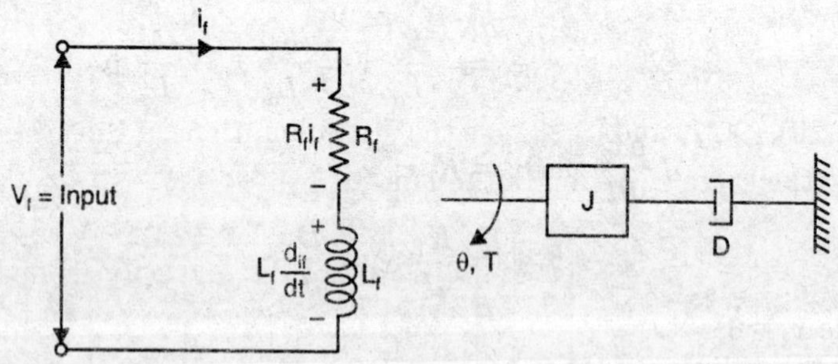

Fig. 9.7

By applying KVL, we get

$$V_f = R_f i_f + L_f \frac{di_f}{dt} \tag{9.16}$$

Torque of the DC motor is proportional to product of flux and armature current.

$$\therefore \qquad T \propto i_f \qquad (\because\ \theta \propto i_f)$$

$$T = K_{tf} i_f \tag{9.17}$$

The differential equation governing the mechanical system of the motor is given by

$$J\frac{d^2\theta}{dt^2} + D\frac{d\theta}{dt} = T \tag{9.18}$$

From eqns. (9.17) and (9.18),

$$J\frac{d^2\theta}{dt^2} + D\frac{d\theta}{dt} = K_{tf} i_f \tag{9.19}$$

Consider i_f, ω and θ as state variables to model the field controlled DC motor

$$x_1 = i_f,\ x_2 = \frac{d\theta}{dt} = \omega \text{ and } x_3 = \theta$$

and $V_f = u$ = input to the system

Substituting the state variables in eqns. (9.16) and (9.19)

$$R_f x_1 + L_f\frac{dx_1}{dt} = u \tag{9.20}$$

or

$$\dot{x}_1 = -\frac{R_f}{L_f}x_1 + \frac{1}{L_f}u$$

and

$$J\frac{dx_2}{dt} + D\,x_2 = K_{tf}x_1$$

or

$$\dot{x}_2 = \frac{K_{tf}}{J}x_1 - \frac{D}{J}x_2 \tag{9.21}$$

The state variable,

$$x_3 = \theta$$

$$\frac{dx_3}{dt} = \frac{d\theta}{dt} = x_2$$

$$\therefore \qquad \dot{x}_3 = x_2 \tag{9.22}$$

Equations (9.20), (9.21) and (9.22) are the state equations of the system.

On arranging the state equation in the matrix form

$$\begin{bmatrix}\dot{x}_1\\ \dot{x}_2\\ \dot{x}_3\end{bmatrix} = \begin{bmatrix}\dfrac{-R_f}{L_f} & \dfrac{-K_b}{L_a} & 0\\ \dfrac{K_{tf}}{J} & \dfrac{-D}{J} & 0\\ 0 & 1 & 0\end{bmatrix}\begin{bmatrix}x_1\\ x_2\\ x_3\end{bmatrix} + \begin{bmatrix}\dfrac{1}{L_f}\\ 0\\ 0\end{bmatrix}u$$

Let output,

$$y = \theta = x_3$$

The output equation in the matrix form is

$$y = \begin{bmatrix} 0 & 0 & 1 \end{bmatrix} \begin{bmatrix} x_1 \\ x_2 \\ x_3 \end{bmatrix}$$

The state equation and the output equation together constitute state model of the field controlled DC motor.

Problem 9.3. *Determine the state model of the electrical network shown in Fig. 9.8.*

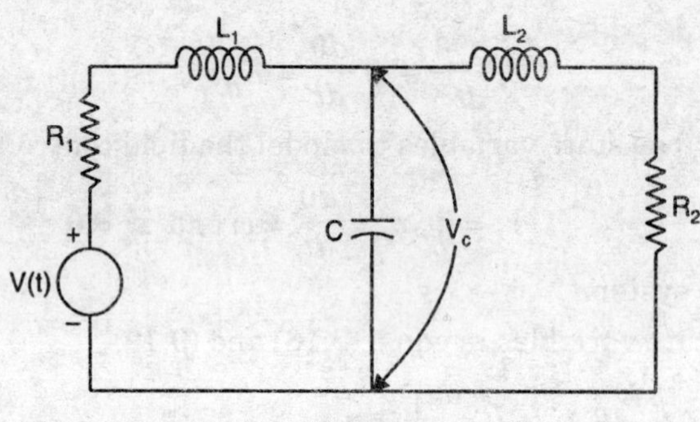

Fig. 9.8

Solution: Apply KVL for loop i_1 :

$$V(t) = R_1 i_1 + L_1 \frac{di_1}{dt} + V_c$$

or
$$\frac{di_1}{dt} = \frac{-R_1}{L_1} i_1 - \frac{1}{L_1} V_c + \frac{1}{L_1} V(t) \qquad (9.22a)$$

for loop i_2 :

$$R_2 i_2 + L_2 \frac{di_2}{dt} + V_c = 0$$

or
$$\frac{di_2}{dt} = \frac{-R_2}{L_2} i_2 - \frac{1}{L_2} V_c \qquad (9.23)$$

and current flowing through capacitor is

$$C \frac{dV_c}{dt} = i_1 + i_2$$

or
$$\frac{dV_c}{dt} = \frac{I}{C} i_1 + \frac{I}{C} i_2 \qquad (9.24)$$

Consider the state variables are

$$x_1 = i_1$$
$$x_2 = i_2$$
$$x_3 = V_c$$

Input, $u = V(t)$

Output, $y = i_2 R_2 = x_2 R_2 \qquad (9.25)$

Substituting the above state variables in eqns. (9.22*a*), (9.23) and (9.24)

$$\therefore \qquad \dot{x}_1 = \frac{R_1}{L_1} x_1 - \frac{1}{L_1} x_3 + \frac{1}{L_1} u$$

$$\dot{x}_2 = \frac{R_2}{L_2} x_2 - \frac{1}{L_2} x_3$$

And
$$\dot{x}_3 = \frac{1}{C} x_1 + \frac{1}{C} x_2$$

The above three state equations are written in matrix form,

$$\begin{bmatrix} \dot{x}_1 \\ \dot{x}_2 \\ \dot{x}_3 \end{bmatrix} = \begin{bmatrix} \frac{-R_1}{L_1} & \frac{-K_b}{L_1} & 0 \\ 0 & \frac{-R_2}{L_2} & \frac{-1}{L_2} \\ 0 & 1 & 0 \end{bmatrix} \begin{bmatrix} x_1 \\ x_2 \\ x_3 \end{bmatrix} + \begin{bmatrix} \frac{1}{L_1} \\ 0 \\ 0 \end{bmatrix} u$$

and the output eqn. (9.25) in matrix form

$$y = [0 \;\; R_2 \;\; 0] \begin{bmatrix} x_1 \\ x_2 \\ x_3 \end{bmatrix}$$

Problem 9.4. *Determine the state model of the electrical network shown in Fig. 9.9.*

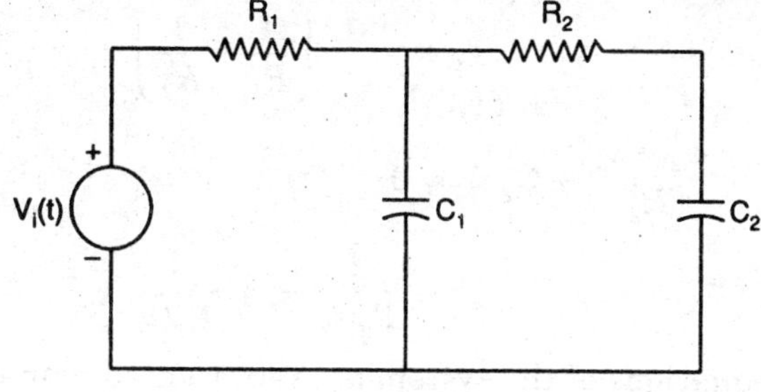

Fig. 9.9

Solution:

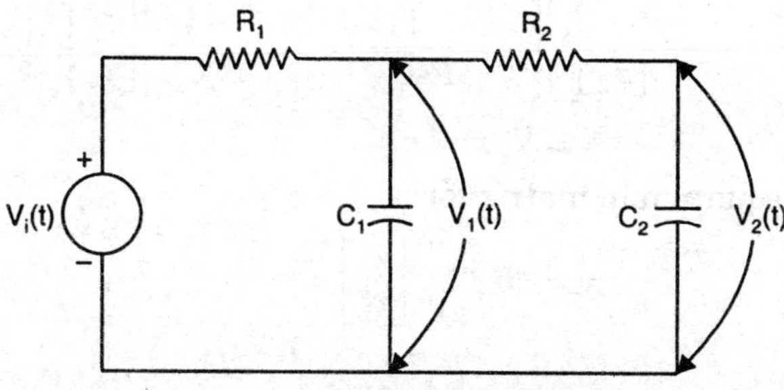

Fig. 9.10

In the network shown in Fig. 9.10,

apply KCL at node 1 :

$$\frac{V_1(t) - V_i(t)}{R_1} + \frac{C_1 dV_1(t)}{dt} + \frac{V_1(t) - V_2(t)}{R_2} = 0 \quad (9.26)$$

At node 2 :

$$\frac{V_2(t) - V_1(t)}{R_2} + \frac{C_2 dV_2}{dt} = 0 \quad (9.27)$$

The number of state variables = number of energy stored components in the network shown in Fig. 9.10. Two capacitors, hence voltage across two capacitors are chosen as state variables.

i.e., $V_1(t) = x_1$

$V_2(t) = x_2$

Input, $V_i(t) = u$

And output, $V_2(t) = x_2 = y$

Substituting the above variables in eqns. (9.26) and (9.27)

$$\frac{x_1 - u}{R_1} + C_1 \dot{x}_1 + \frac{x_1 - x_2}{R_2} = 0$$

or
$$C_1 \dot{x}_1 = -\left(\frac{1}{R_1} + \frac{1}{R_2}\right) x_1 + \frac{1}{R_2} x_2 + \frac{1}{R_1} u$$

or
$$\dot{x}_1 = -\frac{1}{C_1}\left(\frac{1}{R_1} + \frac{1}{R_2}\right) x_1 + \frac{1}{R_2 C_1} x_2 + \frac{1}{R_1 C_1} u$$

and
$$\frac{x_2 - x_1}{R_2} + C_2 \dot{x}_2 = 0$$

or
$$\dot{x}_2 = \frac{1}{R_2 C_2} x_1 - \frac{1}{R_2 C_2} x_2$$

The above state equations of the system are represented in matrix form as

$$\begin{bmatrix} \dot{x}_1 \\ \dot{x}_2 \end{bmatrix} = \begin{bmatrix} \frac{-1}{C_1}\left(\frac{1}{R_1} + \frac{1}{R_2}\right) & \frac{1}{R_2 C_1} \\ \frac{1}{R_2 C_2} & \frac{-1}{R_2 C_2} \end{bmatrix} \begin{bmatrix} x_1 \\ x_2 \end{bmatrix} + \begin{bmatrix} \frac{1}{R_2 C_1} \\ 0 \end{bmatrix} u$$

The output $y = V_2(t) = x_2$

$\therefore$ The output equation in matrix form

$$y = [0 \quad 1] \begin{bmatrix} x_1 \\ x_2 \end{bmatrix}$$

Problem 9.5. *Determine the state space representation of the network shown in Fig. 9.11.*

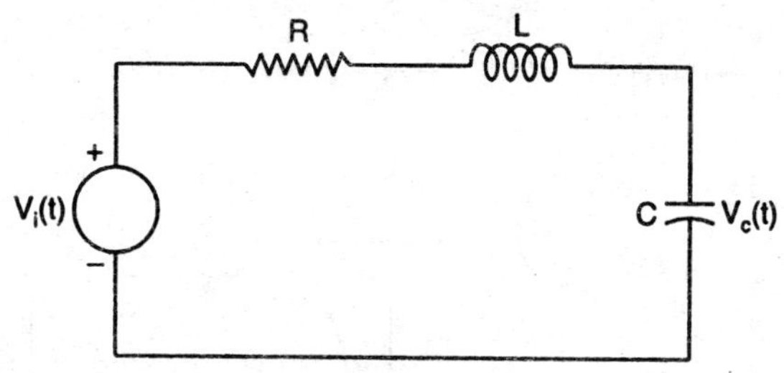

Fig. 9.11

Solution: There are 2 energy storage elements in the network shown in Fig. 9.11. Let us choose 2 state variables.

i.e., $x_1 = i(t)$ = current through L

$x_2 = V_c(t)$ = voltage across C

$u = V_i(t)$ = input voltage of the system

and output, $y = V_c(t) = x_2$

Using KVL

$$V_i(t) = R_i(t) + \frac{Ld_i(t)}{dt} + V_c(t) \tag{9.28}$$

and
$$i(t) = \frac{CdV_c(t)}{dt} \tag{9.29}$$

Substituting the state variables in eqns. (9.28) and (9.29)

$$Rx_1 + L\dot{x}_1 + x_2 = u(t)$$

or
$$\dot{x}_1 = \frac{-R}{L}x_1 - \frac{1}{L}x_2 + \frac{u(t)}{L} \tag{9.30}$$

and
$$x_1 = C\,\dot{x}_2$$

or
$$\dot{x}_2 = \frac{1}{C}x_1 \tag{9.31}$$

The state eqns. (9.30) and (9.31) are written in matrix form

$$\begin{bmatrix} \dot{x}_1 \\ \dot{x}_2 \end{bmatrix} = \begin{bmatrix} \frac{-R}{L} & \frac{-1}{L} \\ \frac{1}{C} & 0 \end{bmatrix} \begin{bmatrix} x_1 \\ x_2 \end{bmatrix} + \begin{bmatrix} \frac{1}{L} \\ 0 \end{bmatrix} u(t)$$

The output in matrix form

$$y = [0 \quad 1] \begin{bmatrix} x_1 \\ x_2 \end{bmatrix}$$

Problem 9.6. *Obtain the state space model of the electric circuit shown in Fig. 9.12.*

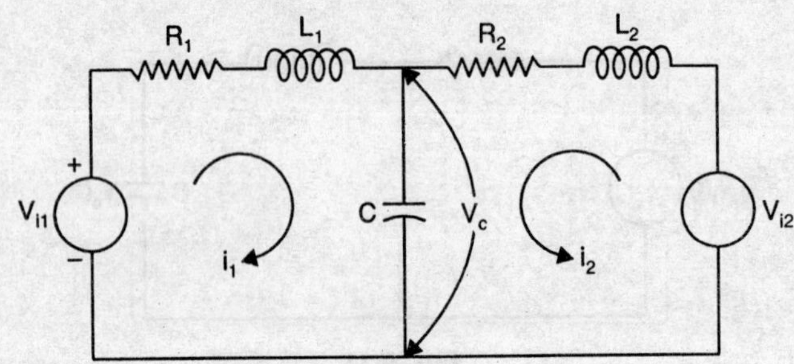

Fig. 9.12

Solution: Applying KVL to the network, we have

$$V_{i1} = R_1\, i_1 + L_1\, \frac{di_1}{dt} + V_c \tag{9.32}$$

$$V_{i2} = R_2\, i_2 + L_2\, \frac{di_2}{dt} + V_c \tag{9.33}$$

$$i + i_2 = C\, \frac{dV_c}{dt} \tag{9.34}$$

There are 3 energy storage elements in the network shown in Fig. 9.12. Let us choose 3 state variables

i.e., $x_1 = i_1$ = current flowing through L_1

$x_2 = i_2$ = current flowing through L_2

$x_3 = V_c$ = voltage across the capacitor

and inputs, $u_1 = V_{1i}$

$u_2 = V_{2i}$

Substituting the state variables in eqns. (9.32), (9.33) and (9.34), we get

$$u_1(t) = R_1 x_1 + L_1\, \dot{x}_1 + x_3$$

or
$$\dot{x}_1 = -\frac{R_1}{L_1}\, x_1 - \frac{1}{L_1}\, x_3 + \frac{u_1(t)}{L_1}$$

$$u_2(t) = R_2 x_2 + L_2\, \dot{x}_2 + x_3$$

or
$$\dot{x}_2 = -\frac{R_2}{L_2}\, x_1 - \frac{1}{L_2}\, x_3 + \frac{1}{L_2}\, u_2(t)$$

and
$$\dot{x}_3 = \frac{1}{C}\, x_1 + \frac{1}{C}\, x_2$$

On arranging the above 3 state equations in matrix form

$$\begin{bmatrix} \dot{x}_1 \\ \dot{x}_2 \\ \dot{x}_3 \end{bmatrix} = \begin{bmatrix} \frac{R_1}{L_1} & 0 & \frac{1}{L_1} \\ 0 & \frac{R_2}{L_2} & \frac{1}{L_2} \\ \frac{1}{C} & \frac{1}{C} & 0 \end{bmatrix} \begin{bmatrix} x_1 \\ x_2 \\ x_3 \end{bmatrix} + \begin{bmatrix} \frac{1}{L_1} \\ \frac{1}{L_2} \\ 0 \end{bmatrix} \begin{bmatrix} u_1 \\ u_2 \\ 0 \end{bmatrix}$$

Problem 9.7. *Determine the state equations of the network shown in Fig. 9.13.*

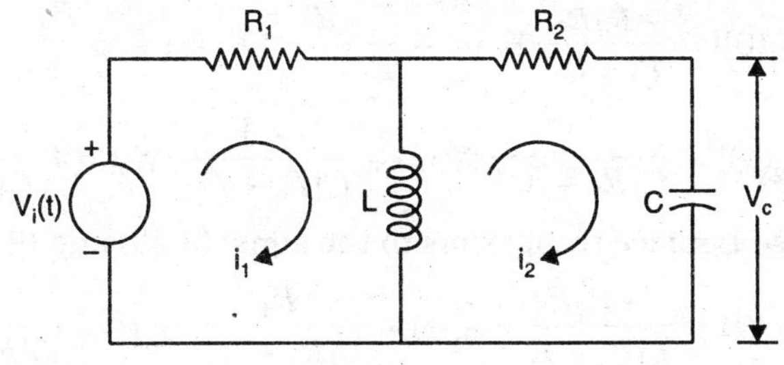

Fig. 9.13

Solution: There are 2 energy storage elements in the network shown in Fig. 9.13.

Let us choose 2 state variables

i.e., $x_1 = i_1 - i_2 =$ current flowing through the inductance L.

and $x_2 = V_c =$ voltage across the capacitor.

And input, $u = V_i(t)$

Apply the KVL for i_1 loop,

$$L\left[\frac{di_1}{dt} - \frac{di_2}{dt}\right] + R_1 i_1 = V_i(t) \tag{9.35}$$

and for i_2 loop,

$$\frac{1}{C}\int i_2\,dt + R_2 i_2 + L\left[\frac{di_2}{dt} - \frac{di_1}{dt}\right] = 0 \tag{9.36}$$

So, from the definition of x_2,

$$x_2 = \frac{1}{C}\int i_2\,dt \tag{9.37}$$

$$\dot{x}_2 = \frac{1}{C} i_2$$

From eqn. (9.35),

$$L\frac{dx_1}{dt} + R_1[x_1 + i_2] = V_i(t) \tag{9.38}$$

and from eqn. (9.36),

$$x_2 + R_2 i_2 - L\frac{dx_1}{dt} = 0$$

From eqn. (9.38), substituting $L\dfrac{dx_1}{dt} = -R_1[x_1 + i_2] + V_i(t)$ in above equation.

$$x_2 + R_2 i_2 + R_1(x_1 + i_2) = V_i(t) \tag{9.39}$$

Taking the Laplace transform for the eqns. (9.37), (9.38) and (9.39)

$$L[sX_1(s) - x_1(0)] + R_1[X_1(s) + I_2(s)] = V_i(s) \tag{9.40}$$

$$X_2(s) + R_2 I_2(s) + R_1[X_1(s) + I_2(s)] = V_i(s) \tag{9.41}$$

and

$$[sX_2 - x_2(0)] = \frac{1}{C} I_2(s) \tag{9.42}$$

Eliminating $I_2(s)$ in eqns. (9.40) and (9.42) from eqn. (9.41), we get

$$L[sX_1(s) - x_1(0)] = \frac{-R1R_2}{R_1 + R_2} X_1(s) + \frac{-R_1}{R_1 + R_2} X_2(s) + \frac{R_2}{R_1 + R_2} V_i(s) \quad (9.43)$$

And $$[sX_2(s) - x_2(0)] = \frac{-R_1}{C[R_1 + R_2]} X_1(s) + \frac{-1}{C[R_1 + R_2]} X_2(s) + \frac{1}{C[R_1 + R_2]} V_i(s) \quad (9.44)$$

Taking the inverse Laplace transforms to the eqns. (9.43) and (9.44)

$$\dot{x}_1(t) = \frac{-R_1R_2}{L[R_1 + R_2]} x_1(t) + \frac{R_1}{L[R_1 + R_2]} x_2(t) + \frac{R_2}{L[R_1 + R_2]} V_i(t)$$

$$\dot{x}_2(t) = \frac{-R_1}{C[R_1 + R_2]} x_1(t) + \frac{-1}{C[R_1 + R_2]} x_2(t) + \frac{1}{C[R_1 + R_2]} V_i(t)$$

The above state equation in the matrix form

$$\begin{bmatrix} \dot{x}_1 \\ \\ \dot{x}_2 \end{bmatrix} = \begin{bmatrix} \frac{-R_1R_2}{L[R_1 + R_2]} & \frac{R_1}{L[R_1 + R_2]} \\ \frac{-R_1}{C[R_1 + R_2]} & \frac{-1}{C[R_1 + R_2]} \end{bmatrix} \begin{bmatrix} x_1 \\ \\ x_2 \end{bmatrix} + \begin{bmatrix} \frac{R_2}{L[R_1 + R_2]} \\ \frac{-1}{C[R_1 + R_2]} \end{bmatrix} V_i(t)$$

Problem 9.8. *Determine the state equations in matrix form for the circuit shown in Fig. 9.14.*

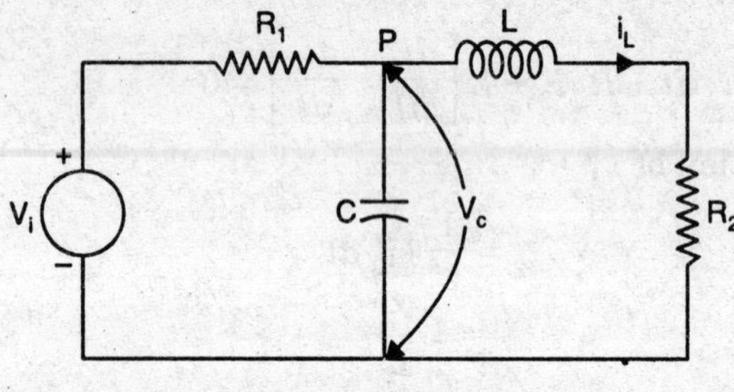

Fig. 9.14

Solution: Taking KVL for right loop

$$L\frac{di_L}{dt} + I_L R_2 = V_c$$

or $$L\frac{di_L}{dt} = \frac{-R_2}{L} i_L + \frac{V_c}{L} \quad (9.45)$$

Taking KCL at point P

and $$C\frac{dV_c}{dt} = \frac{V_i - V_c}{R_1} - i_L$$

$$\frac{dV_c}{dt} = \frac{V_i - V_c}{R_1 C} - \frac{i_L}{C}$$

$$= \frac{1}{C} I_L - \frac{V_c}{R_1 C} + \frac{V_i}{R_1 C} \quad (9.46)$$

writing the state eqns. (9.45) and (9.46) in matrix form

$$\begin{bmatrix} \dot{i}_L \\ \dot{V}_c \end{bmatrix} = \begin{bmatrix} \frac{-R_2}{L} & \frac{1}{L} \\ \frac{-1}{C} & \frac{-1}{R_1 C} \end{bmatrix} \begin{bmatrix} i_L \\ V_c \end{bmatrix} + \begin{bmatrix} 0 \\ \frac{1}{R_1 C} \end{bmatrix} V_i$$

Problem 9.9. *Determine the A, B, C and D matrices of the circuit shown in Fig. 9.15.*

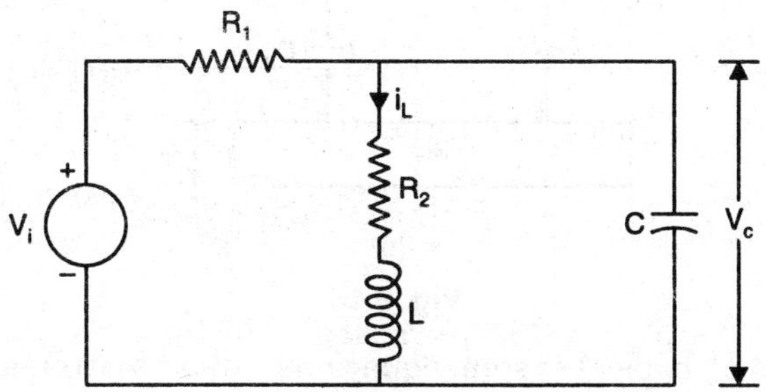

Fig. 9.15

Solution: Apply KCL at junction P,

$$\frac{V_i - V_c}{R_1} = i_L + C\frac{dV_c}{dt}$$

or

$$\frac{dV_c}{dt} = -\frac{V_c}{CR_1} - \frac{i_L}{C} + \frac{V_i}{CR_1} \qquad (9.47)$$

and KVL to the right loop

$$L\frac{di_L}{dt} + i_L R_2 = V_c$$

$$\frac{di_L}{dt} = \frac{V_c}{L} - \frac{i_L R_2}{L} \qquad (9.48)$$

The state equations (9.47) and (9.48) are written in matrix form

$$\begin{bmatrix} \dot{V}_c \\ \dot{i}_L \end{bmatrix} = \begin{bmatrix} \frac{-1}{CR_1} & \frac{-1}{C} \\ \frac{1}{L} & \frac{-R_2}{L} \end{bmatrix} \begin{bmatrix} V_c \\ i_L \end{bmatrix} + \begin{bmatrix} \frac{-1}{CR_1} \\ 0 \end{bmatrix} V_i$$

$$A = \begin{bmatrix} \frac{-1}{CR_1} & \frac{-1}{C} \\ \frac{1}{L} & \frac{-R_2}{L} \end{bmatrix} \quad \text{and} \quad B = \begin{bmatrix} \frac{-1}{CR_1} \\ 0 \end{bmatrix}$$

The output equation is

$$Y = [1\ 0]\begin{bmatrix} V_c \\ i_L \end{bmatrix} + [0]\ V_i$$

$$C = [1\ 0] \quad D = [0]$$

Problem 9.10. *Determine the state equations of the mechanical system shown in Fig. 9.16.*

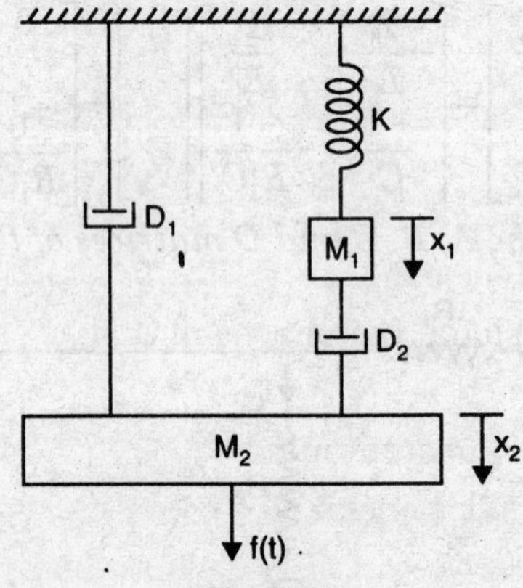

Fig. 9.16

Solution: The basic physical system equations can be written as

$$M_2 \frac{d^2 x_2}{dt^2} + (D_1 + D_2) \frac{dx_2}{dt} - D_2 \frac{dx_1}{dt} = f(t) \tag{9.49}$$

and
$$M_1 \frac{d^2 x_1}{dt^2} + D_1 \frac{dx_1}{dt} + Kx_1 - D_2 \frac{dx_2}{dt} = 0 \tag{9.50}$$

Let us consider

$$x_3 = \frac{dx_1}{dt} \tag{9.51}$$

$$x_4 = \frac{dx_2}{dt} \tag{9.52}$$

Substituting x_3 and x_4 values in eqns. (9.49) and (9.50), we get

$$M_2 \frac{dx_4}{dt} + (D_1 + D_2)\, x_4 - D_2 x_3 = f(t)$$

or
$$\frac{dx_4}{dt} = \frac{D_2}{M_2} x_3 - \frac{(D_1 + D_2)}{M_2} x_4 + \frac{1}{M_2} f(t) \tag{9.53}$$

and
$$M_1 \frac{dx_3}{dt} + D_2\, x_3 + Kx_1 - D_2 x_4 = 0$$

or
$$\frac{dx_3}{dt} = \frac{-K}{M_1} x_1 - \frac{D_2}{M_1} x_3 + \frac{D_2}{M_1} x_4 \tag{9.54}$$

The state equations (9.51) to (9.54) are written in matrix form

$$\begin{bmatrix} \dot{x}_1 \\ \dot{x}_2 \\ \dot{x}_3 \\ \dot{x}_4 \end{bmatrix} = \begin{bmatrix} 0 & 0 & 1 & 0 \\ 0 & 0 & 1 & 0 \\ \dfrac{-K}{M_1} & 0 & \dfrac{-D_2}{M_1} & \dfrac{D_2}{M_1} \\ 0 & 0 & \dfrac{D_2}{M_2} & \left(\dfrac{(D_1 + D_2)}{M_2}\right) \end{bmatrix} \begin{bmatrix} x_1 \\ x_2 \\ x_3 \\ x_4 \end{bmatrix} + \begin{bmatrix} 0 \\ 0 \\ 0 \\ \dfrac{-1}{M_2} \end{bmatrix} f(t)$$

9.8 DIAGONALIZATION (OR) CANONICAL FORM (OR) SIMILAR TRANSFORMATION

Let us consider the state model of linear time invariant system equations as

$$\dot{x} = Ax + Bu \quad (9.55)$$

and
$$y = Cx + Du \quad (9.56)$$

where, x is an n-dimensional state vector.

Consider another n-dimensional vector z, such that

$$x = Pz \quad (9.57)$$

where, $P = n \times n$ is a non singular constant matrix

Taking derivative on both sides of eqn. (9.57), we get

$$\therefore \quad \dot{x} = P\dot{z} \quad (\because \ P \text{ is constant matrix}) \quad (9.58)$$

Substituting x and $\dot{x}$ from eqns. (9.57) and (9.58) respectively in equation (9.55)

$$\therefore \quad P\dot{z} = APz + Bu$$

or
$$\dot{z} = P^{-1}APz + P^{-1}Bu \quad (9.59)$$

Similarly

$$y = CPz + Du \quad (9.60)$$

If the matrix P can be selected such that $P^{-1}AP$ is a diagonal matrix, Λ, then this model given by eqns. (9.59) is a **canonical state model.**

The eqns. (9.59) can be written as

$$\dot{z} = \Lambda z + B'u$$

and
$$y = C'z + Du$$

where $\Lambda = P^{-1}AP$ = diagonal matrix

$B' = P^{-1}B$ and $C' = CP$

The matrix P can be determined by use of the eigen vectors of matrix A.

i.e.,
$$P = [P_1 \, P_2 \, P_3 \, ...P_n]$$

where P_k denotes the eigen vectors that is associated with the eigen value λ_k. If matrix A is of phase variable canonical form like

$$A = \begin{bmatrix} 0 & 1 & 0 & \dots & 0 \\ 0 & 0 & 1 & \dots & 0 \\ \dots & \dots & \dots & \dots & \dots \\ 0 & 0 & 0 & \dots & 1 \\ -a_n & -a_{n-1} & \dots & \dots & -a_1 \end{bmatrix}$$

Then matrix P which diagonalizes matrix A is the Vander Monde matrix.

$$\therefore \quad P = \begin{bmatrix} 1 & 1 & \dots & 1 \\ \lambda_1 & \lambda_2 & \dots & \lambda_n \\ \lambda_1^2 & \lambda_2^2 & \dots & \lambda_n^2 \\ \lambda_1^{n-1} & \lambda_2^{n-2} & \dots & \lambda_n^{n-1} \end{bmatrix}$$

where $\lambda_1, \lambda_2, \dots, \lambda_n$ are the eigen values of matrix A.

Problem 9.11. *Reduce the matrix A to diagonal matrix,* $A = \begin{bmatrix} 3 & -2 \\ -1 & 2 \end{bmatrix}$

Solution: The characteristic equation is

$$|\lambda I - A| = \begin{vmatrix} \lambda & 0 \\ 0 & \lambda \end{vmatrix} - \begin{vmatrix} 3 & -2 \\ -1 & 2 \end{vmatrix} = \begin{vmatrix} \lambda - 3 & 2 \\ 1 & \lambda - 2 \end{vmatrix} = 0$$

or $$(\lambda - 3)(\lambda - 2) = 0$$

$$\lambda^2 - 5\lambda + 4 = 0$$

$$(\lambda - 1)(\lambda - 4) = 0$$

The eigen values of A matrix are $\lambda_1 = 1$ and $\lambda_2 = -4$. The given A matrix is not of phase canonical form and hence P matrix will not be the Vander Monde matrix. Therefore, matrix P can be determined from the eigen vectors of A.

Let the eigen vector associated with $\lambda_1 = 1$ be represented by $P_1 = \begin{vmatrix} P_{11} \\ P_{12} \end{vmatrix}$.

Then $$[\lambda_1 I - A]\, P_1 = 0$$

$\Rightarrow$ $$\begin{bmatrix} -2 & 2 \\ 1 & -1 \end{bmatrix} \begin{bmatrix} P_{11} \\ P_{12} \end{bmatrix} = 0$$

$$-2P_{11} + 2P_{12} = 0$$

and $$P_{11} - P_{12} = 0$$

Let $P_{11} = 1$, then $$P_{12} = 1$$

$\therefore$ $$P_1 = \begin{bmatrix} 1 \\ 1 \end{bmatrix}$$

The eigen vector corresponding to $\lambda_2 = 4$ is $[\lambda_2 I - A]\, P_2 = 0$

$\Rightarrow$ $$\begin{bmatrix} 1 & 2 \\ 1 & 2 \end{bmatrix} \begin{bmatrix} P_{21} \\ P_{22} \end{bmatrix} = 0$$

$$P_{21} + 2P_{22} = 0$$

Let $P_{21} = 2$, then $$P_{22} = -1$$

$\therefore$ $$P_2 = \begin{bmatrix} 2 \\ -1 \end{bmatrix}$$

$\therefore$ $$P = \begin{bmatrix} 1 & 2 \\ 1 & -1 \end{bmatrix}$$

The diagonal matrix, $$\Lambda = P^{-1}AP = \begin{bmatrix} 1 & 0 \\ 0 & 4 \end{bmatrix}$$

Problem 9.12. *Reduce the matrix A to diagonal matrix,* $A = \begin{bmatrix} 0 & 1 & 0 \\ 0 & 0 & 1 \\ -6 & -11 & -6 \end{bmatrix}$.

Solution: The eigen values can be determined as follows:

The characteristic equation is

$$|\lambda I - A| = \begin{vmatrix} \lambda & 0 & 0 \\ 0 & \lambda & 0 \\ 0 & 0 & \lambda \end{vmatrix} - \begin{vmatrix} 0 & 1 & 0 \\ 0 & 0 & 1 \\ -6 & -11 & -6 \end{vmatrix} = \begin{vmatrix} \lambda & -1 & 0 \\ 0 & \lambda & -1 \\ 6 & 11 & \lambda+6 \end{vmatrix} = 0$$

or $$\lambda^3 + 6\lambda^2 + 11\lambda + 6 = 0$$

or $$(\lambda + 1)(\lambda + 2)(\lambda + 3) =$$

The eigen values of matrix A are $\lambda_1 = -1$, $\lambda_2 = -2$ and $\lambda_3 = -3$. The given matrix A is a phase variable canonical form. Then matrix P will be Vander Monde matrix

$$\therefore \qquad P = \begin{bmatrix} 1 & 1 & 1 \\ \lambda_1 & \lambda_2 & \lambda_3 \\ \lambda_1^2 & \lambda_2^2 & \lambda_3^2 \end{bmatrix} = \begin{vmatrix} 1 & 1 & 1 \\ -1 & -2 & -3 \\ 1 & 4 & 9 \end{vmatrix}$$

$\therefore$ The diagonal matrix, $\Lambda = P^{-1} A P = \begin{vmatrix} 1 & 0 & 0 \\ 0 & -2 & 0 \\ 0 & 0 & -3 \end{vmatrix}$.

Problem 9.13. *Reduce the matrix A to diagonal matrix,* $A = \begin{vmatrix} 1 & 1 & -1 \\ -6 & -11 & 6 \\ -6 & -11 & 5 \end{vmatrix}$

Solution: The characteristic equation is

$$|\lambda I - A| = \begin{vmatrix} \lambda & 0 & 0 \\ 0 & \lambda & 0 \\ 0 & 0 & \lambda \end{vmatrix} - \begin{vmatrix} 0 & 1 & -1 \\ -6 & -11 & 6 \\ -6 & -11 & 5 \end{vmatrix} = 0$$

$$= \begin{vmatrix} \lambda & -1 & 1 \\ 6 & \lambda+11 & -6 \\ 6 & -11 & \lambda-5 \end{vmatrix} = 0 \qquad (9.61)$$

or $$\lambda^3 + 6\lambda^2 + 11\lambda + 6 = 0$$

$$(\lambda + 1)(\lambda + 2)(\lambda = 3) = 0$$

$\therefore$ The eigen values of matrix A are $\lambda_1 = -1$, $\lambda_2 = -2$ and $\lambda_3 = -3$.

The given A matrix is not in phase-canonical form and hence P matrix will not be the Vander Monde matrix. Therefore, matrix P can be determined from the eigen vectors of A.

Let the eigen vector associated with $\lambda_1 = -1$ be represented by $P_1 = \begin{bmatrix} P_{11} \\ P_{12} \\ P_{13} \end{bmatrix}$,

then $$[\lambda I - A] P_1 = 0$$

Substituting $\lambda_1 = -1$ in eqn. (8.61)

$$\therefore \qquad [\lambda I - A] P_1 = \begin{bmatrix} -1 & -1 & 1 \\ 6 & 10 & -6 \\ 6 & -11 & -6 \end{bmatrix} \begin{vmatrix} P_{11} \\ P_{12} \\ P_{13} \end{vmatrix} = 0$$

The solution of the above equation is

$$P_1 = \begin{vmatrix} 1 \\ 0 \\ 1 \end{vmatrix}$$

Similarly, $$P_2 = \begin{vmatrix} 1 \\ 2 \\ 4 \end{vmatrix} \quad \text{and} \quad P_3 = \begin{vmatrix} 1 \\ 6 \\ 9 \end{vmatrix}$$

$\therefore$ $$P = \begin{vmatrix} 1 & 0 & 1 \\ 1 & 2 & 1 \\ 1 & 6 & 9 \end{vmatrix}$$

The diagonal matrix, $$\Lambda = P^{-1} AP = \begin{vmatrix} -1 & 0 & 0 \\ 0 & -2 & 0 \\ 0 & 0 & -3 \end{vmatrix}$$

9.9 STATE VARIABLE REPRESENTATION USING PHASE VARIABLES

Phase variables are defined as those particular state variables which are obtained from one of the system variables and its derivatives. Usually system output and their derivatives of the outputs are chosen as state variables. The number of state variables will be equal to the order of the differential equation.

The transfer function of physical system is expressed as ratio of two polynomials, which are rational

i.e., $$G(s) = \frac{P(s)}{Q(s)} = \frac{b_0 S^m + b_1 S^{m-1} + \ldots\ldots + b^{m-1} S + b_m}{a_0 S^n + a_1 S^{n-1} + \ldots\ldots + a^{n-1} S + a_n} \tag{9.62}$$

where the degree of the numerator polynomial is 'm' and that of the denominator is 'n', for a proper transfer function, which is realizable physically $n \geq m$.

Case I: Transfer function having no zeros

If $G(s)$ has no zeros at all and only poles, then $n = 0$ and let the polynomial $Q(s)$ be non-valued.

Then

$$G(s) = \frac{b}{S^n + a_1 S^{n-1} + \ldots\ldots + a^{n-1} S + a_n} = \frac{Y(s)}{U(s)}$$

(9.63)

where $Y(s)$ and $U(s)$ are Laplace transformed output and input respectively.

Then eqn.(9.63), represented in state variable model as follows:

$$y^{n\cdot} + a_1 y^{(n-1)\cdot} + a_2 y^{(n-2)\cdot} + \ldots\ldots + a_{n-2}\, \ddot{y} + a_{n-1}\, \dot{y} + a_n y = bu \tag{9.64}$$

Let us choose the output 'y' and their derivatives as state variables.

$$x_1 = y$$

$$x_2 = \dot{y} = \dot{x}_1$$

$$\ldots \quad \ldots \quad \ldots$$

$$x_n = y^{(n-1)} = \dot{x}_{n-1}$$

Substituting the above state variables in the eqn. (9.64)

$$\therefore \quad \dot{x}_n + a_1 x_n + a_2 x_{n-1} + \ldots\ldots\ldots + a_n x_1 = bu$$

or

$$\dot{x}_n = -a_1 x_n - a_2 x_{n-1} - \ldots\ldots\ldots - a_n x_1 + bu$$

State equations of the system are

$$\dot{x}_1 = x_2$$

$$\dot{x}_2 = x_3$$

$$\ldots \quad \ldots$$

$$\dot{x}_n = -a_1 x_n - a_2 x_{n-1} - \ldots\ldots\ldots - a_n x_1 + bu$$

Writing the above state equations in matrix form

$$\begin{bmatrix} \dot{x}_1 \\ \dot{x}_2 \\ \cdot \\ \cdot \\ \cdot \\ \dot{x}_n \end{bmatrix} = \begin{bmatrix} 0 & 1 & 0 & \ldots & 0 \\ 0 & 0 & 1 & \ldots & 0 \\ \ldots & \ldots & \ldots & \ldots & \ldots \\ \ldots & \ldots & \ldots & \ldots & \ldots \\ -a_n & -a_{n-1} & \ldots a_1 & \cdot & \cdot \end{bmatrix} \begin{bmatrix} x_1 \\ x_2 \\ \cdot \\ \cdot \\ x_n \end{bmatrix} + \begin{bmatrix} 0 \\ 0 \\ \cdot \\ \cdot \\ b \end{bmatrix} u$$

In vector form, $\dot{x} = Ax + Bu$

Here, matrix A is in the upper off-diagonal. Last row is comprised of negative of the coefficients of the original differential equation and all other elements are zero. This form of matrix A is known as **bush form** or **comparison form**.

The output equation given by,

$$y = [1\ 0\ \ldots\ldots\ 0] \begin{bmatrix} x_1 \\ x_2 \\ \ldots \\ x_n \end{bmatrix}$$

In vector form, $\quad y = Cx$

Advantage

State model can be written directly by inspection from the differential equation governing the system.

Problem 9.14. *Given* $G(s) = \dfrac{2}{s^2 + 5s + 6}$, *obtain the state space model of the system in the diagonal canonical form.*

Solution: The given transfer function, $G(s) = \dfrac{Y(s)}{U(s)} = \dfrac{2}{s^2 + 5s + 5}$

By taking inverse $L.T.$ the given transfer function can be written as

$$\ddot{y} + 5\dot{y} + 6y = 2u$$

Let us choose the state variable as

$$x_1 = y$$

$$x_2 = \dot{y} = \dot{x}_1$$

and y = output variable

$$\therefore \qquad \dot{x}_2 = -6x_1 - 5x_2 + 2u$$

The above state equations in matrix form

$$\begin{bmatrix} \dot{x}_1 \\ \dot{x}_2 \end{bmatrix} = \begin{bmatrix} 0 & 1 \\ -6 & -5 \end{bmatrix} \begin{bmatrix} x_1 \\ x_2 \end{bmatrix} + \begin{bmatrix} 0 \\ 2 \end{bmatrix} u$$

and output, $\quad y = [1 \quad 0] \begin{bmatrix} x_1 \\ x_2 \end{bmatrix}$

Problem 9.15. *Determine the state model of the system for the following transfer functions*

$$\frac{Y(s)}{U(s)} = \frac{8}{s^3 + 3s^2 + 2s + 5}$$

Solution: By taking inverse L.T, the given transfer functions can be written as

$$\dddot{y} + 3\ddot{y} + 2\dot{y} + 5y = 8u$$

Let us choose the state variable

$$y = x_1,$$

$$x_2 = \dot{y} = \dot{x}_1$$

and $\qquad x_3 = \ddot{y} = \dot{x}_2$

$$\therefore \qquad \dot{x}_3 = -5x_1 - 2x_2 - 3x_3 + 8u$$

The above state equations represents in matrix form

$$\begin{bmatrix} \dot{x}_1 \\ \dot{x}_2 \\ \dot{x}_3 \end{bmatrix} = \begin{bmatrix} 0 & 1 & 0 \\ 0 & 0 & 1 \\ -5 & -2 & -3 \end{bmatrix} \begin{bmatrix} x_1 \\ x_2 \\ x_3 \end{bmatrix} + \begin{bmatrix} 0 \\ 0 \\ 8 \end{bmatrix} u$$

and the output in matrix form

$$y = [1 \quad 0 \quad 0] \begin{bmatrix} x_1 \\ x_2 \\ x_3 \end{bmatrix}$$

Problem 9.16. *Determine the state model of the system characterized by the differential equation* $(s^4 + 2s^2 + 8s^3 + 4s + 3)\, Y(s) = 10\, U(s)$

Solution: The given differential equation is converted into time domain as

$$\frac{d^4y(t)}{dt^4} + 2\frac{d^3y(t)}{dt^3} + 8\frac{d^2y(t)}{dt^2} + 4\frac{dy(t)}{dt} + 3y(t) = 10\, u(t)$$

Let us choose the state variable as

$$x_1 = y(t),$$

$$x_2 = \dot{y}(t) = \dot{x}_1$$

$$x_3 = \ddot{y}(t) = \dot{x}_2$$

$$x_4 = \dddot{y}(t) = \dot{x}_3$$

and $y(t) = x_1$ = output variable.

$$\therefore \qquad \dot{x}_4 = -3x_1 - 4x_2 - 8x_3 - 2x_4 + 10\,u(t)$$

Represent the above state equation in matrix form

$$\dot{x} = \begin{bmatrix} 0 & 1 & 0 & 0 \\ 0 & 0 & 1 & 0 \\ 0 & 0 & 0 & 1 \\ -3 & -4 & -8 & -2 \end{bmatrix} x + \begin{bmatrix} 0 \\ 0 \\ 0 \\ 10 \end{bmatrix} U(t)$$

and $$y = [1 \quad 0 \quad 0 \quad 0]\, x$$

Case II: Transfer function having zeros

For all practical realizable systems, consider the eqn.(9.62) with $G(s) = \dfrac{Y(s)}{U(s)}$ and the numerator degree $m \le n$, for physically realizable system.

Let us take $n = m = 3$. Thus we have

$$\frac{Y(s)}{U(s)} = \frac{b_0 s^3 + b_1 s^2 + b_2 s + b_3}{s^3 + a_1 s^2 + a_2 s + a_3} \tag{9.65}$$

Equation (9.65) can be written as

$$\dddot{y} + a_1 \ddot{y} + a_2 \dot{y} + a_3 y = b_0 \dddot{u} + b_1 \ddot{u} + b_2 \dot{u} + b_3 u$$

Taking L.T on both sides, assuming initial conditions are zero.

$$\frac{Y(s)}{U(s)} = \frac{b_0 s^3 + b_1 s^2 + b_2 s + b_3}{s^3 + a_1 s^2 + a_2 s + a_3}$$

The above equation is modified as

$$\frac{Y(s)}{U(s)} = \frac{Y(s)}{P(s)} \frac{P(s)}{U(s)}$$

where $$\frac{P(s)}{U(s)} = \frac{1}{s^3 + a_1 s^2 + a_2 s + a_3}$$

Let $\dddot{P} + a_1 \ddot{P} + a_2 \dot{P} + a_3 P = u$

$$P = x_1$$

$$\dot{P} = x_2 = \dot{x}_1$$

and $$\ddot{P} = x_3 = \dot{x}_2$$

$$\therefore \qquad \dot{x}_3 = u - a_1 x_3 - a_2 x_2 - a_3 x_1$$

The above state equations written in matrix form

$$\begin{bmatrix} \dot{x}_1 \\ \dot{x}_2 \\ \dot{x}_3 \end{bmatrix} = \begin{bmatrix} 0 & 1 & 0 \\ 0 & 0 & 1 \\ -a_3 & -a_2 & -a_1 \end{bmatrix} \begin{bmatrix} x_1 \\ x_2 \\ x_3 \end{bmatrix} + \begin{bmatrix} 0 \\ 0 \\ 1 \end{bmatrix} u$$

and $$\frac{Y(s)}{P(s)} = b_0 s^3 + b_1 s^2 + b_2 s + b_3$$

$$y = b_0 (u - a_1 x_3 - a_2 x_2 - a_3 x_1) + b_1 x_3 + b_2 x_2 + b_3 x_1$$

$$y = b_0 u + x_1 [b_1 - b_0 a_3] + x_2 [b_2 - b_0 a_2 0 + x_3 [b_1 - a_1 b_0]$$

or $$y = (b_3 - b_0 a_3) x_1 + (b_2 - b_0 a_2) x_2 + (b_1 - b_0 a_1) x_3 + b_0 u$$

On arranging the above equation in matrix form

$$y = [b_3 - b_0 a_3 \quad b_2 - b_0 a_2 \quad b_1 - b_0 a_1] \begin{bmatrix} x_1 \\ x_2 \\ x_3 \end{bmatrix} + [b_0] u$$

Generalization for n^{th} order differential equations

$$\begin{bmatrix} \dot{x}_1 \\ \dot{x}_2 \\ \cdots \\ \dot{x}_{n-1} \\ \dot{x}_n \end{bmatrix} = \begin{bmatrix} 0 & 1 & 0 & \ldots\ldots & 0 \\ 0 & 0 & 1 & \ldots\ldots & 0 \\ \cdots & . & . & . & \cdots \\ 0 & 0 & 0 & \ldots\ldots & 1 \\ -a_n & -a_{n-1} & \ldots\ldots & \ldots\ldots & -a_1 \end{bmatrix} \begin{bmatrix} x_1 \\ x_2 \\ . \\ . \\ x_{n-1} \\ x_n \end{bmatrix} + \begin{bmatrix} 0 \\ 0 \\ . \\ . \\ 0 \\ 1 \end{bmatrix} u$$

and $$y = [(b_n - b_0 a_n)(b_{n-1} - b_0 a_{n-1}) \ldots\ldots (b_2 - b_0 a)(b_1 - b_0 a_1)] \begin{bmatrix} x_1 \\ x_2 \\ \cdots \\ x_n \end{bmatrix} + [b_0] u$$

Advantage

It provides a link between the transfer functions design approach and time domain design approach.

Problem 9.17. *Determine the state model of the system for the following transfer function*

$$\frac{Y(s)}{U(s)} = \frac{2s^2 + s + 5}{s^3 + 6s^2 + 11s + 4}$$

Solution:

Let $$\frac{Y(s)}{U(s)} = \frac{Y(s)}{P(s)} \frac{P(s)}{U(s)}$$

where $$\frac{P(s)}{U(s)} = \frac{1}{s^3 + 6s^2 + 11s + 4}$$

or $$\dddot{P} + 6\ddot{P} + 11\dot{P} + 4P = u \qquad (9.66)$$

Choose state variables

$$x_1 = P,$$

$$x_2 = \dot{P} = \dot{x}_1$$

and $$x_3 = \ddot{P} = \dot{x}_2$$

Substituting the state variables in eqn. (9.66)

$\therefore$ $$\dot{x}_3 = -4x_1 - 11x_2 - 6x_3 + u$$

The above state equations are written in matrix form

$$\begin{bmatrix} \dot{x}_1 \\ \dot{x}_2 \\ \dot{x}_3 \end{bmatrix} = \begin{bmatrix} 0 & 1 & 0 \\ 0 & 0 & 1 \\ -4 & -11 & -6 \end{bmatrix} \begin{bmatrix} x_1 \\ x_2 \\ x_3 \end{bmatrix} + \begin{bmatrix} 0 \\ 0 \\ 1 \end{bmatrix} u$$

and $$\frac{Y(s)}{P(s)} = 2s^2 + s + 5$$

$$y = 2\ddot{P} + \dot{P} + 5P$$

$$y = 2x_3 + x_2 + 5x_1$$

or $$y = 5x_1 + x_2 + 2x_3$$

and the output in matrix form

$$y = [5 \quad 1 \quad 2] \begin{bmatrix} x_1 \\ x_2 \\ x_3 \end{bmatrix}$$

9.10 SOLUTION OF STATE EQUATIONS

Case-1: Solution of homogeneous state equations

In this case, the input term $u(t)$ is not present. When input is not presented, eqn. (9.3) can be modified as

$$\dot{x} = Ax \tag{9.67}$$

Taking $L.T.$ on both sides of eqn. (9.67)

$$sX(s) - x_{(0)} = AX(s)$$

or $$(sI - A)\,X(s) = x_{(0)}$$

or $$X(s) = (sI - A)^{-1} x_{(0)} \tag{9.68}$$

Taking inverse $L.T.$ on both sides of eqn. (9.68)

$$x(t) = L^{-1}\left[(sI - A)^{-1}\right] x_{(0)} \tag{9.69}$$

$$= e^{At}\, x_{(0)}$$

where $$e^{At} = I + At + \frac{A^2t^2}{2!} + \ldots\ldots + \frac{A^nt^n}{n!}$$

or $$x(t) = \phi(t)\, x_{(0)}$$

where $$\phi(t) = e^{At} = L^{-1}\left[(sI - A)^{-1}\right]$$

The matrix e^{At} is called state transition matrix **(STM)** and is denoted by $\phi(t)$. From the solution of the state equation it is observed that the initial state x_0 *at* $t = 0$ is driven to state $x(t)$ at the time t by state transition matrix.

Case-2: Solution of non-homogeneous state equations

In this case, input term or forced function $u(t)$ is taken into consideration. The eqn. (9.3) is

$$\dot{x}(t) = Ax(t) + Bu(t)$$

Taking L.T. on both sides of above equation.

$$sX(s) - x_{(0)} = AX(s) + BU(s)$$

or
$$(sI - A)\, X(s) = x_{(0)} + B\ U(s)$$

or
$$X(s) = (sI - A)^{-1} x_{(0)} + (sI - A)^{-1} BU(s) \tag{9.70}$$

Taking inverse $L.T.$ on both sides of eqn. (8.70)

$$\therefore \quad x(t) = L^{-1}\left[(sI - A)^{-1}\right] x_{(0)} + L^{-1}\left[(sI - A)^{-1}\, BU(s)\right]$$

$$= e^{At}\, x_{(0)} + \int_0^t e^{A\,(t-T)}\; Bu(T)\, dT \tag{9.71}$$

First part and second part of eqn. (9.71) are the **homogeneous solution and forced solution** respectively.

The eqn. (9.71) is the solution of state equation, when the initial conditions are known at $t = 0$. If the initial conditions are known at $t = t_0$ then the solution of state equation is given by

$$x(t) = e^{A(t-t_0)}\; x(t_0) + \int_{t_0}^{t} e^{A(t-T)}\; Bu(T)\, dT$$

9.10.1 Properties of State Transition Matrix

1. $\phi_{(0)} = e^{A0} = I$
2. $\phi^{-1}\,(t) = \phi\,(-t)$

Proof: $\phi(t) = e^{At}$

Post multiply both sides by e^{-At}

$$\phi(t)\, e^{-At} = e^{At}\, .\, e^{-At} = I$$

$$e^{-At} = \phi^{-1}\,(t)$$

$$\therefore \quad \phi^{-1}\,(t) = \phi\,(-t) = e^{-At}$$

3. $\phi(t_1 + t_2) = e^{A(t_1+t_2)} = (e^{At_1})\,(e^{At_2}) = \phi\,(t_1)\,\phi\,(t_2)$
4. $[\phi(t)]^k = \phi\,(kt)$

Problem 9.18. *Determine the time response of the following system*

$$[\dot{x}] = \begin{bmatrix} 1 & 0 \\ 1 & 2 \end{bmatrix} [x] + \begin{bmatrix} 1 \\ 1 \end{bmatrix} u$$

where $u(t)$ is a unit step occurring at $t = 0$ and $x^T(0) = [1 \quad 0]$

Solution: Time response of system when step input is zero is $x(t) = \phi(t)\, x_{(0)}$

Where $\phi(t) = L^{-1}[(sI - A) - 1]$,

$$[sI - A] = \begin{bmatrix} s & 0 \\ 0 & s \end{bmatrix} - \begin{bmatrix} 1 & 0 \\ 1 & 2 \end{bmatrix} = \begin{bmatrix} s-1 & 0 \\ -1 & s-2 \end{bmatrix}$$

$$|sI - A| = (s-1)(s-2)$$

$$(sI - A)^{-1} = \begin{bmatrix} \dfrac{(s-2)}{(s-1)(s-2)} & 0 \\ \dfrac{1}{(s-1)(s-2)} & \dfrac{s-1}{(s-1)(s-2)} \end{bmatrix}$$

$$\therefore \qquad \phi(t) = \begin{bmatrix} e^t & 0 \\ e^{2t} - e^t & e^{2t} \end{bmatrix}$$

$$\therefore \qquad x(t) = \begin{bmatrix} e^t & 0 \\ e^{2t} - e^t & e^{2t} \end{bmatrix} \begin{bmatrix} 1 \\ 0 \end{bmatrix} = \begin{bmatrix} e^t \\ e^{2t} - e^t \end{bmatrix}$$

Problem 9.19. *The state equation of LTIV system is given by* $\begin{bmatrix} \dot{x}_1 \\ \dot{x}_2 \end{bmatrix} = \begin{bmatrix} 0 & 1 \\ -1 & 0 \end{bmatrix} x + \begin{bmatrix} 0 \\ 1 \end{bmatrix} u.$ *Determine the STM.*

Solution : STM, $\phi(t) = L^{-1}[(sI - A)^{-1}]$

$$[sI - A] = \begin{bmatrix} s & 0 \\ 0 & s \end{bmatrix} - \begin{bmatrix} 0 & 1 \\ -1 & 0 \end{bmatrix} \begin{bmatrix} s & -1 \\ 1 & s \end{bmatrix}$$

$$|sI - A| = s^2 + 1$$

$$(sI - A)^{-1} = \begin{bmatrix} \dfrac{s}{s^2+1} & \dfrac{1}{s^2+1} \\ \dfrac{-1}{s^2+1} & \dfrac{s}{s^2+1} \end{bmatrix}$$

$$\therefore \qquad \phi(t) = \begin{bmatrix} \cos t & \sin t \\ -\sin t & \cos t \end{bmatrix}$$

Problem 9.20. *The state equations of LTIV is represented by* $\dot{x} = Ax + Bu$. *Find the state transition matrix.*

$$A = \begin{bmatrix} -2 & 0 & 0 \\ 0 & -1 & 1 \\ 0 & 0 & -1 \end{bmatrix}, B = \begin{bmatrix} 0 \\ 1 \\ 0 \end{bmatrix}$$

Solution: State transition matrix is $\phi(t) = L^{-1}[(sI - A)^{-1}]$, it does not depend on B.

$$[sI - A] = \begin{bmatrix} s & 0 & 0 \\ 0 & s & 0 \\ 0 & 0 & s \end{bmatrix} - \begin{bmatrix} 2 & 0 & 0 \\ 0 & -1 & 1 \\ 0 & 0 & -1 \end{bmatrix} = \begin{bmatrix} s+2 & 0 & 0 \\ 0 & s+1 & -1 \\ 0 & 0 & s+1 \end{bmatrix}$$

and $$(sI-A)^{-1} = \frac{Adj(sI-A)}{|sI-A|}$$

$$Adj\,(sI-A) = \begin{bmatrix} (s+1)^2 & 0 & 0 \\ 0 & (s+2)(s+1) & (s+2) \\ 0 & 0 & (s+2)(s+1) \end{bmatrix}$$

$$\therefore \quad (sI-A)^{-1} = \begin{bmatrix} \frac{(s+1)^2}{(s+2)(s+1)^2} & 0 & 0 \\ 0 & \frac{(s+2)(s+1)}{(s+2)(s+1)^2} & \frac{(s+2)}{(s+2)(s+1)^2} \\ 0 & 0 & \frac{(s+2)(s+1)}{(s+2)(s+1)^2} \end{bmatrix}$$

$$\phi\,(t) = L^{-1} \begin{bmatrix} \frac{1}{(s+2)} & 0 & 0 \\ 0 & \frac{1}{(s+1)} & \frac{1}{(s+1)^2} \\ 0 & 0 & \frac{1}{(s+1)} \end{bmatrix}$$

$$\therefore \quad \phi(t) = \begin{bmatrix} e^{-2t} & 0 & 0 \\ 0 & e^{-t} & te^{-t} \\ 0 & 0 & e^{-t} \end{bmatrix}$$

Problem 9.21. *The state equation of a LTIV system is given by*

$$\begin{bmatrix} \dot{x}_1 \\ \dot{x}_2 \end{bmatrix} = \begin{bmatrix} -2 & 0 \\ 1 & -1 \end{bmatrix} \begin{bmatrix} x_1 \\ x_2 \end{bmatrix} + \begin{bmatrix} 0 \\ 1 \end{bmatrix} u(t).$$ *Determine the following :*

(i) STM and

(ii) state equation for unit step input under zero initial condition.

Solution:

(*i*) STM, $\phi\,(t) = L^{-1}\,[(sI-A)^{-1}]$

$$(sI-A) = \begin{bmatrix} s & 0 \\ 0 & s \end{bmatrix} - \begin{bmatrix} -2 & 0 \\ 1 & -1 \end{bmatrix} = \begin{bmatrix} s+2 & 0 \\ -1 & s+1 \end{bmatrix}$$

$$|\,(sI-A)\,| = (s+2)\,(s+1)$$

$$Adj\,(sI-A) = \begin{bmatrix} s+1 & 0 \\ 1 & (s+2) \end{bmatrix}$$

$$(sI-A)^{-1} = \begin{bmatrix} \frac{1}{s+2} & 0 \\ \frac{1}{(s+1)(s+2)} & \frac{1}{s+1} \end{bmatrix}$$

$$\therefore \qquad \phi(t) = L^{-1}[(sI - A)^{-1}] = \begin{bmatrix} e^{-2t} & 0 \\ e^{-t} - e^{-2t} & e^{-t} \end{bmatrix}$$

(*ii*) The solution of state equation is

$$x(t) = e^{At}x(0) + \int_0^t e^{A(t-T)} Bu(T)dT$$

$$= \begin{bmatrix} e^{-2t} & 0 \\ e^{-t} - e^{-2t} & e^{-t} \end{bmatrix}\begin{bmatrix} x_{1(0)} \\ x_{2(0)} \end{bmatrix} + \int_0^t \begin{bmatrix} e^{-2t(t-T)} & 0 \\ e^{-(t-T)} - e^{-2(t-T)} & e^{-(t-T)} \end{bmatrix}\begin{bmatrix} 0 \\ 1 \end{bmatrix}[1]\, dT$$

$$= 0 + \int_0^t e^{-(t-T)}\, dT = 1 - e^{-t} \text{ (since initial condition, } x_0 = 0)$$

Problem 9.22. *The state equation of the LTIV system are given by*

$$\begin{bmatrix} \dot{x}_1 \\ \dot{x}_2 \end{bmatrix} = \begin{bmatrix} -2 & 0 \\ 0 & -1 \end{bmatrix}\begin{bmatrix} x_1 \\ x_2 \end{bmatrix} + \begin{bmatrix} 0 \\ 1 \end{bmatrix}u\,;\; y = [1 \quad 0]\begin{bmatrix} x_1 \\ x_2 \end{bmatrix}$$

(*i*) *determine the STM*

(*ii*) *find the solution for y(t) and*

(*iii*) *If a unit step is given to the input, what will be the behaviour of the output?*

Solution:

Given $\qquad A = \begin{bmatrix} -2 & 0 \\ 0 & -1 \end{bmatrix}, B = \begin{bmatrix} 0 \\ 1 \end{bmatrix} C = [1 \quad 0]$

(*i*) STM, $\qquad \phi(t) = L^{-1}[(sI - A)^{-1}]$

$$[sI - A] = \begin{bmatrix} s & 0 \\ 0 & s \end{bmatrix} - \begin{bmatrix} -2 & 0 \\ 0 & -1 \end{bmatrix} = \begin{bmatrix} s+2 & 0 \\ 0 & s+1 \end{bmatrix}$$

$$|sI - A| = (s+2)(s+1)$$

$$\therefore \qquad \phi(t) = L^{-1}\begin{bmatrix} \dfrac{(s+1)}{(s+2)(s+1)} & 0 \\ 0 & \dfrac{(s+2)}{(s+2)(s+1)} \end{bmatrix} = \begin{bmatrix} e^{-2t} & 0 \\ 0 & e^{-t} \end{bmatrix}$$

(*ii*) The transfer function, $\dfrac{Y(s)}{U(s)} = C[(sI - A)^{-1}B + D$

$$(sI - A)^{-1} = \begin{bmatrix} \dfrac{1}{(s+2)} & 0 \\ 0 & \dfrac{1}{(s+1)} \end{bmatrix}$$

$$(sI - A)^{-1}B = \begin{bmatrix} \dfrac{1}{(s+2)} & 0 \\ 0 & \dfrac{1}{(s+1)} \end{bmatrix}\begin{bmatrix} 0 \\ 1 \end{bmatrix} = \begin{bmatrix} 0 \\ \dfrac{1}{(s+1)} \end{bmatrix}$$

$$\frac{Y(s)}{U(s)} = [1 \quad 0]\begin{bmatrix} 0 \\ \dfrac{1}{(s+1)} \end{bmatrix} = 0$$

Hence, $Y(s) = 0$

$\therefore$ $Y(t) = 0$

(*iii*) since, $Y(s) = 0$ for any input, the output is zero.

Problem 9.23. *Solve for y(t) for the following system represented in state space, where u(t) is the unit step*

$$\dot{x} = \begin{bmatrix} -4 & 1 & 0 \\ 0 & -5 & 1 \\ 0 & 0 & -2 \end{bmatrix} x + \begin{bmatrix} 0 \\ 0 \\ 1 \end{bmatrix} u$$

$$y = [1 \;\; 1 \;\; 0]\, x, \; x_{(0)} = 0$$

Solution: The response of the state equation is $x(t) = \phi(t)\, x_{(0)} + \int_0^t \phi(t-T)\, Bu(T)\, dT$

When $x_{(0)} = 0$, the response is

$$x(t) = 0 + \int_0^t \phi(t-T)\, Bu(T)\, dT$$

$$= L^{-1}\,[(sI - A)^{-1} BU(s)]$$

$$(sI - A) = \begin{bmatrix} s & 0 & 0 \\ 0 & s & 0 \\ 0 & 0 & s \end{bmatrix} - \begin{bmatrix} -4 & 1 & 0 \\ 0 & -5 & 1 \\ 0 & 0 & -2 \end{bmatrix} = \begin{bmatrix} s+4 & -1 & 0 \\ 0 & s+5 & -1 \\ 0 & 0 & s+2 \end{bmatrix}$$

$$|\, sI - A \,| = (s+4)\,[(s+2)\,(s+5)]$$

$$= (s+2)\,(s+4)\,(s+5)$$

$$[sI - A]^{-1} = \frac{\text{Adj}[sI - A]}{|sI - A|}$$

$$= \frac{1}{(s+2)(s+4)(s+5)} \begin{bmatrix} (s+5)(s+2) & (s+2) & 1 \\ 0 & (s+4)(s+2) & (s+4) \\ 0 & 0 & (s+4)(s+5) \end{bmatrix}$$

$$[sI - A]^{-1} B = \frac{1}{(s+2)(s+4)(s+5)} \begin{bmatrix} (s+5)(s+2) & (s+2) & 1 \\ 0 & (s+4)(s+2) & (s+4) \\ 0 & 0 & (s+4)(s+5) \end{bmatrix} \begin{bmatrix} 0 \\ 0 \\ 1 \end{bmatrix}$$

$$= \frac{1}{(s+2)(s+4)(s+5)} \begin{bmatrix} 1 \\ s+4 \\ (s+4)(s+5) \end{bmatrix}$$

and $$(sI - A)^{-1} B\, U(s) = \frac{1}{(s+2)(s+4)(s+5)} \begin{bmatrix} 1 \\ s+4 \\ (s+4)(s+5) \end{bmatrix} \left[\frac{1}{s}\right] \qquad (\because \; U(s) = \text{unit step input})$$

$$= \begin{bmatrix} \dfrac{1}{s(s+2)(s+4)(s+5)} \\ \dfrac{1}{s(s+2)(s+5)} \\ \dfrac{1}{s(s+2)} \end{bmatrix}$$

$$\therefore \qquad x(t) = L^{-1}\begin{bmatrix} \dfrac{1}{s(s+2)(s+4)(s+5)} \\ \dfrac{1}{s(s+2)(s+5)} \\ \dfrac{1}{s(s+2)} \end{bmatrix} = \begin{bmatrix} \dfrac{1}{40} - \dfrac{1}{5} - e^{-5t} - \dfrac{1}{12} e^{-2t} - \dfrac{1}{8} e^{-4t} \\ \dfrac{1}{10} + \dfrac{1}{15} e^{-5t} - \dfrac{1}{6} e^{-2t} \\ \dfrac{1}{2} - \dfrac{1}{2} e^{-2t} \end{bmatrix}$$

The output $y(t) = C\,x(t)$

$$= [1 \;\; 1 \;\; 0]\begin{bmatrix} \dfrac{1}{40} - \dfrac{1}{5} - e^{-5t} - \dfrac{1}{12} e^{-2t} - \dfrac{1}{8} e^{-4t} \\ \dfrac{1}{10} + \dfrac{1}{15} e^{-5t} - \dfrac{1}{6} e^{-2t} \\ \dfrac{1}{2} - \dfrac{1}{2} e^{-2t} \end{bmatrix}$$

$$= \frac{1}{40} - \frac{1}{5} e^{-5t} - \frac{1}{12} e^{-2t} - \frac{1}{8} e^{-4t} + \frac{1}{10} + \frac{1}{15} e^{-5t} - \frac{1}{6} e^{-2t}$$

$$= \frac{1}{8} + \frac{2}{15} e^{-5t} - \frac{1}{8} e^{-4t} - \frac{1}{4} e^{-2t}.$$

9.11 DERIVATION OF TRANSFER FUNCTION FROM STATE MODEL

Consider the state model as

$$\dot{x}(t) = A\,x(t) + Bu(t)$$

$$y(t) = C\,x(t) + Du(t)$$

Taking L.T. for the above equations and according to definition of transfer function the initial conditions on x are all zero.

$$\therefore \qquad sX(s) = AX(s) + BU(s)$$

or $$(sI - A)\,X(s) = BU(s)$$

or $$X(s) = (sI - A)^{-1} BU(s) \qquad (9.72)$$

and $$Y(s) = CX(s) + DU(s)$$

Substituting $X(s)$ from eqn.(9.72) in above equation

$$Y(s) = C\,(sI - A)^{-1} BU(s) + DU(s)$$

or $$Y(s) = [C\,(sI - A)^{-1} B + D]U(s)$$

Transfer function is

$$\frac{Y(s)}{U(s)} = C\,(sI - A)^{-1} B + D.$$

Problem 9.24. *The state equations of the LTIV system are given by* $[\dot{x}] = \begin{bmatrix} 1 & 0 \\ 0 & -1 \end{bmatrix}[x] + \begin{bmatrix} 1 \\ 0 \end{bmatrix}$ *u and* $y = [1\ 1]\,[x]$. *Determine the transfer function of the system.*

Solution: The transfer function, $\dfrac{Y(s)}{U(s)} = C\,[sI - A]^{-1} B + D$

$$[sI - A] = \begin{bmatrix} s & 0 \\ 0 & s \end{bmatrix} - \begin{bmatrix} 1 & 0 \\ 0 & -1 \end{bmatrix} = \begin{bmatrix} s-1 & 0 \\ 0 & s+1 \end{bmatrix}$$

$$|\,sI - A\,| = (s^2 - 1)$$

$$[sI - A]^{-1} = \frac{1}{(s^2 - 1)}\begin{bmatrix} s+1 & 0 \\ 0 & s-1 \end{bmatrix}$$

$$[SI - A]^{-1} B = \frac{1}{(s^2 - 1)}\begin{bmatrix} s+1 & 0 \\ 0 & s-1 \end{bmatrix}\begin{bmatrix} 0 \\ 1 \end{bmatrix} = \frac{1}{(s^2 - 1)}\begin{bmatrix} 0 \\ s-1 \end{bmatrix}$$

$$\therefore \quad \text{Transfer function, } \frac{Y(s)}{U(s)} = [1\quad 1]\frac{1}{(s^2 - 1)}\begin{bmatrix} 0 \\ \\ s-1 \end{bmatrix} = \frac{1}{s+1}$$

Problem 9.25. *Given the state model*

$$\dot{x} = \begin{bmatrix} 1 & 2 \\ 3 & 4 \end{bmatrix} x + \begin{bmatrix} 1 \\ 1 \end{bmatrix} U$$

$$y = \begin{bmatrix} 1 & 0 \\ 0 & 2 \end{bmatrix} x + \begin{bmatrix} 1 \\ 0 \end{bmatrix} U.$$ *Determine the transfer function.*

Solution : Given $A = \begin{bmatrix} 1 & 2 \\ 3 & 4 \end{bmatrix}, B = \begin{bmatrix} 1 \\ 1 \end{bmatrix}, C = \begin{bmatrix} 1 & 0 \\ 0 & 2 \end{bmatrix}$ and $D = \begin{bmatrix} 1 \\ 0 \end{bmatrix}$

The transfer function, $\dfrac{Y(s)}{U(s)} = C\,[sI - A]^{-1} B + D$

$$[sI - A] = \begin{bmatrix} s & 0 \\ 0 & s \end{bmatrix} - \begin{bmatrix} 1 & 2 \\ 3 & 4 \end{bmatrix} = \begin{bmatrix} s-1 & -2 \\ -3 & s-4 \end{bmatrix}$$

$$|\,sI - A\,| = (s - 1)(s - 4) - 6 = s^2 - 5s - 2$$

$$(sI - A)^{-1} = \frac{1}{(s^2 + 5s - 2)}\begin{bmatrix} s-4 & 2 \\ 3 & s-1 \end{bmatrix}\begin{bmatrix} 1 \\ 1 \end{bmatrix}$$

$$= \frac{1}{(s^2 + 5s - 2)}\begin{bmatrix} s-2 \\ s+2 \end{bmatrix}$$

$$C\,(sI - A)^{-1} B = \begin{bmatrix} 1 & 0 \\ 0 & 2 \end{bmatrix}\frac{1}{(s^2 + 5s - 2)}\begin{bmatrix} s-2 \\ s+2 \end{bmatrix}$$

$$= \frac{1}{(s^2 + 5s - 2)}\begin{bmatrix} s-2 \\ 2(s+2) \end{bmatrix}$$

$$\therefore \quad \frac{Y(s)}{U(s)} = C\,(sI-A)^{-1}\,B + D = \frac{1}{(s^2+5s-2)}\begin{bmatrix} s-2 \\ 2(s+2) \end{bmatrix} + \begin{bmatrix} 1 \\ 0 \end{bmatrix}$$

$$= \begin{bmatrix} \dfrac{(s-2)+(s^2-5s-2)}{s^2-5s-2} \\ \dfrac{2(s+2)+0}{(s^2-5s-2)} \end{bmatrix} = \begin{bmatrix} \dfrac{s^2-4s-4}{s^2-5s-2} \\ \dfrac{2s+4}{s^2-5s-2} \end{bmatrix}$$

Problem 9.26. *The state variable formulation of a system is given by*

$$[\dot{x}] = \begin{bmatrix} -3 & 1 \\ -2 & 0 \end{bmatrix}[x] + \begin{bmatrix} 0 \\ 1 \end{bmatrix} u \text{ and } y = [1 \quad 0]\,[x]. \text{ Determine the following}$$

(i) transfer function of the system

(ii) state transaction matrix and

(iii) state equation for a unit step input under zero initial condition.

Solution:

$$A = \begin{bmatrix} -3 & 1 \\ -2 & 0 \end{bmatrix}, \qquad B = \begin{bmatrix} 0 \\ 1 \end{bmatrix}, \qquad C = [1 \quad 0]$$

Transfer function, $\dfrac{Y(s)}{U(s)} = C\,(sI-A)^{-1}\,B + D$

$$(sI-A) = \begin{bmatrix} s & 0 \\ 0 & s \end{bmatrix} - \begin{bmatrix} -3 & 1 \\ -2 & 0 \end{bmatrix} = \begin{bmatrix} s+3 & -1 \\ 2 & s+2 \end{bmatrix}$$

$$|\,sI-A\,| = s(s+3) + 2 = s^2 + 3s + 2$$

$$(sI-A)^{-1} = \frac{1}{s^2+3s+2}\begin{bmatrix} s & 1 \\ -2 & s+3 \end{bmatrix}$$

$$(sI-A)^{-1} = \frac{1}{s^2+3s+2}\begin{bmatrix} s & 1 \\ -2 & s+3 \end{bmatrix}\begin{bmatrix} 0 \\ 1 \end{bmatrix}$$

$$= \frac{1}{s^2+3s+2}\begin{bmatrix} 1 \\ s+3 \end{bmatrix}$$

$$\therefore \quad \frac{Y(s)}{U(s)} = [1 \quad 0]\frac{1}{s^2+3s+2}\begin{bmatrix} 1 \\ s+3 \end{bmatrix} = \frac{1}{s^2+3s+2}$$

(*ii*) STM, $\phi(t) = L^{-1}\,[(sI-A)^{-1}]$

$$\therefore \quad \phi(t) = L^{-1}\begin{bmatrix} \dfrac{s}{(s+2)(s+1)} & \dfrac{1}{(s+2)(s+1)} \\ \dfrac{-2}{(s+2)(s+1)} & \dfrac{s+3}{(s+2)(s+1)} \end{bmatrix}$$

$$= \begin{bmatrix} -e^{-t} + 2e^{-2t} & e^{-t} - e^{-2t} \\ -2e^{-t} + 2e^{-2t} & 2e^{-t} - e^{-2t} \end{bmatrix}$$

(*iii*) State equation for a unit step input under zero initial condition

$$x(t) = e^{At}\, x_{(0)} + \int_0^t e^{A(t-T)}\, Bu(T)\, dT$$

$$= 0 + \int_0^t \begin{bmatrix} -e^{-(t-T)} + 2e^{-2(t-T)} & e^{-(t-T)} - e^{-2(t-T)} \\ -2e^{-(t-T)} + 2e^{-2(t-T)} & 2e^{-(t-T)} - e^{-2(t-T)} \end{bmatrix}$$

$$= \int_0^t \begin{bmatrix} e^{-(t-T)} & -e^{-2(t-T)} \\ 2e^{-(t-T)} & -e^{-2(t-T)} \end{bmatrix} dT$$

$$= \begin{bmatrix} 1-e^{-t} & -\frac{1}{2}+\frac{1}{2} \quad e^{-2t} \\ 2-2e^{-t} & -\frac{1}{2}+\frac{1}{2} \quad e^{-2t} \end{bmatrix} = \begin{bmatrix} \frac{1}{2} - e^{-t} & +\frac{1}{2} \quad e^{-2t} \\ \frac{3}{2} - 2e^{-t} & +\frac{1}{2} \quad e^{-2t} \end{bmatrix}$$

Problem 9.27. *A second order linear system is described by*

$$\dot{x}_1 = -3x_1 + x_2 + u$$

$$\dot{x}_2 = -x_1 - x_2 + u$$

and $\quad y = x_1 + x_2.$

Determine the transfer function and also calculate the zero input response of $x_{1(0)} = 1$ *and* $x_{2(0)} = -1$.

Solution:

$$\begin{bmatrix} \dot{x}_1 \\ \dot{x}_2 \end{bmatrix} = \begin{bmatrix} -2 & 1 \\ -1 & -1 \end{bmatrix} \begin{bmatrix} x_1 \\ x_2 \end{bmatrix} + \begin{bmatrix} 1 \\ 1 \end{bmatrix} u$$

$$y = [1 \quad 1] \begin{bmatrix} x_1 \\ x_2 \end{bmatrix} \quad \text{and} \quad \begin{bmatrix} x_{1(0)} \\ x_{2(0)} \end{bmatrix} = \begin{bmatrix} 1 \\ -1 \end{bmatrix}$$

Transfer function, $\dfrac{Y(s)}{U(s)} = C\,[sI - A]^{-1} B + D$

$$[sI - A] = \begin{bmatrix} s & 0 \\ 0 & s \end{bmatrix} - \begin{bmatrix} -2 & 1 \\ -1 & -1 \end{bmatrix} = \begin{bmatrix} s+2 & -1 \\ 1 & s+1 \end{bmatrix}$$

$$|\, sI - A \,| = (s + 3)(s + 1) + 1 = s^2 + 4s + 4 = (s + 2)^2$$

$$(sI - A)^{-1} = \frac{1}{(s+2)^2} \begin{bmatrix} s+1 & 1 \\ -1 & s+3 \end{bmatrix}$$

$$[sI - A]^{-1} B = \frac{1}{(s+2)^2} \begin{bmatrix} s+1 & 1 \\ -1 & s+3 \end{bmatrix} \begin{bmatrix} 1 \\ 1 \end{bmatrix} = \frac{1}{(s+2)^2} \begin{bmatrix} s+2 \\ s+2 \end{bmatrix} = \frac{1}{s+2} \begin{bmatrix} 1 \\ 1 \end{bmatrix}$$

$$\therefore \quad \frac{Y(s)}{U(s)} = [1 \quad 1] \frac{1}{s+2} \begin{bmatrix} 1 \\ 1 \end{bmatrix} = \frac{2}{s+2}$$

(*ii*) The solution of state equation with zero input is $x(t) = \phi(t)\, x_{(0)}$

$$\therefore \quad \phi(t) = L^{-1}\, [(sI - A)^{-1}]$$

$$= L^{-1}\begin{bmatrix} \dfrac{s+1}{(s+2)^2} & \dfrac{1}{(s+2)^2} \\ \dfrac{-1}{(s+2)^2} & \dfrac{s+3}{(s+2)^2} \end{bmatrix} = \begin{bmatrix} -te^{-2t}+e^{-2t} & te^{-2t} \\ -te^{-2t} & te^{-2t}+e^{-2t} \end{bmatrix}$$

$$\therefore \qquad x(t) = \begin{bmatrix} -te^{-2t}+e^{-2t} & te^{-2t} \\ -te^{-2t} & te^{-2t}+e^{-2t} \end{bmatrix}\begin{bmatrix} 1 \\ -1 \end{bmatrix}$$

$$x_1(t) = -te^{-2t} + e^{-2t} - te^{-2t} = e^{-2t} - 2te^{-2t}$$

$$x_2(t) = -te^{-2t} - te^{-2t} - e^{-2t} = -(e^{-2t} + te^{-2t})$$

Problem 9.28. *The state equations of a LTIV system are given by*

$$\begin{bmatrix} \dot{x}_1 \\ \dot{x}_2 \\ \dot{x}_3 \end{bmatrix} = \begin{bmatrix} -2 & 0 & 1 \\ 1 & -3 & 0 \\ 1 & 1 & -1 \end{bmatrix}\begin{bmatrix} x_1 \\ x_2 \\ x_3 \end{bmatrix} + \begin{bmatrix} 1 \\ 0 \\ 1 \end{bmatrix} u \;;\; y = \begin{bmatrix} 2 & 1 & -1 \end{bmatrix}\begin{bmatrix} x_1 \\ x_2 \\ x_3 \end{bmatrix}.$$ *Determine the transfer function.*

Solution: The transfer function, $\dfrac{Y(s)}{U(s)} = C\,[sI - A]^{-1}B + D$

$$A = \begin{bmatrix} -2 & 0 & 1 \\ 1 & -3 & 0 \\ 1 & 1 & -1 \end{bmatrix},\; B = \begin{bmatrix} 1 \\ 0 \\ 1 \end{bmatrix},\; C = \begin{bmatrix} 2 & 1 & -1 \end{bmatrix}$$

$$[sI - A] = \begin{bmatrix} s & 0 & 0 \\ 0 & s & 0 \\ 0 & 0 & s \end{bmatrix} - \begin{bmatrix} -2 & 0 & 1 \\ 1 & -3 & 0 \\ 1 & 1 & -1 \end{bmatrix} = \begin{bmatrix} s+2 & 0 & -1 \\ -1 & s+3 & 0 \\ -1 & -1 & s+1 \end{bmatrix}$$

$$|\,sI - A\,| = (s+2)\,[(s+3)(s+1) + 0] - 0 + 1\,(1 + s + 3)$$
$$= s^3 + 6s^2 + 12s + 10$$

$$[sI - A]^{-1} = \frac{Adj[sI - A]}{|sI - A|}$$

$$\therefore \qquad [sI - A]^{-1} = \frac{1}{(s^3 + 6s^2 + 12s + 100)}\begin{bmatrix} (s+1)(s+3) & 1 & (s+3) \\ 0 & s^2+3s+1 & 1 \\ s+4 & 1 & (s+2)(s+3) \end{bmatrix}$$

$$(sI\text{–}A)^{-1}B = \frac{1}{(s^3 + 6s^2 + 12s + 100)}\begin{bmatrix} (s+1)(s+3) & 1 & (s+3) \\ 0 & s^2+3s+1 & 1 \\ s+4 & 1 & (s+2)(s+3) \end{bmatrix}\begin{bmatrix} 1 \\ 0 \\ 1 \end{bmatrix}$$

$$= \frac{1}{(s^3 + 6s^2 + 12s + 100)}\begin{bmatrix} (s+1)(s+3) + 0 + s + 3 \\ 1 \\ s + 4 + (s+2)(s+3) \end{bmatrix}$$

$$= \frac{1}{(s^3 + 6s^2 + 12s + 100)}\begin{bmatrix} s^2 + 5s + 6 \\ 1 \\ s^2 + 6s + 10 \end{bmatrix}$$

$$\therefore \quad \frac{Y(s)}{U(s)} = C\,(sI - A)^{-1} = \begin{bmatrix} 2 & 1 & -1 \end{bmatrix} \frac{1}{(s^3 + 6s^2 + 12s + 100)} \begin{bmatrix} s^2 + 5s + 6 \\ 1 \\ s^2 + 6s + 10 \end{bmatrix}$$

$$= \frac{1}{(s^3 + 6s^2 + 12s + 100)} [2(s^2 + 5s + 6) + 1 - 1\,(s^2 + 6s + 10)]$$

$$= \frac{s^2 + 4s + 2}{s^3 + 6s^2 + 12s + 100}.$$

9.12 CALEY-HAMILTON THEOREM

Consider the transfer function as

$$G(s) = \frac{Y(s)}{U(s)} = C\,[sI - A]^{-1}\,B + D$$

where $$(sI - A)^{-1} = \frac{Adj[sI - A]}{|sI - A|}$$

The denominator polynomial $|sI - A|$ is called the characteristic polynomial and the roots of the characteristic polynomial are the eigen values of the matrix A. The characteristic polynomial is of nth polynomial, if A is $(n \times n)$ square matrix. If λ is an eigen value of A, then λ is obtained by solving the polynomial as

$$q(\lambda) = |\lambda I - A| = 0$$

$$= \lambda^n I + a_1 \lambda^{n-1} + \ldots.. + a_{n-1} \lambda + a_n = 0 \tag{9.73}$$

According to the Caley Hamilton theorem, every square matrix A satisfies its own characteristic polynomial

i.e., $$q(A) = A^n + a_1 A^{n-1} + \ldots.. + a_{n-1} A + a_0 = 0$$

$q(A)$ is a matrix polynomial. Evaluation of such polynomial involves computation of matrix powers. Using Caley-Hamilton theorem a matrix polynomial is computed by considering a scalar polynomial which is just similar to the computation of state transition matrix and involves evaluation of matrix polynomial e^{At}.

The procedure is as follows:

Given a square matrix A. Let $\lambda_1, \lambda_2, \ldots\ldots, \lambda_n$ be eigen values. The matrix polynomial is

$$f(A) = K_0 I + K_1 A + K_2 A^2 + \ldots\ldots + K_i A^i \tag{9.74}$$

can be computed by considering

$$f(\lambda) = K_0 + K_1 \lambda + K_2 \lambda^2 + \ldots.. + K_i \lambda^i \tag{9.75}$$

Dividing $f(\lambda)$ by $q(\lambda)$ of eqn. (9.73) gives

$$\frac{f(\lambda)}{q(\lambda)} = Q(\lambda) + \frac{R(\lambda)}{q(\lambda)}$$

or $$f(\lambda) = Q(\lambda)\, q(\lambda) + R(\lambda)$$

where $R(\lambda)$ is the remainder polynomial having the form

$$\therefore \quad R(\lambda) = \alpha_o I + \alpha_1 \lambda^1 + \alpha_2 \lambda^2 + \ldots.. + \alpha_{n-1} \lambda^{n-1} \tag{9.76}$$

The n coefficients of $R(\lambda)$ can be obtained by substituting the 'n' eigen values *i.e.*, $\lambda_1, \lambda_2 \ldots \lambda_n$ in eqn. (9.76)

i.e., $$f(\lambda)_i = R(\lambda)_i \; , i = 1, 2, \ldots\ldots n$$

On similar lines, we have

$$f(A) = Q(A)\, q(A) + R(A)$$

Since $q(A) = 0$, we have

$$f(A) = R(A)$$

$$= \alpha_o I + \alpha_1 A + \alpha_2 A^2 + \ldots\ldots + \lambda_{n-1} A^{n-1} \quad (9.77)$$

If A has repeating eigen values, say λ_k of order m, then

$$f(\lambda_k) = R(\lambda_k)$$

results only the coefficient of $R(\lambda)$. The remaining coefficients are obtained by differentiating equation (9.77) w.r.t. λ_j, $j = 1, 2, \ldots\ldots, m - 1$

$$\therefore \quad \frac{d^j q(\lambda)}{d\lambda^j} = \frac{d^j R(\lambda)}{d\lambda^j} \; , j = 0, 1, 2, \ldots. m - 1 \quad (9.78)$$

Problem 9.29. *Using Caley-Hamilton method, determine STM for A, given*

$$A = \begin{bmatrix} 0 & 2 \\ -2 & -4 \end{bmatrix}.$$

Solution: $$e^{At} = \alpha_o I + \alpha_1 A \quad (9.79)$$

Eigen values are given by

$$|\lambda I - A| = \begin{vmatrix} \lambda & 0 \\ 0 & \lambda \end{vmatrix} - \begin{vmatrix} 0 & 2 \\ -2 & -4 \end{vmatrix} = \begin{vmatrix} \lambda & -2 \\ 2 & \lambda+4 \end{vmatrix} = 0$$

or $$\lambda(\lambda + 4) + 4 = 0$$

or $$\lambda^2 + 4\lambda + 4 = 0$$

or $$(\lambda + 2)^2$$

$\therefore \quad \lambda_1 = -2$ is an eigen value repeated twice

$$\therefore \quad e^{\lambda_1 t} = \alpha_o + \alpha_1 \lambda_1 \quad (9.80)$$

Differentiating the above equation w.r.t. λ_1

$$te^{\lambda_1 t} = \alpha_1 \quad (9.81)$$

Substituting $\lambda_1 = -2$ in eqns. (9.80) and (9.81)

$$e^{-2t} = \alpha_o - 2\alpha_1$$

and $$te^{-2t} = \alpha_1$$

$$\therefore \quad \alpha_o = e^{-2t} + 2t\, e^{-2t}$$

$$= (1 + 2t)\, e^{-2t}$$

Hence, from eqn. (9.79)

$$e^{At} = (1 + 2t)\, e^{-2t} \begin{bmatrix} 1 & 0 \\ 0 & 1 \end{bmatrix} + t\, e^{-2t} \begin{bmatrix} 0 & 2 \\ -2 & -4 \end{bmatrix} = \begin{bmatrix} (1+2t)\, e^{-2t} & 2te^{-2t} \\ -2te^{-2t} & (1-2t)\, e^{-2t} \end{bmatrix}$$

9.13 CONTROLLABILITY AND OBSERVABILITY

Let us consider LTIV system equations as

$$\dot{x} = Ax + Bu \tag{9.82}$$

$$y = Cx + Du \tag{9.83}$$

Controllability

A system is said to be completely state controllable if it is possible to transfer the system state from any initial state to any other desired state in a finite time.

The eqn. (9.82) (*i.e.*, $\dot{x} = Ax + Bu$) is completely controllable if and only if the rank of the composite matrix Q_c is n or determinant of $Qc \neq 0$.

$$\therefore \quad Q_c = [B : AB \;........ A^{n-1}B] \tag{9.84}$$

Observability

A system is said to be completely observable, if every state can be completely identified by measurements of the output in a finite time.

The eqns. (8.82) and (8.83) are completely observable if and only if the rank of composite matrix Q_o is n or $|\ Q_o| \neq 0$.

$$\therefore \quad Q_o = [C^T : A^T C^T : \;......\; : (A^T)^{n-1} C^T] \tag{9.85}$$

Problem 9.30. *The state equation of a LTIV system is given by*

$\begin{bmatrix} \dot{x}_1 \\ \dot{x}_2 \end{bmatrix} = \begin{bmatrix} -2 & 0 \\ 1 & -1 \end{bmatrix} \begin{bmatrix} x_1 \\ x_2 \end{bmatrix} + \begin{bmatrix} 0 \\ 1 \end{bmatrix} u$. *Determine whether the system is controllable.*

Solution :

$$A = \begin{bmatrix} -2 & 0 \\ 1 & -1 \end{bmatrix}, \quad B = \begin{bmatrix} 1 \\ 0 \end{bmatrix}$$

For controllability, $Q_c = [B \;:\; AB]$ must be non-singular.

$$\therefore \quad [A \;\; B] = \begin{bmatrix} -2 & 0 \\ 1 & -1 \end{bmatrix} \begin{bmatrix} 0 \\ 1 \end{bmatrix} = \begin{bmatrix} 0 \\ -1 \end{bmatrix}$$

$$\therefore \quad Q_c = [B \;:\; AB] = \begin{bmatrix} 0 & 0 \\ 1 & -1 \end{bmatrix} = 0$$

Hence, the given system is not state controllable since det (Q_c) is zero.

Problem 9.31. *Determine the state controllability and observability of the following system*

$$\text{(i) } A = \begin{bmatrix} -1 & 1 & 0 \\ 0 & -3 & 2 \\ 0 & 0 & -8 \end{bmatrix}, B = \begin{bmatrix} 0 \\ 0 \\ 1 \end{bmatrix}, C = [1 \;\; 0 \;\; 1]$$

Solution: Controllability $Q_c = [B \quad AB \quad A^2B]$

$$A^2 = A.\,A = \begin{bmatrix} -1 & 1 & 0 \\ 0 & -3 & 2 \\ 0 & 0 & -8 \end{bmatrix}\begin{bmatrix} -1 & 1 & 0 \\ 0 & -3 & 2 \\ 0 & 0 & -8 \end{bmatrix}$$

$$= \begin{bmatrix} 1 & -4 & 2 \\ 0 & 9 & -22 \\ 0 & 0 & 64 \end{bmatrix}$$

$$A^2B = \begin{bmatrix} 1 & -4 & 2 \\ 0 & 9 & -22 \\ 0 & 0 & 64 \end{bmatrix}\begin{bmatrix} 0 \\ 0 \\ 1 \end{bmatrix} = \begin{bmatrix} 2 \\ -22 \\ 64 \end{bmatrix}$$

$$AB = \begin{bmatrix} -1 & 1 & 0 \\ 0 & -3 & 2 \\ 0 & 0 & -8 \end{bmatrix}\begin{bmatrix} 0 \\ 0 \\ 1 \end{bmatrix} = \begin{bmatrix} 0 \\ 2 \\ 8 \end{bmatrix}$$

$$\therefore \qquad Q_c = \begin{bmatrix} 0 & 0 & 2 \\ 0 & 2 & -22 \\ 1 & -8 & 64 \end{bmatrix}$$

Determinant of $\quad Q_c = 0 - 0 + 2\,(-2) = -4$

$\neq 0$

And rank of Q_c = order of A

$\therefore$ Given system is controllable.

Observability $\quad Q_o = \left[C' \quad A'C' \quad (A')^2C'\right]$

$$C' = \begin{bmatrix} 1 \\ 0 \\ 1 \end{bmatrix}$$

$$A'C' = \begin{bmatrix} -1 & 0 & 0 \\ 1 & -3 & 0 \\ 0 & 2 & -8 \end{bmatrix}\begin{bmatrix} 1 \\ 0 \\ 1 \end{bmatrix} = \begin{bmatrix} -1 \\ 1 \\ -8 \end{bmatrix}$$

$$(A')^2 = \begin{bmatrix} -1 & 0 & 0 \\ 1 & -3 & 0 \\ 0 & 2 & -8 \end{bmatrix}\begin{bmatrix} -1 & 0 & 0 \\ 1 & -3 & 0 \\ 0 & 2 & -8 \end{bmatrix} = \begin{bmatrix} 1 & 0 & 0 \\ -4 & 9 & 0 \\ 2 & -22 & 64 \end{bmatrix}$$

$$(A')^2C' = \begin{bmatrix} 1 & 0 & 0 \\ -4 & 9 & 0 \\ 2 & -22 & 64 \end{bmatrix}\begin{bmatrix} 1 \\ 0 \\ 1 \end{bmatrix} = \begin{bmatrix} 1 \\ -4 \\ 66 \end{bmatrix}$$

$$\therefore \qquad Q_o = \begin{bmatrix} 1 & -1 & 1 \\ 0 & 1 & -4 \\ 1 & -8 & 66 \end{bmatrix}$$

Determinant of $Q_o = 1\,(66 - 32) + 1(4) + 1\,(-1)$

$$= 34 + 4 - 1 = 37$$

$$\neq 0$$

And rank of Q_o = order of A

$\therefore$ Given system is observable.

Problem 9.32. *Determine the state controllability of the following system*

$$A = \begin{bmatrix} 0 & 1 & 0 \\ 0 & 0 & 1 \\ -2 & -4 & -5 \end{bmatrix}, B = \begin{bmatrix} 0 \\ 0 \\ 1 \end{bmatrix}, C = [1 \quad 1 \quad 0]$$

Solution: Controllability $Q_c = [B \quad AB \quad A^2B]$

$$AB = \begin{bmatrix} 0 & 1 & 0 \\ 0 & 0 & 1 \\ -2 & -4 & -5 \end{bmatrix} \begin{bmatrix} 0 \\ 0 \\ 1 \end{bmatrix} = \begin{bmatrix} 0 \\ 1 \\ -5 \end{bmatrix}$$

$$A^2 = \begin{bmatrix} 0 & 1 & 0 \\ 0 & 0 & 1 \\ -2 & -4 & -5 \end{bmatrix} \begin{bmatrix} 0 & 1 & 0 \\ 0 & 0 & 1 \\ -2 & -4 & -5 \end{bmatrix} = \begin{bmatrix} 0 & 0 & 1 \\ -2 & -4 & 5 \\ 10 & 19 & 21 \end{bmatrix}$$

$$A^2B = \begin{bmatrix} 0 & 0 & 1 \\ -2 & -4 & 5 \\ 10 & 18 & 21 \end{bmatrix} \begin{bmatrix} 0 \\ 0 \\ 1 \end{bmatrix} = \begin{bmatrix} 1 \\ -5 \\ 21 \end{bmatrix}$$

$$Q_c = \begin{bmatrix} 0 & 0 & 1 \\ 0 & 1 & -5 \\ 1 & -5 & 21 \end{bmatrix}$$

Determinant of $Q_c = -1$ and rank is 3

$\therefore$ Given system is controllable.

Problem 9.33. *Determine the state controllability and observability of the following system:*

$$A = \begin{bmatrix} -1 & 0 \\ 0 & -4 \end{bmatrix}, B = \begin{bmatrix} 0 \\ 1 \end{bmatrix}, \quad C = [1 \quad 3]$$

Solution: Controllability $Q_c = [B \quad AB \quad A^2B]$

$$AB = \begin{bmatrix} -1 & 0 \\ 0 & -4 \end{bmatrix} \begin{bmatrix} 0 \\ 1 \end{bmatrix} = \begin{bmatrix} 0 \\ 0 \\ -4 \end{bmatrix}$$

$$A^2 = \begin{bmatrix} -1 & 0 \\ 0 & -4 \end{bmatrix} \begin{bmatrix} -1 & 0 \\ 0 & -4 \end{bmatrix} = \begin{bmatrix} 1 & 0 \\ 0 & 16 \end{bmatrix}$$

$$A^2B = \begin{bmatrix} 1 & 0 \\ 0 & 16 \end{bmatrix} \begin{bmatrix} 0 \\ 1 \end{bmatrix} = \begin{bmatrix} 0 \\ 16 \end{bmatrix}$$

$$\therefore \qquad Q_c = \begin{bmatrix} 0 & 0 & 0 \\ 1 & -4 & 16 \end{bmatrix}$$

Determinant of $Q_c = 0$,

Therefore the given system is not controllable.

Observability $\qquad Q_o = \begin{bmatrix} C' & A'C' & (A')^2 C' \end{bmatrix}$

$$C' = \begin{bmatrix} 1 \\ 3 \end{bmatrix}$$

$$A'C' = \begin{bmatrix} -1 & 0 \\ 0 & -4 \end{bmatrix}\begin{bmatrix} 1 \\ 3 \end{bmatrix} = \begin{bmatrix} -1 \\ -12 \end{bmatrix}$$

$$(A')^2 = \begin{bmatrix} -1 & 0 \\ 0 & -4 \end{bmatrix}\begin{bmatrix} -1 & 0 \\ 0 & -4 \end{bmatrix} = \begin{bmatrix} 1 & 0 \\ 0 & 16 \end{bmatrix}$$

$$C'(A')^2 = \begin{bmatrix} 1 \\ 3 \end{bmatrix}\begin{bmatrix} 1 & 0 \\ 0 & 16 \end{bmatrix} = \begin{bmatrix} 1 \\ 48 \end{bmatrix}$$

$$Q_o = \begin{bmatrix} 1 & -1 & 1 \\ 3 & -12 & 48 \end{bmatrix}$$

Determinant of $Q_o \neq 0$

Therefore the given system is observable.

SHORT QUESTIONS AND ANSWERS

1. **What are the advantages of state variable techniques?**
   - This approach can be applied to linear or nonlinear, time variant or time invariant systems.
   - It is easier to apply where the Laplace transform cannot be applied.
   - n^{th} order differential equations can be expressed as 'n' equation of first order whose solutions are easier.
   - It is a time domain approach.

2. **What is state?**

   The state of a dynamic system is the minimal set of variables (known as state variables) such that the knowledge of these variables at the instant $t = t_0$ together with the knowledge of the inputs for $t \geq t_0$, completely determines the behavior of the system for any time $t > t_0$.

3. **What are state variables?**

   State variables are the minimum set of variables, which determine the state of dynamic system.

4. **What is state vector?**

   The state vector $x(t)$ is the vector sum of all the state variables.

5. What is state space?

The space whose coordinate axis consist of x_1 axis, x_2 axis,......, x_n axis is called a State Space.

6. What are state equations?

The system equations written in the form of first order differential equations in time domain are known as State Equations.

7. What is state diagram?

The pictorial representation of the state model of the system is called as State Diagram.

8. What are the properties of state transition matrix?

- $\phi(0) = e^{A0} = I$
- $\phi^{-1}(t) = \phi(-t)$
- $\phi(t_1 + t_2) = \phi(t_1)\,\phi(t_2)$
- $[\phi(t)]^K = \phi(kt)$

9. What is controllability?

A system is said to be completely state controllable if it is possible to transfer the system state from any initial state to any other desired state in a finite time.

10. What is observability?

A system is said to be completely observable, if every state can be completely identified by measurements of the output in a finite time.

OBJECTIVE TYPE QUESTIONS

1. State space analysis is applicable to

   (*a*) time variant system (*b*) Time invariant system

   (*c*) (*b*), (*c*) and (*d*) (*d*) Multi input *N*–multi output system

2. Solution for non homogeneous system

   (*a*) $x_o e^{-At}$ (*b*) $x_o e^{At}$

   (*c*) $[SI-A]^{-1}$ (*d*) $L^{-1}[SI-A]^{-1}$

3. State space analysis state transmission matrix $\phi(t)$ is equal to

   (*a*) $L^{-1}[SI-A]^{-1}$ (*b*) $[SI-A]^{-1}$

   (*c*) e^{-At} (*d*) Both (*a*) and (*c*)

4. Given a state variable model $x = Ax + bu$ and $y = cx + du$ under this transformation $x = P\overline{x}$; P is a nonsingular matrix, the model becomes

   (*a*) $\overline{A} = PAP^{-1}$; $\overline{b} = P^{-1}b$; $\overline{c} = cP$ (*b*) $\overline{A} = P^{-1}AP$; $\overline{b} = P^{-1}b$; $\overline{c} = cP$

   (*c*) $\overline{A} = P^{-1}AP$; $\overline{b} = Pb$; $\overline{c} = cP$ (*d*) $\overline{A} = P^{-1}AP$; $\overline{b} = P^{-1}b$; $\overline{c} = cP^{-1}$

5. A state variable formation of a system is given by the equations

$$\begin{pmatrix} \dot{x}_1 \\ x \end{pmatrix} = \begin{pmatrix} -1 & 0 \\ 0 & 3 \end{pmatrix}\begin{pmatrix} x_1 \\ x_2 \end{pmatrix} + \begin{pmatrix} 1 \\ 1 \end{pmatrix} u$$

$y = [1 \quad 0]\begin{pmatrix} x_1 \\ x_2 \end{pmatrix}$. The transfer function of the system is

(*a*) $\dfrac{1}{(s+1)(s+3)}$ (*b*) $\dfrac{1}{s+1}$

(*c*) $\dfrac{1}{s+3}$ (*d*) None of these.

6. A state variable formulation of a system is given by the equations

$$\begin{pmatrix} \dot{x}_1 \\ \dot{x} \end{pmatrix} = \begin{pmatrix} -1 & 0 \\ 0 & 3 \end{pmatrix}\begin{pmatrix} x_1 \\ x_2 \end{pmatrix} + \begin{pmatrix} 1 \\ 1 \end{pmatrix} u \; ; x_{1(0)} = x_{2(0)} = 0$$

$y = [1 \quad 0]\begin{pmatrix} x_1 \\ x_2 \end{pmatrix}$. The response $y(t)$ to unit-step input is

(*a*) $1 + e^{-t}$ (*b*) $(1/3)\,[1 - e^{-3t}]$

(*c*) $1 - e^{-t}$ (*d*) None of the answers is correct

7. The eigen values of the matrix $A = \begin{pmatrix} 0 & 1 & 0 \\ 0 & 0 & 1 \\ 0 & -3 & -4 \end{pmatrix}$ are

(*a*) 0, – 1, – 3 (*b*) 0, – 3, – 4

(*c*) 0, 0, – 4 (*d*) None of these.

8. Given the system

$$\dot{x} = \begin{pmatrix} 0 & 0 & -20 \\ 1 & 0 & -24 \\ 0 & 1 & -9 \end{pmatrix} x + \begin{pmatrix} 3 \\ 1 \\ 0 \end{pmatrix} u$$

$$y = [0 \quad 0 \quad 1]\, x$$

The characteristic equation of the system is

(*a*) $s^3 + 20s^2 + 24s + 9 = 0$ (*b*) $s^3 + 9s^2 + 24s + 20 = 0$

(*c*) $s^3 + 24s^2 + 9s + 20 = 0$ (*d*) None of the answers is correct

9. A state variable model of a system is given by

$$\begin{pmatrix} \dot{x}_1 \\ \dot{x}_2 \end{pmatrix} = \begin{pmatrix} 1 & 1 \\ -2 & -1 \end{pmatrix}\begin{pmatrix} x_1 \\ x_2 \end{pmatrix} + \begin{pmatrix} 0 \\ 1 \end{pmatrix} u$$

$y = [1 \quad 0]\begin{pmatrix} x_1 \\ x_2 \end{pmatrix}$. The system is

(*a*) controllable and observable (*b*) controllable but unobservable

(*c*) observable and uncontrollable (*d*) uncontrollable and unobservable

10. The transfer function $G(s) = C\,(sI - A)^{-1}\,B$ of the system

$$\dot{x} = Ax + Bu\ ;\ y = Cx + Du$$

has pole-zero cancellation. The system is

(*a*) uncontrollable and unobservable (*b*) Observable but uncontrollable

(*c*) controllable but unobservable (*d*) May be any one of (*a*), (*b*), and (*c*)

11. Consider the system

$$A = \begin{pmatrix} 0 & -2 \\ 1 & -3 \end{pmatrix}\ ;\ b = \begin{pmatrix} 1 \\ 1 \end{pmatrix}\ ;\ c = [0 \quad 1]$$

The transfer function of the system has pole-zero cancellation. The system is

(*a*) controllable and observable (*b*) Uncontrollable and unobservable

(*c*) controllable but unobservable (*d*) Observable but uncontrollable

12. For all values of '*t*', the matrix exponential e^{At} is nonsingular for

(*a*) Singular *A*

(*b*) Non-singular *A*

(*c*) All *A*

(*d*) Nothing can be said, in general, about non-singularity of e^{At} for a given *A*.

KEY

1. (*a*)	**2.** (*a*)	**3.** (*d*)	**4.** (*b*)	**5.** (*b*)	**6.** (*c*)
7. (*d*)	**8.** (*b*)	**9.** (*a*)	**10.** (*d*)	**11.** (*a*)	**12.** (*d*)

EXERCISE

1. What are the merits of state variable techniques?
2. Define the following:
   (*i*) State
   (*ii*) State variables
   (*iii*) State vector.
3. Develop the state model of linear time invariant systems.
4. State the properties of state transaction matrix.
5. Derive the state transition matrix that is required in the solution of the state equation using Laplace transform.
6. State and explain the controllability and observability of a system.
7. Derivation of Transfer Function from State Model.
8. Explain the Caley-Hamilton theorem.

9. Write the state equation in matrix form for the circuit shown in Fig. 9.17.

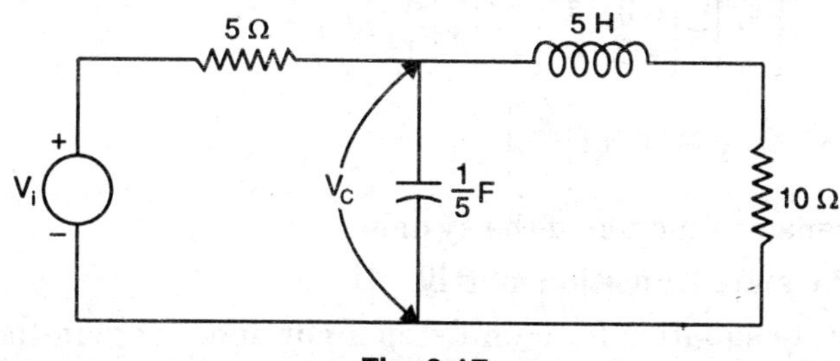

Fig. 9.17

10. Find the eigen values of RLC series circuit which is source free.

11. Write the state equations and eigen values of the circuit shown in Fig. 9.18.

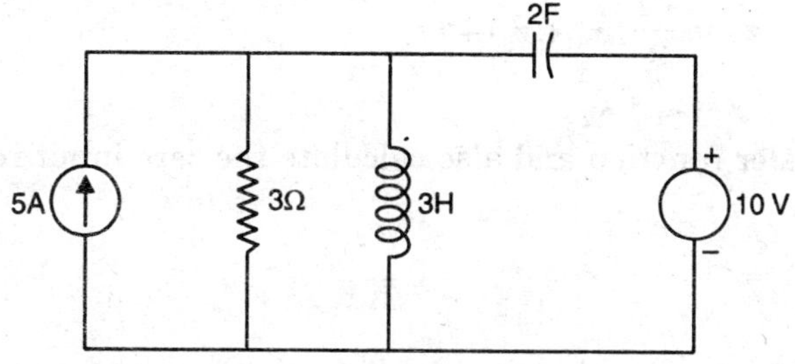

Fig. 9.18

12. Find the state transition matrix of the system, given

$$A = \begin{bmatrix} -2 & 0 & 0 \\ 0 & -1 & 1 \\ 0 & 0 & -1 \end{bmatrix}, B = \begin{bmatrix} 0 \\ 1 \\ 0 \end{bmatrix}$$

13. The state equation of a system is given by

$$\begin{bmatrix} \dot{x}_1 \\ \dot{x}_2 \end{bmatrix} = \begin{bmatrix} -3 & 1 \\ -2 & 0 \end{bmatrix} \begin{bmatrix} x_1 \\ x_2 \end{bmatrix} + \begin{bmatrix} 0 \\ 1 \end{bmatrix} u(t), t > 0$$

(*a*) Is the system controllable?

(*b*) Compute the state transition matrix

(*c*) Compute $x_1(t)$ under zero initial condition and a unit step input

14. State variable formulation of a system is given by the expression

$$\begin{bmatrix} \dot{x}_1 \\ \dot{x}_2 \end{bmatrix} = \begin{bmatrix} 1 & 0 \\ 0 & 1 \end{bmatrix} \begin{bmatrix} x_1 \\ x_2 \end{bmatrix} + \begin{bmatrix} 0 \\ 1 \end{bmatrix} u(t)$$

$$y = [1 \quad 0] \begin{bmatrix} x_1 \\ x_2 \end{bmatrix}$$

What is the transfer function of the system?

15. A system is characterized by the following state equation

$$\begin{bmatrix} \dot{x}_1 \\ \dot{x}_2 \end{bmatrix} = \begin{bmatrix} -3 & 1 \\ -2 & 0 \end{bmatrix} \begin{bmatrix} x_1 \\ x_2 \end{bmatrix} + \begin{bmatrix} 0 \\ 1 \end{bmatrix} u(t),\ t > 0$$

$$y = [1 \quad 0] \begin{bmatrix} x_1 \\ x_2 \end{bmatrix}$$

(*a*) Find the transfer function of the system

(*b*) Compute the state transition matrix

(*c*) Solve the state equation for a unit step input under zero initial condition

16. A second order system is described by

$$\dot{x}_1 = \frac{-3}{2} x_1 + \frac{1}{2} x_2 + u$$

$$\dot{x}_2 = \frac{-1}{2} (x_1 + x_2) + u$$

$$y = x_1 + x_2$$

Obtain the transfer function and also calculate the zero input response $y(t)$ of $x_{1(0)} = 1$ and $x_{2(0)} = -1$.

Appendix

Example A.1: *Finding the roots of quadratic equation of $ax^2 + bx + c = 0$.*

Program:

```
a=input('enter the value a');
b=input('enter the value b');
c=input('enter the value c');
x1=(-b+sqrt(b^2-4*a*c))/(2*a);
x2=(-b-sqrt(b^2-4*a*c))/(2*a);
fprintf('roots of quadratic expression are ')
disp(x1)
disp(x2)
```

Output:

```
%enter the value a = 5
%enter the value b = 8
%enter the value c = 7
%roots of the quadratic expression are
% -0.800+0.8718i
%  -0.8000-0.8718i
```

Example A.2: *To perform the block diagram reduction using MATLAB program.*

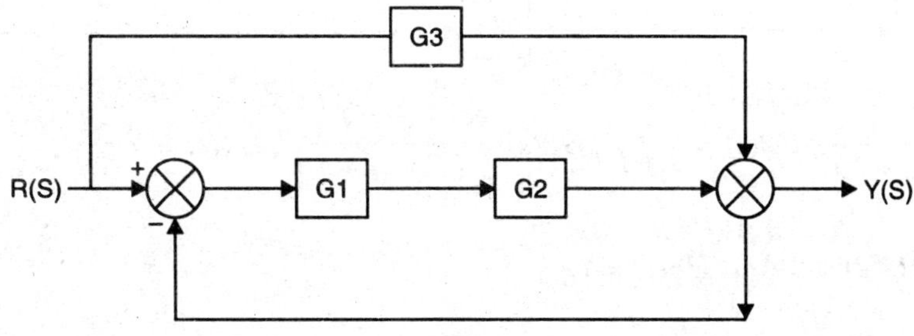

where $G1 = \dfrac{1}{s+1}$, $G2 = \dfrac{1}{s}$, $G3 = 2$ *and* $H1 = 1$.

Program:

```
clear   clc
num1=[0 1];
den1=[1 1];
num2=[0 1];
den2=[1 0];
num3=[2];
den3=[1];
num5=[1];
den5=[1];
[num4,den4]=series(num1,den1,num2,den2);
[num4,den4]=feedback(num4,den4,num5,den5,+1);
[num,den]=parallel(num4,den4,num3,den3);
Printsys (num,den)
```

Output :

$$\% \text{ num/den} = \frac{2s^2 + 2s - 1}{s^2 + s - 1}$$

Example A.3: *To perform the block diagram reduction using MATLAB program*

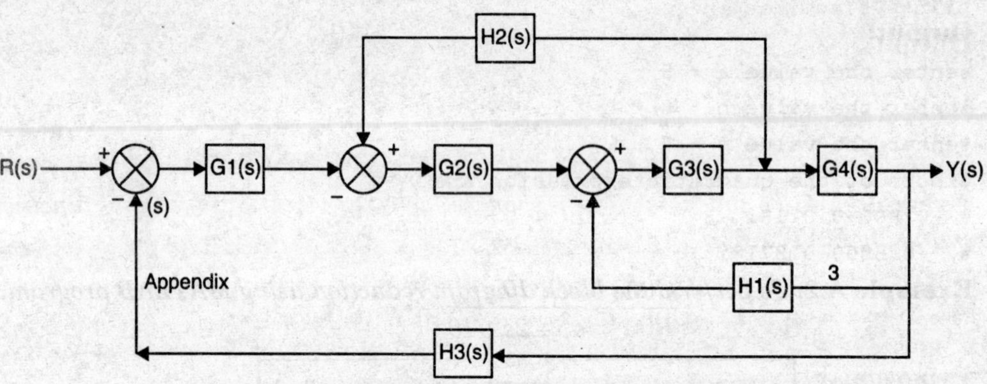

Where $G1(s) = \dfrac{1}{s+10}$, $G2(s) = \dfrac{1}{s+1}$, $G3(s) = \dfrac{s^2+1}{s^2+4s+10}$, $G4(s) = \dfrac{s+1}{s+6}$,

$H1(s) = \dfrac{s+1}{s+2}$, *H2(s) = 2 And H3(s) = 1.*

Program:

```
clear
clc
num1=[0 1];
den1=[1 10];
num2=[0 1];
```

```
den2=[1 1];
num3=[1 0 1];
den3=[1 4 4];
num4=[1 1];
den4=[1 6];
num5=[1 1];
den5=[1 2];
num6=[2];
den6=[1];
num7=[1];
den7=[1];
num8=[2 12];
den8=[1 1];
[num9,den9]=series(num3,den3,num4,den4);
[num10,den10]=feedback(num9,den9,num5,den5,+1);
[num11,den11]=series(num10,den10,num2,den2);
[num12,den12]=feedback(num11,den11,num8,den8);
[num13,den13]=series(num12,den12,num1,den1);
[num,den]=feedback(num13,den13,num7,den7);
printsys(num,den)
```

Output :

```
                s^5 + 4 s^4 + 6 s^3 + 6 s^2 + 5s + 2
      -------------------------------------------------------------
      12 s^6 + 205 s^5 + 1066 s^4 + 2517 s^3 + 3128 s^2 + 2196 s + 712
```

Example A.4: *To obtain the time response specifications of given transfor function with step input using MATLAB.*

$$\frac{C(s)}{R(s)} = \frac{10}{s^2 + 2s + 10}.$$

Program:

```
clc
clear
t=0:.05:10;
num1=[0 10]
den1=[1 2 10]
step(num1,den1,t)
title('step response')
xlabel('time(sec'))
ylabel('output')
```

Output:

```
delay time = 0.371
rise time = 0.63
```

```
peak time = 1.07
settling time  = 4.28
```

Output Responce

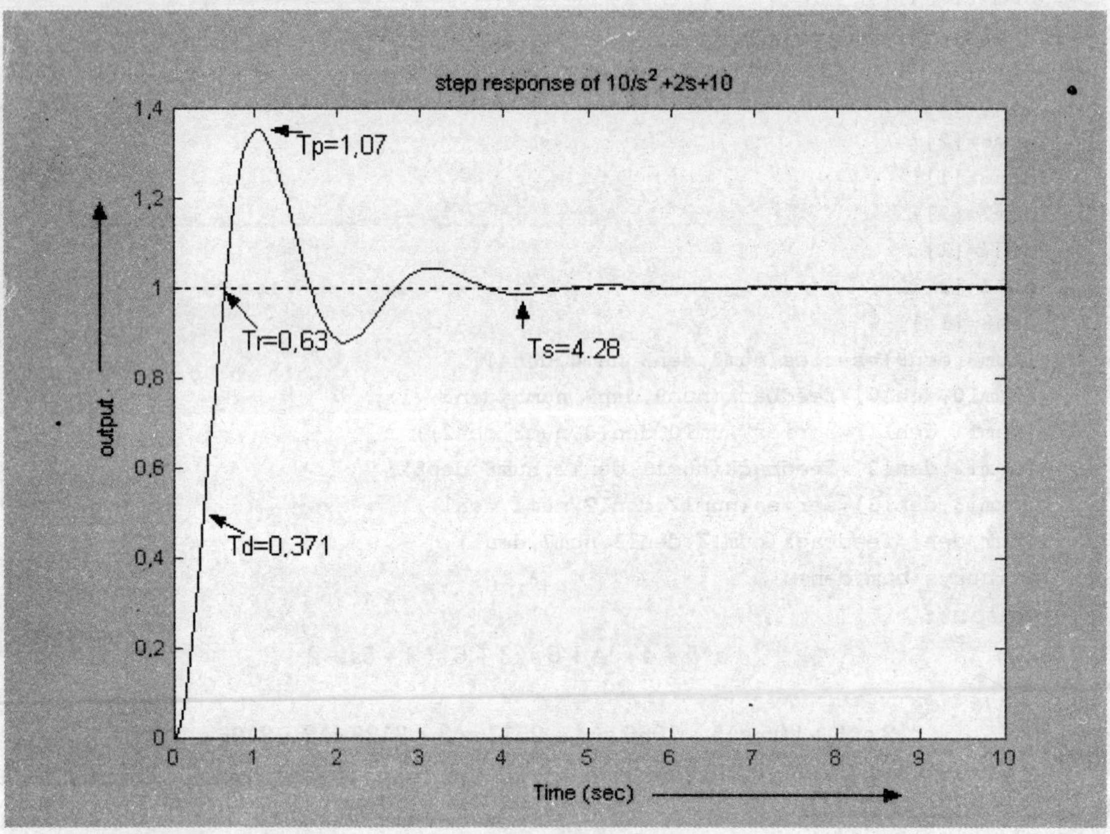

Example A.5: *To obtain the unit step response for transfer function using MATLAB. Given transfer function*

$$\frac{C(s)}{R(s)} = \frac{2s+4}{s^3+5s^2+6s+4}.$$

Program:

```
clc
num=[2 4];
den=[1 5 6 4];
trf=tf(num,den);
step(trf);
title('step response of given system')
```

Output Response:

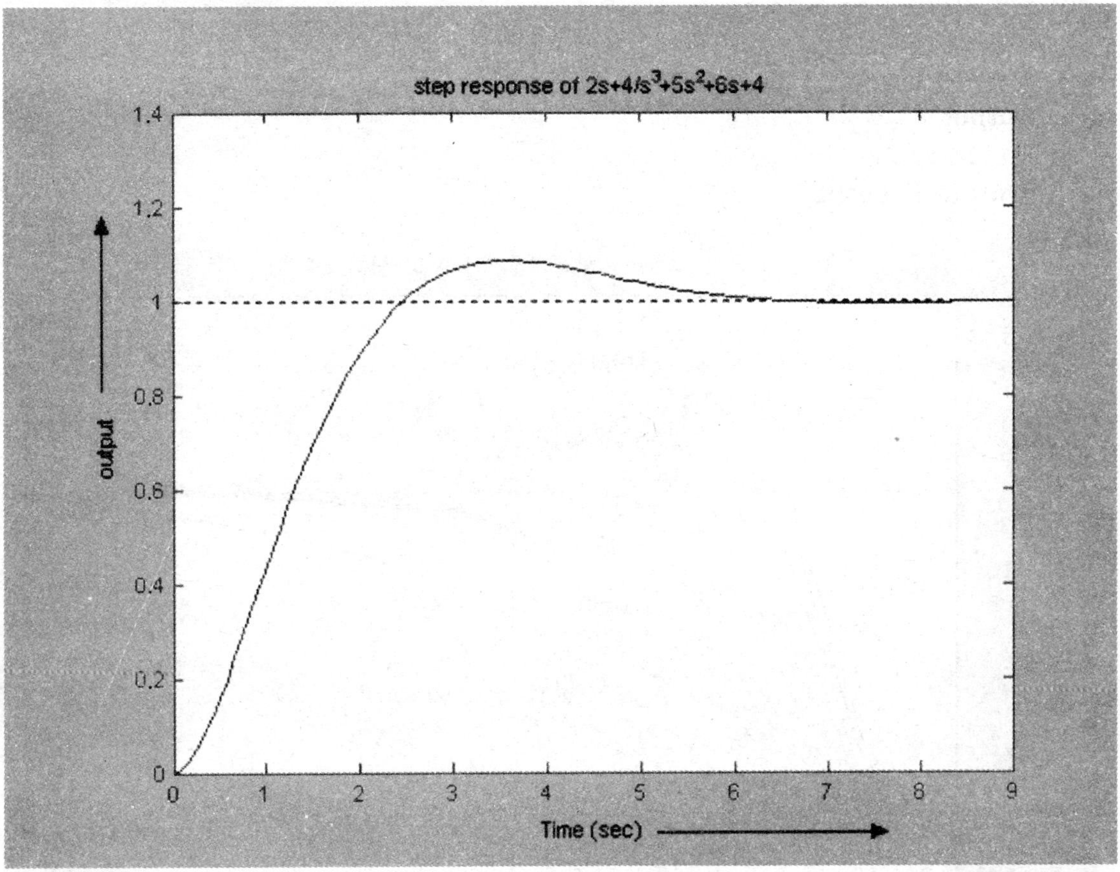

Example A.6: *To obtain the step response of system for different values of open loop gain K*

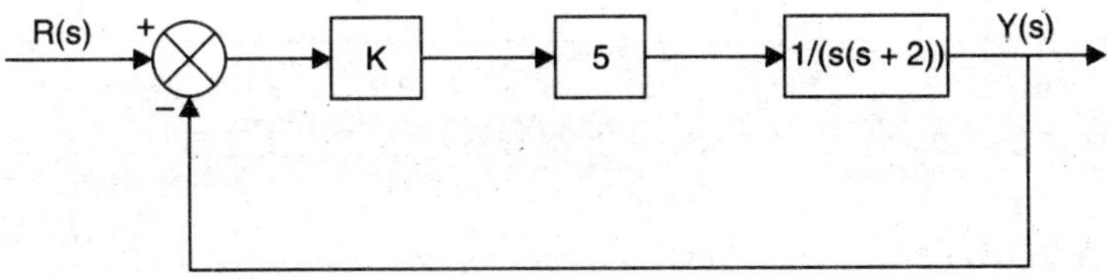

Program:

```
clc
clear
t=[0:0.05:0.5]
k=[20 30 40 50 60 70 80 90 100]
for i=1:9
    num=[0 0 5*k(i)];
    den=[1 20 5*k(i)];
    [y(:,i),x,t]=step(num,den,t);
end
```

```
plot(t,y,'-',t,y,'= '- -');
title('step response');
xlabel('time')
ylabel('output')
```

Output

```
% k=20 30 40 50 60 70 80 90 100
```

Output Response:

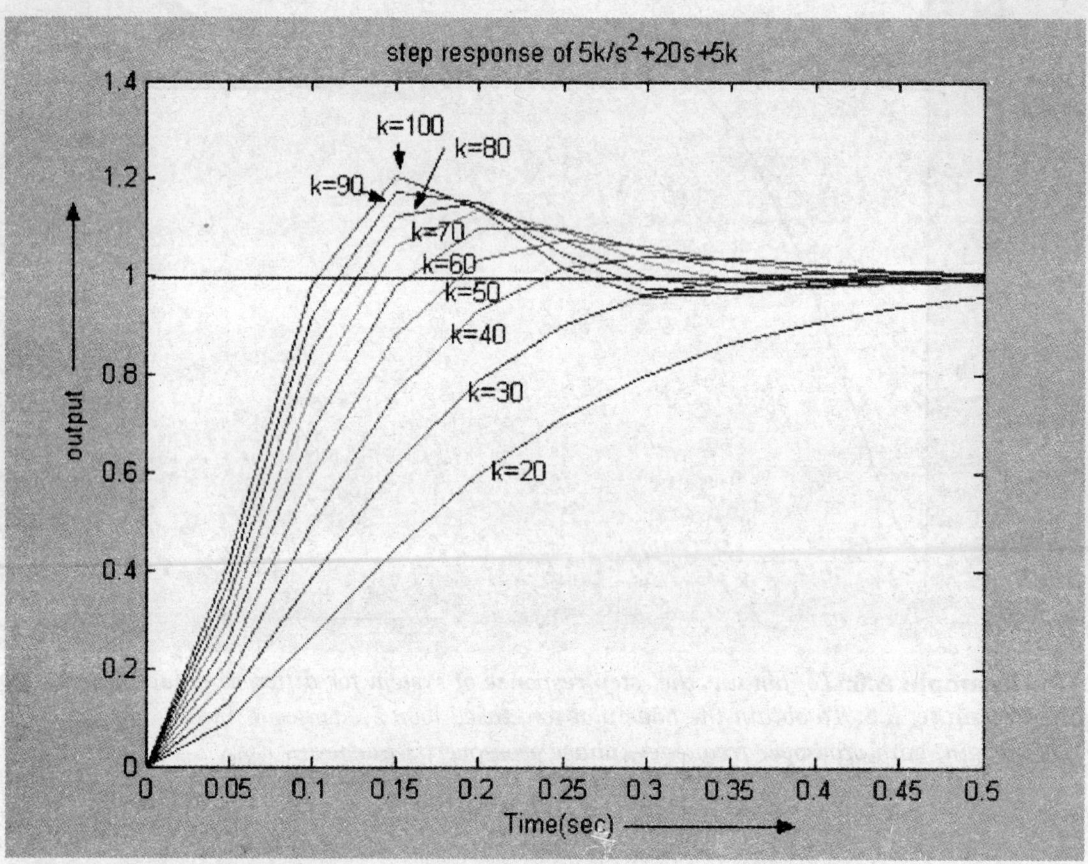

Example A.7: *To obtain the Root locus for given system is* $\frac{K}{s(s^2 + 4s + 13)}$

Program:

```
clc
k=[0:.01:100]
num=[1 ]
den=[1 4 13 0]
to=tf(num,den)
rlocus(to,k)
grid on
title('Root locus of  1/s(s^2+4s+13 ')
```

```
xlabel('real axis')
ylabel('imaginary axis')
```

Output Response:

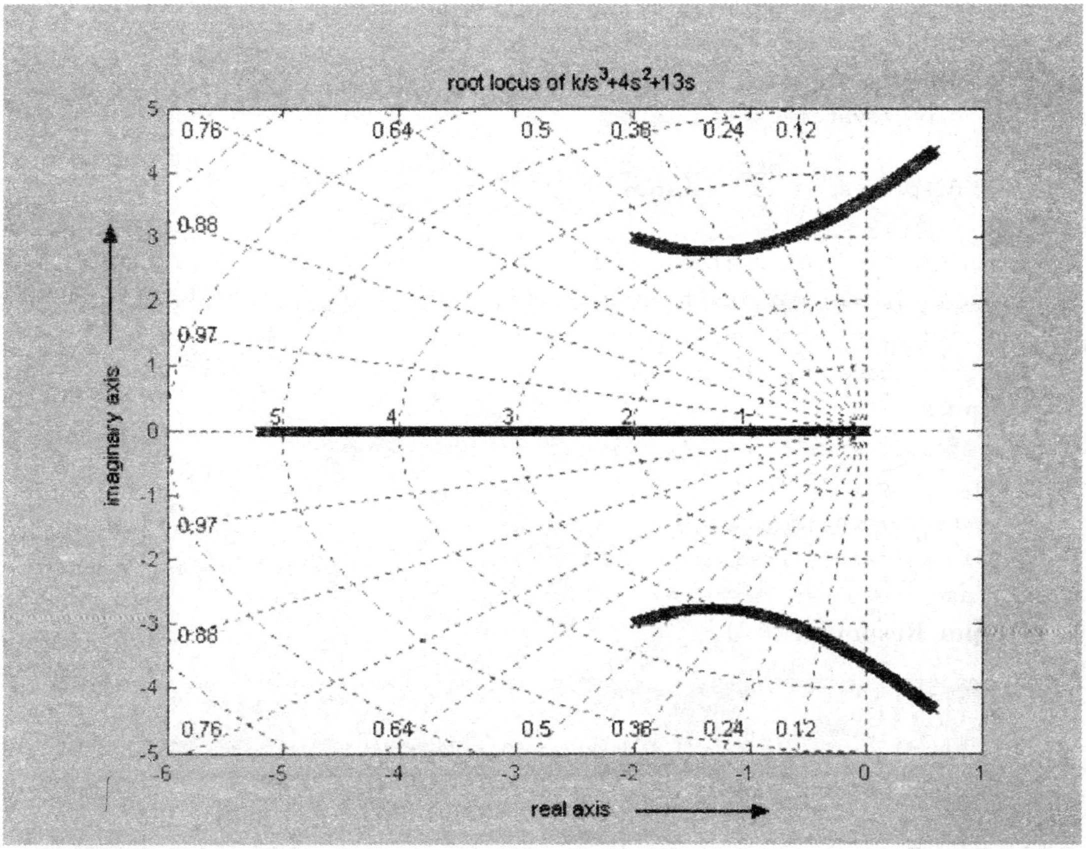

Example A.8: *To obtain the bode plot for closed loop system and finding gain margin, phase margin, gain crossover frequency, phase crossover frequency.*

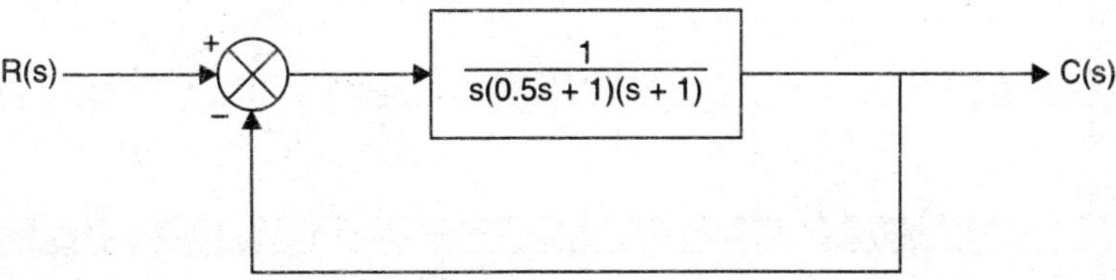

Program:

```
clear
clc
nump=[0 0 0 1]
denp=[0.5 1.5 1 0]
sysp=tf(nump,denp)
w=logspace(-1,1,10);
bode(sysp,w)
```

```
title('bode plot of 1/s(0.5s+1)(s+1)')
xlabel('frequency ')
grid on;
[gm,pm,wcp,wcg]=margin(sysp);
gmdb=20*log(gm);
if(gmdb>=0&&pm>=0)
    disp('system is stable');
else
    disp('system is unstable');
end
fprintf('gain margin=%f \n',gmdb)
fprintf('[phase margin=%f \n',pm);
fprintf('gain crossover frequency =%f \n',wcg);
fprintf('phase crossover frequency=%f \n',wcp);
```

Output:

```
% system is stable
% gain margin=21.97 db
% phase margin=32.61 deg
% gain crossover frequency =0.74 rad/sec
% phase crossover frequency=1.41 rad/sec
```

Output Response:

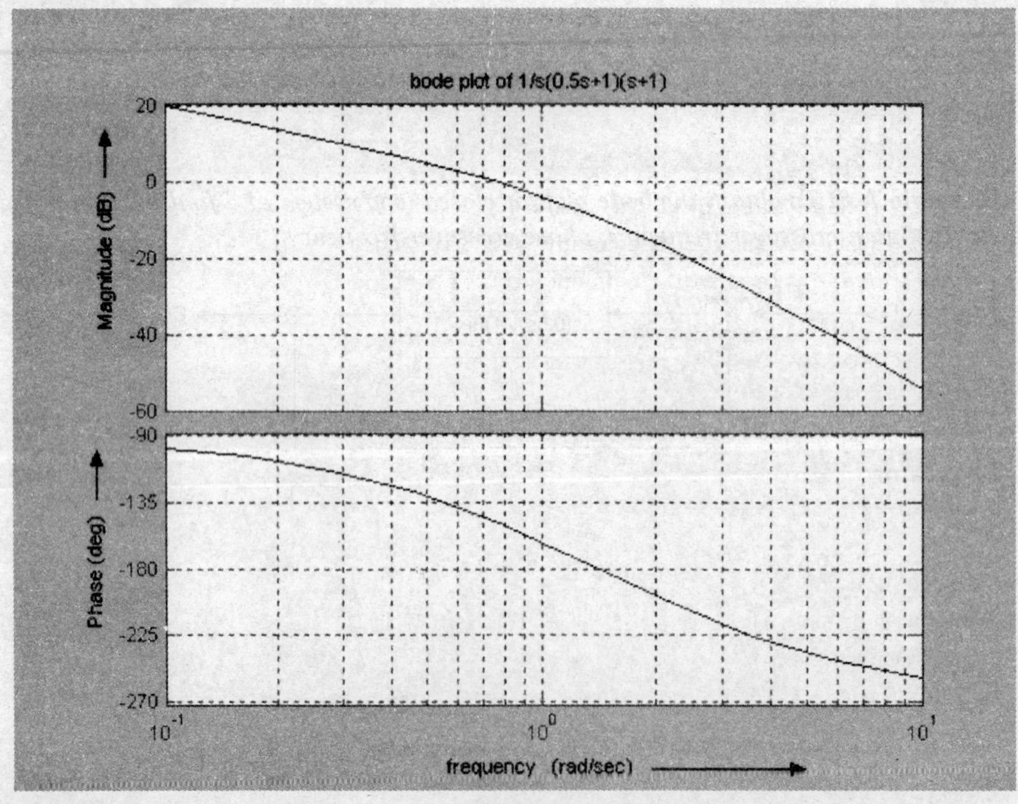

Example A.9: *To obtain the bode plot for closed loop system and finding gain margin, phase margin, gain cross over frequency, phase crossover frequency.*

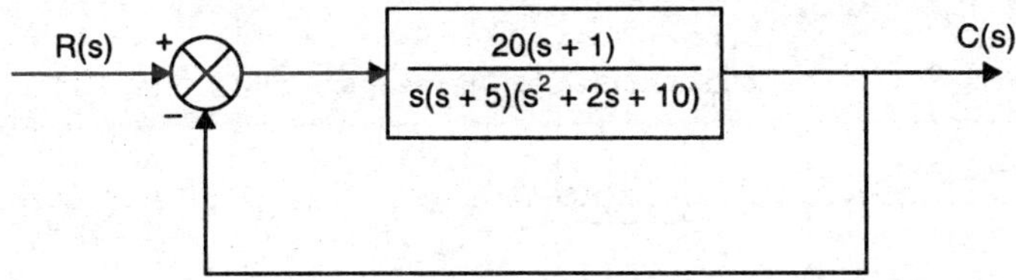

Program:

```
clear
clc
num=[0 0 0 20 20];
den=conv([1 5 0],[1 2 10]);
sys=tf(num,den);
w=logspace(-1,2,10);
bode(sys,w);
title('bode plot of  20s(s+1)/s(s+5)(s^2+2s+10)')
xlabel('frequency')
grid on;
[gm,pm,wcp,wcg]=margin(sys);
gmdb=20*log(gm);
if(gmdb>=0&&pm>=0)
    disp('system is stable');
else
    disp('system is unstable');
end
fprintf('gain margin=%f \n',gmdb);
fprintf('[phase margin=%f \n',pm);
fprintf('gain crossover frequency =%f \n',wcg);
fprintf('phase crossover frequency=%f \n',wcp);
```

Output :

```
% system is stable
% gain margin=22.86 db
% phase margin=103.65 deg
% gain crossover frequency=0.44 rad/sec
% phase crossover frequency=4.013 rad/sec
```

Output Response:

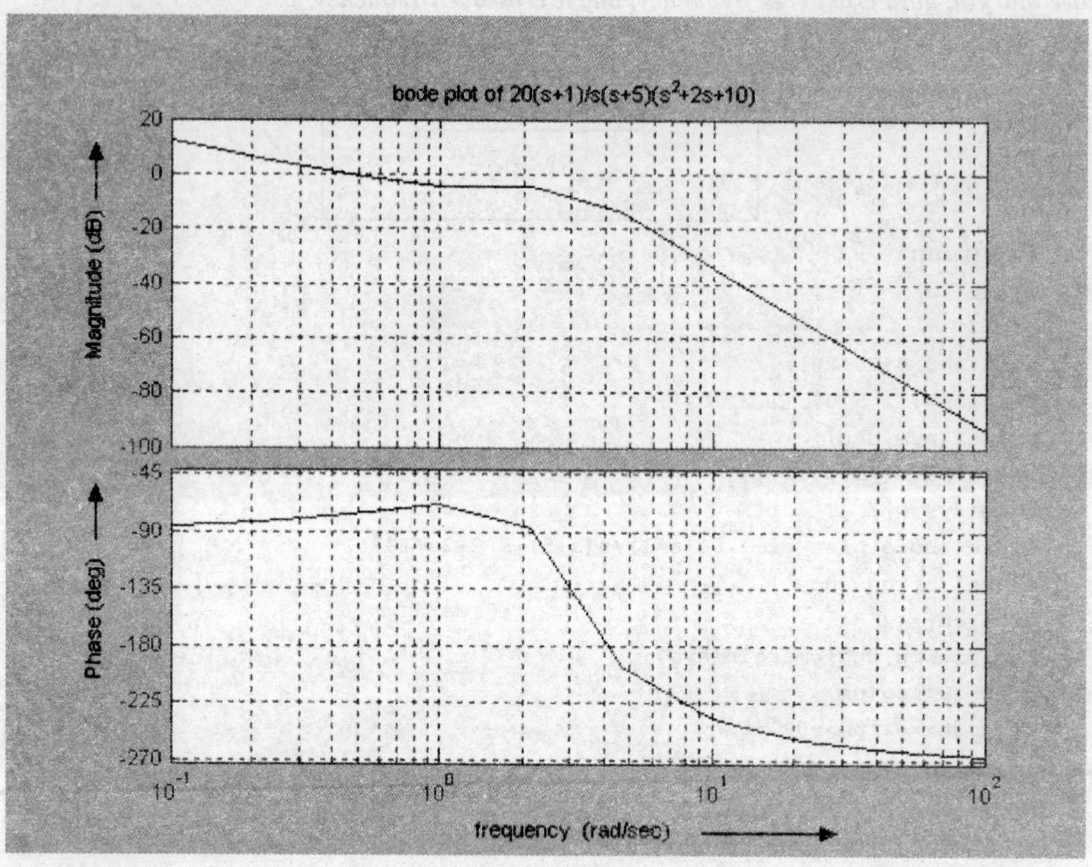

Example A.10: *To obtain the Nyquist plot for given system of* $G(s) = \dfrac{(2s^2 + 5s + 1)}{(s^2 + 2s + 3)}$

Program

```
clc
num=[2 5 1]
den=[1 2 3]
to=tf(num,den)
nyquist(to)
title('nyquist plot of   2s^2 +5s+1/s^2+2s+3  ')
grid on
```

Output Response:

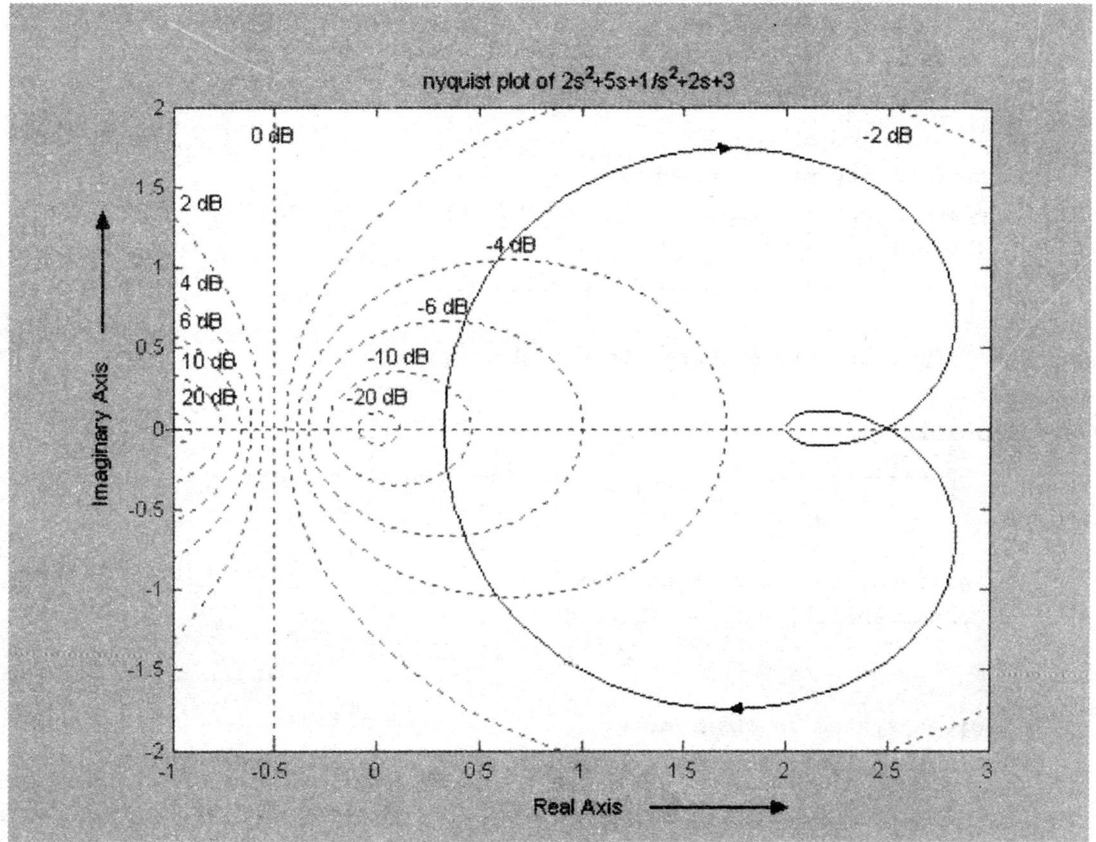

Example A.11: *To determine the controllability and observability of given state space system. Given matrix are*

$$\dot{x} = Ax + Bu$$

$$y = Cx$$

Where $A = \begin{bmatrix} -1 & 1 & 0 \\ 0 & -4 & 2 \\ 0 & 0 & -10 \end{bmatrix}$, $B = \begin{bmatrix} 1 \\ 0 \\ -1 \end{bmatrix}$ *and* $C = [1 \quad 0 \quad 1]$.

Program:

```
clear
clc
a=[-1 1 0;0 -4 2;0 0 -10]
b=[1;0;-1]
c=[1 0 -1]
d=[0]
cm=ctrb(a,b)
cond=length(a)-rank(cm)
```

```
if (cond==0)
    fprintf('given system is controllable');
else
    fprintf('given is uncontrollable');
end
    % observability
observability matrix =obsv(a,c)
ob=length(a)-rank(om)
if(ob==0)
    fprintf('given system is observable')
else
    fprintf('given system is not observable')
end
```

Output :

```
% conrollable matrix =   1    -1    -1
%                        0    -2    28
%                       -1    10    -100
%  given system is controllable
%  observability matrix = 1    0    -1
%                        -1    1    10
%                         1   -5    -98
% given system is observable
```

Example A.12: *Finding transfer function, poles, zeros of a given system*

$$A = \begin{bmatrix} -4.2 & -17.4 & -30.8 & -60 \\ 1 & 0 & 0 & 0 \\ 0 & 1 & 0 & 0 \\ 0 & 0 & 1 & 0 \end{bmatrix}, B = \begin{bmatrix} 1 \\ 0 \\ 0 \\ 0 \end{bmatrix}, C = [48 \quad 30 \quad 78 \quad 60] \text{ and } D = [0].$$

Program:

```
clc
a=[-4.2 -17.4 -30.8 -60;1 0 0 0;0 1 0 0;0 0 1 0];
b=[1;0;0;0];
c=[4 8.4 30.78 60];
d=0;
s=ss(a,b,c,d)
t=tf(s)
p=pole(t)
z=zero(t)
iopzmap(t)
sgrid
```

Output:

```
% transfer function  =   4s^3 + 8.4s^2 + 30.78s + 60
                        ----------------------------------
                        s^4 + 4.2s^3 + 17.4s^2 + 30.8s + 60
%
%
%    p =   -1.8475+2.2192i
%             -1.8475-2.2192i
%             -0.2525+2.6706i
%             -0.2525-2.6706i
%     z = -0.0496+2.7376i
%             -0.0496-2.7376i
%             -2.0009
```

Output Response:

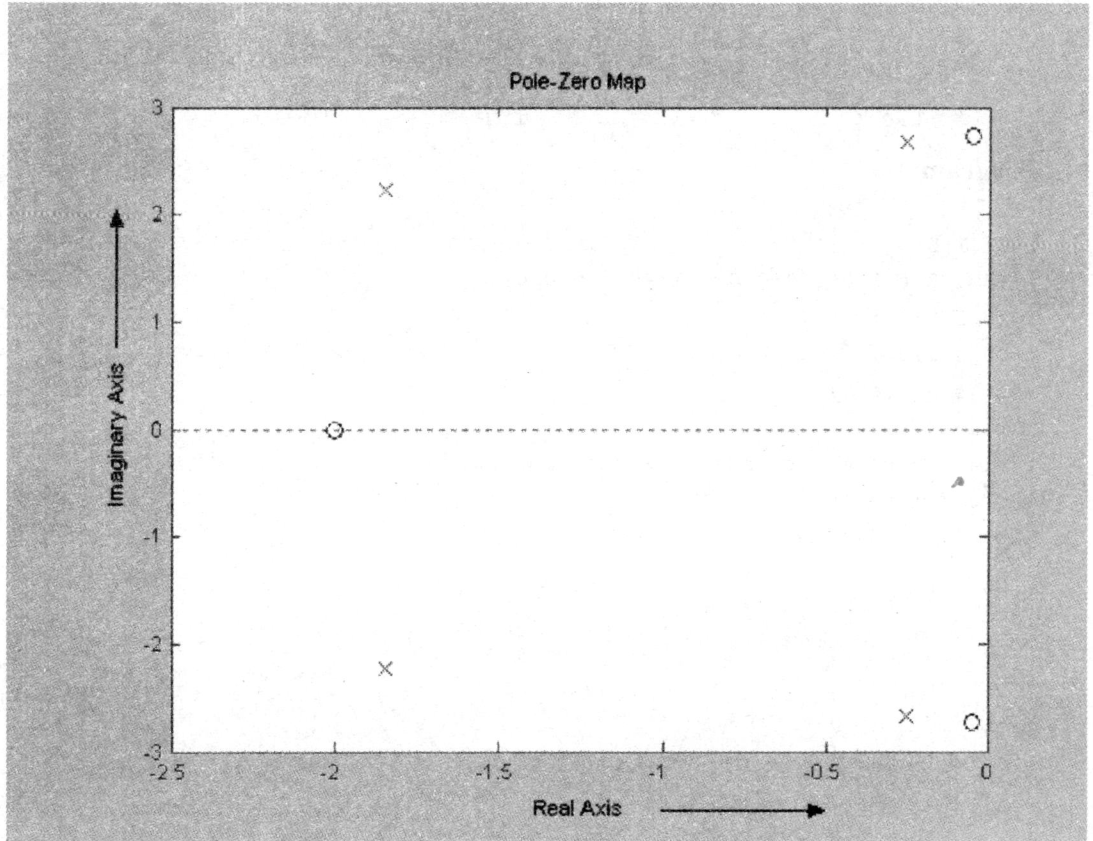

Example A.13: *To obtain the feedback gain matrix K using pole placement method*

$$A = \begin{bmatrix} 0 & 1 & 0 & 0 \\ 1 & 0 & 0 & 0 \\ 0 & 0 & 0 & 1 \\ -0.5 & 0 & 1 & 0 \end{bmatrix}, B = \begin{bmatrix} 1 \\ 0 \\ 0 \\ 0 \end{bmatrix}, C = [0 \quad 0 \quad 1 \quad 0] \text{ and } D = [0]$$

Program:

```
clc
clear
A=[0 1 0 0;1 0 0 0;0 0 0 1;-0.5 0 0 0];
B=[0;1;0;0];
C=[0 0 1 0];
D=[0];
p=[1 -1 -1+i -1-i]
K=place(A,B,p)
```

Output:

```
% k =2.00 2.00  4.00  4.00
```

Example A.14: *To obtain the impulse response of a given state model using pole placement method:*

$$A = \begin{bmatrix} 0 & 1 & 0 & 0 \\ 1 & 0 & 0 & 0 \\ 0 & 0 & 0 & 1 \\ -0.5 & 0 & 1 & 0 \end{bmatrix}, B = \begin{bmatrix} 1 \\ 0 \\ 0 \\ 0 \end{bmatrix}, C = [0 \quad 0 \quad 1 \quad 0] \quad and\ D = [0]$$

Program:

```
clc
t=0:.1:5
a=[0 1 0 0;1 0 0 0;0 0 0 1;-0.5 0 0 0];
b=[0;1;0;0];
c=[0 0 1 0];
sys=ss(a,b,c,0)
impulse(sys,t)
title('impulse response of state space model')
xlabel('time')
ylabel('amplitude')
```

Output Response:

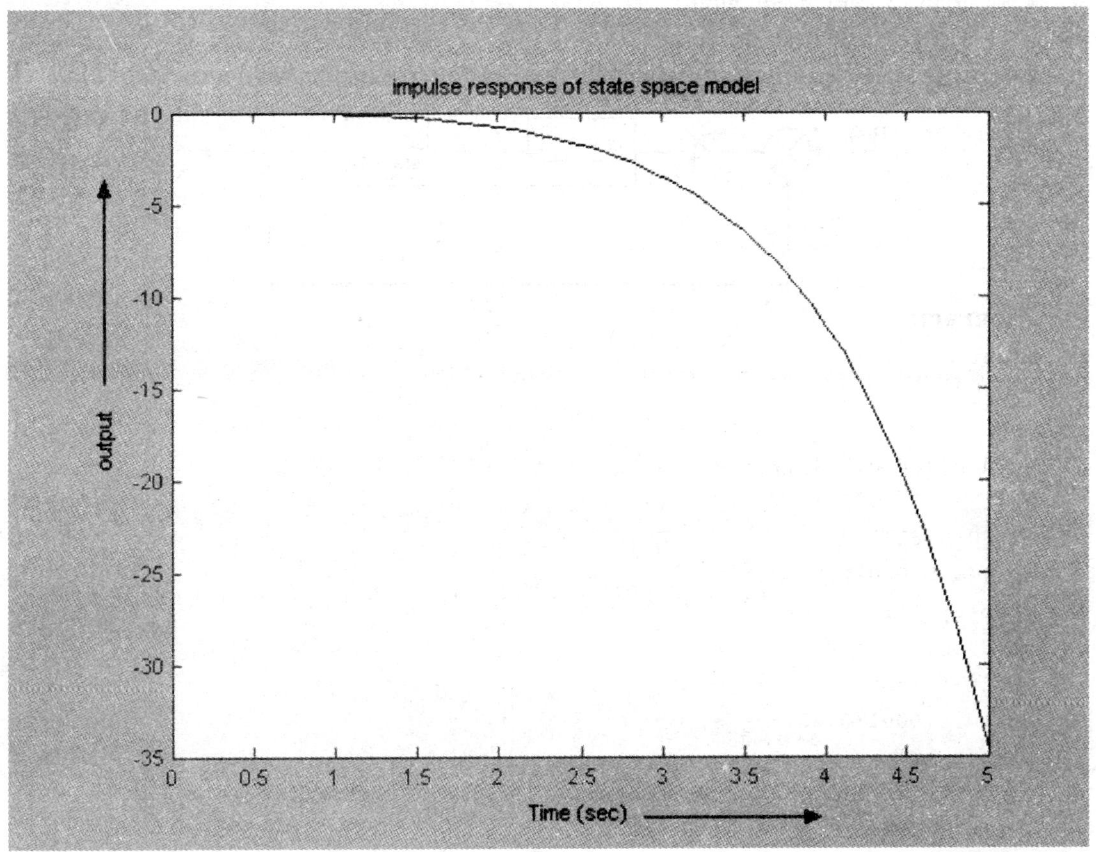

Example A.15: *To obtain the feedback gain matrix K using place and Acker commands in pole placement method*

$$A = \begin{bmatrix} 0 & 1 & 0 & 0 \\ 1 & 0 & 0 & 0 \\ 0 & 0 & 0 & 1 \\ -0.5 & 0 & 1 & 0 \end{bmatrix}, B = \begin{bmatrix} 1 \\ 0 \\ 0 \\ 0 \end{bmatrix}, C = [0 \quad 0 \quad 1 \quad 0] \quad and\ D = [0]$$

Program:

```
clc
clear
A=[0 1 0 0;1 0 0 0;0 0 0 1;-0.5 0 0 0];
B=[0;1;0;0];
C=[0 0 1 0];
D=[0];
p=[-1.5 0 -1+i -1-i]
K=place(A,B,p)
T=acker(A,B,p)
```

Output:

```
% K= 6.00 3.500 0 -0.6000
% T= 6.00 3.500 0 -0.6000
```

Example A.16: *To obtain the time response of discrete time system with step input*

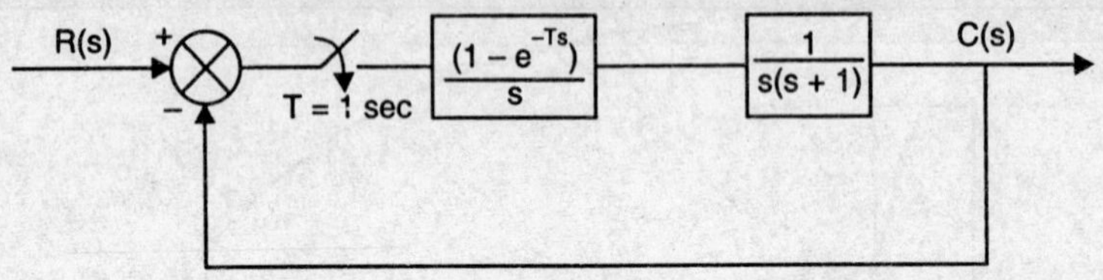

Program:

```
clear
clc
pack
a=[0 0.368 0.2644];
b=[1 -1 0.6322];
t1=[0:0.01:0.25];
t=tf(a,b,0.01);
step(t,t1);
hold on;
h=d2c(t,'zoh');
t2=(0:0.0001:0.25);
step(h,t2)
title('discrete  time response of  0.368 Z+ 0.2644/Z^2-Z+0.6322')
xlabel('time')
ylabel('output')
end
```

Output Response:

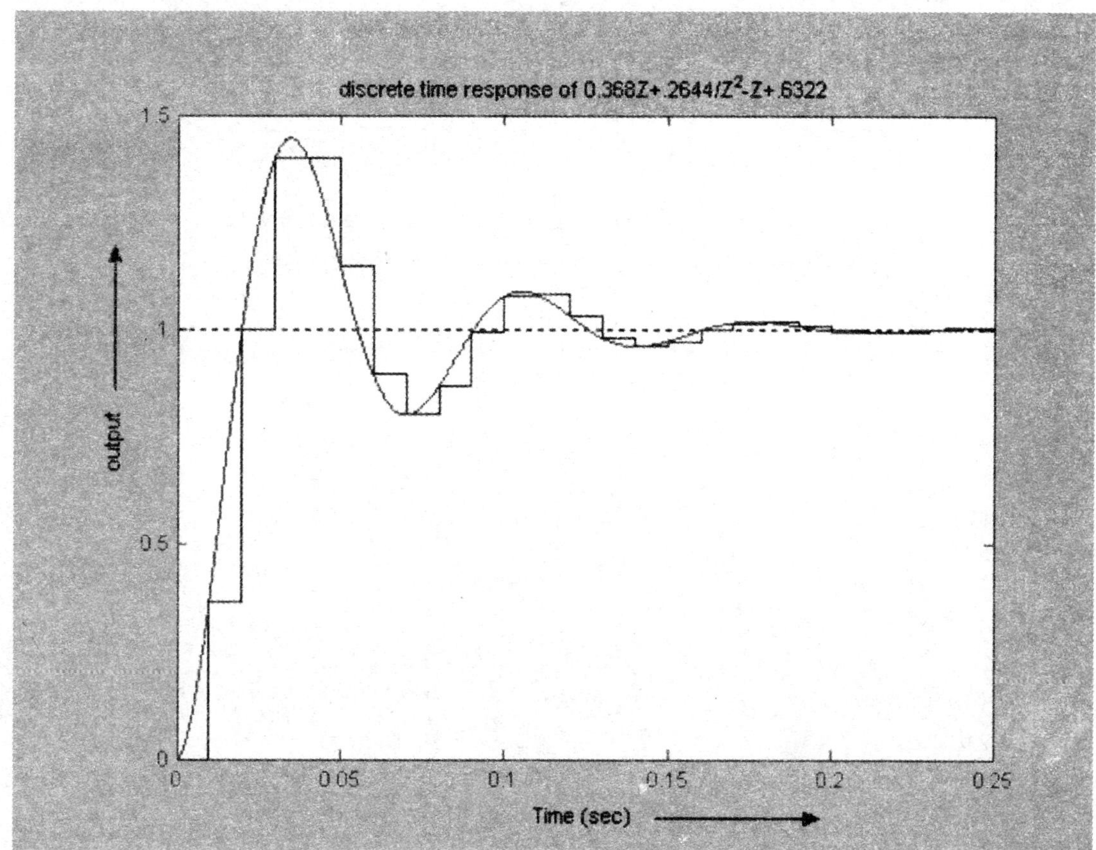

Index

G

H

I

L

M

N

O

P

R

S

T

U